# Advances in Cellulose-Based Hydrogels

# Advances in Cellulose-Based Hydrogels

Editors

**Christian Demitri**
**Lorenzo Bonetti**
**Laura Riva**

MDPI • Basel • Beijing • Wuhan • Barcelona • Belgrade • Manchester • Tokyo • Cluj • Tianjin

*Editors*

Christian Demitri
Department of Engineering
for Innovation
University of Salento
Lecce
Italy

Lorenzo Bonetti
Department of Chemistry,
Materials and Chemical
Engineering "G. Natta"
Politecnico di Milano
Milano
Italy

Laura Riva
Department of Chemistry,
Materials and Chemical
Engineering "G. Natta"
Politecnico di Milano
Milano
Italy

*Editorial Office*
MDPI
St. Alban-Anlage 66
4052 Basel, Switzerland

This is a reprint of articles from the Special Issue published online in the open access journal *Gels* (ISSN 2310-2861) (available at: www.mdpi.com/journal/gels/special_issues/Cellulose_Hydrogels).

For citation purposes, cite each article independently as indicated on the article page online and as indicated below:

LastName, A.A.; LastName, B.B.; LastName, C.C. Article Title. *Journal Name* **Year**, *Volume Number*, Page Range.

**ISBN 978-3-0365-7111-9 (Hbk)**
**ISBN 978-3-0365-7110-2 (PDF)**

© 2023 by the authors. Articles in this book are Open Access and distributed under the Creative Commons Attribution (CC BY) license, which allows users to download, copy and build upon published articles, as long as the author and publisher are properly credited, which ensures maximum dissemination and a wider impact of our publications.

The book as a whole is distributed by MDPI under the terms and conditions of the Creative Commons license CC BY-NC-ND.

# Contents

**About the Editors** . . . . . . . . . . . . . . . . . . . . . . . . . . . . . . . . . . . . . . . . . . . . . . . . . . . . . . . . . . . . vii

**Preface to "Advances in Cellulose-Based Hydrogels"** . . . . . . . . . . . . . . . . . . . . . . . . . . . . ix

**Lorenzo Bonetti, Christian Demitri and Laura Riva**
Editorial on the Special Issue "Advances in Cellulose-Based Hydrogels"
Reprinted from: *Gels* 2022, *8*, 790, doi:10.3390/gels8120790 . . . . . . . . . . . . . . . . . . . . . . . 1

**Daniela Filip, Doina Macocinschi, Mihaela Balan-Porcarasu, Cristian-Dragos Varganici, Raluca-Petronela Dumitriu and Dragos Peptanariu et al.**
Biocompatible Self-Assembled Hydrogen-Bonded Gels Based on Natural Deep Eutectic Solvents and Hydroxypropyl Cellulose with Strong Antimicrobial Activity
Reprinted from: *Gels* 2022, *8*, 666, doi:10.3390/gels8100666 . . . . . . . . . . . . . . . . . . . . . . . 5

**Roko Blažic, Dajana Kučić Grgić, Marijana Kraljić Roković and Elvira Vidović**
Cellulose-*g*-poly(2-(dimethylamino)ethylmethacrylate) Hydrogels: Synthesis, Characterization, Antibacterial Testing and Polymer Electrolyte Application
Reprinted from: *Gels* 2022, *8*, 636, doi:10.3390/gels8100636 . . . . . . . . . . . . . . . . . . . . . . . 31

**Frederik L. Zitzmann, Ewan Ward and Avtar S. Matharu**
Use of Carbotrace 480 as a Probe for Cellulose and Hydrogel Formation from Defibrillated Microalgae
Reprinted from: *Gels* 2022, *8*, 383, doi:10.3390/gels8060383 . . . . . . . . . . . . . . . . . . . . . . . 55

**Pengfei Zou, Jiaxin Yao, Ya-Nan Cui, Te Zhao, Junwei Che and Meiyan Yang et al.**
Advances in Cellulose-Based Hydrogels for Biomedical Engineering: A Review Summary
Reprinted from: *Gels* 2022, *8*, 364, doi:10.3390/gels8060364 . . . . . . . . . . . . . . . . . . . . . . . 65

**Xinkuan Liu, Mingxin Zhang, Wenliang Song, Yu Zhang, Deng-Guang Yu and Yanbo Liu**
Electrospun Core (HPMC–Acetaminophen)–Shell (PVP–Sucralose) Nanohybrids for Rapid Drug Delivery
Reprinted from: *Gels* 2022, *8*, 357, doi:10.3390/gels8060357 . . . . . . . . . . . . . . . . . . . . . . . 95

**Subin Jin, Yewon Kim, Donghee Son and Mikyung Shin**
Tissue Adhesive, Conductive, and Injectable Cellulose Hydrogel Ink for On-Skin Direct Writing of Electronics
Reprinted from: *Gels* 2022, *8*, 336, doi:10.3390/gels8060336 . . . . . . . . . . . . . . . . . . . . . . . 111

**Hamid M. Shaikh, Arfat Anis, Anesh Manjaly Poulose, Niyaz Ahamad Madhar and Saeed M. Al-Zahrani**
Development of Bigels Based on Date Palm-Derived Cellulose Nanocrystal-Reinforced Guar Gum Hydrogel and Sesame Oil/Candelilla Wax Oleogel as Delivery Vehicles for Moxifloxacin
Reprinted from: *Gels* 2022, *8*, 330, doi:10.3390/gels8060330 . . . . . . . . . . . . . . . . . . . . . . . 125

**Sayaka Fujita, Toshiaki Tazawa and Hiroyuki Kono**
Preparation and Enzyme Degradability of Spherical and Water-Absorbent Gels from Sodium Carboxymethyl Cellulose
Reprinted from: *Gels* 2022, *8*, 321, doi:10.3390/gels8050321 . . . . . . . . . . . . . . . . . . . . . . . 141

**Lorenzo Bonetti, Andrea Fiorati, Agnese D'Agostino, Carlo Maria Pelacani, Roberto Chiesa and Silvia Farè et al.**
Smart Methylcellulose Hydrogels for pH-Triggered Delivery of Silver Nanoparticles
Reprinted from: *Gels* 2022, *8*, 298, doi:10.3390/gels8050298 . . . . . . . . . . . . . . . . . . . . . . . 157

**Hang Yang, Xianyu Lan and Yuzhu Xiong**
In Situ Growth of Zeolitic Imidazolate Framework-L in Macroporous PVA/CMC/PEG Composite Hydrogels with Synergistic Antibacterial and Rapid Hemostatic Functions for Wound Dressing
Reprinted from: *Gels* **2022**, *8*, 279, doi:10.3390/gels8050279 . . . . . . . . . . . . . . . . . . . . . . 173

**Jinyu Yang, Dongliang Liu, Xiaofang Song, Yuan Zhao, Yayang Wang and Lu Rao et al.**
Recent Progress of Cellulose-Based Hydrogel Photocatalysts and Their Applications
Reprinted from: *Gels* **2022**, *8*, 270, doi:10.3390/gels8050270 . . . . . . . . . . . . . . . . . . . . . . 189

**Chaehyun Jo, Sam Soo Kim, Balasubramanian Rukmanikrishnan, Srinivasan Ramalingam, Prabakaran D. S. and Jaewoong Lee**
Properties of Cellulose Pulp and Polyurethane Composite Films Fabricated with Curcumin by Using NMMO Ionic Liquid
Reprinted from: *Gels* **2022**, *8*, 248, doi:10.3390/gels8040248 . . . . . . . . . . . . . . . . . . . . . . 205

**Esubalew Kasaw Gebeyehu, Xiaofeng Sui, Biruk Fentahun Adamu, Kura Alemayehu Beyene and Melkie Getnet Tadesse**
Cellulosic-Based Conductive Hydrogels for Electro-Active Tissues: A Review Summary
Reprinted from: *Gels* **2022**, *8*, 140, doi:10.3390/gels8030140 . . . . . . . . . . . . . . . . . . . . . . 219

**Kyuha Park, Heewon Choi, Kyumin Kang, Mikyung Shin and Donghee Son**
Soft Stretchable Conductive Carboxymethylcellulose Hydrogels for Wearable Sensors
Reprinted from: *Gels* **2022**, *8*, 92, doi:10.3390/gels8020092 . . . . . . . . . . . . . . . . . . . . . . 237

**Laura Riva, Angelo Davide Lotito, Carlo Punta and Alessandro Sacchetti**
Zinc- and Copper-Loaded Nanosponges from Cellulose Nanofibers Hydrogels: New Heterogeneous Catalysts for the Synthesis of Aromatic Acetals
Reprinted from: *Gels* **2022**, *8*, 54, doi:10.3390/gels8010054 . . . . . . . . . . . . . . . . . . . . . . 249

# About the Editors

**Christian Demitri**

Christian Demitri is an associate professor of Industrial Bioengineering, specializing in the synthesis and characterization of hydrogel polymers based on natural and synthetic polymers for biomedical applications. In particular, these materials have been applied in the field of regenerative medicine as cartilage substitutes or as devices for the treatment of metabolic diseases such as obesity. They have also been applied in agriculture as vehicles for the controlled release of moisture in arid soils. Demitri has conducted research at Italian and foreign institutions on the synthesis and characterization of natural polymers for several applications. During this research, he explored techniques for detecting DNA using analytical techniques such as mass spectrometry, HPLC, fluorescence microscopy, and electron microscopy (TEM and SEM). Over the years, he has also been involved in evaluating the cytotoxicity of materials using cell assays to identify specific markers. The research activities focus on the property-structure relationships of biomaterials, with a specific focus on the cell recognition processes of natural polymer-based scaffolds. He continues to work on the development and industrialization of research results concerning hydrogels and biodegradable polymers in the fields of metabolic disease treatment, regenerative medicine, and biotechnology, through collaboration with various companies in the sector.

**Lorenzo Bonetti**

Lorenzo Bonetti is a Post-doctoral Research Fellow at the Department of Chemistry, Materials and Chemical Engineering "G. Natta", Politecnico di Milano (Milan, Italy). After a M.Eng. degree cum laude in Biomedical Engineering (2017), he obtained a Ph.D. in Materials Engineering (2021) with the highest grade from Politecnico di Milano. While completing his Ph.D., he had experience as a Visiting Researcher (2019) at the Faculty of Mechanical Engineering at Ottawa University (ON, CA). He is currently a Post-doctoral Research Fellow at Politecnico di Milano and his research is focused on the design and characterization of stimuli-responsive hydrogels, based on natural polymers, for tissue engineering and regenerative medicine. In this framework, he developed innovative responsive hydrogels, based on methylcellulose and chitosan, as advanced platforms for drug delivery, cell sheet engineering, and 3D bioprinting. Dr. Bonetti received the Julia Polak European Doctorate Award in 2022 from the European Society for Biomaterials (ESB), as a recognition for his outstanding doctoral track record. He currently serves as a Guest Editor for *Gels* and is a member of the Topical Advisory Panel of the same journal. With an H-index of 8, he is co-author of 15 research articles.

**Laura Riva**

Laura Riva graduated in 2018 in Medicinal Chemistry and Pharmaceutical Technology cum laude at the University of Milan, and in 2021, she obtained a Ph.D. in Industrial Chemistry and Chemical Engineering with the highest grade at Politecnico di Milano. Since March 2023, she is a junior researcher (RTD-A) in the Department of Chemistry, Materials and Chemical Engineering, Politecnico di Milano. During her Ph.D., conducted in collaboration with Innovhub Stazioni Sperimentali per l'Industria – area Carta e Cartone and focused on the production and applications of high value-added nanodimensional materials from cellulose-containing sources, she did part of her research in France, at the INP PAGORA GRENOBLE - LGP2 LABORATORY (Laboratory of Pulp and Paper Science). Co-author of 12 scientific papers, one book chapter and one patent application, her research activities mainly focus on production, characterization and applications of nanocellulose obtained from virgin and waste sources, combined with the synthesis of cellulose-based nanostructured organic materials used in environmental decontamination (water treatment for removal of dyes, heavy metals and pesticides), in the field of sensing and heterogeneous catalysis. The focus of her research is also on the environmental sustainability of the materials produced and the processes used to produce them, studied through Life Cycle Assessment analyses conducted in collaboration with other groups at the Politecnico di Milano, with a focus also on the eco-safety of materials, developed in collaboration with the faculty of eco-toxicology at the University of Siena. She has numerous collaborations with various national and international research groups, with which she is currently conducting scientific research in all the areas previously described.

# Preface to "Advances in Cellulose-Based Hydrogels"

Cellulose is the most abundant natural biopolymer on Earth. With an estimated annual production of $1.5 \times 10^{12}$ tons globally, and the possibility of its extraction even from waste sources, it is considered an almost inexhaustible source of raw material capable to make up for the growing demand for environmentally friendly and biocompatible products. Within this framework, cellulose-based hydrogels usually combine hydrophilicity, biodegradability, non-toxicity, and biocompatibility together with low costs and massive availability, which make them extremely attractive in both academic and industrial fields. Possible application fields include biomedical engineering (e.g., tissue engineering and regenerative medicine, drug/cell delivery systems, and 3D printing and bioprinting), progress in smart systems (e.g., sensors, actuators, and soft robotics) and stimuli-responsive systems (e.g., pH- or thermo-responsive hydrogels), the agricultural sector (e.g., soil conditioning, nutrient carriers, and water reservoirs), and water purification.

This volume collects the recent progress in cellulose-based hydrogels, including gels prepared from natural cellulose and its derivatives, cellulose graft co-polymers, and composite gels based on cellulose, covering key aspects of cellulose-based hydrogels, including design, characterization, as well as application-focused research.

We want to deeply thank all contributing authors whose expert contributions made the publication of this Special Issue possible. We would also like to express our deepest appreciation to the editorial team, especially Ms. Miranda Song at MDPI for her encouragement, technical guidance, editing, and publication of this Special Issue.

**Christian Demitri, Lorenzo Bonetti, and Laura Riva**
*Editors*

*Editorial*

# Editorial on the Special Issue "Advances in Cellulose-Based Hydrogels"

Lorenzo Bonetti [1,*], Christian Demitri [2] and Laura Riva [1]

1. Department of Chemistry, Materials and Chemical Engineering "G. Natta", Politecnico di Milano, Via Luigi Mancinelli 7, 20131 Milan, Italy
2. Department of Engineering for Innovation, Campus University Ecotekne, Università del Salento, Via per Monteroni, 73100 Lecce, Italy
* Correspondence: lorenzo.bonetti@polimi.it; Tel.: +39-02-2399-4741

Cellulose is one of the most ubiquitous and naturally abundant biopolymers found on Earth and is primarily obtained from plants and other biomass sources. It is considered to be a nearly limitless raw material supply that is able to meet the rising demand for ecologically friendly and biocompatible products, with a huge annual production worldwide (~$1.5 \times 10^{12}$ tons) and a capacity for extraction even from waste sources. Cellulose-based hydrogels are particularly appealing for both the academic and industrial fields, since they often combine hydrophilicity, biodegradability, non-toxicity, and biocompatibility with a low cost and wide availability. Biomedical engineering, advances in smart systems (such as sensors, actuators, and soft robotics), stimuli-responsive systems (such as pH- or thermo-responsive hydrogels), the agricultural industry (such as soil conditioning, nutrient carriers, and water reservoirs), and water purification are just some of the potential areas of application.

This field of research has continuously and exponentially grown over the past decade, as demonstrated by the rising number of annual publications on cellulose-based hydrogels, which amount to only 8 in the 2000s versus 497 in 2022 (Scopus database). This book includes 15 papers and reviews covering a wide variety of cellulose-based hydrogel applications, ranging from drug delivery to heterogeneous catalysis. Together, these papers provide a valuable overview of the fields engaged in research on cellulose-based hydrogels, offering a perspective on the challenges and opportunities for future development in this area.

In recent years, cellulose-based hydrogels have attracted significant attention in the biomedical field owing to both their outstanding intrinsic properties (e.g., biocompatibility and degradability) and the possibility of obtaining designable functions through different preparation methods and structure designs. In this regard, the review conducted by Zou et al. [1] covers the recent advances in research on cellulose-based hydrogels, introducing their applications and future developments in the field. In the same vein, the paper of Filip et al. describes self-assembled gels based on natural deep eutectic solvents and hydroxypropyl cellulose, with strong antibacterial and antifungal activities [2]. These systems were tested in vitro in the presence of *S. aureus*, *E. coli*, *P. aeruginosa*, and *C. albicans*, confirming their antimicrobial activity and showing a good biocompatibility with primary cells (human gingival fibroblasts). In their work, Blažic et al. synthesized and fully characterized cellulose-g-poly (2-(dimethylamino)ethylmethacrylate) hydrogels [3]. The obtained hydrogels showed a weak antibacterial activity in vitro against *E. coli*, *P. aeruginosa*, and *B. subtilis*. However, cyclic voltammetry (CV) and electrochemical impedance spectroscopy (EIS) were also performed, demonstrating that these systems, despite their weak antibacterial performance, have good characteristics as supercapacitors. A novel hydrogel based on polyvinyl alcohol (PVA), carboxymethyl cellulose (CMC), and polyethylene glycol (PEG) was designed by Yang et al. for the growth of zeolitic imidazolate framework-L (ZIF-L) in situ [4]. The release of $Zn^{2+}$ and imidazolyl groups elicited a synergistic antibacterial activity against *S. aureus* in vitro. At the same time, the obtained hydrogel promoted cell proliferation at an early stage, enhancing its coagulation efficiency. A rat liver injury model was further employed to confirm the rapid hemostasis capacity of the developed hydrogel.

Cellulose-based hydrogels have successfully been employed as delivery systems to improve the controlled release of particles and (bio)molecules, even as a function of physiological and/or pathological stimuli (e.g., temperature and pH variations). In this regard, Bonetti et al. [5] designed a novel methylcellulose (MC)-based hydrogel for the controlled release of silver nanoparticles (AgNPs) in infected chronic wounds, characterized by a local pH increase. The obtained MC hydrogels showed swelling and degradation behaviors that were dependent on both the pH and temperature, and a noteworthy pH-triggered release of AgNPs (release ~10 times higher at pH = 12 than pH = 4). In their work, Liu et al. [6] fabricated innovative electrospun core–shell nanohybrids composed of hydroxypropyl methylcellulose (HPMC) and acetaminophen (AAP) in the core sections and composites of polyvinylpyrrolidone (PVP) and sucralose in the shell sections. Interestingly, HMPC gelation ensured a faster release of AAP, which could be beneficial for potential orodispersible drug delivery applications. Shaik et al. [7] reported an oleogel-based bigel derived from a guar gum hydrogel and a sesame oil/candelilla wax, which they further tailored using date palm-derived cellulose nanocrystals (dp-CNC). The addition of dp-CNC as a novel reinforcing agent allowed these bigels to act as carriers of moxifloxacin hydrochloride (MH).

The biomedical applications of cellulose-based hydrogels are not limited to the above-mentioned studies but can also be extended to the field of conductive hydrogels. In this regard, Kasaw Gebeyehu et al. [8] present a review of the current state of the art of conductive hydrogel manufacturing based on cellulosic materials used for tissue engineering, in addition to reporting the current scenario and the possible future developments in the field. In this regard, Jin et al. [9] report on conductive hydrogel inks consisting of CMC, tannic acid, and metal ions ($HAuCL_4$) developed for on-skin direct printing. These hydrogels showed self-healing properties and reversible conductivity under cyclic strain, and they were successfully on-skin printed, achieving the continuous electrical flow of the electronic circuit under the conditions of skin deformation. In their work, Park et al. [10] present a soft stretchable conductive hydrogel composite consisting of alginate (Alg), CMC, polyacrylamide (PAM), and silver flakes (AgF). The obtained hydrogel fully supported the operation of a light-emitting diode under the conditions of mechanical deformation and successfully enabled the measurement of electromyogram signals.

Another important area in which cellulose-based hydrogels have found application is photo-catalysis, with a focus on water remediation. Photocatalytic systems hold great promise as innovative alternatives to common environmental decontamination systems, and the review conducted by Yang et al. [11] extensively describes the recent advances in cellulose-based hydrogel photo-catalysts, first discussing the properties and preparation methods and then classifying these systems according to the type of catalyst and the research progress in different fields. In the context of sustainable technological development, the work of Riva et al. [12] reports on hydrogel-derived nano-sponges that are capable of capturing heavy metal ions from aqueous solutions. Based on their observations of this interesting property, the authors describe the application of these materials in the field of heterogeneous catalysis, successfully catalyzing the formation of aromatic acetals using nanosponges loaded with $Cu^{2+}$ and $Zn^{2+}$ metals and achieving extremely high yields and a very good selectivity for the desired products.

Cellulose-based hydrogels can also find application in other industrial fields, such as food packaging and the agricultural sectors. In this regard, the work conducted by Fujita et al. [13] investigates the production of a biodegradable alternative to polyacrylic-acid-based superabsorbent spheres, producing spherical hydrogel particles from CMC that could be useful in industrial and agricultural applications. Meanwhile, Jo et al. [14] report on the preparation of cellulose pulp (CP), polyurethane (PU), and curcumin-based composite films using a cost-effective method based on the use of N-methylmorpholine N-oxide (NMMO) as a solvent. The obtained composite films presented good antioxidant activities and the absence of cytotoxic effects on a HaCat cell line in vitro, thus lending themselves well to packaging and biomedical applications.

Lastly, Zitzmann et al. [15] propose a novel analytical tool for visualizing the cellulose content in defibrillated cellulose derived from microalgae using Carbotrace 480. Exploiting the distinctive fluorescent properties of the optotracer, this study provides a means through which to correlate the cellulose content in the samples with the successful hydrogel formation.

**Funding:** This research received no external funding.

**Conflicts of Interest:** The authors declare no conflict of interest.

# References

1. Zou, P.; Yao, J.; Cui, Y.N.; Zhao, T.; Che, J.; Yang, M.; Li, Z.; Gao, C. Advances in Cellulose-Based Hydrogels for Biomedical Engineering: A Review Summary. *Gels* **2022**, *8*, 364. [CrossRef] [PubMed]
2. Filip, D.; Macocinschi, D.; Balan-Porcarasu, M.; Varganici, C.D.; Dumitriu, R.P.; Peptanariu, D.; Tuchilus, C.G.; Zaltariov, M.F. Biocompatible Self-Assembled Hydrogen-Bonded Gels Based on Natural Deep Eutectic Solvents and Hydroxypropyl Cellulose with Strong Antimicrobial Activity. *Gels* **2022**, *8*, 666. [CrossRef] [PubMed]
3. Blažic, R.; Kučić Grgić, D.; Kraljić Roković, M.; Vidović, E. Cellulose-g-poly(2-(dimethylamino)ethylmethacrylate) Hydrogels: Synthesis, Characterization, Antibacterial Testing and Polymer Electrolyte Application. *Gels* **2022**, *8*, 636. [CrossRef] [PubMed]
4. Yang, H.; Lan, X.; Xiong, Y. In Situ Growth of Zeolitic Imidazolate Framework-L in Macroporous PVA/CMC/PEG Composite Hydrogels with Synergistic Antibacterial and Rapid Hemostatic Functions for Wound Dressing. *Gels* **2022**, *8*, 279. [CrossRef] [PubMed]
5. Bonetti, L.; Fiorati, A.; D'agostino, A.; Pelacani, C.M.; Chiesa, R.; Farè, S.; De Nardo, L. Smart Methylcellulose Hydrogels for pH-Triggered Delivery of Silver Nanoparticles. *Gels* **2022**, *8*, 298. [CrossRef] [PubMed]
6. Liu, X.; Zhang, M.; Song, W.; Zhang, Y.; Yu, D.G.; Liu, Y. Electrospun Core (HPMC–Acetaminophen)–Shell (PVP–Sucralose) Nanohybrids for Rapid Drug Delivery. *Gels* **2022**, *8*, 357. [CrossRef] [PubMed]
7. Shaikh, H.M.; Anis, A.; Poulose, A.M.; Madhar, N.A.; Al-Zahrani, S.M. Development of Bigels Based on Date Palm-Derived Cellulose Nanocrystal-Reinforced Guar Gum Hydrogel and Sesame Oil/Candelilla Wax Oleogel as Delivery Vehicles for Moxifloxacin. *Gels* **2022**, *8*, 330. [CrossRef] [PubMed]
8. Gebeyehu, E.K.; Sui, X.; Adamu, B.F.; Beyene, K.A.; Tadesse, M.G. Cellulosic-Based Conductive Hydrogels for Electro-Active Tissues: A Review Summary. *Gels* **2022**, *8*, 140. [CrossRef] [PubMed]
9. Jin, S.; Kim, Y.; Son, D. Tissue Adhesive, Conductive, and Injectable Cellulose Hydrogel Ink for On-Skin Direct Writing of Electronics. *Gels* **2022**, *8*, 336. [CrossRef] [PubMed]
10. Park, K.; Choi, H.; Kang, K.; Shin, M.; Son, D. Soft Stretchable Conductive Carboxymethylcellulose Hydrogels for Wearable Sensors. *Gels* **2022**, *8*, 92. [CrossRef] [PubMed]
11. Yang, J.; Liu, D.; Song, X.; Zhao, Y.; Wang, Y.; Rao, L.; Fu, L.; Wang, Z.; Yang, X.; Li, Y.; et al. Recent Progress of Cellulose-Based Hydrogel Photocatalysts and Their Applications. *Gels* **2022**, *8*, 270. [CrossRef]
12. Riva, L.; Lotito, A.D.; Punta, C.; Sacchetti, A. Zinc- and Copper-Loaded Nanosponges from Cellulose Nanofibers Hydrogels: New Heterogeneous Catalysts for the Synthesis of Aromatic Acetals. *Gels* **2022**, *8*, 54. [CrossRef] [PubMed]
13. Fujita, S.; Tazawa, T.; Kono, H. Preparation and Enzyme Degradability of Spherical and Water-Absorbent Gels from Sodium Carboxymethyl Cellulose. *Gels* **2022**, *8*, 321. [CrossRef] [PubMed]
14. Jo, C.; Kim, S.S.; Rukmanikrishnan, B.; Ramalingam, S.; Prabakaran, D.S.; Lee, J. Properties of Cellulose Pulp and Polyurethane Composite Films Fabricated with Curcumin by Using NMMO Ionic Liquid. *Gels* **2022**, *8*, 248. [CrossRef] [PubMed]
15. Zitzmann, F.L.; Ward, E.; Matharu, A.S. Use of Carbotrace 480 as a Probe for Cellulose and Hydrogel Formation from Defibrillated Microalgae. *Gels* **2022**, *8*, 383. [CrossRef] [PubMed]

*Article*

# Biocompatible Self-Assembled Hydrogen-Bonded Gels Based on Natural Deep Eutectic Solvents and Hydroxypropyl Cellulose with Strong Antimicrobial Activity

Daniela Filip [1], Doina Macocinschi [1], Mihaela Balan-Porcarasu [2,*], Cristian-Dragos Varganici [3], Raluca-Petronela Dumitriu [1], Dragos Peptanariu [3], Cristina Gabriela Tuchilus [4] and Mirela-Fernanda Zaltariov [5,*]

1. Laboratory of Physical Chemistry of Polymers, "Petru Poni" Institute of Macromolecular Chemistry, Aleea Gr. Ghica Voda 41 A, 700487 Iasi, Romania
2. Laboratory of Polycondensation and Thermostable Polymers, "Petru Poni" Institute of Macromolecular Chemistry, Aleea Gr. Ghica Voda 41 A, 700487 Iasi, Romania
3. Centre of Advanced Research in Bionanoconjugates and Biopolymers, "Petru Poni" Institute of Macromolecular Chemistry, Aleea Gr. Ghica Voda 41 A, 700487 Iasi, Romania
4. Microbiology Department, Faculty of Medicine, "Gr. T. Popa" University of Medicine and Pharmacy, 16 Universitatii Street, 700115 Iasi, Romania
5. Department of Inorganic Polymers, "Petru Poni" Institute of Macromolecular Chemistry, Aleea Gr. Ghica Voda 41 A, 700487 Iasi, Romania

* Correspondence: mihaela.balan@icmpp.ro (M.B.-P.); zaltariov.mirela@icmpp.ro (M.-F.Z.)

**Abstract:** Natural deep eutectic solvents (NADES)-hydroxypropyl cellulose (HPC) self-assembled gels with potential for pharmaceutical applications are prepared. FT-IR, [1]HNMR, DSC, TGA and rheology measurements revealed that hydrogen bond acceptor–hydrogen bond donor interactions, concentration of NADES and the water content influence significantly the physico-chemical characteristics of the studied gel systems. HPC-NADES gel compositions have thermal stabilities lower than HPC and higher than NADES components. Thermal transitions reveal multiple glass transitions characteristic of phase separated systems. Flow curves evidence shear thinning (pseudoplastic) behavior. The flow curve shear stress vs. shear rate were assessed by applying Bingham, Herschel–Bulkley, Vocadlo and Casson rheological models. The proposed correlations are in good agreement with experimental data. The studied gels evidence thermothickening behavior due to characteristic LCST (lower critical solution temperature) behavior of HPC in aqueous systems and a good biocompatibility with normal cells (human gingival fibroblasts). The order of antibacterial and antifungal activities (*S.aureus*, *E.coli*, *P. aeruginosa* and *C. albicans*) is as follows: citric acid >lactic acid > urea > glycerol, revealing the higher antibacterial and antifungal activities of acids.

**Keywords:** deep eutectic solvents; hydroxypropyl cellulose; rheology; antimicrobial gels; biocompatibility

## 1. Introduction

Natural deep eutectic solvents (NADES) belong to a new generation of solvents [1–3] comprising mixtures of cheap and readily obtainable components which generate eutectics with melting points significantly lower that of their individual components due to ion–dipole interaction or hydrogen bonding. They are ionic green solvents which are not volatile, not flammable and a low-cost alternative to room temperature ionic liquids. Their use opens the possibility of replacing toxic imidazolium ionic liquids by more sustainable compounds that can be applied in the development of biocompatible and biodegradable drug delivery responsive systems [4].

Their unique properties recommend them in varied practical applications ranging from extraction and biocatalysis to biomedical ones. In biomedical field NADES can be used as biopolymer modifiers, acting as template delivery compounds also knows as "therapeutic deep eutectic solvents", being able to solubilize and stabilize different pharmaceutical products [5].

NADES are interesting liquid-like gels in which H-bonds mediate anion binding, and are able to produce major changes in bulk material characteristics. The field of "eutectogels" is less developed than that of ionogels based on classical ionic liquids. Thus, NADES can be used as template for supramolecular gelators [6], being environmentally friendly and providing low preparation costs, non-inflammability, increased thermal and chemical stability, low toxicity and biodegradability. They are developed as a good alternative for poor soluble drugs, being used as excipients in the pharmaceutical industry [7].

In NADES, choline chloride is the hydrogen bond acceptor (HBA) and the other natural components such as alcohols, acids, amides, amines and sugars are hydrogenbond donors (HBD). Choline chloride is a substituted quaternary ammonium salt, a human and animal nutrient [8]. It is found in high quantities in eggs, liver, peanuts, meats and vegetables. This advantage promises great possibilities for drug-delivery systems, bone-therapy scaffolds, cosmetic, pharmaceutical and food-related applications [9]. It was suggested that NADES represent the third liquid phase in organisms [2]. Their increased viscosities constitute a disadvantage but adding water is employed to modulate physicochemical properties and facilitate their applications [10].

Water-based NADES can be an alternative to DESs of high viscosities, poor conductivities and higher toxicities [11]. The organic solvents are not adequate for pharmaceutical applications, therefore NADES are preferred to solubilize the hydrophobic drugs. Interest in NADES for pharmaceutical applications is recent [12,13]. They possess intrinsic antimicrobial activity, can promote absorption and diminish phenomena like polymorphism or degradation.

Hydroxypropyl cellulose (HPC) is a cellulose ether derivative which is obtained by hydroxypropylation of hydroxyl groups of the cellulose backbone, i.e., by reacting alkali cellulose with propylene oxide at elevated pressure and temperature. The hydroxyl groups as well as the incorporated hydroxypropyl groups are able to donate hydrogen bonds to active pharmaceutical ingredients with hydrogen accepting groups [14]. HPC serves well as an excipient for pharmaceuticals. It is also used as emulsifier, stabilizer and thickener in pharmaceutical applications, additives in cosmetics, formulation aid and texturizer for foods.

In this study, we investigate the properties of hydrophilic polymer matrices comprising HPC and four different NADES (NADES gels): choline chloride: urea 1:2 (ChCl-U), choline chloride: citric acid 1:1 (ChCl-CA), choline chloride: lactic acid 1:1 (ChCl-LA), choline chloride: glycerol 1:2 (ChCl-Gly) in aqueous systems. The rheological behavior of HPC gels at various concentrations of NADES is investigated. The thermal transitions and thermal stability of HPC-NADES compositions are investigated by DSC and TGA analyses, respectively. The strength of hydrogen bonding in the HPC-NADES gels is investigated by means of FT-IR and $^1$HNMR spectroscopy. The antimicrobial activities of pure NADES and NADES-HPC gels were studied against Gram positive bacteria (*Staphylococcus aureus* ATCC 25923), Gram negative bacteria (*Escherichia coli* ATCC 25922, *Pseudomonas aeruginosa* ATCC 27853) and pathogenic yeasts (*Candida albicans* ATCC 90028). The biocompatibility of NADES gels was evaluated on HGF (human gingival fibroblast) cell line.

## 2. Results and Discussion
### 2.1. FT-IR Spectroscopy

The HPC-NADES gels are prepared by physical interactions, mainly H-bonding starting from original NADES containing two components: choline chloride (ChCl) as hydrogen acceptor and four different hydrogen donors: urea, glycerol, lactic acid and citric acid. Based on the IR and NMR data it was possible to establish the HPC spectral changes by addition of various amounts of NADES: 17% and 29%, mainly in the O-H spectral region by involving of O-H groups in the H-bonding with NADES components. The IR comparative spectra of HPC solution 14%, original NADES in solution and two

HPC-NADES mixtures with 17% and 29% NADES containing water as third component can be seen in Figure 1.

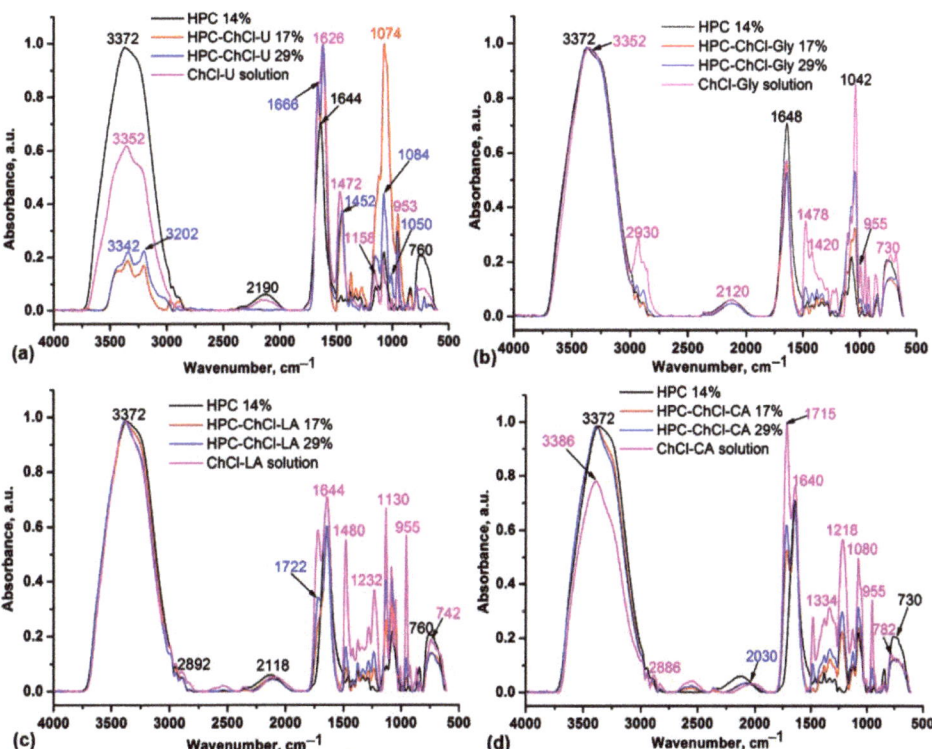

**Figure 1.** Comparative IR spectra in ATR (Attenuated Total Reflectance) mode of the HPC-ChCl-U (a), HPC-ChCl-Gly (b), HPC-ChCl-LA (c) and HPC-ChCl-CA (d) at 17% and 29% NADES concentrations compared with the starting components HPC solution 14% and original NADES in solution.

It is evident that the –OH stretching bands of the HPC-NADES aq. solutions are blueshifted by 30 cm$^{-1}$ as compared with HPC solution 14% and by 10 cm$^{-1}$ as compared with original NADES (ChCl-U) in solution (Figure 1a). In the HPC-ChCl-Gly, HPC-ChCl-LA and HPC-ChCl-CA mixtures the O-H stretches are close to those found in HPC solution 14% (Figure 1b–d). These are redshifted by 20 cm$^{-1}$ compared to the corresponding original NADES in solution and confirm their involvement in the H-bonding interactions. The C-H stretches are present at 2962–2886 cm$^{-1}$, while the broad bands at 2540–2030 cm$^{-1}$ are characteristic for H-bonding O-H groups. In the 1800–1500 cm$^{-1}$ spectral range specific stretching vibrations for carboxylic C=O groups in HPC-ChCl-LA and HPC-ChCl-CA can be seen at 1722 and 1715 cm$^{-1}$, respectively (Figure 1c,d).

The C=O stretching band of aq. HPC 14% located at 1682 (sh) and 1644 cm$^{-1}$ is explained with deformative water molecules [15]. The C=O stretching band of HPC-NADES aq. solutions are close to that of HPC solution 14% and influenced by C=O stretching bands of studied NADES. The characteristic bands associated with ChCl located at about 1480 cm$^{-1}$ attributed to ρCH$_3$ and the band at 955 cm$^{-1}$ attributed to ammonium structure identity of the NADES (Ch)$^+$ are distinct from the bands of HPC aq. solution and decrease their intensities with decreasing concentration of NADES [16]. The C-O-C stretching bands of HPC-NADES aq. solutions located in the 1200–1000 cm$^{-1}$ region are close to those of HPC 14% $w/v$ and influenced by the bands of NADES.

NADES have similar bands to the starting HBD and HBA molecules [17,18]. For ChCl-CA 17% and ChCl-Gly 17% samples the peaks at 955 cm$^{-1}$ vanish (Figure 1b,d). This result is explained by the strength of hydrogen bonding and less phase separation in comparison with ChCl-LA and ChCl-U NADES. In general, ChCl-U, ChCl-Gly, ChCl-LA and ChCl-CA are high viscosity liquid systems, with high hygroscopicity mainly due to the ChCl component, so that these can be isolated with lower water (adsorbed moisture from surrounding atmosphere) content in their composition. In binary mixtures of studied NADES there is a clear contribution of HBD and HBA role in the system, while in the ternary and quaternary systems, resulting from the addition of water and HPC, there is a more complex hydrogen bonding network in the system.

We studied the effect of water addition in the original NADES (prepared in solid state by simple plastering the components) by IR spectroscopy in order to highlight that water can stabilize HBD components in the liquid phase and mediate the interaction with choline chloride (HBA component). In Figure 2 the IR spectra of ChCl-U during the addition of different amounts of water are shown. One can observe the redshift of the O-H and N-H stretches by 33–58 cm$^{-1}$ (Figure 2a), respectively, and of O-H and N-H deformations (Figure 2b) at 1662 and 1610 cm$^{-1}$ by adding water as a third component in the NADES system. These spectral changes confirmed that the strong H-bond interactions between ChCl and urea gradually decrease in the presence of water, also leading to the decrease in the intensity of the characteristic vibrations at 1480 and 955 cm$^{-1}$ (Figure 2c) and the appearance of broad bands at 800–600 cm$^{-1}$, specifically for different H-bonding water molecules (Figure 2d).

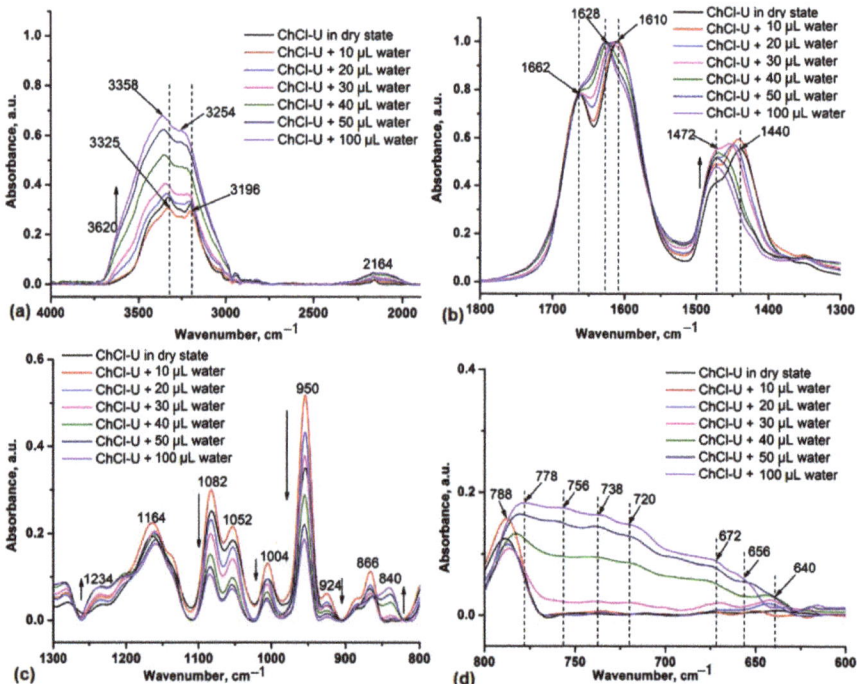

**Figure 2.** IR spectra of ChCl-U during the addition of water in the 4000–1800 cm$^{-1}$ (a), 1800–1300 cm$^{-1}$ (b), 1300–800 cm$^{-1}$ (c) and 800–600 cm$^{-1}$ (d) spectra range. The vertical dashed lines highlight the shifts of the main absorptions in the presence of different amounts of water.

For all HPC-based NADES and water/NADES systems, there was a decrease in H-bond energy by addition of both, water and HPC solution 14%, as compared with these systems in dry state (Table 1). This process is accompanied by the reorganization of the H bonding interactions between both components of NADES, as can be seen in a schematic representation (Scheme 1), by involving water and HPC chains.

Table 1. The energy and the distance of hydrogen bonds in NADES and HPC-NADES samples.

| Sample | HPC-NADES (Dry State) | | HPC-NADES (in Solution) | |
|---|---|---|---|---|
| | $E_H$ (kJ) | R(Å) | $E_H$ (kJ) | R(Å) |
| HPC | 27.832 | 2.753 | 15.318 | 2.792 |
| HPC-ChCl-U 17% | 20.401 (N-H) | 2.717 (N-H) | 14.885 (N-H) | 2.851 (N-H) |
| | 22.151 (O-H) | 2.769 (O-H) | 20.281 (O-H) | 2.776 (O-H) |
| HPC-ChCl-U 29% | 20.552 (N-H) | 2.714 (N-H) | 15.860 (N-H) | 2.827 (N-H) |
| | 22.366 (O-H) | 2.770 (O-H) | 20.568 (O-H) | 2.775 (O-H) |
| HPC-ChCl-Gly 17% | 21.863 | 2.771 | 20.065 | 2.777 |
| HPC-ChCl-Gly 29% | 20.281 | 2.776 | 20.136 | 2.776 |
| HPC-ChCl-LA17% | 19.633 | 2.778 | 19.921 | 2.777 |
| HPC-ChCl-LA 29% | 18.482 | 2.782 | 19.489 | 2.778 |
| HPC-ChCl-CA 17% | 18.770 | 2.781 | 18.627 | 2.782 |
| HPC-ChCl-CA 29% | 16.397 | 2.788 | 18.770 | 2.781 |
| Original NADES | in dry state | | in solution | |
| ChCl-U | 21.006 (N-H) | 2.703 (N-H) | 16.623 (N-H) | 2.808 (N-H) |
| | 23.373 (O-H) | 2.766 (O-H) | 21.001 (O-H) | 2.774 (O-H) |
| ChCl-Gly | 24.092 | 2.764 | 22.150 | 2.770 |
| ChCl-LA | 22.798 | 2.768 | 20.424 | 2.775 |
| ChCl-CA | 21.575 | 2.772 | 18.771 | 2.781 |

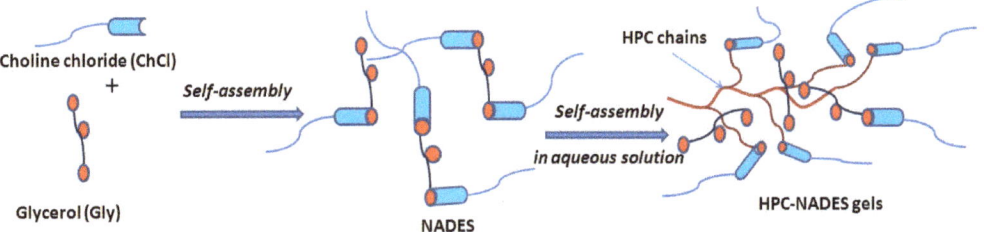

Scheme 1. Schematic representation of the reorganization of H-bonding (donor–acceptor) by addition of HPC.

The H-bond distance also highlighted this aspect, by slightly increasing values of its value, supporting the interconnectivity between NADES, HPC and/or water components. Moreover, one can see from Table 1 that the H bonds distance of HPC-NADES gels differ slightly in solution and in solid state as compared with original NADES. Higher differences appear to initial HPC, the values of H-bond distance being larger than the HPC in solution. This suggests that original H-bonding NADES co-exist with H-bonded HPC-NADES in the gel mixture. Thus, it was found that the presence of HBD (hydrogen bond donor) molecules favors the interaction with polymer and diminishes the interaction with water molecules and therefore the solute–solvent interactions in the ternary solutions are similar to those in binary solutions. The NADES complex separated into its precursors can be treated as a pseudo-compound when the HBA (hydrogen bond acceptor): HBD molar ratio is maintained and the system is pseudo-ternary. Stronger hydrogen bonding of NADES determines favorable interactions and increased solvation capability of the carbohydrates.

H-bonds energy ($E_H$) and distance (R) of original NADES and HPC-NADES solutions (Table 1) have been estimated by using Sederholm and Struszczyk methods by a previously published procedure [19].

One can observe that in dry state in the original NADES there is a strong interaction between both components, the higher energy of the H bonds being found for ChCl-Gly, followed by ChCl-U, ChCl-LA and ChCl-CA, while in water these interactions considerably decrease. For HPC-NADES mixture in dry state, the energy of the H bonds increases at lower concentration (17%) of NADES. A strong interaction was observed for HPC-ChCl-U 29%, which is maintained inclusively in solution of 14% HPC. A lower interaction was found for HPC-ChCl-CA both in dry state and in solution. The value of the distance of H bonds is lower for HPC-ChCl-U and HPC-ChCl-Gly at both concentrations, also proving a strong interaction within the ternary systems.

## 2.2. $^1$H NMR and ROESY Spectra

There are several NMR studies concerning the molecular interactions in NADES based on ChCl and CA, LA, U and Gly, respectively, and their mixtures with water [20–24]. These studies show that adding more than 10% water to the NADES leads to weakening of the hydrogen bond network which is reflected in the ROESY spectra by the disappearance of the intermolecular correlation peaks and in the $^1$H NMR spectra by the collapse of all the peaks for labile protons and water into one peak, due to the fast exchange that takes place in presence of water.

The same trend was observed in the case of our mixtures, where the ROESY spectra (Figures S1–S8) recorded for mixtures of ChCl-U-$H_2O$ (1:2:6 molar ratio) (Figures S1 and S2), ChCl-Gly-$H_2O$ (1:2:11 molar ratio) (Figures S3 and S4), ChCl-LA-$H_2O$ (1:1:7 molar ratio) (Figures S5 and S6) and ChCl-CA-$H_2O$ (1:1:10 molar ratio) (Figures S7 and S8), and show only intramolecular correlation peaks and no intermolecular interactions.

In the $^1$H NMR spectrum for ChCl-U 100% (Figure 3a) we can observe the following peaks (δ, ppm): 6.08 (U-*NH$_2$*), 5.35 (ChCl-*OH*), 4.42 ($H_2O$), 3.94 (ChCl-*CH$_2$*-OH), 3.49 (ChCl-*CH$_2$*-N), 3.18 (ChCl-N-(*CH$_3$*)$_3$). This eutectic mixture also contains intrinsic water (adsorbed water during the preparation) and the molar ratio between ChCl:U:$H_2O$, as calculated from the integral ratio, is 1:2:0.6. In the $^1$H NMR spectra of HPC-ChCl-U 29% and HPC-ChCl-U 17% from Figure 3b,c, we can observe the peaks from HPC (1.1 ppm), 3–4.5 ppm, overlapped with the peaks from ChCl) and the peaks for U-*NH$_2$* and ChCl-*OH* are still observed at 5.75–5.77 ppm and 5.32–5.36 ppm, respectively, but their intensity is diminished due to the exchange with deuterium from the solvent. Urea is known to form very strong hydrogen bonds which leads to a slower exchange rate, thus the peaks for the labile protons are still observed in the $^1$H NMR spectra even at higher water contents. A ROESY experiment (Figure S9) was performed for HPC-ChCl-U 17% in order to evaluate if the supramolecular structure of ChCl-U is maintained after mixing it with HPC and water. No intermolecular correlation peaks were observed between the three components, meaning that the hydrogen bond network considerably weakened.

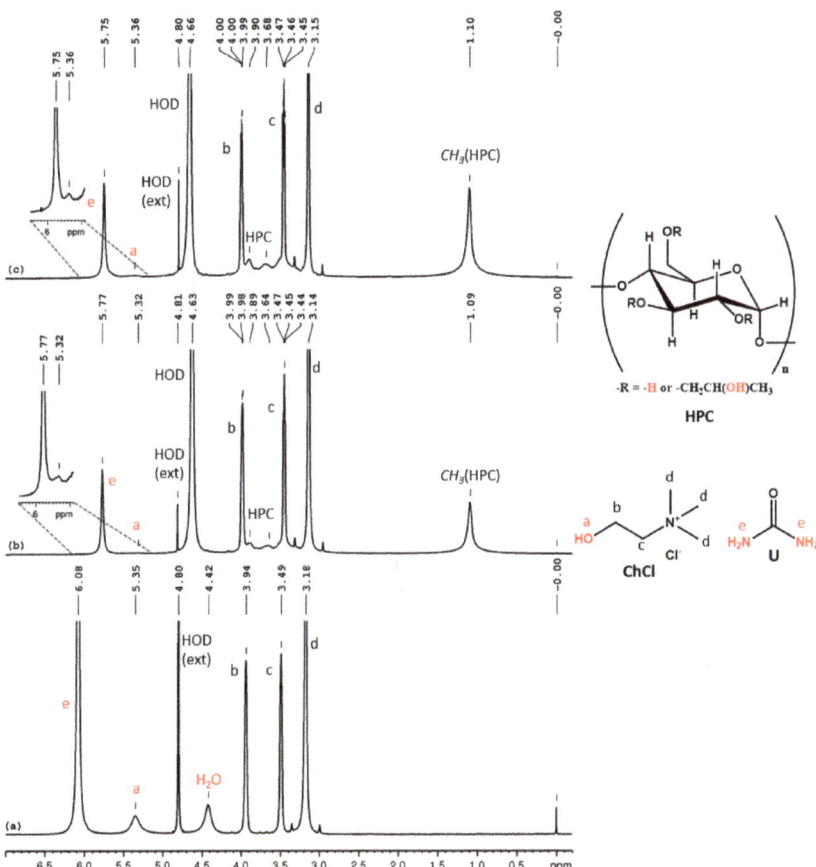

**Figure 3.** $^1$H NMR spectra of ChCl-U 100% (a), HPC-ChCl-U 29% (b) and HPC-ChCl-U 17% (c). In red color are marked the exchangeable protons.

In the $^1$H NMR spectrum of ChCl-Gly 100% (Figure 4a) we can observe the following peaks (δ, ppm): 5.16 (ChCl-*OH*),4.89 (Gly-CH-*OH*), 4.80 (Gly-CH$_2$-*OH*, partially overlapped with external HOD), 4.33 (H$_2$O), 3.87 (ChCl-*CH$_2$*-OH), 3.55 (ChCl-*CH$_2$*-N), 3.46–3.35 (Gly-*CH$_2$*- and Gly-*CH*-) and 3.11 (ChCl-N-(*CH$_3$*)$_3$). Although no water was added during the preparation of ChCl-Gly, the $^1$H NMR spectrum shows the presence of a small content of water. The molar ratio of the ChCl:Gly:H$_2$O ternary mixture, calculated from the integral values, is 1:2:2.5. Figure 4b,c shows the $^1$H NMR spectra for two mixtures of different proportions of ChCl-Gly, HPC and D$_2$O. We can observe the peak for the methyl group from the 2-hydroxypropyl substituent of cellulose at 1.10 ppm but the rest of the peaks of HPC, which resonate between 3 and 4.5 ppm, are overlapped with the peaks from ChCl and Gly. The hydroxylic protons from ChCl and Gly appear with the HOD peak due to fast exchange with the water protons.

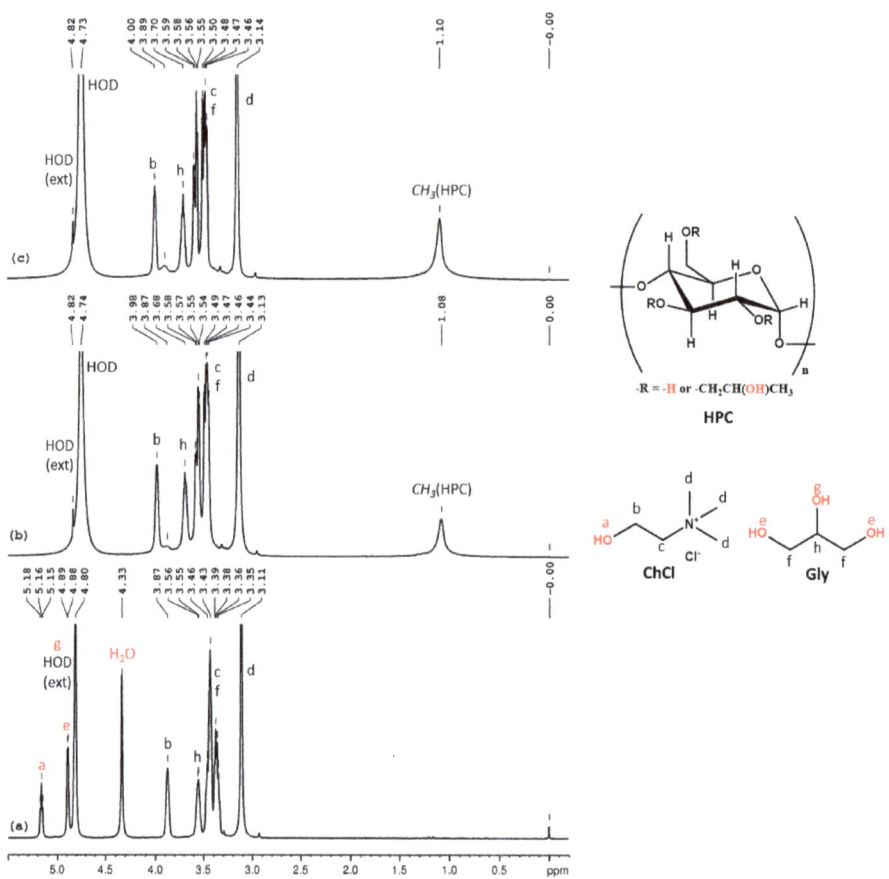

**Figure 4.** $^1$H NMR spectra of ChCl-Gly (**a**), HPC-ChCl-Gly 29% (**b**) and HPC-ChCl-Gly 17% (**c**). In red color are marked the exchangeable protons.

The assignments of the peaks from the $^1$H NMR spectrum of neat ChCl:LA (Figure 5a) are as follows (δ, ppm): 6.09 (LA-*OH*, LA-*COOH*, ChCl-*OH*, H$_2$O), 4.28 (LA-*CH*-), 3.96 (ChCl-*CH$_2$*-OH), 3.58 (ChCl-*CH$_2$*-N), 3.26 (ChCl-N-(*CH$_3$*)$_3$) and 1.31 (LA-*CH$_3$*). Since carboxylic protons are more acidic than alcohols, a faster exchange occurs in the case of ChCl-LA and a single, time-averaged signal for all the labile protons and intrinsic water is observed in the $^1$H NMR spectrum. In Figure 5b,c the $^1$H NMR spectra for HPC-ChCl-U 17% and HPC-ChCl-U 29% are depicted. Since the addition of water increases the rate of exchange, we can observe the displacement of the peak for the labile and water protons from 6.09 ppm to 4.8 ppm and the presence of the HPC peaks (1.10 ppm and 3–4.5 ppm, overlapped with the ChCl and LA peaks).

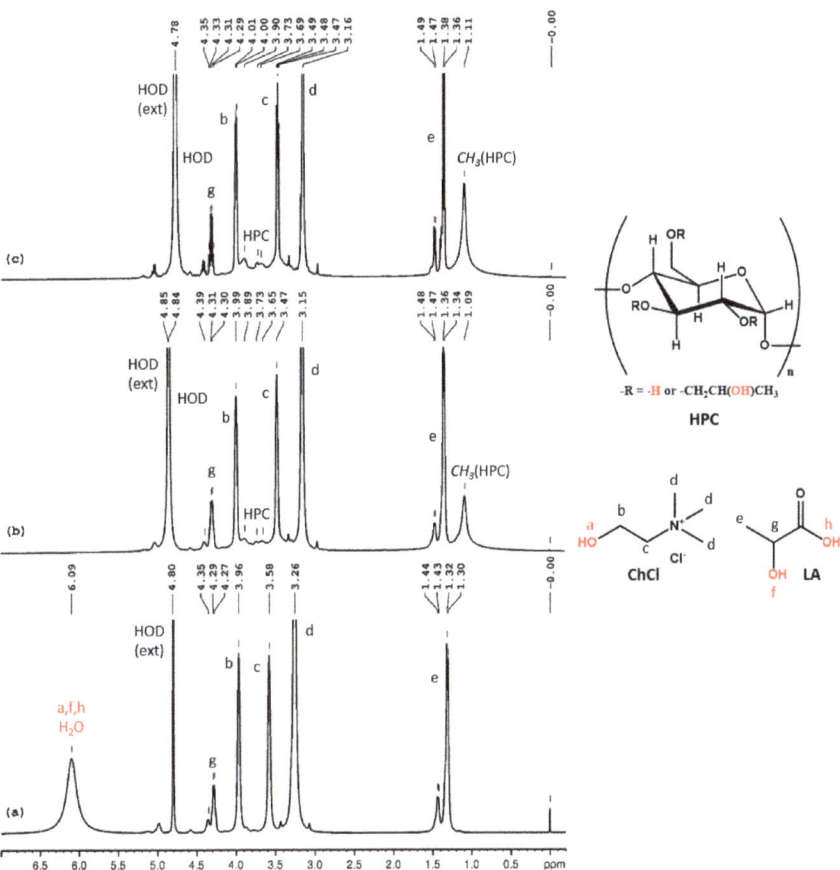

**Figure 5.** $^1$H NMR spectra of ChCl-LA (**a**), HPC-ChCl-LA 29% (**b**) and HPC-ChCl-LA 17% (**c**). In red color are marked the exchangeable protons.

ChCl-CA 100% could not be analyzed by NMR due to its high viscosity. We recorded the $^1$H NMR spectrum (Figure 6a) for a less viscous mixture of 80% ChCl-CA and 20% D$_2$O. The assignments of the peaks are (δ, ppm): 6.16 (CA-*COOH*, CA-*OH*, ChCl-*OH*, HOD), 3.95 (CHCl-*CH$_2$*-OH), 3.42 (CHCl-*CH$_2$*-N), 3.09 (ChCl-N-(*CH$_3$*)$_3$), 2.94 (CA-*CH$_2$*) and 2.79 (CA-*CH$_2$*). Since D$_2$O was used for the sample preparation, the intrinsic water content of this mixture could not be calculated from NMR data. In the $^1$H NMR spectra for 14% HPC, 29% ChCl-CA and 57% D$_2$O (*w/w/w*) and 16% HPC, 16% ChCl-CA and 66% D$_2$O (*w/w/w*) the peaks for HPC are observed at 1.10 ppm and 3–4.5 ppm (overlapped with ChCl and CA peaks). The peak for HOD and for the labile protons shifts from 6.16 in ChCl-CA to 4.96 and 4.86 ppm in the two samples (Figure 6b,c) due to increasing rate of exchange determined by increasing the amount of water from the system.

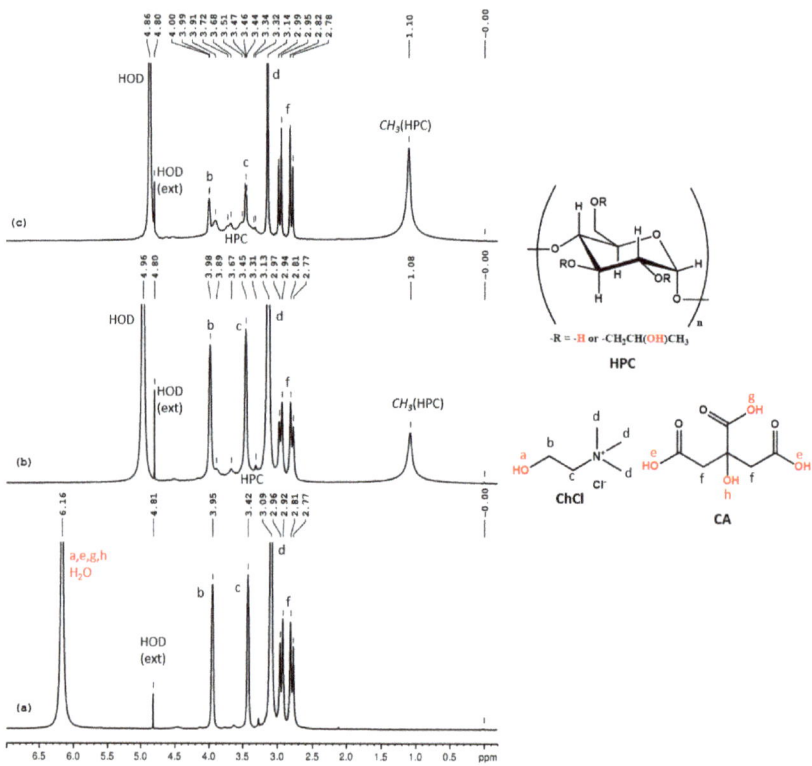

**Figure 6.** $^1$H NMR spectra of ChCl-CA (**a**), HPC-ChCl-CA 29% (**b**) and HPC-ChCl-CA 17% (**c**). In red color are marked the exchangeable protons.

### 2.3. Thermal Stability

The thermal stabilities and kinetic parameters of pure studied NADES and their gels with HPC are investigated using thermogravimetry and derivative thermogravimetry under dynamic conditions of temperature. The thermogravimetric curves of the studied pure NADES and gel mixtures with HPC (17 and 29% $w/w$) are shown in Figure 7. The thermal characteristics obtained from TG and DTG curves are summarized in Table 2.

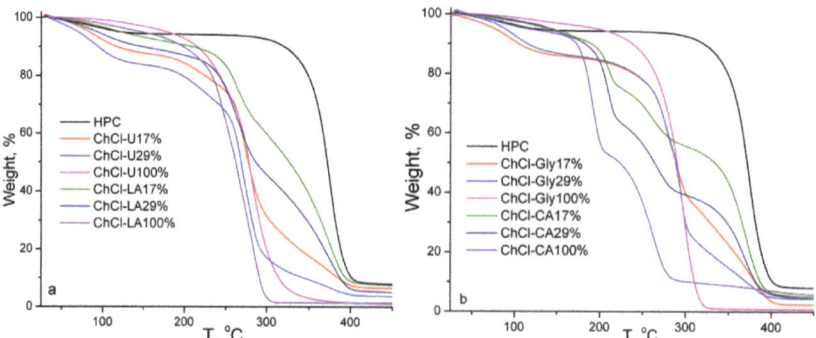

**Figure 7.** Cont.

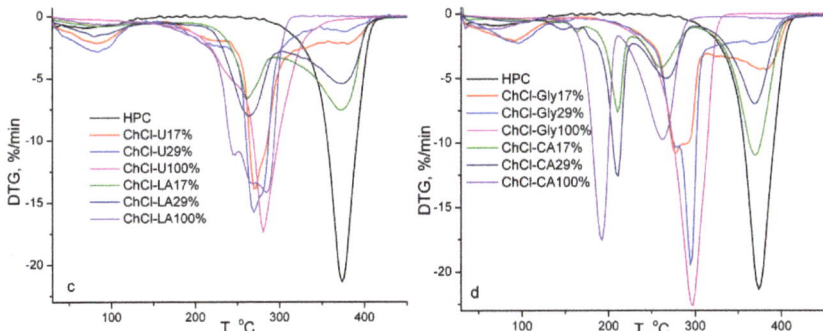

**Figure 7.** TG and DTG curves of HPC, NADES and HPC-NADES gels (17% and 29% w/w).

**Table 2.** The thermal characteristics from TG and DTG curves of HPC, NADES and HPC-NADES gels (17% and 29% w/w).

| Sample | Stages, °C | $T_{max}$ [a], °C | $\Delta w$ [b], % |
|---|---|---|---|
| ChCl-CA 100% | I 36.8–121.6 | 61.7 | 6.8 |
|  | II 171.7–200.9 | 191.6 | 40.9 |
|  | III 225.5–277.2 | 262.5 | 41.2 |
|  | IV 307.8–419.0 | 364.3 | 3.2 |
| ChCl-Gly 100% | I 109.7–176.9 | 128.0 | 5.5 |
|  | II 228.2–318.9 | 296.3 | 92.5 |
| ChCl-LA 100% | I 74.5–143 | 89.1 | 4.87 |
|  | II 171.3–232.9 | 206.1 | 14.5 |
|  | III 232.9–315.0 | 284.6 | 78.1 |
| ChCl-U 100% | I 127.9–223.2 | 179.0 | 10.9 |
|  | II 223.2–269.3 | 237.0 | 24.8 |
|  | III 269.3–320.8 | 280.7 | 62.4 |
| HPC | I 26.7–122.7 | 72.6 | 6.7 |
|  | II 306.3–399.7 | 374.0 | 85.1 |
| ChCl-CA 17% | I 157.3–177.2 | 162.8 | 8.9 |
|  | II 194.4–216.2 | 210.4 | 16.6 |
|  | III 236.6–282.7 | 260.5 | 18.2 |
|  | IV 328.2–391.6 | 369.5 | 51.8 |
| ChCl-Gly 17% | I 43.5–120.4 | 90.8 | 17.2 |
|  | II 249.3–301.5 | 277.3 | 47.8 |
|  | III 334.7–398.6 | 382.2 | 31.9 |
| ChCl-LA 17% | I 61.6–143.3 | 89.8 | 9.48 |
|  | II 231.4–280.0 | 261.6 | 27.3 |
|  | III 301.3–396.3 | 372.2 | 55.8 |
| ChCl-U 17% | I 36.3–121.6 | 88.0 | 12.8 |
|  | II 184.3–259.1 | 214.0 | 14.4 |
|  | III 259.1–298.2 | 270.5 | 39.6 |
|  | IV 298.2–404.3 | 380.1 | 25.7 |
| ChCl-CA 29% | I 69.7–135.5 | 90.0 | 3.4 |
|  | II 135.5–156.8 | 147.2 | 4.95 |
|  | III 191.7–216.8 | 210.1 | 28.8 |
|  | IV 238.5–285.3 | 266.1 | 23.0 |
|  | V 320.7–391.6 | 368.9 | 35.3 |
| ChCl-Gly 29% | I 55.5–126.9 | 95.8 | 15.9 |
|  | II 249.7–287.0 | 278.2 | 31.1 |
|  | III 287.0–303.5 | 295.0 | 26.7 |
|  | IV 303.5–393.7 | 366.7 | 20.6 |

Table 2. Cont.

| Sample | Stages, °C | $T_{max}$ [a], °C | $\Delta w$ [b], % |
|---|---|---|---|
| ChCl-LA 29% | I 40.9–125.8 | 79.8 | 12.1 |
| | II 223.3–281.1 | 264.4 | 40.2 |
| | III 298.5–398.6 | 374.2 | 42.3 |
| ChCl-U 29% | I 38.9–116.6 | 83.7 | 15.5 |
| | II 182.1–247.5 | 206.9 | 13.5 |
| | III 254.6–292.6 | 269.4 | 57.3 |
| | IV 334.9–402.9 | 377.6 | 8.4 |

[a] Temperature corresponding to maximum rate of decomposition. [b] Weight loss percentage corresponding to degradation stage.

Generally, the values of initial decomposition temperatures for NADES are between those of pure constituents, HBD and HBA. The thermal stability of HBDs determines the thermal stabilities of the resulting NADES, i.e., the thermal stabilities of NADES increase compared to pure HBDs and become worse compared to HBAs [25–27]. The thermal stability of NADES is influenced by hydrogen bonding between HBA and HBD molecules. The degradation stages below 150 °C are associated with evaporation of water. It is evident from Figure 7 and Table 2 that the thermal stability of HPC is superior to those of pure NADES and their mixtures with HPC because the onset degradation temperature of HPC is higher. The thermal degradation is complex because the main degradation peaks are comprised of several processes of thermal degradation.

The non-isothermal kinetic parameters of the thermal degradation for overlapped processes following water evaporation were evaluated in terms of the Coats–Redfern [28], Flynn–Wall [29] van Krevelen [29] and Urbanovici–Segal [30] integral methods (Table 3). The activation energy of thermal degradation is regarded as a semi-quantitative factor of thermal stability. It is evident from Table 3 that the values of activation energy and order of reaction are found higher for NADES and their mixtures with HPC (the degradation stage corresponding to HPC component) based on citric acid and glycerol. This result confirms the importance of the hydrogen bonding network in dictating the thermal stability which is in connection with the FT-IR, DSC results and rheology measurements.

Table 3. Kinetic parameters of thermal degradation of HPC, NADES and HPC-NADES gels (17 and 29% $w/w$).

| Sample | Activation Energy (kJ/mol)/ln A/Order of Reaction | | | |
|---|---|---|---|---|
| | Coats–Redfern | Flynn–Wall | Van Krevelen | Urbanovici–Segal |
| ChCl-CA 100% | II 367.1/92.13/1.7 | 356.2/90.35/1.7 | 382.2/96.0/1.7 | 354.9/88.8/1.5 |
| | III 211.2/43.69/1.3 | 209.0/43.26/1.3 | 233.0/48.72/1.4 | 214.2/44.43/1.3 |
| ChCl-Gly 100% | 163.98/30.17/1.1 | 164.5/30.32/1.1 | 182.8/34.2/1.1 | 165.67/30.6/1.1 |
| ChCl-LA 100% | 96.8/16.31/0.9 | 100.2/17.38/0.9 | 120.8/21.74/1.0 | 96.1/16.18/0.8 |
| ChCl-U 100% | 51.3/4.86/0.1 | 55.8/6.84/0.0 | 70.4/9.42/0.4 | 53.4/5.55/0.2 |
| HPC | 222.2/36.71/1.1 | 221.1/36.53/1.1 | 251.3/42.23/1.2 | 224.0/37.1/1.1 |
| ChCl-CA 17% [a] | 305.7/53.15/1.4 | 300.6/52.37/1.4 | 337.8/59.25/1.5 | 309.3/53.89/1.4 |
| ChCl-Gly 17% [a] | 235.8/39.55/1.4 | 234.3/39.3/1.4 | 263.7/44.84/1.5 | 239.9/40.37/1.4 |
| ChCl-LA 17% [a] | 160.4/25.64/1.3 | 156.8/24.96/1.2 | 178.6/29.09/1.3 | 157.2/25.0/1.2 |
| ChCl-U 17% [a] | 186.6/30.47/1.4 | 187.4/30.67/1.4 | 213.9/35.7/1.5 | 189.8/31.16/1.4 |
| ChCl-CA 29% [a] | 192.4/31.60/1.2 | 192.8/31.70/1.2 | 214.0/35.65/1.2 | 194.4/32.03/1.2 |
| ChCl-Gly 29% [a] | 108.1/15.57/1.1 | 112.6/16.78/1.1 | 122.6/18.36/1.1 | 111.2/16.26/1.1 |
| ChCl-LA 29% [a] | 150.8/23.79/1.3 | 153.2/24.36/1.3 | 168.9/27.22/1.3 | 153.3/24.35/1.3 |
| ChCl-U 29% [a] | 234.0/39.27/1.5 | 232.6/39.04/1.5 | 252.7/42.71/1.5 | 237.3/39.95/1.5 |

[a] The kinetic parameters were evaluated for thermal degradation stage corresponding to HPC component.

## 2.4. DSC Analysis

In Figure 8 the DSC thermograms (second heating runs) of the HPC, NADES and HPC-NADES gel mixtures (17% and 29% $w/w$) are shown. The thermal characteristics resulting from DSC are summarized in Table 4.

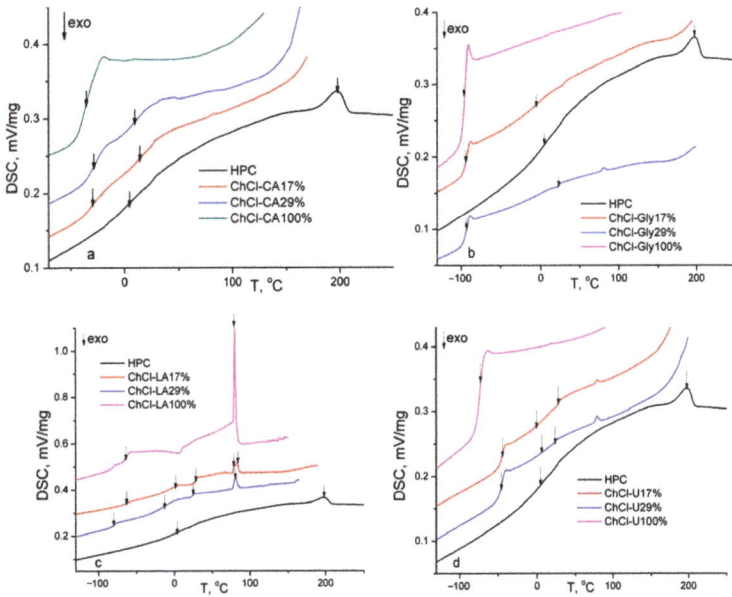

**Figure 8.** DSC thermograms (second heating runs) of HPC, NADES and HPC-NADES gels (17% and 29% $w/w$).

**Table 4.** DSC results (second heating runs) of HPC, NADES and HPC-NADES gels (17% $w/w$).

| Sample | $T_g$, °C | $T_m$, °C | Sample | $T_g$, °C | $T_m$, °C |
|---|---|---|---|---|---|
| ChCl-CA 100% | −34.1 | - | ChCl-LA 100% | −64.7 | -[a] |
| HPC-ChCl-CA 17% | −28.7; 13.9 | - | HPC-ChCl-LA 17% | −63.5;0.8; 28.2 | -[b] |
| HPC-ChCl-CA 29% | −27.6; 9.9 | - | HPC-ChCl-LA 29% | −79.5; −14.5; 24.2 | - |
| ChCl-Gly 100% | −96.2 | - | ChCl-U 100% | −73.7 | - |
| HPC-ChCl-Gly 17% | −93.3; −5.5 | - | HPC-ChCl-U 17% | −44.8; −0.2; 28.1 | - |
| HPC-ChCl-Gly 29% | −93.9; 23.7 | - | HPC-ChCl-U 29% | −45.7; 5.0; 23.5 | - |
| HPC | 4.8 | 197.2 | - | - | - |

[a] 80.0 °C attributed to solid-solid transition; [b] 79.5/83.1 °C attributed to solid-solid transition.

The pure NADES evidence mainly glass transitions, this variant being called low transition temperature mixtures [3]. It is speculated that the low amounts of water determine the increased viscosity and decreased molecular mobility, promoting glass formation instead of crystal [31]. Only ChCl-U reveals melting endotherm at 15 °C (first heating run). Pure ChCl-LA reveals at around 80 °C a solid–solid transition found in case of choline chloride-urea systems for choline chloride rich composition ($\chi_{ChCl}$ =0.5–0.67) [31,32]. Multiple glass transitions are observed which is characteristic for phase separated systems. Each phase has its own glass transition. For the ChCl-CA and ChCl-Gly mixtures two glass transitions corresponding to the NADES and HPC component are observed. In the case of ChCl-LA and ChCl-U mixtures three glass transitions are observed. The Tg at highest temperature is found around 28 °C for both mixtures and is attributed to a phase composed mainly of HPC. The Tg of HPC (powder) is found at 21 °C (second heating run, 10 °C/min). The three values of the glass transitions indicate higher degree of phase segregation for ChCl-LA and ChCl-U mixtures due to less strength of the hydrogen bonding. The values of Tgs in mixtures (for the phase rich in NADES) are higher than those of pure NADES suggesting the formation of the hydrogen bonding network. Some polymer complexes have higher Tg values because hydrogen bonds act as physical crosslinks [33]. It was found in literature that the values of $T_g$ of mixtures of biomaterials are in connection with the number of hydroxyl groups per molecule [34] intermolecular hydrogen bonding and molecular packing. The decrease of the Tg corresponding to HPC component indicates a plasticization of HPC by NADES.

## 2.5. Rheological Properties

The study of rheological properties of pharmaceutical systems (simple liquids, ointments, creams, pastes, suppositories, suspensions and colloidal dispersing, emulsifying and suspending agents) are very important as a quality control instrument in order to assure product quality and diminished batch-to-batch discrepancies. These studies contribute to the characterization of manufacturing, storage and transport processes, or the behavior during pharmaceutical products' administration or therapeutic outcome [35–37].

### 2.5.1. Flow Curves

In Figure 9a,b the flow behavior of NADES-HPC gel solutions (17% and 29% $w/w$ NADES) is illustrated. The rheological plots (log-log scale) of NADES-HPC gel solutions show (Figure 9a) non-Newtonian, shear thinning (pseudoplastic fluid) behavior. The viscosities decrease by adding NADES in solutions in the following order: HPC 14% > ChCl-CA 29% > ChCl-Gly 29% > ChCl-U 29% > ChCl-LA 29%. The rheological behavior of these systems is governed by the formation of an extensive network of inter- and intramolecular interactions such as hydrogen bonding, assuring high cohesive forces in the bulk liquid. The strength of hydrogen bonding and viscosities of the starting NADES influences the resulting viscosity values of the NADES-HPC aqueous solutions. The hydrogen bond density of the NADES is determined by the number of hydrogen bonds donated. The hydrogen bond donor count values for citric acid (4) (https://pubchem.ncbi.nlm.nih.gov/compound/Citric-acid), glycerol (3) (https://pubchem.ncbi.nlm.nih.gov/compound/Glycerol), urea (2) (https://pubchem.ncbi.nlm.nih.gov/compound/Urea) and lactic acid (2) https://pubchem.ncbi.nlm.nih.gov/compound/Lactic-acid) are in agreement with viscosity, thermal stability, DSC and FT-IR results.

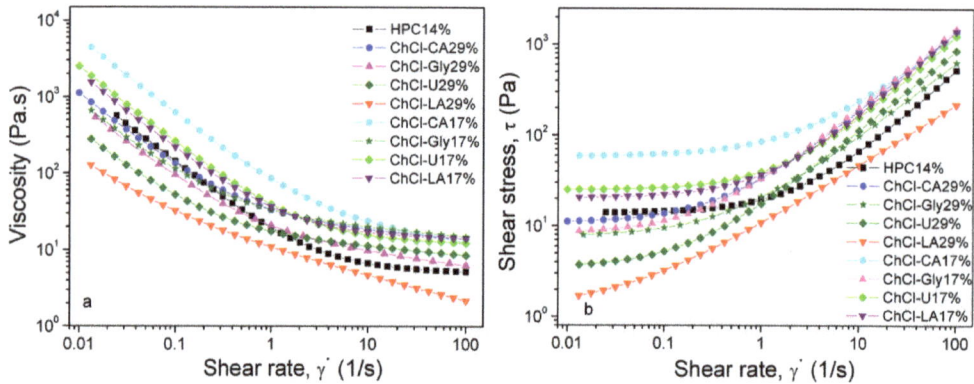

**Figure 9.** The flow behavior of the HPC (solution 14%) and HPC-based NADES gels (17% and 29%): viscosity (**a**) and shear stress (**b**).

The addition of water or polymers changes the flow behavior and internal resistance of the resulting systems. This result can be corroborated with IR, DSC and TGA results. In case of NADES or ionic liquids shear thinning is associated with breaking of the hydrogen bonds. Electrostatic interactions and the re-arrangement of the ions could contribute to shear thinning as well. In case of polymers the disentanglement of the polymer coils and alignment of the polymer chain into the flow direction are also associated with shear thinning phenomenon.

From Figure 9b it is evident that the studied solutions exhibit yield stress ($\tau_0$) which must be exceeded prior to deformation or flowing of the fluid [38]. The yield stress is obtained through extrapolation of the flow curve at low shear rates to zero shear rate. For shear stress values below the yield stress values the fluid behaves as a rigid solid. Usually, the yield stress is considered as the transition stress between elastic solid-like behavior and viscous liquid-like behavior and is connected to the characteristic network structure. When external stress is greater than yield stress the flow curve does not pass through origin and may be linear or non-linear. For a non-Newtonian fluid the shear stress vs. shear rate curve is non-linear or does not pass through the origin. Shulman model [39] was chosen to characterize the shear stress vs. shear rate curves:

$$\tau = \left[\tau_0^{1/n} + (\mu\gamma\cdot)^{1/m}\right]^n$$

where $\tau$ is shear stress, $\tau_0$ is the yield stress, $\mu$ is the plastic viscosity (Pa..s), $K$ is consistency index (Pa.s$^n$), $\gamma\cdot$ is shear rate and $m$, $n$ are power exponents related to material properties. The following simpler rheological models derived by reducing the coefficients were applied for the studied shear stress-shear rate curves:

1. Bingham model [40]:

$$\tau = \tau_0 + \mu\gamma \quad (1)$$

2. Herschel–Bulkley model [41]:

$$\tau = \tau_0 + K\gamma\cdot^n \quad (2)$$

3. Vocadlo model [42]

$$\tau = \left(\tau_0^{1/n} + K\gamma\cdot\right)^n \quad (3)$$

4. Casson model [43]

$$\tau^{0.5} = \tau_0^{0.5} + (\mu\gamma\cdot)^{0.5} \quad (4)$$

When the flow index (dimensionless) $n = 1$ the rheological models reduce to the Bingham model. For $n < 1$ the system is non-Newtonian pseudoplastic (shear thinning) whereas for $n > 1$ (unusual) [44] the system is shear thickening (dilatant). A lower value of $n$ indicates a more non-Newtonian shear thinning fluid (increased pseudoplasticity). The behavior of NADES is mainly Newtonian and non-Newtonian when they are similar to ionic liquids [45]. The rheology of HPC gels is well described by Herschel-Bulkley model [46]. The values of $n$ close to 1 (Herschel Bulkley model in the Table 5) indicate that Bingham model is more appropriate. The values of $K$ serve as the viscosity indices of the systems.

The value of $\tau_0$ refers to the amount of minimum stress necessary for disrupting the networked structure in order to initiate the flow. The yield stress gives information on chain rigidity, hydrogen bonding and molecular weight being connected to viscosity values [47]. The calculated values of $\tau_0$ are in good agreement with the experimental ones. The values of $\tau_0$ are in the same order as viscosity for the studied samples (Figure 9b).

By increasing the value of exponent >0.5 in the Casson equation approaching the Bingham model the fitting results are closer to those experimentally found.

Table 5. The shear stress-shear rate curves coefficients of HPC (solution 14%) and HPC-NADES 17% and 29%.

| Sample | Bingham | Herschel-Bulkley | Vocadlo | Casson | $\tau_{0exp}$, Pa |
|---|---|---|---|---|---|
| HPC 14% | $\tau_0$ = 14.78; $\mu$ = 5.11; R = 0.999 | $\tau_0$ = 13.90; $\mu$ = 4.93; n = 0.97; R = 0.9997 | $\tau_0$ = 13.89; $\mu$ = 6.40; n = 0.96; R = 0.9999 | $\tau_0$ = 14.00; $\mu$ = 3.70; R = 0.9957 | 14.18 |
| ChCl-Gly 29% | $\tau_0$ = 18.18; $\mu$ = 6.44; R = 0.99648 | $\tau_0$ = 15.00; $\mu$ = 6.49; n = 1.04; R = 0.98892 | $\tau_0$ = 7.81; $\mu$ = 34.06; n = 0.79; R = 0.99989 | $\tau_0$ = 5.77; $\mu$ = 5.38; R = 0.99964 | 10.62 |
| ChCl-CA 29% | $\tau_0$ = 23.92; $\mu$ = 14.22; R = 0.99859 | $\tau_0$ = 24.40; $\mu$ = 11.35; n = 1.09; R = 0.9902 | $\tau_0$ = 11.74; $\mu$ = 30.10; n = 0.91; R = 0.99971 | $\tau_0$ = 6.67; $\mu$ = 12.29; R = 0.99974 | 13.70 |
| ChCl-U 29% | $\tau_0$ = 12.07; $\mu$ = 8.72; R = 0.99848 | $\tau_0$ = 3.40; $\mu$ = 14.30; n = 0.88; R = 0.99998 | $\tau_0$ = 3.63; $\mu$ = 21.76; n = 0.88; R = 0.99998 | $\tau_0$ = 2.21; $\mu$ = 7.96; R = 0.99976 | 5.38 |
| ChCl-LA 29% | $\tau_0$ = 10.55; $\mu$ = 2.31; R = 0.98363 | $\tau_0$ = 1.10; $\mu$ = 9.96; n = 0.66; R = 0.99994 | $\tau_0$ = 1.53; $\mu$ = 38.74; n = 0.65; R = 0.9999 | $\tau_0$ = 2.88; $\mu$ = 2.02; R = 0.9924 | 3.33 |
| ChCl-CA 17% | $\tau_0$ = 79.95; $\mu$ = 12.57; R = 0.9961 | $\tau_0$ = 62.70; $\mu$ = 17.92; n = 0.97; R = 0.9912 | $\tau_0$ = 59.01; $\mu$ = 208.06; n = 0.71; R = 0.9996 | $\tau_0$ = 80.00; $\mu$ = 8.17; R = 0.9989 | 64.53 |
| ChCl-Gly 17% | $\tau_0$ = 23.81; $\mu$ = 15.22; R = 0.9984 | $\tau_0$ = 11.50; $\mu$ = 17.18; n = 1.02; R = 0.9926 | $\tau_0$ = 8.87; $\mu$ = 39.49; n = 0.88; R = 0.9999 | $\tau_0$ = 23.00; $\mu$ = 13.59; R = 0.9998 | 11.68 |
| ChCl-U 17% | $\tau_0$ = 28.67; $\mu$ = 12.17; R = 0.9998 | $\tau_0$ = 27.50; $\mu$ = 9.65; n = 1.09; R = 0.9936 | $\tau_0$ = 24.82; $\mu$ = 19.27; n = 0.94; R = 0.9999 | $\tau_0$ = 28.00; $\mu$ = 9.15; R = 0.9969 | 26.24 |
| ChCl-LA 17% | $\tau_0$ = 24.38; $\mu$ = 13.99; R = 0.9998 | $\tau_0$ = 22.60; $\mu$ = 11.43; n = 1.09; R = 0.9944 | $\tau_0$ = 20.45; $\mu$ = 19.78; n = 0.95; R = 0.9999 | $\tau_0$ = 24.00; $\mu$ = 11.16; R = 0.9978 | 22.56 |

### 2.5.2. Dynamic Oscillatory Measurements

Viscoelastic properties of HPC-NADES solutions were studied by oscillation. The elastic modulus is linked with the energy stored in elastic deformation whereas the viscous modulus is linked with viscous dissipation effects.

In Figure 10 the dependences of complex viscosity and dynamic moduli, $G'$ and $G''$ on angular frequency for the studied solutions are illustrated.

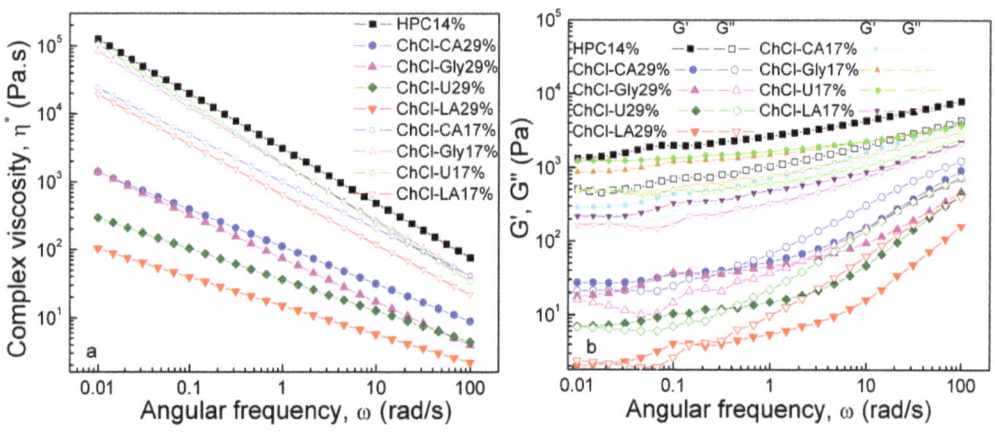

Figure 10. Complex viscosity and dynamic moduli, $G'$ and $G''$ as functions of angular frequency at 20 °C.

Oscillatory measurements are preferred instead of steady shear measurements in order to avoid the disruption of the networks [48]. The departure from the Cox–Merz rule [49]

$$\eta(\gamma \cdot) \cong |\eta * (\omega)|_\omega = \gamma \quad (5)$$

can be explained by the structural damage caused by the excessive shear during flow curves measurements. The complex viscosity decreases at increased concentration of NADES.

The property of viscoelasticity is important in polymeric solutions because they show both liquid and solid-like properties. The extent of intermolecular association/aggregation and chain entanglement determine the relative contributions of the elastic and viscous elements.

The viscoelastic response of the samples is influenced both by composition and concentration of NADES with visible differences in behavior between the two series of samples. It can be noticed that HPC 14% has an elastic (solid-like) behavior ($G' > G''$) over the entire frequency domain studied. The capability of the polymer network to store the imposed energy increases and the behavior is more elastic (solid-like). A similar pattern of behavior is observed for all the mixtures of 17% concentration, with high, but quite close values measured for $G'$ and $G''$.

In the case of 29% concentration series, the storage ($G'$) and loss ($G''$) moduli reveal a modified behavior and much lower values as compared with HPC and the 17-coded samples series. All 29%-compositions show cross-over points when the elastic component outweighs the viscous ones, which indicate transition from a predominantly liquid-like behavior ($G'' > G'$) at increased frequencies towards an elastic/solid-like response at low $\omega$. The cross-over point is located at characteristic frequency the reciprocal of which represents a measure of the relaxation time of the polymer network. In case of polymeric systems the relaxation time $\lambda$ under dynamic shear can be determined by means of the following equation [50,51]:

$$\lambda = \frac{G'}{\eta^* \omega^2} \quad (6)$$

where $G'$ is the storage modulus, $\omega$ is the angular frequency and $\eta^*$ is complex viscosity. The systems with high values of complex viscosity correspond to early relaxations of polymer chains (short relaxation times).

Increasing trend of $\omega$ and $G' = G''$ values at crossover point:
ChCl-LA 29% — $\omega$ = 0.1668//0.02132 rad/s and $G' = G''$ = 4.036//2.186 Pa
ChCl-U 29% — $\omega$ = 0.3683 rad/s and $G' = G''$ = 12.20 Pa
ChCl-CA 29% — $\omega$ = 0.3706 rad/s and $G' = G''$ = 41.24 Pa
ChCl-Gly 29% — $\omega$ = 2.145 rad/s and $G' = G''$ = 54.88 Pa

### 2.5.3. Effect of Temperature

The influence of temperature was investigated by time-temperature sweep measurements performed from 10 to 50 °C with a heating rate of 2 °C/min (Figure 11). The storage modulus ($G'$), loss modulus ($G''$) and complex viscosity ($\eta^*$) were recorded at frequency of 1 Hz, applying a constant stress ($\tau$) value corresponding for each sample in the regime of linear viscoelasticity.

All samples show thermothickening properties. At low temperatures, generally below 20 °C, the mixtures show fluid/viscous-like behavior with a weak elastic contribution. Above 20 °C a sol-gel transition takes place, with increase of elasticity, mainly between 20–50 °C. The transition is sharper for HPC 14%, HPC-ChCl-Gly 29% and HPC-ChCl-U 29% and smoother for HPC-ChCl-CA29% and HPC-ChCl-LA29% depending on the composition of the sample. Solutions behave as elastic gels above 40 °C, but a new crossover point occurred around a dissolution temperature ($T_{dis}$) for HPC-ChCl-U 29% and HPC-ChCl-Gly 29%.

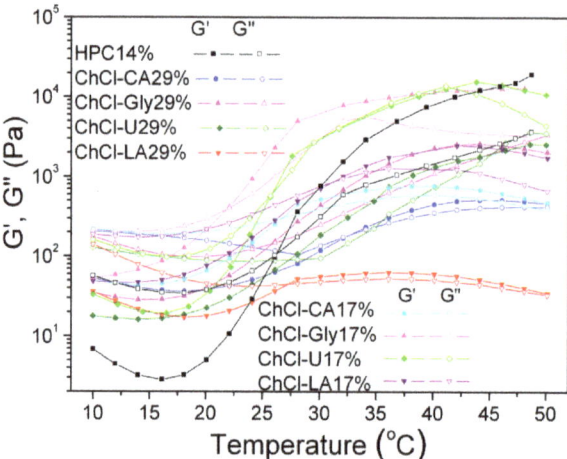

**Figure 11.** Dynamic moduli, $G'$ and $G''$ versus temperature for studied HPC-NADES solutions.

The following cross-over points are observed for the studied solutions:
HPC-$T_{gel}$ = 28.6 °C; $G' = G'' = 642.6$ Pa
ChCl-CA 17% $T_{gel}$ = 26.43 °C; $G' = G'' = 304.87$ Pa
ChCl-Gly 17% $T_{gel}$ = 21.87 °C; $G' = G'' = 384.00$ Pa
ChCl-U 17% $T_{gel}$ = 25.47 °C; $G' = G'' = 575.71$ Pa
ChCl-LA 17% $T_{gel}$ = 31.19 °C; $G' = G'' = 888.78$ Pa
ChCl-CA 29% $T_{gel}$ = 32.1 °C; $G' = G'' = 186.9$ Pa
ChCl-LA 29% $T_{gel}$ = 27.3 °C; $G' = G'' = 43.1$ Pa
ChCl-Gly 29% $T_{gel}$ = 26.2 °C; $G' = G'' = 142.8$ Pa; $T_{dis}$ = 47.2 °C; $G' = G'' = 2611.3$ Pa
ChCl-U 29% $T_{gel}$ = 28.3 °C; $G' = G'' = 94.8$ Pa; $T_{dis}$ = 45.4 °C; $G' = G'' = 2059.8$ Pa

It is well known that HPC exhibits a lower critical solution *temperature* (LCST) at 41–43 °C in aqueous media [52,53].

The phase separation from solution and solidification above a certain temperature, LCST or "cloud point" when water becomes a poor solvent and the solution becomes cloudy is characteristic for thermoresponsive hydrogels (negative temperature-sensitive hydrogels). Below LCST the polymers are soluble and above LCST they become more and more hydrophobic and insoluble with gel formation. Hydrogen bonding and hydrophobic interactions are dependent on temperature and create phase separation. The increase of moduli with increasing temperature is correlated with the structure formation as phase separation and gelation occur nearly at the same time [54]. The LCST behavior is modified for NADES-HPC solutions. Moreover, $T_{dis}$ is associated with upper critical solution temperature behavior (UCST-positive temperature-sensitive hydrogels). Above UCST the polymers are soluble.

### 2.6. Antibacterial and Antifungal Activities

It is crucial to characterize the biological properties of NADES prior to their industrial applications especially in drug delivery systems, pharmaceutical and food-related applications. Recently there have been reported studies on the development of deep eutectic solvents with antibacterial properties [55–57]. Depending on the NADES's composition different antibacterial and antifungal activities are found. The components of NADES give the antibacterial properties to the complex and can manifest synergistic effects [55–61]. The increased antibacterial effect of NADES with acids was reported [55,62]. Table 6 evidences the antibacterial and antifungal activities of the tested NADES towards common pathogen bacteria and yeast strain which were determined by means of disc diffusion method [63,64].

The diameters of the inhibition zones (in mm) corresponding to the tested compounds are shown in Table 6. Results are expressed as means ± SD. It is obvious from Table 6 that the order of antibacterial and antifungal activities is as follows: ChCl-CA > ChCl-LA > ChCl-U > ChCl-Gly, evidencing the higher antibacterial and antifungal activities given by acids. The antibacterial and antifungal activities in the mixtures with HPC give lower values of the diameters of the inhibition zones.

**Table 6.** Antibacterial and Antifungal Activities of the Tested Compounds.

| Compounds | Diameter of Inhibition Zones (mm) | | | |
|---|---|---|---|---|
| | S. aureus ATCC 25923 | E. coli ATCC 25922 | Pseudomonas aeruginosa ATCC 27853 | C. albicans ATCC 90028 |
| ChCl-CA 29% | 20.30 ± 0.57 | 19.00 ± 0.00 | 0.00 | 17.50 ± 0.50 |
| ChCl-Gly 29% | 0.00 | 0.00 | 0.00 | 0.00 |
| ChCl-LA 29% | 16.00 ± 0.00 | 15.10 ± 0.05 | 0.00 | 18.30 ± 0.57 |
| ChCl-U 29% | 10.10 ± 0.05 | 0.00 | 0.00 | 12.00 ± 0.00 |
| ChCl-CA 100% | 29.00 ± 0.00 | 26.00 ± 0.00 | 30.00 ± 0.00 | 28.00 ± 0.00 |
| ChCl-Gly 100% | 0.00 | 0.00 | 0.00 | 13.00 ± 0.00 |
| ChCl-LA100% | 20.00 ± 0.00 | 23.00 ± 0.00 | 15.00 ± 0.00 | 21.10 ± 0.05 |
| ChCl-U 100% | 12.00 ± 0.00 | 0.00 | 0.00 | 15.00 ± 0.00 |
| Ciprofloxacin (5 µg/disc) | 26.70 ± 0.06 | 30.00 ± 0.00 | 32.00 ± 0.00 | * NT |
| Fluconazol (25 µg/disc) | NT * | NT * | NT * | 28.00 ± 0.00 |
| Voriconazol (1 µg/disc) | NT * | NT * | NT * | 32.50 ± 0.50 |

* NT—not tested.

The inhibition zones of the studied samples against S. aureus, E. coli, Pseudomonas aeruginosa and C. albicans are shown in Figure 12.

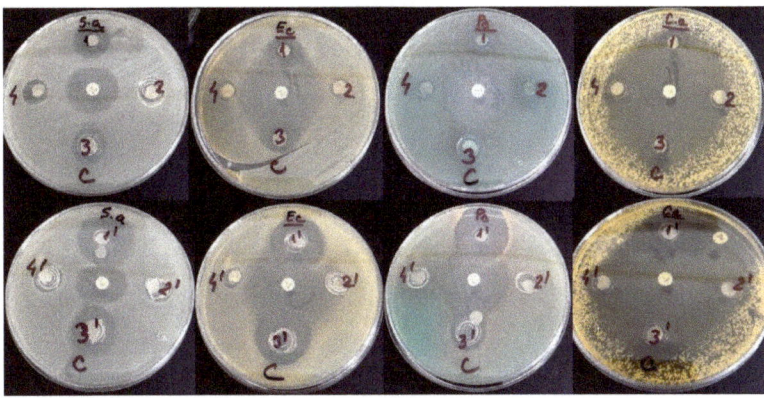

**Figure 12.** Antimicrobial activity of tested compounds HPC-ChCl-CA 29% (1) and ChCl-CA 100% (1′); HPC-ChCl-Gly 29% (2) and ChCl-Gly 100% (2′); HPC-ChCl-LA 29% (3) and ChCl-LA 100% (3′); HPC-ChCl-U 29% (4) and ChCl-U 100% (4′) against Staphylococcus aureus ATCC 25923, E. coli ATCC 25922, Pseudomonas aeruginosa ATCC 27853 and Candida albicans ATCC 90028.

### 2.7. Evaluation of Biocompatibility

Biocompatibility of NADES gels mixture depends on the structure of the mixture components. In general, choline chloride showed a lower cytotoxicity than many other ionic liquids, such as imidazolium or pyridinium [65]. Based on the principle of green chemistry, two characteristics, biocompatibility and biodegradability, are necessary to be investigated before that they can be named "green solvents" or co-solvents for pharmaceutical applications. NADES possess a high potential to improve drug development and release. For use as pharmaceutical excipients these systems must be investigated for their toxicological effects by evaluation of cytotoxicity in human cell line [66]. In order to evaluate the biocompatibility with normal cells (human gingival fibroblast) of the obtained HPC-NADES gels, MTS assay was performed, which allows the estimation of the cell viability and proliferation (Figure 13). The compounds were tested at three concentrations: 1 µg/mL, 10 µg/mL and 100 µg/mL. The results indicated a very good compatibility at all studied concentrations, the cell viability being almost 100%, excepting the NADES gels based on ChCl-CA with a cell viability of 90%. The MTS test is an indirect colorimetric method for estimating cell viability and proliferation. The reagent contains a tetrazolium compound that is transformed by the mitochondria of healthy cells into a colored formazan. The more living cells there are, or the more intense the mitochondrial metabolism, the more formazan will be obtained. Thus, in our case, a stronger signal than in the untreated sample may mean a more intense proliferation or a stimulation of the mitochondrial metabolism. Choline chloride component is mainly responsible for the higher viability, this being a part of the vitamin B complex.

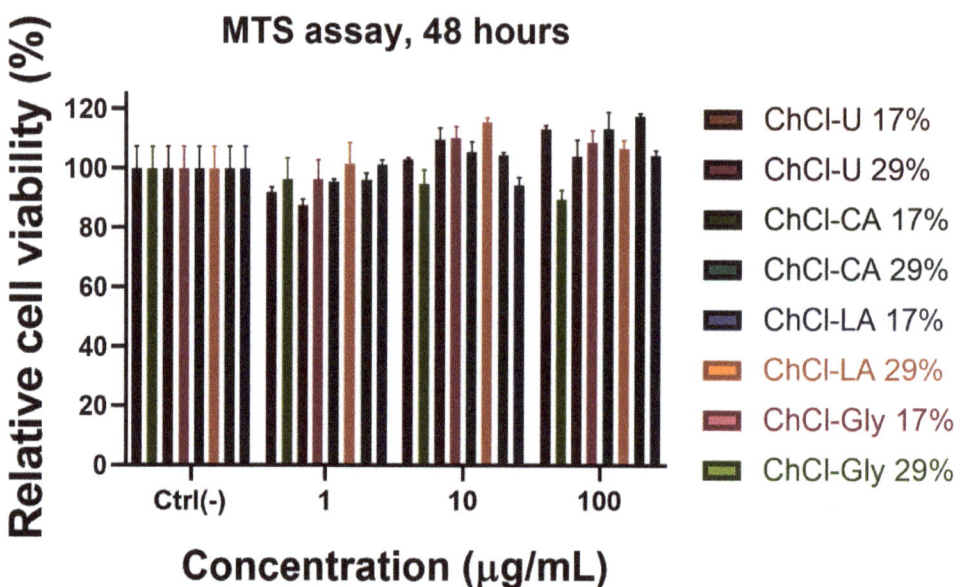

**Figure 13.** MTS assay on human gingival fibroblast of the prepared HPC-NADES gels.

## 3. Conclusions

In this work NADES-HPC gels were prepared. FT-IR, $^1$HNMR, DSC, TGA and rheology data revealed that hydrogen bond acceptor–hydrogen bond donor interactions of the studied NADES, their concentration and water content considerably influence the physico-chemical characteristics of the studied systems. The peak at 955 cm$^{-1}$ in the IR spectra attributed to ammonium structure in NADES vanishes for lower content of NADES in solutions for the systems with increased strength of hydrogen bonding. HPC-NADES gel compositions have thermal stabilities lower than HPC and higher than NADES components. The thermal stability was investigated by means of activation energy and order of reaction. Thermal transitions reveal multiple glass transitions characteristic for phase separated systems. Increased strength of hydrogen bonding network was obtained for citric acid and glycerol-based systems compared to urea and lactic acid-based systems. Flow curves evidence shear thinning behavior. Bingham, Herschel–Bulkley, Vocadlo and Casson rheological models were employed to fit the rheological data. The studied solutions evidence thermothickening behavior because of specific LCST behavior of HPC in aqueous solutions. All prepared NADES-HPC gels proved to have a very good biocompatibility with HGF normal cell line. S. aureus, E. coli, P. aeruginosa and C. albicans were tested to assess the antibacterial and antifungal activities of the investigated systems. The order of antibacterial and antifungal activities is as follows: citric acid >lactic acid > urea > glycerol content in NADES-HPC gels.

## 4. Experimental Part

### 4.1. Materials

Hydroxypropyl cellulose (Klucel LF $M_w$ 95,000 Da) was purchased from Aqualon, Hercules Inc., Wilmington, NC, USA and used as received. Choline chloride, $\geq$98%, was purchased from (Sigma-Aldrich, Merck Romania SRL, Bucharest, Romania, an affiliate of Merck KGaA, Darmstadt, Germany ), urea was purchased from (Sigma-Aldrich, Merck Romania SRL, Bucharest, Romania, an affiliate of Merck KGaA, Darmstadt, Germany), citric acid (99%)was purchased from (Sigma-Aldrich, Merck Romania SRL, Bucharest, Romania, an affiliate of Merck KGaA, Darmstadt, Germany), DL-lactic acid (90%)was purchased from (Sigma-Aldrich, Merck Romania SRL, Bucharest, Romania, an affiliate of Merck KGaA, Darmstadt, Germany) and glycerol was purchased from Merck Romania SRL, Bucharest, Romania, an affiliate of Merck KGaA, Darmstadt, Germany). Choline chloride and urea were dried under vacuum for several days before use.

### 4.2. Preparation of Aqueous HPC-NADES Gel Solutions

HPC solutions (14% $w/v$) were prepared by solving HPC in distilled water. The solutions were kept on a magnetic stirrer at room temperature and were stirred at moderate speed for several hours. Original NADES were prepared by mixing the two initial compounds (choline chloride with the corresponding HBD component: urea, glycerol, lactic acid or citric acid) with added small amounts of water at 60–80°C until homogeneous liquid was obtained, then the mixtures were dried in oven at 25 °C. HPC-NADES aqueous solutions 17% and 29% were obtained by gradual addition of NADES at RT to HPC solution 14%. The compositions (wt%) of the studied HPC-NADES aqueous solutions are shown in Table 7.

Table 7. The composition of the studied NADES and HPC-NADES systems.

| HPC-NADES Components | HPC-NADES Systems | |
| --- | --- | --- |
| | HPC-NADES 17% | HPC-NADES 29% |
| HPC | 1.36 g | 1.12 g |
| ChCl-U 100% | 0.73 g choline chloride + 0.63 g urea (1:2 molar ratio) | 1.25 g choline chloride + 1.07 g urea (1:2 molar ratio) |
| ChCl-Gly 100% | 0.58 g choline chloride + 0.77 g glycerol (1:2 molar ratio) | 1.00 g choline chloride + 1.32 g glycerol (1:2 molar ratio) |
| ChCl-LA 100% | 0.83 g choline chloride + 0.53 g lactic acid (1:1 molar ratio) | 1.41 g choline chloride + 0.91 g lactic acid (1:1 molar ratio) |
| ChCl-CA 100% | 0.57 g choline chloride + 0.78 g citric acid (1:1 molar ratio) | 0.98 g choline chloride + 1.34 g citric acid (1:1 molar ratio) |
| water | 5.28 g | 4.56 g |

### 4.3. Measurements

ATR (attenuated total reflection infrared)-IR (infrared) spectra were measured with an equipment Bruker Vertex 70 having an ATR accessory module equipped with a ZnSe crystal. The IR spectra were performed at room temperature with accumulation of 32 scans. The resolution of the registered spectra was 4 cm$^{-1}$. The O-H and N-H spectral regions in the 3800–3000 cm$^{-1}$ spectral range were deconvoluted with the curve-fitting function accessible from OPUS 6.5 software (Bruker, Ettlingen, Germany). The maxima peak's position was determined by second derivative of the spectra and used to calculate the energy of the hydrogen bonding and the H-bond distance according to a procedure already described [19].

NMR spectra were recorded on a Bruker Avance NEO 400 MHz Spectrometer equipped with a 5-mm QNP direct detection probe and z-gradients, using standard parameter sets provided by Bruker. The $^1$H NMR and ROESY spectra were recorded at room temperature in presence of external $D_2O$ containing TSP (sodium salt of trimethylsilyl propionic acid) and were calibrated on the TSP peak (0 ppm). NMR sample preparation: 0.6 mL of each pure NADES mixture were transferred into NMR tubes and prior to recording the spectra, in each tube a capillary containing $D_2O$ and TSP was added. For HPC-NADES 17% mixtures: a HPC solution in $D_2O$ (14% $w/w$) was prepared by dissolving 0.68 g HPC in 3 mL $D_2O$; in 0.85 g of this HPC in $D_2O$ solution, 0.68 g of NADES were slowly added, stirred for homogenization and transferred into NMR tubes. Prior to recording the spectra, in each tube a capillary containing $D_2O$ and TSP was added. For each of the HPC-NADES 29% mixtures: 0.43 g mixture was weighted and then dissolved in 0.52 mL $D_2O$. The resulting solutions were transferred into NMR tubes and prior to recording the spectra capillaries containing $D_2O$ and TSP were added.

STA 449F1 Jupiter NETZSCH (NETZSCH Analysing and Testing, Netzsch, Germany) equipment for investigation of thermal stability of the HPC/NADES blends was used. The measurements were conducted in the 30–700 °C temperature range, in 50 mL min$^{-1}$ nitrogen flow. The heating rate was 10 °C min$^{-1}$.

DSC 200 F3 Maia equipment (Netzsch, Germany) operating under nitrogen 50 mL min$^{-1}$ flow in heating/cooling rates of 10 °C min$^{-1}$ was used to analyze the NADES/HPC blends. About 10 mg of each sample were heated in aluminum crucibles with pierced and pressed lids from $-150$ °C to 250 °C. Before TGA and DSC analyses, the HPC-NADES aqueous solutions were dehydrated at room temperature for several days followed by drying under vacuum for several days.

The rheological behavior study was accomplished by using an Anton PaarPhysica MCR 301 Rheometer (Anton Paar, Austria), equipped with a 50 mm diameter cone-plate geometry with a 1° angle. Steady shear flow and dynamic oscillatory measurements on HPC-NADES aqueous solutions were carried out at 25 ± 0.1 °C. A Peltier heating system was used for precise temperature control. Flow measurements were performed over the shear rate range 0.05–200 1/s. For the oscillatory shear tests, the logarithmic frequency sweeps were carried out over the angular frequency range 200–0.1/0.05 rad/s. Preliminary strain sweep tests done at 10 rad/s over the strain range 0.01–200%confirmed that the tests were in the linear viscoelastic regime (LVE). Temperature sweep measurements were carried out at a constant shear rate of 20 1/s, over the temperature range between 10 to 50 °C at a heating rate of 2 °C/min. Prior each measurement, after loading, the sample was held for certain time, as previously tested, to permit stress relaxation and temperature equilibration.

4.3.1. Antimicrobial Susceptibility Tests

The antimicrobial activity was studied using Gram positive bacteria (*Staphylococcus aureus* ATCC 25923), Gram negative bacteria (*Escherichia coli* ATCC 25922, *Pseudomonas aeruginosa* ATCC 27853) and a pathogenic yeast (*Candida albicans* ATCC 90028).

The antimicrobial activity was assessed by employing the disk diffusion methods [63,64].

*Disc-diffusion method.* Mueller–Hinton agar (Oxoid) and Mueller–Hinton agar Fungi (Biolab) were inoculated with the suspensions of the tested microorganisms: *Staphylococcus aureus* ATCC 25923, *Escherichia coli* ATCC 25922, *Pseudomonas aeruginosa* ATCC 27853 and *Candida albicans* ATCC 90028. Sterile stainless-steel cylinders (5 mm internal diameter; 10 mm height) were applied on the agar surface in Petri plates. Then, 100 µL of the tested compounds were added into cylinders. The plates were left 10 min at room temperature to ensure the equal diffusion of the compound in the medium and then incubated at 35 °C for 24 h. As reference antimicrobial drugs commercially available discs were used containing Ciprofloxacin (5 µg/disk), Fluconazole (25 µg/disk) and Voriconazole (1 µg/disk). After incubation, the diameters of inhibition were measured. All assays were carried out in triplicate.

### 4.3.2. MTS Assay

Human gingival fibroblast (HGF) cells were cultured in a complete medium containing alpha-MEM, 10% FBS and 1% penicillin-streptomycin-amphotericin B mixture. For cell culture the cells were maintained in a humidified environment with 5% $CO_2$ at 37 °C. After the cells were multiplied sufficiently, the culture medium was removed and the cells were washed with phosphate buffer and then detached with Tryple.

For MTS assay, $25 \times 10^5$ cells/well were seeded into 24-well plates and incubated over night. The next day, the HPC-NADES mixtures (30 mg) were diluted at three concentrations and then placed on the top of each well previously seeded with cells. The thus prepared plates were returned to the incubator.

After 48 h, plates were treated with MTS reagent according to manufacturer's protocol and the obtained formazan was quantified after 2 h by measuring the absorbance at 490 nm with a plate reader (BMG LABTECH, Ortenberg, Germany). The cell viability was estimated as the absorbance of the samples as a percentage of the absorbance of the untreated cells.

**Supplementary Materials:** The following supporting information can be downloaded at: https://www.mdpi.com/article/10.3390/gels8100666/s1, Figure S1. $^1$H NMR spectrum of ChCl-U-$H_2O$ (1:2:6); Figure S2. 2D ROESY spectrum of ChCl-U-$H_2O$ (1:2:6 molar ratio); Figure S3. $^1$H NMR spectrum of ChCl-Gly-H2O (1:2:11); Figure S4. 2D ROESY spectrum of ChCl-Gly-$H_2O$ (1:2:11 molar ratio); Figure S5. $^1$H NMR spectrum of ChCl-LA-$H_2O$ (1:1:7); Figure S6. 2D ROESY spectrum of ChCl-LA-$H_2O$ (1:1:7 molar ratio); Figure S7. $^1$H NMR spectrum of ChCl-CA-$H_2O$ (1:1:10); Figure S8. 2D ROESY spectrum of ChCl-CA-$H_2O$ (1:1:10 molar ratio); Figure S9. 2D ROESY spectrum of HPC-ChCl-U 17%.

**Author Contributions:** Conceptualization, D.F. and M.-F.Z.; data curation, D.F., D.M., M.B.-P., R.-P.D., D.P., C.G.T. and M.-F.Z.; formal analysis, D.F., D.M., C.-D.V., R.-P.D., D.P., C.G.T. and M.-F.Z.; funding acquisition, M.-F.Z.; investigation, D.M., M.B.-P., C.-D.V., R.-P.D., D.P., C.G.T. and M.-F.Z.; methodology, D.M. and C.-D.V.; resources, D.P., C.G.T. and M.-F.Z.; validation, D.F., M.B.-P. and M.-F.Z.; visualization, M.-F.Z.; writing—original draft, D.F., D.M., M.B.-P. and M.-F.Z. All authors have read and agreed to the published version of the manuscript.

**Funding:** This research was funded by the grants from the Ministry of Research, Innovation and Digitization, CNCS/CCCDI-UEFISCDI: PN-III-P4-ID-PCCF-2016-0050, within PNCDI III.

**Institutional Review Board Statement:** Not applicable.

**Informed Consent Statement:** Not applicable.

**Data Availability Statement:** Not applicable.

**Conflicts of Interest:** The authors declare no conflict of interest.

## Abbreviation

| | | | |
|---|---|---|---|
| NADES | Natural Deep Eutectic Solvents | HBD | Hydrogen-Bond Donor |
| HPC | Hydroxypropyl cellulose | HBA | Hydrogen-Bond Acceptor |
| LCST | Lower Critical Solution Temperature | HGF | Human Gingival Fibroblast |
| ChCl | Choline chloride | DSC | Differential Scanning Calorimetry |
| U | Urea | TGA | Thermogravimetric analysis |
| CA | Citric acid | DTG | Differential thermogravimetric analysis |
| LA | Lactic acid | FT-IR | Fourier-Transform Infrared |
| Gly | Glycerol | NMR | Nuclear Magnetic Resonance |

## References

1. Abbott, A.P.; Capper, G.; Davies, D.L.; Rasheed, R.K. Ionic Liquid Analogues Formed from Hydrated Metal Salts. *Chem. Eur. J.* **2004**, *10*, 3769–3774. [CrossRef] [PubMed]
2. Choi, Y.H.; van Spronsen, J.; Dai, Y.; Verberne, M.; Hollmann, F.; Arends, I.W.C.E.; Witkamp, G.-J.; Verpoorte, R. Are natural deep eutectic solvents the missing link in understanding cellular metabolism and physiology? *Plant Physiol.* **2011**, *156*, 1701–1705. [CrossRef] [PubMed]
3. Francisco, M.; van den Bruinhorst, A.; Kroon, M.C. Low-Transition-Temperature Mixtures (LTTMs): A New Generation of Designer Solvents. *Angew. Chem. Int. Ed.* **2013**, *52*, 3074–3085. [CrossRef]
4. Dias, A.M.A.; Cortez, A.R.; Barsan, M.M.; Santos, J.B.; Brett, C.M.A.; de Sousa, H.C. Development of Greener Multi-Responsive Chitosan Biomaterials Doped with Biocompatible Ammonium Ionic Liquids. *ACS Sustain. Chem. Eng.* **2013**, *1*, 1480–1492. [CrossRef]
5. Marques Silva, J.M.; Reis, R.L.; Paiva, A.; Duarte, A.R.C. Design of functional therapeutic deep eutectic solvents based on choline chloride and ascorbic acid. *ACS Sustain. Chem. Eng.* **2018**, *6*, 10355–10363. [CrossRef]
6. Ruiz-Olles, J.; Slavik, P.; Whitelaw, N.K.; Smith, D.K. Self-Assembled Gels formed in Deep Eutectic Solvents—Supramolecular Eutectogels with High Ionic Conductivities. *Angew. Chem. Int. Ed.* **2019**, *58*, 4173–4178. [CrossRef] [PubMed]
7. Lomba, L.; García, C.B.; Ribate, M.P.; Giner, B.; Zuriaga, E. Applications of Deep Eutectic Solvents Related to Health,Synthesis, and Extraction of Natural Based Chemicals. *Appl. Sci.* **2021**, *11*, 10156. [CrossRef]
8. Blusztajn, J.K. Choline, A Vital Amine. *Science* **1998**, *281*, 794–795. [CrossRef]
9. Paiva, A.; Craveiro, R.; Aroso, I.; Martins, M.; Reis, R.L.; Duarte, A.R.C. Natural deep eutectic solvents-solvents for the 21st century. *ACS Sustain. Chem. Eng.* **2014**, *2*, 1063–1071. [CrossRef]
10. Dai, Y.; Witkamp, G.-J.; Verpoorte, R.; Choi, Y.H. Tailoring properties of natural deep eutectic solvents with water to facilitate their applications. *Food Chem.* **2015**, *187*, 14–19. [CrossRef] [PubMed]
11. Hayyan, M.; Mbous, Y.P.; Looi, C.Y.; Wong, W.F.; Hayyan, A.; Salleh, Z.; Mohd-Ali, O. Natural deep eutectic solvents: Cytotoxic profile. *Springerplus* **2016**, *5*, 913–924. [CrossRef]
12. Agatemor, C.; Ibsen, K.N.; Tanner, E.E.L.; Mitragotri, S. Ionic liquids for addressing unmet needs in healthcare. *Bioeng. Transl. Med.* **2018**, *3*, 7–25. [CrossRef]
13. Chowdhury, M.R.; MdMoshikur, R.; Wakabayashi, R.; Tahara, Y.; Kamiya, N.; Moniruzzaman, M.; Goto, M. In vivo biocompatibility, pharmacokinetics, antitumor efficacy, and hypersensitivity evaluation of ionic liquid-mediated paclitaxel formulations. *Int. J. Pharm.* **2019**, *565*, 219–226. [CrossRef]
14. Majee, S.B. (Ed.) *Emerging Concepts in Analysis and Applications of Hydrogels*; IntechOpen: Rijeka, Croatia, 2016.
15. Alharbi, N.D.; Guirguis, O.W. Macrostructure and optical studies of hydroxypropyl cellulose in pure and Nano-composites forms. *Results Phys.* **2019**, *15*, 102637. [CrossRef]
16. Yue, D.; Jia, Y.; Yao, Y.; Sun, J.; Jing, Y. Structure and electrochemical behavior of ionic liquid analogue based on choline chloride and urea. *Electrochim. Acta* **2012**, *65*, 30–36. [CrossRef]
17. Aissaoui, T. Novel Contribution to the Chemical Structure of Choline Chloride Based Deep Eutectic Solvents. *Pharm. Anal. Acta* **2015**, *6*, 1000448. [CrossRef]
18. Azevedo, A.M.O.; Costa, S.P.F.; Dias, A.F.P.; Marques, A.H.O.; Pinto, P.C.A.G.; Bica, K.; Ressmann, A.K.; Passos, M.L.C.; Araújo, A.R.T.S.; Reis, S.; et al. Anti-inflammatory choline based ionic liquids: Insights into their lipophilicity, solubility and toxicity parameters. *J. Mol. Liq.* **2017**, *232*, 20–26. [CrossRef]
19. Filip, D.; Macocinschi, D.; Zaltariov, M.F.; Ciubotaru, B.I.; Bargan, A.; Varganici, C.D.; Vasiliu, A.L.; Peptanariu, D.; Balan-Porcarasu, M.; Timofte-Zorila, M.M. Hydroxypropyl Cellulose/Pluronic-Based Composite Hydrogels as Biodegradable Mucoadhesive Scaffolds for Tissue Engineering. *Gels* **2022**, *8*, 519. [CrossRef]
20. López, N.; Delso, I.; Matute, D.; Lafuente, C.; Artal, M. Characterization of xylitol or citric acid:choline chloride:water mixtures: Structure, thermophysical properties, and quercetin solubility. *Food Chem.* **2020**, *306*, 125610. [CrossRef] [PubMed]
21. Jangir, A.K.; Mandviwala, H.; Patel, P.; Sharma, S.; Kuperkar, K. Acumen into the effect of alcohols on choline chloride: L-lactic acid-based natural deep eutectic solvent (NADES): A spectral investigation unified with theoretical and thermophysical characterization. *J. Mol. Liq.* **2020**, *317*, 113923. [CrossRef]
22. Posada, E.; López-Salas, N.; Jiménez Riobóo, R.J.; Ferrer, M.L.; Gutiérrez, M.C.; del Monte, F. Reline aqueous solutions behaving as liquid mixtures of H-bond co-solvents: Microphase segregation and formation of co-continuous structures as indicated by Brillouin and $^1$H NMR spectroscopies. *Phys. Chem. Chem. Phys.* **2017**, *19*, 17103–17110. [CrossRef]
23. Delso, I.; Lafuente, C.; Muñoz-Embid, J.; Artal, M. NMR study of choline chloride-based deep eutectic solvents. *J. Mol. Liq.* **2019**, *290*, 111236. [CrossRef]
24. Liu, P.; Pedersen, C.M.; Zhang, J.; Liu, R.; Zhang, Z.; Hou, X.; Wang, Y. Ternary deep eutectic solvents catalyzed D-glucosamine self-condensation todeoxyfructosazine: NMR study. *Green Energy Environ.* **2021**, *6*, 261–270. [CrossRef]
25. Delgado-Mellado, N.; Larriba, M.; Navarro, P.; Rigual, V.; Ayuso, M.; Garcia, J.; Rodriguez, F. Thermal stability of choline chloride deep eutectic solvents by TGA/FTIR-ATR analysis. *J. Mol. Liq.* **2018**, *260*, 37–43. [CrossRef]
26. Hammond, O.S.; Mudring, A.-V. Ionic liquids and deep eutectics as a transformative platform for the synthesis of nanomaterials. *Chem. Comm.* **2022**, *58*, 3865–3892. [CrossRef] [PubMed]

27. Florindo, C.; Oliveira, F.S.; Rebelo, L.P.N.; Fernandes, A.M.; Marrucho, I.M. Insights into the synthesis and properties of deep eutectic solvents based on chlinium chloride and carboxylic acids. *ACS Sustain. Chem. Eng.* **2014**, *2*, 2416–2425. [CrossRef]
28. Coats, A.W.; Redfern, J.T. Kinetics parameters from thermogravimetric data. *Nature* **1964**, *201*, 68–69. [CrossRef]
29. Flynn, J.H.; Wall, L.A. A Quick, Direct Method for the Determination of Activation Energy from Thermogravimetric Data. *Polym. Lett.* **1966**, *4*, 323–328. [CrossRef]
30. Urbanovici, E.; Segal, E. Is classical non-isothermal kinetics with constant heating rate actually non-isothermal kinetics with quasi-constant heating rate. *Thermochim. Acta* **1990**, *159*, 369–372. [CrossRef]
31. Gilmore, M.; Swadzba-Kwasny, M.; Holbrey, J.D. Thermal Properties of Choline Chloride/Urea System Studied under Moisture-Free Atmosphere. *J. Chem. Eng. Data* **2019**, *64*, 5248–5255. [CrossRef]
32. Maneffa, A.J.; Harrison, A.B.; Radford, S.J.; Whitehouse, A.S.; Clark, J.H.; Matharu, A.S. Deep Eutectic Solvents Based on Natural Ascorbic Acid Analogues and Choline Chloride. *Chem. Open* **2020**, *9*, 559–567. [CrossRef] [PubMed]
33. Wang, L.F.; Pearce, E.M.; Kwei, T.K. Glass transitions in Hydrogen-Bonded Polymer Complexes. *J. Polym. Sci. Part B Polym. Phys.* **1991**, *29*, 619–626. [CrossRef]
34. van der Sman, R.G.M. Predictions of Glass Transition Temperature for Hydrogen Bonding Biomaterials. *J. Phys. Chem. B* **2013**, *117*, 16303–16313. [CrossRef]
35. Tamburic, S.; Craig, D.Q.M. The effects of ageing on the rheological,dielectric and mucoadhesive properties of poly(acrylic acid) gel systems. *Pharm. Res.* **1996**, *13*, 279–283. [CrossRef] [PubMed]
36. Thorgeirsdottir, T.O.; Kjoniksen, A.L.; Knudsen, K.D.; Kristmundsdottir, T.; Nystrom, B. Viscoelastic and Structural Properties of Pharmaceutical Hydrogels Containing Monocaprin. *Eur. J. Pharm. Biopharm.* **2005**, *59*, 333–342. [CrossRef]
37. Walicka, A.; Falicki, J.; Iwanowska-Chomiak, B. Rheology of drugs for topical and transdermal delivery. *Int. J. Appl. Mech. Eng.* **2019**, *24*, 179–198. [CrossRef]
38. Chhabra, R.P.; Richardson, J.F. *Non-Newtonian Flow in Process Industries*; Butterworth-Heinemann: Oxford, UK, 1999.
39. Gallegos, C.; Franco, J.M. Rheology of food, cosmetics and pharmaceuticals. *Curr. Opin. Colloid Interface* **1999**, *4*, 288–293. [CrossRef]
40. Bingham, E.C. *Fluidity and Plasticity*; MC Graw-Hill: New York, NY, USA, 1922.
41. Hershel, W.H.; Bulkley, R. Konsistenzmessungen von Gummi Benzollosungen. *Kolloid Z.* **1926**, *39*, 291–300. [CrossRef]
42. Vocadlo, J.J.; Charles, M.E. Characterization and Laminar Flow of Fluid-Like Viscoplastic Substances. *Can. J. Chem.* **1973**, *51*, 116–121. [CrossRef]
43. Casson, N. *The Rheology of Disperse Systems*; Pergamon Press: London, UK, 1959.
44. Patural, L.; Marchal, P.; Govin, A.; Grosseau, P.; Ruot, B.; Deves, O. Cellulose ethers influence on water retention and consistency in cement-based mortars. *Cem. Concr. Res.* **2011**, *41*, 46–55. [CrossRef]
45. Elhamarnah, Y.A.; Nasser, M.; Qiblawey, H.; Benamor, H.; Atilhan, A.; Aparicio, M. A comprehensive review on the rheological behavior of imidazolium based ionic liquids and natural deep eutectic solvents. *J. Mol. Liq.* **2019**, *277*, 932–958. [CrossRef]
46. Ramachandran, S.; Chen, S.; Etzler, F. Rheological Characterization of Hydroxypropyl cellulose Gels. *Drug. Dev. Ind. Pharm.* **1999**, *25*, 153–161. [CrossRef] [PubMed]
47. Altamash, T.; Nasser, M.S.; Elhamarnah, Y.; Magzoub, M.; Ullah, R.; Anaya, B.; Aparicio, S.; Atilhan, M. Gas Solubility and Rheological Behavior of Natural Deep Eutectic Solvents (NADES) via Combined Experimental and Molecular Simulation *Tech. Chem. Select* **2017**, *2*, 7278–7295.
48. Tao, R.; Simon, S.L. Rheology of Imidazolium-Based Ionic Liquids with Aromatic Functionality. *J. Phys. Chem. B* **2015**, *119*, 11953–11959. [CrossRef] [PubMed]
49. Cox, W.P.; Merz, E.H. Correlation of dynamic and steady flow viscosities. *J. Polym. Sci.* **1958**, *28*, 619–622. [CrossRef]
50. Wissbrun, K.F.; Griffin, A.C. Rheology of a Thermotropic Polyester in the Nematic and Isotropic States. *J. Polym. Sci. Polym. Phys.* **1982**, *20*, 1835–1845. [CrossRef]
51. Tan, L.; Wan, A.; Pan, D. Viscoelasticity of concentrated polyacrylonitrile solutions: Effects of solution composition and temperature. *Polym. Int.* **2011**, *60*, 1047–1052. [CrossRef]
52. Xia, X.; Tang, S.; Lu, X.; Hu, Z. Formation and Volume Phase Transition of Hydroxypropyl cellulose Microgels in Salt Solution. *Macromolecules* **2003**, *36*, 3695–3698. [CrossRef]
53. Gao, J.; Haidar, G.; Lu, X.; Hu, Z. Self-Association of Hydroxypropyl cellulose in Water. *Macromolecules* **2001**, *34*, 2242–2247. [CrossRef]
54. Fairclough, J.P.A.; Yu, H.; Kelly, O.; Ryan, A.J.; Sammler, R.L.; Radler, M. Interplay between gelation and phase separation in aqueous solutions of methylcellulose and hydroxypropyl methylcellulose. *Langmuir* **2012**, *28*, 10551–10557. [CrossRef]
55. Silva, J.M.; Silva, E.; Reis, R.L.; Duarte, A.R.C. A closer look in the antimicrobial properties of deep eutectic solvents based on fatty acids. *Sustain. Chem. Pharm.* **2019**, *14*, 100192. [CrossRef]
56. Wikene, K.O.; Rukke, H.V.; Bruzell, E.; Tonnesen, H.H. Investigation of the antimicrobial effect of natural deep eutectic solvents (NADES) as solvents in antimicrobial photodynamic therapy. *J. Photochem. Photobiol. B Biol.* **2017**, *171*, 27–33. [CrossRef] [PubMed]
57. Radosevic, K.; Canak, I.; Panic, M.; Markov, K.; Bubalo, M.C.; Frece, J.; Srcek, V.G.; Redovnikovic, I.R. Antimicrobial, cytotoxic and antioxidative evaluation of natural deep eutectic solvents. *Environ. Sci. Pollut. Res.* **2018**, *25*, 14188–14196. [CrossRef]
58. Burel, C.; Kala, A.; Purevdorj-Gage, L. Impact of pH on citric acid antimicrobial activity against Gram-negative bacteria. *Lett. Appl. Microbiol.* **2021**, *72*, 332–340. [CrossRef]

59. Wang, C.; Chang, T.; Yang, H.; Cui, M. Antibacterial mechanism of lactic acid on physiological and morphological properties of Salmonella Enteritidis, Escherichia Coli and Listeria monocytogenes. *Food Control* **2015**, *47*, 231–236. [CrossRef]
60. Piquero-Casals, J.; Morgado-Carrasco, D.; Granger, C.; Trullas, C.; Jesus-Silva, A.; Krutmann, J. Urea in Dermatology: A Review of its Emollient, Moisturizing, Keratolytic, Skin Barrier Enhancing and Antimicrobial Properties. *Dermatol. Ther.* **2021**, *11*, 1905–1915. [CrossRef] [PubMed]
61. Linser, A. Glycerine as fungicide or bactericide active substance. U.S. Patent 02/069708, 12 September 2002.
62. Zhao, B.Y.; Xu, P.; Yang, F.X.; Wu, H.; Zong, M.H.; Lou, W.Y. Biocompatible deep eutectic solvents based on choline chloride: Characterization and application to the extraction of rutin from Sophora Japonica. *ACS Sustain. Chem. Eng.* **2015**, *3*, 2746–2755. [CrossRef]
63. Clinical and Laboratory Standard Institute. *Method for Antifungal Disk Diffusion Susceptibility Testing of Yeasts*, Approved Guideline, 2nd ed.; Clinical and Laboratory Standard Institute: Wayne, PA, USA, 2009.
64. Clinical and Laboratory Standards Institute. *Performance Standards for Antimicrobial Susceptibility Testing*, CLSI Supplement M100, 30th ed.; Clinical and Laboratory Standard Institute: Wayne, PA, USA, 2020.
65. Egorova, K.S.; Gordeev, E.G.; Ananikov, V.P. Biological Activity of Ionic Liquids and Their Application in Pharmaceutics and Medicine. *Chem. Rev.* **2017**, *117*, 7132–7189. [CrossRef]
66. Moshikur, R.M.; Chowdhury, M.R.; Moniruzzaman, M.; Goto, M. Biocompatible ionic liquids and their application in pharmaceutics. *Green Chem.* **2020**, *22*, 8116–8139. [CrossRef]

Article

# Cellulose-*g*-poly(2-(dimethylamino)ethylmethacrylate) Hydrogels: Synthesis, Characterization, Antibacterial Testing and Polymer Electrolyte Application

Roko Blažic, Dajana Kučić Grgić, Marijana Kraljić Roković and Elvira Vidović *

Faculty of Chemical Engineering and Technology, University of Zagreb, Marulićev trg 19, 10000 Zagreb, Croatia
* Correspondence: evidov@fkit.hr

**Abstract:** Hydrogels have been investigated due to their unique properties. These include high water content and biocompatibility. Here, hydrogels with different ratios of poly(2-(dimethylamino) ethylmethacrylate) (PDMAEMA) were grafted onto cellulose (Cel-*g*-PDMAEMA) by the free radical polymerization method and gamma-ray radiation was applied in order to increase crosslinking and content of PDMAEMA. Gamma irradiation enabled an increase of PDMAEMA content in hydrogels in case of higher ratio of 2-(dimethylamino)ethyl methacrylate in the initial reaction mixture. The swelling of synthesized hydrogels was monitored in dependence of pH (3, 5.5 and 10) during up to 60 days. The swelling increased from 270% to 900%. Testing of antimicrobial activity of selected hydrogel films showed weak inhibitory activity against *Escherichia coli*, *Pseudomonas aeruginosa*, and *Bacillus subtilis*. The results obtained by the cyclic voltammetry (CV) and electrochemical impedance spectroscopy (EIS) indicate that chemically synthesized hydrogels have good characteristics for the supercapacitor application.

**Keywords:** antimicrobial activity; 2-(dimethylamino)ethyl methacrylate; gamma irradiation; hydrogel; hydrogel electrolytes

## 1. Introduction

Hydrogels are three-dimensional polymer networks of natural or synthetic materials capable of adsorbing and retaining significant amounts of water without being dissolved [1]. A number of naturally occurring materials exhibit a hydrogel structure. The fact that the extracellular matrix (ECM) hydrogels are water-swollen fibrillary three-dimensional (3D) networks speaks of the importance and prevalence of such a structure. Naturally-derived hydrogels can be classified into three groups: protein-based materials, polysaccharide-based materials and those derived from decellularized tissue [2]. These types of hydrogels include ones derived from collagen, gelatin, elastin, fibrin and silk fibroin. Elastin and fibrin are widely found proteins in the ECM structure, and collagen is the major component inside ECM. They give ECM the required strength and elasticity to function properly, making them very promising materials for tissue engineering and cell culture systems [2,3]. The first applications of synthetic hydrogels after their synthesis were for medical reasons such as optical lenses [4,5]. They were later used for wound dressings [6], implantable medical devices, artificial blood vessels and heart valves [2,3], matrices for bone and cartilage tissue engineering [7,8] as well as drug delivery systems [9–11]. Due to their properties, they were also used widely in pharmacy [12–14], agriculture [15–17] and water retention [18,19]. Hydrogels are able to retain a large amount of water or biological fluids under physiological conditions and are characterized by a high degree of flexibility. Their consistency is similar to living tissues making them an ideal substance for a variety of applications. The characteristic properties of hydrogels, such as desired functionality, biocompatibility and in some cases reversibility or sterilizability meet important requirements to treat or replace tissues and organs, or interact safely with the biological system [20,21]. Recently,

**Citation:** Blažic, R.; Kučić Grgić, D.; Kraljić Roković, M.; Vidović, E. Cellulose-g-poly(2-(dimethylamino)ethylmethacrylate) Hydrogels: Synthesis, Characterization, Antibacterial Testing and Polymer Electrolyte Application. *Gels* **2022**, *8*, 636. https://doi.org/10.3390/gels8100636

**Academic Editors:** Christian Demitri, Lorenzo Bonetti and Laura Riva

Received: 15 August 2022
Accepted: 5 October 2022
Published: 7 October 2022

**Publisher's Note:** MDPI stays neutral with regard to jurisdictional claims in published maps and institutional affiliations.

**Copyright:** © 2022 by the authors. Licensee MDPI, Basel, Switzerland. This article is an open access article distributed under the terms and conditions of the Creative Commons Attribution (CC BY) license (https://creativecommons.org/licenses/by/4.0/).

their application for water treatments, i.e., as adsorbent for removing heavy metal ions from wastewater has been particularly intensively researched [22–25]. In addition, the preparation of hydrogels with antibacterial properties is often achieved by incorporating silver particles, which are known to have good antimicrobial properties. In the preparation of these hydrogel–silver particle composites, the free spaces between the polymer molecules in the hydrogels act as nanoreactors, allowing control of the size and shape of the silver nanoparticles [26,27].

Smart hydrogels or stimulus-responsive hydrogels, which can display, for example, antibacterial properties, have dramatic volume changes in response to external environments, such as temperature, pH and certain stimuli, and are considered a special subgroup of materials attracting research attention. The choice of materials for stimuli is limited. One of the commonly chosen stimuli is poly(N-isopropylmethacrylamide [28] or poly(2-(dimethylamino)ethylmethacrylate) (PDMAEMA). They are pH and temperature double-responsive [29] and possess antibacterial activity [30,31]. More than two decades ago, PDMAEMA and its copolymers were synthesized and evaluated as gene transfer agents and carrier systems for DNA [32,33]. Often, they are combined with petrochemical based polymers [34] or biobased polymers [35,36]. Water swelling tests on interpenetrating network (IPN) hydrogels based on nanofibrillated cellulose (NFC) and PDMAEMA prepared via crosslinking free radical polymerization showed that the IPN hydrogels were both pH-sensitive and temperature-sensitive. The swelling of hydrogels was limited at high temperature and in neutral medium. There was an increase in the swelling ratio as the NFC content increased. The synthesized materials were tested for removal of Pb(II) and Cu(II) ions. The achieved adsorption of both ions was better in comparison with other literature data (more than 50% for Cu, and several times higher for Pb). The high porosity and uniform pore structure, among other characteristics, presumably contributed to removal of Pb(II) and Cu(II) ions [25].

In this study, we aimed at preparing hydrogels from PDMAEMA grafted onto cellulose (Cel-g-PDMAEMA) by a free radical polymerization method, chemical synthesis, as well as by applying gamma irradiation in an attempt to increase crosslinking and content of PDMAEMA in hydrogels. Regarding the possible application of hydrogels the antimicrobial activities of selected films that are additionally decorated with silver were also tested. The cyclic voltammetry (CV) and electrochemical impedance spectroscopy (EIS) were used to test potential application of chemically synthesized hydrogels as sheets in the assembled supercapacitor.

## 2. Results and Discussion

### 2.1. Polymerization Reaction

The polymerization of 2-(dimethylamino)ethyl methacrylate (DMAEMA) with N,N-methylenebis (acrylamide) (MBA) in a cellulose solution is a complex process that can lead to different products [37]: (a) polymer grafted onto cellulose, (b) a semi-interpenetrating network formed by crosslinking the polymer, (c) homopolymer or copolymer (branched structure), and (d) polymer microgel. A detailed study of the structure of the prepared Cel-g-PDMAEMA hydrogels with different molar ratios of cellulose to DMAEMA (1:1, 1:3, and 1:5), especially with respect to the content of PDMAEMA microgel, was undertaken in other research [38]. The chemical reactions of grafting PDMAEMA onto cellulose and crosslinking with MBA to form a network with(out) gamma irradiation are shown in Figure 1.

In this work, the focus is on the swelling of Cel-g-PDMAEMA hydrogels over a broader pH range. In addition, the effect of PDMAEMA on the antibacterial properties of copolymer hydrogels as well as on the efficiency of hydrogel decoration with silver was investigated, where improvement is expected due to the presence of nitrogen atoms. The analysis of the electrochemical properties of the selected hydrogels indicates their suitability for application as supercapacitors (Scheme 1). Therefore, the obtained results indicate a possible application of the synthesized materials as separators, for example, in electrochemical devices (capacitors, etc.) or in the medical field (dressings, wraps, polymers for controlled delivery of drugs, etc.).

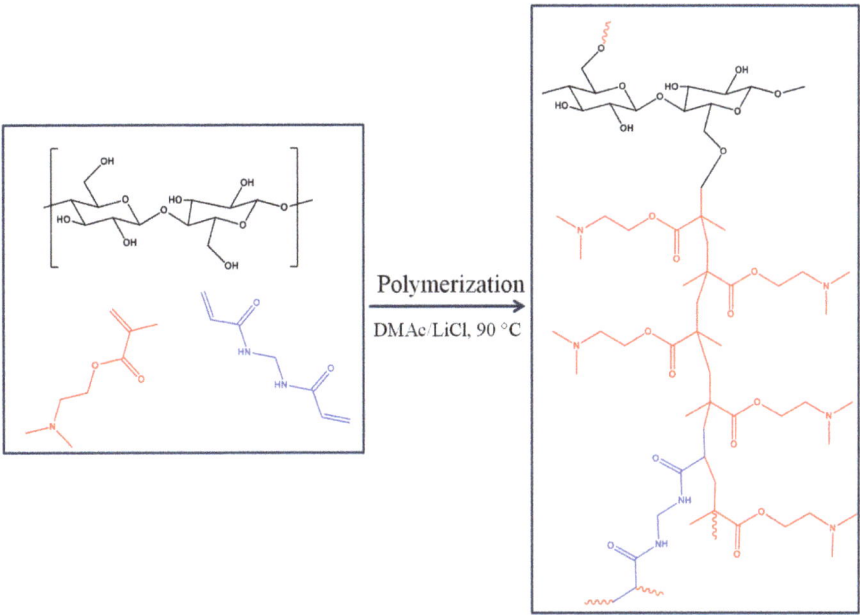

**Figure 1.** The chemical reaction of polymerization and grafting of poly(2-(dimethylamino)ethylmethacrylate) (PDMAEMA) onto cellulose to produce networks Cel-g-PDMAEMA crosslinked with MBA.

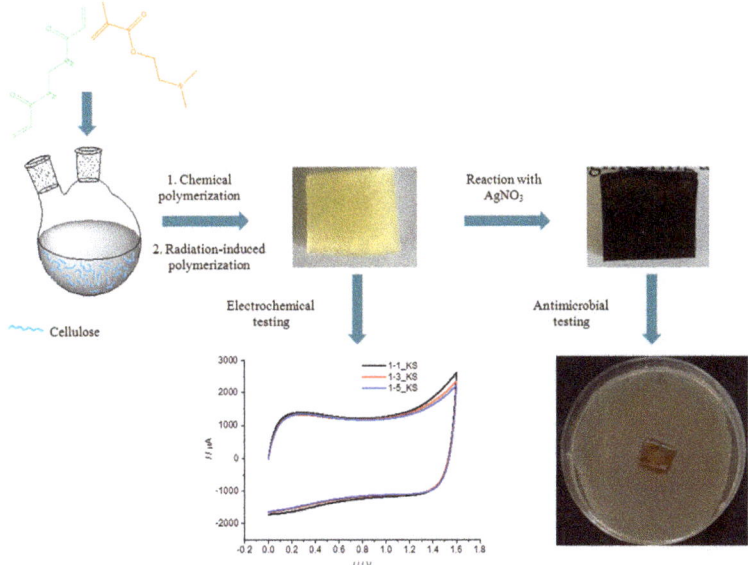

**Scheme 1.** Methodology for Cel-g-PDMAEMA hydrogel preparation and testing regarding potential applications.

33

Cellulose grafted with PDMAEMA and crosslinked with MBA hydrogels are named Cel-g-PDMAEMA. Samples are referred as series, generally x-y_KS or x-y_Z where x-y represents the ratio $n$(cellulose)/$n$(DMAEMA):1-1, 1-3 or 1-5, KS represents hydrogels prepared via chemical synthesis, while Z (10, 30 or 100) represents the irradiation dose in kGy.

## 2.2. Structural Characterization of Synthesized Cel-g-PDMAEMA Hydrogels

### 2.2.1. FTIR Spectroscopy Analysis

The FTIR spectrum of microcellulose is shown in Figure 2. In the wavenumber range between 3000 and 3600 cm$^{-1}$, there is a broad signal with a maximum at 3331 cm$^{-1}$, which is characteristic of the vibrational band of the hydroxyl group in the polysaccharide, which is affected by intra- and intermolecular hydrogen bonding.

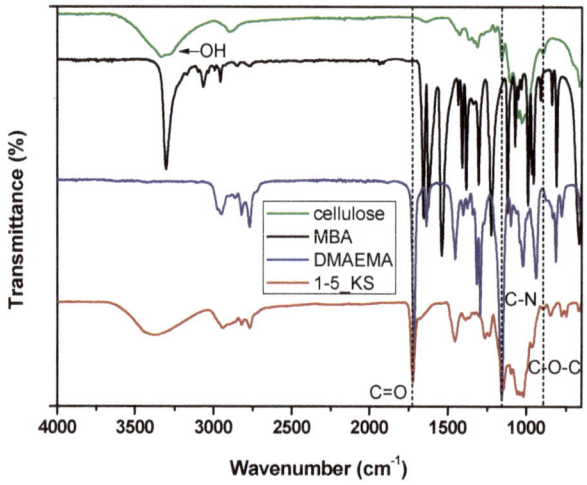

**Figure 2.** FTIR spectra of microcellulose, N,N-methylenebis (acrylamide) (MBA), 2-(dimethylamino)ethyl methacrylate (DMAEMA) and Cel-g-PDMAEMA network crosslinked with MBA sample synthesized chemically.

The band at 2896 cm$^{-1}$ belongs to the vibrations of the CH bonds. The band at 1634 cm$^{-1}$ corresponds to the vibrations of the water molecules absorbed in the cellulose. The bands at 1428 and 1030 cm$^{-1}$ are characteristic of the cellulose and belong to the vibrations of the groups $CH_2$ and OH in the cellulose. The band at 897 cm$^{-1}$ corresponds to the vibration of the characteristic β-(1 → 4) glycosidic bond [39,40].

The FTIR spectrum of MBA presented in Figure 2 shows a band with maxima at 3304 cm$^{-1}$ assigned to N-H stretching vibrations. Bands corresponding to the stretching vibrations of $CH_2$ appear at 3067 and 2956 cm$^{-1}$. The strong band at 1656 cm$^{-1}$ is assigned to the C=O stretching mode (Amide I band), while the strong band at 1535 cm$^{-1}$ is assigned to the N-H deformation (Amide II band). The strong band around 1620 cm$^{-1}$ indicates the presence of a C=C double bond and its stretching mode. Moreover, the medium–strong band at 1428 cm$^{-1}$ is assigned to the in-plane scissoring or bending of $CH_2$. The strong band around 1380 cm$^{-1}$ is assigned to the out-of-plane bending of $CH_2$, while the band at 987 cm$^{-1}$ is assigned to the vibration of $CH_2$, respectively [41,42].

Figure 2 shows the FTIR spectrum of DMAEMA with the maxima at 2949 cm$^{-1}$ corresponding to the sp$^3$ vibration of the CH bond. The signals at 2822 and 2770 cm$^{-1}$ also correspond to sp$^3$ vibrations of the CH bond, but within the $N(CH_3)_2$ group. The signal at 1717 cm$^{-1}$ corresponds to the vibration of the carbonyl group, while at 1638 cm$^{-1}$ the signal belonging to the vibration of the C=C bond is visible. In addition, an adsorption band appeared at 1452 cm$^{-1}$, which is characteristic of the bending of $CH_2$. The adsorption band with a maximum at 1150 cm$^{-1}$ results from the C-N stretching vibrations [39,43].

The infrared spectrum of the 1-5_KS sample with the highest amount of DMAEMA in the reaction mixture is also shown in Figure 2. In this hydrogel apart from a broad peak around 3330 cm$^{-1}$, which is characteristic for cellulose, the signals characteristic for PDMAEMA appear at 2940, 2828 and 2770 cm$^{-1}$. There was a band from the stretching vibrations of the C=O bond, with a maximum at 1724 cm$^{-1}$, which was shifted to higher wavenumbers by 7 units compared to the position of the same band in the FTIR of the DMAEMA monomer (Figure 2). Additionally, the band from stretching vibrations of the C-N and C-O groups with a maximum at 1148 cm$^{-1}$ was shifted by 8 units to lower wavenumbers compare to the position in the DMAEMA. The band at 1455 cm$^{-1}$ from CH$_2$ bending occurred at the same wavenumber. In the 1-5_KS hydrogel spectra, there were no bands from the double C=C bond and from the vinyl group, indicating that the bonding of the DMAEMA monomers and crosslinking with the MBA occurs by breaking the C=C bonds.

Figure 3 shows FTIR spectra of hydrogel films prepared from synthesized Cel-g-PDMAEMA after chemical reaction and additionally irradiated with 100 kGy. The band characteristic of the stretching vibrations of the C=O bond was strong and appeared at the same wavenumber, 1724 cm$^{-1}$, in samples 1-3 and 1-5. In samples 1-1, the C=O signal was weak and shifted to 1731 and 1737 cm$^{-1}$, respectively, while the band of amide band I appeared with similar intensity at 1648 cm$^{-1}$. Intramolecular hydrogen bonds could be formed via the C=O group of PDMAEMA and the N–H group of MBA. The shift of these maxima to lower wavenumbers suggests that the N–H and C=O groups of PDMAEMA and MBA are involved in the hydrogen bond formation. The results of the FTIR analysis are in agreement with the literature data [40–44].

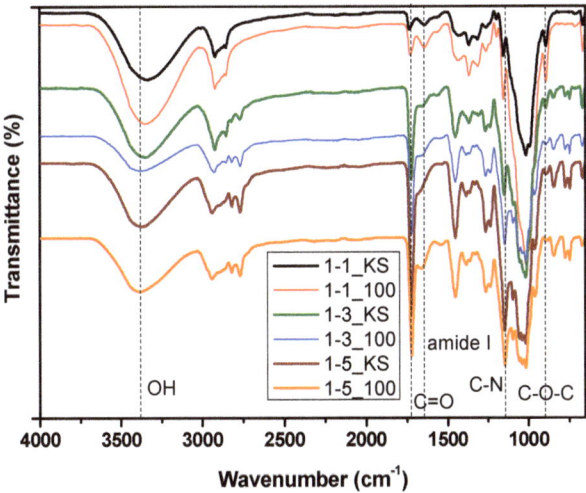

**Figure 3.** FTIR spectra of Cel-g-PDMAEMA hydrogel films (cellulose:DMAEMA = 1:1, 1:3, 1:5) prepared after chemical reaction (KS) and additionally irradiated with 100 kGy.

By comparing the intensity ratio of the band characteristic of PDMAEMA (C=O) and the band characteristic of the cellulose ($\beta$-(1 → 4) glycosidic bond occurring at 897 cm$^{-1}$), it was possible to follow the increase in PDMAEMA content in the hydrogel in agreement with its increased content in the reaction mixture (Table 1).

Table 1. Intensity of characteristic bands and intensity ratio of carbonyl group and glycosidic bond in prepared hydrogels.

| Sample | 2940/cm$^{-1}$ (2925/cm$^{-1}$) | 2850/cm$^{-1}$ (2821/cm$^{-1}$) | 2770/cm$^{-1}$ | 1724/cm$^{-1}$ | 1156/cm$^{-1}$ | 1019/cm$^{-1}$ | 897/cm$^{-1}$ | 851/cm$^{-1}$ | 780/cm$^{-1}$ | 749/cm$^{-1}$ | $I_{1724}/I_{897}$ |
|---|---|---|---|---|---|---|---|---|---|---|---|
| 1-1_KS  | 8.0  | 6.3 | /   | 3.1  | 7.9  | 26.0 | 6.2  | /   | /   | /   | 0.5  |
| 1-1_100 | 11.7 | 9.3 | /   | 5.6  | 13.7 | 43.8 | 10.5 | /   | /   | /   | 0.5  |
| 1-3_KS  | 11.3 | 8.7 | 5.7 | 16.8 | 19.5 | 34.6 | 3.8  | 1.9 | 1.6 | 2.1 | 4.5  |
| 1-3_100 | 6.6  | 4.9 | 5.5 | 17.9 | 19.7 | 22.7 | 1.7  | 2.5 | 2.5 | 3.3 | 10.5 |
| 1-5_KS  | 9.0  | /   | /   | 8.9  | 27.9 | 30.5 | 2.1  | 4.4 | 4.5 | 4.7 | 13.3 |
| 1-5_100 | 7.5  | /   | /   | 6.9  | 21.5 | 23.3 | 0.7  | 3.4 | 3.7 | 4.2 | 32.1 |

The ratio $I_{1725}/I_{895}$ increased from 0.5 in 1-1_KS up to 4.5 and 13.3 in 1-3_KS and 1-5_KS samples. The absorption of irradiation caused further increase of PDMAEMA in hydrogels of series 1-3_100 and 1-5_100 but did not influence sample 1-1_100, as its ratio $I_{1725}/I_{895}$ reveals [45]. The primary radiation damage is the formation of carbon-centered radicals. In the case of cellulose, five different types of radicals produced by H atom elimination from a C–H bond are expected and all the final radiation chemical changes of cellulose are consequences of unimolecular or bimolecular reactions of the radicals. Subsequently, β cleavage of the radical can lead to the breaking of the glucoside bond or opening of the anhydroglucose ring. In both cases carbonyl groups are produced as a result of the cleavage. In case of series 1-1 there is no change in the ratio $I_{1725}/I_{895}$ (Table 1), which would imply that neither radiation damaged cellulose nor enhanced incorporation of PDMAEMA into the hydrogel. The applied dose at 100 kGy does not cause significant damage to cellulose, which is in accordance with the literature where it is mostly claimed that doses of more than 100 kGy cause destruction [46,47]. Irradiation did not increase the content of PDMAEMA, but its presence, even small, affects the structure, porosity and swelling ability of the material, as will be seen below.

It was mentioned earlier that these kinds of reactions, which include chemical synthesis and irradiation, are complex processes that can result in different products: (a) polymer grafted onto cellulose, (b) a semi-interpenetrating network formed through crosslinking of the polymer, (c) homopolymer or copolymer (branched structure), and (d) polymer microgel. Here, as well as during the final design of spheres or films that includes freeze-extraction, numerous influences intertwine that causes phase separation, chain orientation, and different crosslinking densities are possible, and all of this leads to differences in the composition and structure of the material. A detailed study of the influence of an irradiation dose on the structure of Cel-g-PDMAEMA hydrogels was conducted [38,43].

Furthermore, in the spectra in Figure 3, one can see that with the increasing ratio of PDMAEMA, the signals characteristic of the N(CH$_3$)$_2$ group at 2940, 2828 and 2770 cm$^{-1}$ become pronounced. Samples with a higher ratio of PDMAEMA display a strong signal at 1457 cm$^{-1}$ characteristic for bending of CH$_2$ in PDMAEMA. The intensity of the band from stretching vibrations of the CO group, $\nu_s$(C–O), with a maximum at 1157 cm$^{-1}$ relative to the intensity of β-glycosidic linkages (897 cm$^{-1}$) in sample 1-1 and cellulose, are similar. With an increasing amount of PDMAEMA in samples 1-3 and 1-5 the relative increase of the signal at 1157 cm$^{-1}$ against the two signals characteristic of cellulose (897 and 1020 cm$^{-1}$) was significant. It was more conspicuous in the case of irradiated hydrogels. In addition, numerous small signals confirm the presence of methacrylate compounds in the hydrogel. Weakly separated signals at 1235 cm$^{-1}$ and 1265 cm$^{-1}$, i.e., "shoulder" at 1296 cm$^{-1}$, originate from the presence of MBA (1226 cm$^{-1}$) and PDMAEMA (1296 cm$^{-1}$), respectively. Additionally, signals recorded in the fingerprint region (851 cm$^{-1}$ and 780 cm$^{-1}$) originate from MBA and PDMAEMA (Figure 3).

In hydrogels 1-1_KS and 1-1_100 instead of a strong band from the stretching vibrations of the C=O bond, $\nu$(C=O) with a maximum at 1724 cm$^{-1}$ a weak band was recorded. Simultaneously, a relatively stronger intensity amide band I appeared at 1648 cm$^{-1}$. Intramolecular hydrogen bonds could be formed via the C=O group from PDMAEMA and N–H group from MBA. The shifting of these maxima towards lower wavenumbers indicated that the N–H and C=O groups of PDMAEMA and MBA participated in the hydrogen bond formation. Additionally, intensity of the band from stretching vibrations of the CO group, $\nu_s$(C–O), with a maximum at 1157 cm$^{-1}$, displayed an intensity ratio to the β-glycosidic linkages similar to that in pure cellulose (897 cm$^{-1}$). With increasing amount of PDMAEMA these signals both displayed a significant increase relative to the band at 897 cm$^{-1}$, whereat it was more conspicuous in the case of irradiated hydrogels.

### 2.2.2. Scanning Electron Microscopy Analysis

To characterize the morphologies of the Cel-g-PDMAEMA hydrogel sample (both spheres and films) swollen at equilibrium, the SEM micrograms were obtained. Figure 4 shows micrographs of the outer surface and cross-section of the sphere for samples 1-1_KS, 1-3_KS, and 1-5_KS. The surface of the spheres differs in roughness, although not significantly. However, the porosity pattern on the cross-section differs significantly depending on the composition. Drying material by the freeze-extraction method enables the preparation of porous materials because the pores are not crushed by capillary force, therefore all samples prepared by the freeze-extraction method have a porous structure. The size distribution of the pores changes from the outer layer (edge) of the sphere towards its center. In the outer layer, near the edge of the sphere, there are smaller pores (<50 µm), while the pores in the center of the sphere are larger. The smaller pores are the result of the formation of smaller water crystals during the cooling/freezing of the spheres in the cryostat at −40 °C. The slower cooling in the center of the sphere contributed to the formation of larger water crystals and subsequently to the formation of larger pores. The pore size and the distribution of the pores over the cross-section of the spherical samples clearly show the influence of the composition and design protocol of the hydrogels.

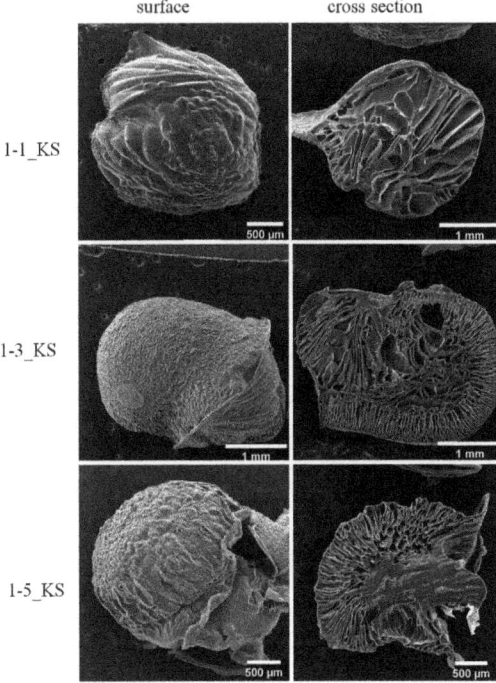

**Figure 4.** SEM micrographs of outer sphere surface and cross-section of sphere for samples 1-1_KS, 1-3_KS and 1-5_KS.

Figure 5 shows SEM micrographs of the Cel-*g*-PDMAEMA hydrogel samples shaped into films. Again, a difference in bulk structure (in cross-section) can be seen depending on the composition of the material. Individually, the three-dimensional structure of the hydrogels looked like a semi-uniform cross-linked network, with a more or less thin layer of different structure formed when the hydrogel films were made, as one side was in contact with the glass surface and the other side was in contact with the air. If necessary, this characteristic could be modified by applying a different film casting procedure. The thickness of the swollen films was 0.15–1 mm. The cross-section shows differences between hydrogels depending on their composition. Hydrogel 1-1_KS has a fairly uniform, dense structure except for the thin porous edge layer. Hydrogel 1-3_KS has a very porous cross-section in which two perpendicular layers with different pore sizes can be seen. A similar structure is present in hydrogel 1-5_KS, which is characterized by high porosity. Here, the difference between the porous layers is less pronounced. Similar to the spheres, this structural organization especially in 1-3_KS and 1-5_KS hydrogels provided a lot of free space within the cross-linked polymer network. Therefore, they appear to be suitable for various applications where fluid sorption within the network is required, including carriers for many active substances.

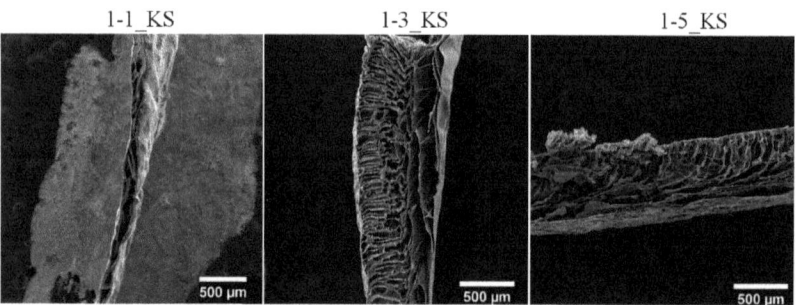

**Figure 5.** SEM micrographs of Cel-*g*-PDMAEMA hydrogel films swollen in the equilibrium state.

Comparing the pore size in samples with different geometries (spheres and films), it can be seen that for spheres 1-1_KS, 1-3_KS and 1-5_KS, the pore size in the swollen state was initially in the range of 40–455 µm, 60–355 µm and 60–180 µm, respectively (Figure 4). The pore size in the layer near the sphere surface was <60 µm for samples 1-3_KS and 1-5_KS. Detailed pore size distribution for initial spheres is presented in Figure S1. At the same time, the pore size of the prepared film samples in the swollen state was in the range of 25–120 µm, 15–185 µm and 15–140 µm, respectively (Figure 5). Detailed pore size distribution for prepared films is presented in Figure S2. Based on the average pore size, the synthesized materials can be classified as macroporous hydrogel [42,48].

Figure 6 shows micrographs of the outer surface of the sphere and the cross-section of the sphere for samples 1-1_KS, 1-3_KS, and 1-5_KS after 60 days of swelling in deionized water. The spheres of 1-1_KS and 1-3_KS exhibited similar porosity accompanied by some shrinkage, while spheres 1-5_KS changed geometry from a spherical to an elliptical shape and porosity disappeared. Their surfaces became somewhat smoother, while the cross-section showed a large change from porous to dense form. This indicates that their behavior and persistence during swelling depend on composition.

The pore size of samples 1-1_KS and 1-3_KS, which were swollen for 60 days, were in the range of 30–315 µm and 60–270 µm, respectively (Figure 6). Comparing the pores in the spheres at the beginning and after 60 days of swelling, it can be seen that their size, shape and distribution changed slightly in the samples with the lower PDMAEMA ratio (1-1_KS and 1-3_KS) (Figure S3). The spheres of the 1-5_KS samples, however, showed a complete change in structure. This can probably be correlated with the method of incorporation of PDMAEMA in high concentrations into the hydrogel and the change in its content during long-term swelling.

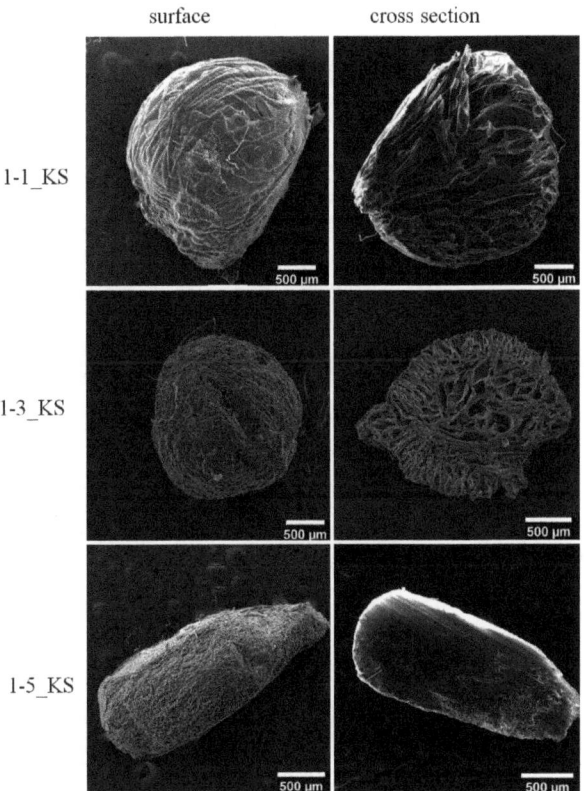

**Figure 6.** SEM micrographs of outer sphere surface and cross-section of sphere for samples 1-1_KS, 1-3_KS and 1-5_KS after 60 days of swelling in deionized water.

Samples 1-1_KS and 1-1_100 as well as 1-3_KS and 1-3_100 after swelling in acidic or alkaline media for 60 days were scanned and SEM micrographs are shown in Figures 7 and 8. They displayed small, up to significant, structural differences. All observed samples prepared by chemical synthesis exhibited porous structure in both media. Samples 1-1_KS and 1-3_KS, which were studied in acidic media, both exhibited a similar pore shape, while in alkaline media they had a significantly different pore structure. If we keep with the 1-1 series, the structure of sample 1-1_100 swollen in alkaline environment is the most different, being crumpled, rough and without the well-defined pore geometry. The samples of series 1-3 showed more pronounced structural differences, the largest being found in sample 1-3_100 swollen in acidic media. It is characterized by a rather compact, smooth structure in which cracks are visible, but no pores are visible. The observed major differences in the structure of the hydrogels do not show a cause-and-effect relationship with their degree of swelling. In particular, the hydrogels 1-3_100, which were swollen in acidic and alkaline media, show the same equilibrium degree of swelling after 60 days, as will be commented on later (see Section 2.3.1).

*2.3. Swelling Study*

The spheres were dried by a freeze-extraction method that enabled the design of porous materials. The behavior of Cel-g-PDMAEMA hydrogels swollen in deionized water (pH 5.5) was monitored at 20 °C. To investigate the influence of pH, the samples were also swollen in acidic (pH = 3) and alkaline (pH = 10) media. The swelling ratio, $\alpha$, was calculated according to Equation (1).

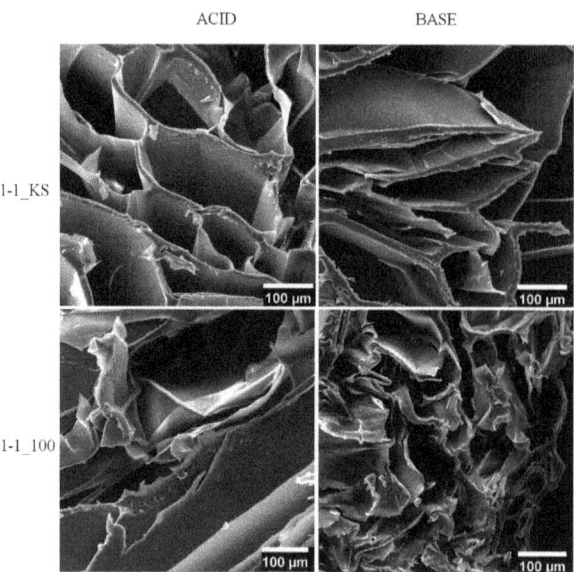

**Figure 7.** SEM micrographs of Cel-*g*-PDMAEMA hydrogel samples, swollen in the equilibrium state for sample 1-1_KS and 1-1_100.

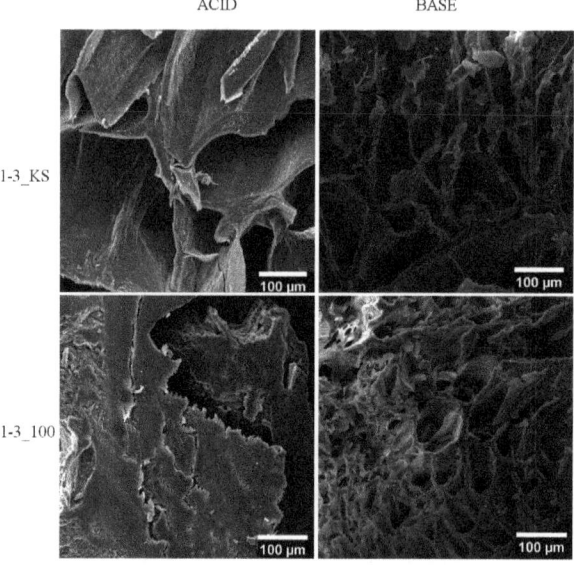

**Figure 8.** SEM micrographs of Cel-*g*-PDMAEMA hydrogel samples, swollen in the equilibrium state for sample 1-3_KS and 1-3_100.

2.3.1. Equilibrium Hydrogel Swelling at 20 °C and pH 5.5

The swelling kinetics of the prepared spheres was studied in deionized water (pH 5.5) at 20 °C for 14 days, and the results are shown in Figure 9. A study of the hydrogels prepared by chemical synthesis showed their fairly extensive swelling. The more porous samples 1-5_KS and especially 1-3_KS reached the maximum degree of swelling in two days, while sample 1-1_KS, which contained predominantly larger elongated pores as

Figure 4 SEM reveals, reached the maximum α in seven days. These results indicate that a higher amount of PDMAEMA affects the structure and porosity of the materials and in this way contributes to faster swelling. After reaching equilibrium samples 1-1_KS and 1-5_KS displayed similar swelling degree of ca. 640% while the highest α of ca. 925% displayed sample 1-3_KS. Perhaps the explanation of the achieved similar swelling is that the sample 1-1_KS, as already mentioned, shows mostly closed pores and the sample 1-5_KS a porous outer zone (layer) but a compact non-porous central part of the sphere. Only the sample 1-3_KS shows a very porous structure throughout the cross-section of the sphere. It is worth mentioning that sample 1-1_KS despite a negligible content of PDMAEMA (Figure 3, FTIR), when compared with the sphere of pure cellulose displayed many times higher degree of swelling. In our preliminary studies we found that the degree of swelling of the cellulose sphere was 80% after 4 h, while the sample of 1-1_KS at the same time showed a degree of swelling of ca. 300%. One can assume that the reason for the much lower swelling of the cellulose sphere is that a porous structure is not obtained despite the same preparation procedure (freeze-extraction method) as for copolymer spheres [43].

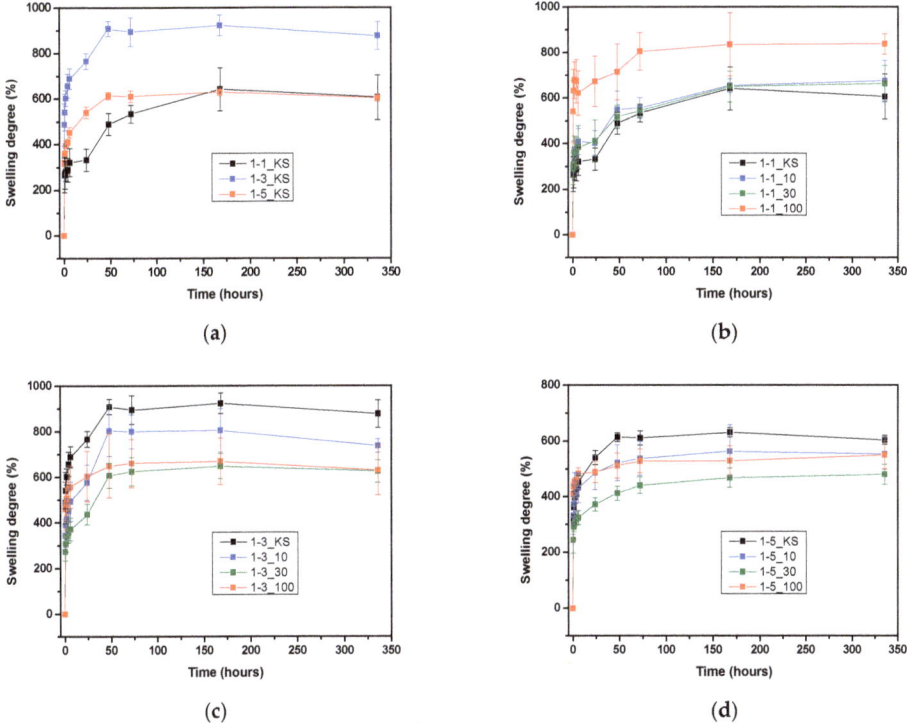

Figure 9. Comparison of swelling kinetic and equilibrium swelling degree for hydrogel samples with different molar ratio of cellulose and DMAEMA. (a) Comparison of swelling kinetic and equilibrium swelling degree for hydrogels after irradiation (b–d).

The highest degree of swelling of sample 1-1_100 among all samples in this series is not likely to be attributed either to increased ratio of PDMAEMA based on FTIR analysis or to increased cellulose degradation. Literature data related to the influence of radiation on the degradation of cellulose differ with regard to the origin and type of cellulose (microcrystalline cellulose, bacterial cellulose, pine wood cellulose, cotton-cellulose, etc.). They generally agree that the weaker radiation does not cause significant changes in the structure. Thus, it was determined that sterilization with gamma irradiation at 25 kGy caused no significant structural changes in the polymer [46] and that irradiation at 10 kGy caused decrease of molecular weight of ca. 12% [47]. On the other hand, doses of 100 kGy and above caused significant changes of cellulose properties such as molecular weight, relative crystallinity, and surface area [49].

Irradiated materials with a higher content of PDMAEMA, series 1-3 and 1-5, displayed a lower degree of swelling in comparison with hydrogel spheres prepared by chemical synthesis (Figure 9). Although there are small differences, it can be said that the swelling ability of these samples decreases with increasing irradiation dose. Due to the higher ratio of DMAEMA in the reaction mixture, and because it seems to be more pliable to irradiation compared to cellulose, its incorporation into hydrogel is enhanced, both by grafting onto cellulose or in the form of microgels. The increased proportion of PDMAEMA was confirmed by FTIR analysis. In this way, the number of possible networking sites is increased. More frequently crosslinked networks swell less. Additionally, smaller deviations from the trend are probably due to variation in porosity. Specifically, samples of series 1-3 displayed swelling degree values in the range between 630% and 925%; with the latter, the largest $\alpha$ was notified in the 1-3_KS after 7 days.

2.3.2. Comparison Equilibrium Hydrogel Swelling at 20 °C and pH = 3.0, pH 5.5 and pH 10

The influence of the pH of the medium on the swelling of Cel-$g$-PDMAEMA hydrogels is shown in Figure 10. Hydrogels prepared by chemical synthesis and those subsequently irradiated with 100 kGy are shown in parallel.

During the first week of swelling, the hydrogels mostly displayed the highest swelling ratios in deionized water (pH 5.5). In the literature, similar hydrogels, i.e., interpenetrating network (IPN) based on nanofibrillated cellulose (NFC) and PDMAEMA prepared via crosslinking free radical polymerization were tested during 24 h, at pH in a range 3 to 11 [25]. They came to the conclusion that the swelling ratios under acidic and alkaline condition are higher than those in a neutral environment, which is the opposite compared to these samples in the initial period. They explained that the phenomenon may be attributed to specific charge properties of PDMAEMA and NFC in aqueous solution. PDMAEMA is a kind of tertiary amine with a pKa of about 7.5 and NFCs have carboxyl groups on the surface with a pKa of about 4.6. Therefore, in acidic medium, the tertiary amine groups can easily be protonated with a positive charge and the polymer chains become stretched due to the electrostatic repulsion. Thus, higher swelling ratios can be achieved because of low resistance of water entering the hydrogel. In alkaline environment, they explained, carboxylic acid groups on NFC gradually shift to the carboxylate anion, which results with weaker hydrogen bonds interaction and enhanced electrostatic repulsion in hydrogels.

Here, the hydrogels displayed higher swelling ratios under acidic and alkaline conditions only after 7 and 14 days, respectively. We can assume that it takes longer due to structural differences that slow the process and restrict stretching in spite of (de)protonation. It is only subsequently that both the repulsive structure and ionization of hydrogels occur and facilitate the diffusion of water molecules and increase the swelling ratios as Salama et al. explained [50]. Furthermore, the swelling ratio of the IPN hydrogels is also associated with the content of the NFCs. With an increase in the NFCs ratio, the swelling ratio of the hydrogel gradually increases [25]. This is also true here, with the exception of the 1-1_KS series. In general, irradiated samples with higher PDMAEMA ratio (1-3_100 and 1-5_100) display a lower degree of swelling in comparison with hydrogels prepared only by chemical reaction, which is in accordance with findings of FTIR. Unlike the latter, hydrogel 1-1_100 showed a higher $\alpha$ in comparison with 1-1_KS. As the previous study showed, the microgelation is very common. The distribution of microgel affects the structure through the density of the crosslinking and its subsequent migration from the loose network is facilitated. It indicates, apart from reaction mixture composition, the influence of different contributions during preparation procedure on materials properties [38]. Almost equal swelling of 1-3 samples in alkaline and acidic medium studied over a longer period makes likely its use for specific applications such as medicine, pharmaceuticals for drug delivery, sensor manufacturing, separators in electrochemical devices (batteries, capacitors and the like), etc. [44].

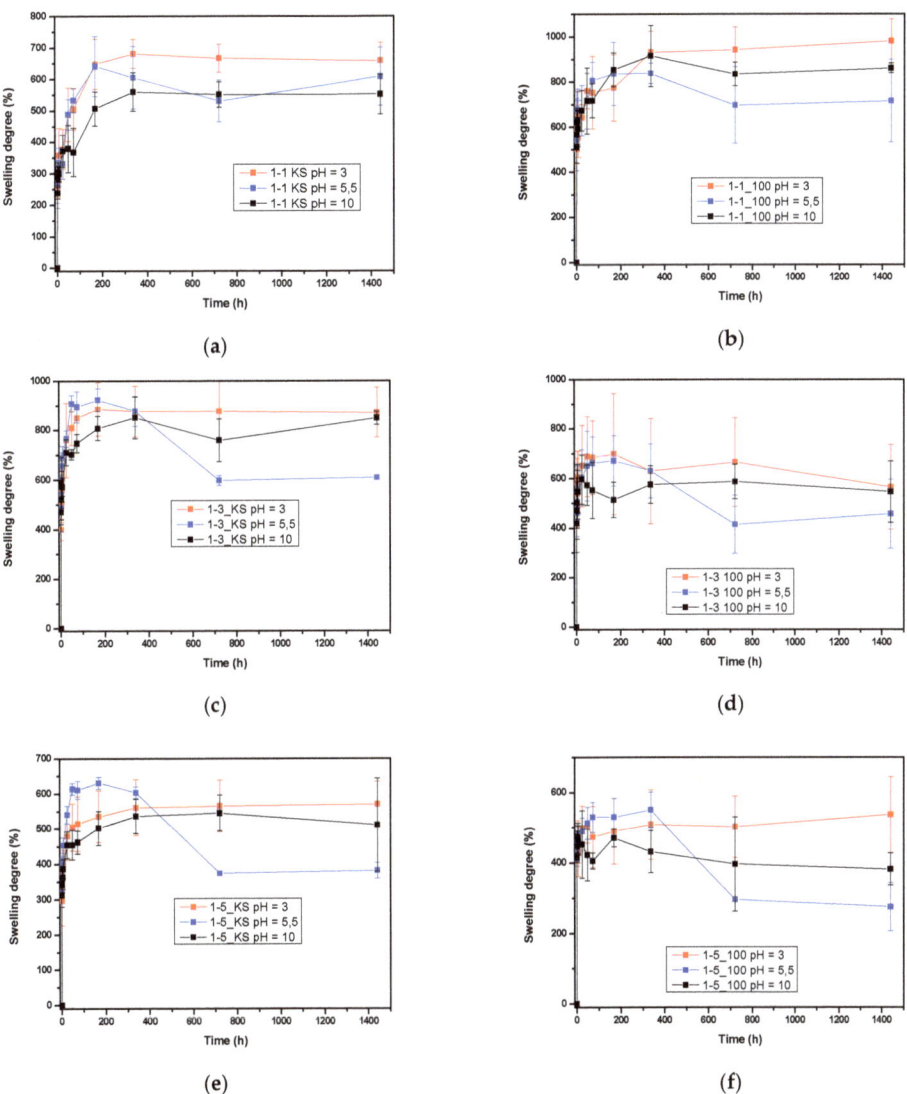

**Figure 10.** Comparison of swelling kinetic and equilibrium swelling degree for hydrogel samples at different pH for initial 1-1_KS, 1-3_KS and 1-5_KS samples (**a,c,e**) and irradiated samples 1-1_100, 1-3_100, 1-5_100 (**b,d,f**).

## 2.4. Decorating Hydrogel Films with Silver Particles

Synthesis of hydrogels with silver was carried out by swelling hydrogels in a silver nitrate solution. Silver atoms are formed by the reduction of silver ions from the complex, which then agglomerate and finally form stable silver particles. Silver is known for its antibacterial activity and can also be used as a catalyst. The introduction of silver into the structure of the hydrogel expands its potential applications, because by combining substances with different properties, new materials with prestigious properties can be created for very different applications, from sensors to various supports and coatings to catalysts, from pharmacy, medicine, water treatment industry and the like [51–54]. Figure 11 shows hydrogels 1-5_KS and 1-5_100 before and after incorporation of silver.

**Figure 11.** Sample 1-5_KS before (**a**) and after (**b**) immersion and sample 1-5_100 before (**c**) and after (**d**) immersion in 0.01 M AgNO$_3$ during 24 h.

It is known that the formation of silver particles in samples is accompanied by a color change, which can be considered a quick and preliminary indicator that a reduction of silver has occurred but the formation of silver particles in solution by Ag$^+$ reduction is commonly and accurately monitored by UV-VIS spectroscopy. However, the formation of silver particles on solid supports, e.g., hydrogels, makes in-situ monitoring of Ag particle formation by UV-VIS spectroscopy impossible, except in rare cases when the media is transparent. At the same time, it is well known from the literature that the presence of Ag$^0$ particles can be confirmed by XRD analysis both in solution and in composites [55,56]. The XRD pattern of samples 1-5_100 + AgNO$_3$ is shown in Figure 12, where broad diffraction lines characteristic of silver (ICDD PDF No. 4-783) can be seen. Moreover, additional diffraction lines in the diffractogram indicate the presence of silver oxide, where the obtained pattern most closely resembles Ag$_2$O$_3$ (ICDD PDF No. 77-607). Due to the large surface area of nanoparticles, contact of silver nanoparticles with aqueous media can lead to oxidation and formation of silver oxide. Silver oxide also shows good antimicrobial properties as reported in the literature [57–59]. Gao et al. showed that silver oxide enhances antibacterial properties of material over a long period of time because the presence of silver oxide modulates the release of Ag$^+$ ions [58].

*2.5. Antimicrobial Testing*

A series of samples for each Cel-*g*-PDMAEMA hydrogel: plain, silver decorated, and those both irradiated and decorated were used to study antibacterial properties against *Bacillus subtilis*, *Escherichia coli*, and *Pseudomonas aeruginosa*. The results of the antimicrobial activity of the tested samples using the disk diffusion method are summarized in Table 2 and Figures S4–S6.

The data in Table 2 show that chemically synthesized hydrogels did not exhibit antibacterial activity, except for sample 1-1_KS, which showed a weak response against *P. aeruginosa* and *B. subtilis*. The samples of all series decorated with silver particles showed weak inhibitory antibacterial activity.

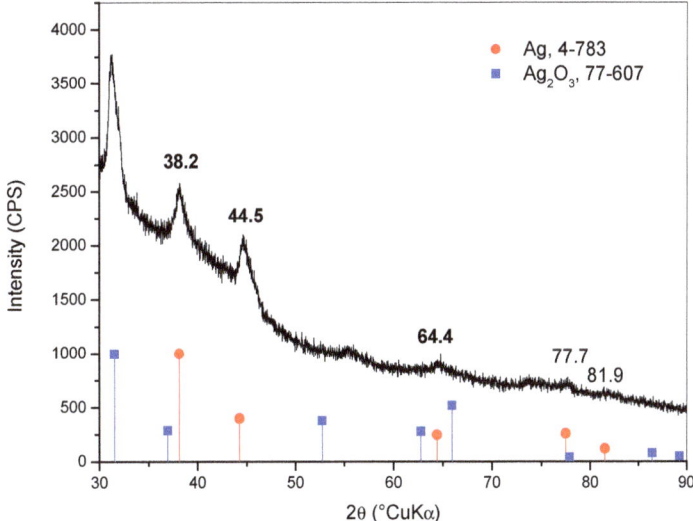

**Figure 12.** XRD diffractogram of formed silver particles in sample 1-5_100+AgNO3 (crystallographic database for silver (ICDD PDF No. 4-783) and $Ag_2O_3$ (ICDD PDF No. 77-607)).

**Table 2.** Antimicrobial activity of tested samples using the agar disk diffusion method.

| Sample | d (Inhibition Zone–E. coli)/cm | d (Inhibition Zone–P. aeruginosa)/cm | d (Inhibition Zone–B. subtilis)/cm |
|---|---|---|---|
| 1-1 KS | 0 | 0.05 | 0.1 |
| 1-1_KS + AgNO3 | 0.1 | 0.15 | 0.5 |
| 1-1_100 + AgNO3 | 0.2 | 0.25 | 0.7 |
| 1-3_KS | 0 | 0 | 0 |
| 1-3_KS + AgNO3 | 0.7 | 0.4 | 0.5 |
| 1-3_100 + AgNO3 | 1 | 0.6 | 0.7 |
| 1-5_KS | 0 | 0 | 0 |
| 1-5_KS + AgNO3 | 0.1 | 0.2 | 0.8 |
| 1-5_100 + AgNO3 | 0.6 | 0.4 | 1 |

Silver nanoparticles have been proven to have antibacterial properties and due to these properties they are used in medical products as well as in consumer products such as textiles with antibacterial properties, cleaning cloths, air filters, food containers, cosmetic products, various coatings, etc. [54]. Their contribution to improving the efficacy of various antibiotics is particularly important. Various studies have shown that they change the activity of antibiotics. For example, the study of the activity of levofloxacin against *E. coli*, *B. subtilis*, *P. aeruginosa* and *S. aureus* has shown a synergistic effect [51]. Such studies are of great importance because of the worrying resistance of bacteria. They not only show a synergistic effect with antibiotics, but also reduce the dose of antibiotic used in therapy. Unlike antibiotics, silver nanoparticles act through a combination of mechanisms rather than a single one. The obtained inhibition zones indicating a stronger antimicrobial effect against all tested bacterial cultures in films decorated with AgNPs, can be explained by their physical properties, where the effect of AgNPs strongly depends on their size, shape and concentration [60]. The AgNPs displayed different activity against various bacteria. They were more active against *E. coli* and *B. subtilis* than against *P. aeruginosa* [61–63]. Here, the irradiated samples show moderately to several times larger zones of inhibition within a single series. This is not surprising since the use of radiation for the purpose of sterilization is well known, e.g., in medicine, food industry, etc., and it has enhanced the incorporation of PDMAEMA into the hydrogel, as determined by FTIR analysis. Studies on the antimicrobial activity of PDMAEMA have shown that it is active against both gram-positive and gram-negative bacteria [31,64]. Therefore, based on the obtained results, it can be concluded that there is a synergistic effect, i.e., higher content of PDMAEMA contributes to higher antibacterial activity, as well as the introduction of AgNPs and applied irradiation [31]. It is important to emphasize that continuous exposure of microorganisms to nanoparticles should be avoided, as a study on E. coli showed that bacteria can become resistant through 225 generations with constant exposure [65]. Unfortunately, silver nanoparticles have certain toxic effects in addition to their antibacterial properties. Many toxic effects of AgNPs have been demonstrated in in vitro studies. Research on the toxicity of AgNPs suggests that size, shape, chemical composition, solubility, surface activity, binding ability, and biological effects such as metabolism and excretion influence their toxicity [66]. The possibility of the AgNPs fixing in order to avoid their aggregation and to prevent their spontaneous release, especially in cases of medical application, seems to be a complementary advantage.

*2.6. Electrochemical Testing*

Aqueous-based polymer electrolytes are commonly used as both separators and ionic conductors in solid-state devices. The advantage of polymer electrolytes over liquid electrolytes is their compact structure, which prevents liquid leakage and electrode displacement. This property is particularly important for flexible and free-standing supercapacitors, which have received considerable attention in recent years [67,68]. Most aqueous-based polymer electrolytes are prepared by blending polymer host materials with ionic conductors, water, and plasticizers [69]. Various fossil-based polymers: PET, aramid nonwovens and PP or PP/PP polyolefin membranes [70] have been investigated as separators for high power lithium ion batteries. In addition, poly(vinyl alcohol), which has excellent properties as a polymer host due to its high hydrophilicity and good film-forming ability is widely used as well as bio-based materials, among which cellulose is leading [71,72]. Recently, separators for high-performance supercapacitors (SCs) have been fabricated from keratin, a natural polymer with excellent wettability, which is an alternative sustainable material [73].

In this work, the polymer electrolytes prepared from 1-1_KS, 1-3_KS, and 1-5_KS hydrogels swollen in 0.5 M $Na_2SO_4$ were used as separators in supercapacitors with reduced graphene oxide/carbon nanotubes (rGO/CNT) as the active material.

Figures 13 and 14 show the CV responses of supercapacitors with different hydrogel electrolytes. The nearly constant current value registered during supercapacitor charging is related to the continuous electrochemical double layer charging by increasing the voltage [74–76]. In each case, a nearly rectangular response was obtained, indicating good capacitive behavior. There was no significant difference between the responses of the different hydrogels. The value of specific capacitance calculated according to Equation (2) was 25.86 F $g^{-1}$ for 1-1_KS, 24.61 F $g^{-1}$ for 1-3_KS and 24.34 F $g^{-1}$ for 1-5_KS. The good capacitive behavior was confirmed by the EIS responses. In the high frequency region, a semicircle with two characteristic resistances was registered. The first one is related to the electrolyte resistance and the second one to the charge transfer resistance between the current collector and the active material. The total resistance represents the equivalent serial resistance (ESR) of the supercapacitor and determines the reversibility of the system and the charge/discharge rate of the supercapacitor. To improve the properties of the supercapacitor, it is of great importance to reduce the ESR. It is obvious that electrolyte resistance values for 1-1_KS and 1-5_KS were 3.2 $\Omega$, while the resistance for 1-3_KS was 5.14 $\Omega$. These values are very similar to those previously reported for supercapacitors assembled by using glass paper fibers wetted with 0.5 mol $dm^{-3}$ $Na_2SO_4$ solution [74]. The low frequency response is related to the capacitance value. From the obtained results, it can be seen that the EIS response of each supercapacitor was similar, which is consistent with the results of CV. From these results, it can be concluded that the prepared hydrogel has good characteristics for supercapacitor application [77].

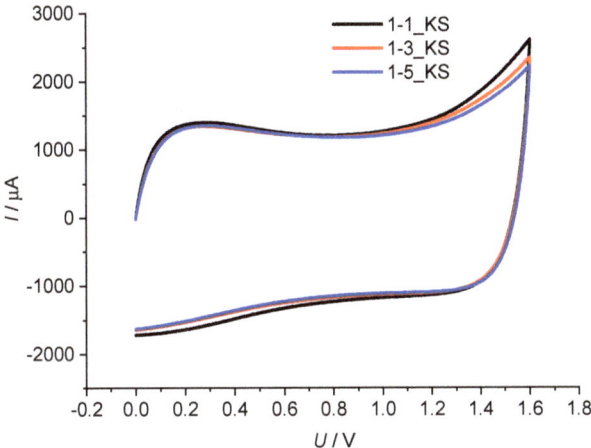

**Figure 13.** Responses obtained by cyclic voltammetry for rGO/CNT supercapacitor with three different hydrogels.

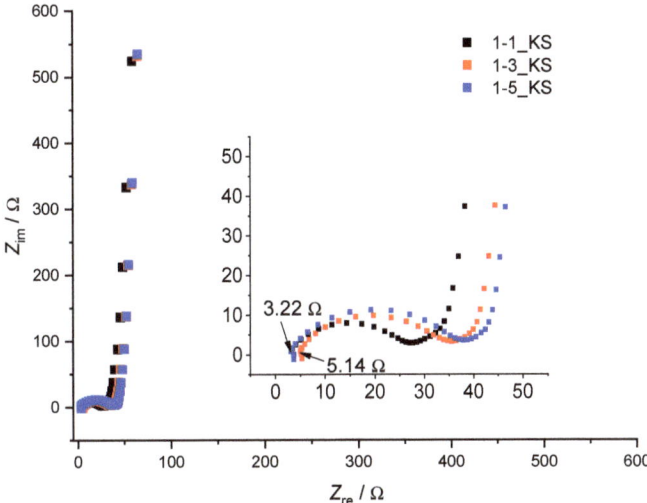

**Figure 14.** Electrochemical impedance spectroscopy responses obtained for rGO/CNT supercapacitor with three different hydrogels.

## 3. Conclusions

The methodology used enables the effective preparation of hydrogels based on cellulose and various ratios of PDMAEMA in the form of porous spheres or films. All prepared hydrogels displayed significant swelling ability. It reveals dependence on the composition of hydrogel, applied irradiation and pH while the achieved porosity of the material has a significant effect, as well. The materials showed steady swelling at acidic and base conditions over a period of two months. The introduction of silver particles and the application of gamma radiation have minimally increased the antibacterial activity of the prepared materials, which opens space for new research such as increasing the silver content or applying a higher dose of gamma radiation. The results from CV showed good capacitive behavior. Therefore prepared hydrogels seem to be promising materials for supercapacitors used as aqueous-based polymer electrolytes in separators in supercapacitors with rGO/CNT as an active material.

## 4. Materials and Methods

### 4.1. Materials

Cellulose ($Mv$ = 25,100 g mol$^{-1}$) was dissolved in the solvent N,N-dimethylacetamide (DMAc) (Fischer Chemical, analytical reagent grade) and lithium chloride (Fischer Chemical, laboratory reagent grade). Before dissolution in DMAc/LiCl solvent, cellulose was dried over phosphorus pentoxide (VWR, purity 99.5%) for 3 h at 90 °C under vacuum. DMAEMA monomer (polymerization grade, Aldrich) was passed through a column of activated basic aluminum oxide (Aldrich) and purged with high-purity nitrogen prior to use. The crosslinking agent MBA (Sigma-Aldrich, St. Louis, MI, USA, purity 99%) was used as received. The initiator of polymerization reaction tert-butylperoxy-2-ethylhexanoate (Trigonox 21, 70 wt% solution, Akzo Chemie, Amsterdam, The Netherlands) was used as received. Absolute ethanol (Gram-mol, p.a.) was used for freeze-extraction method as received. Silver nitrate (Alkaloid, p.a.) was used for the synthesis of silver particles.

### 4.2. Method of Cellulose/Poly(dimethylaminoethyl methacrylate) Hydrogles (Cel-g-PDMAEMA) Synthesis

The samples of cellulose grafted with PDMAEMA were synthesized via free radical polymerization method using MBA as a cross-linker. The polymerization reaction was initiated by adding peroxide initiator. Synthesis was performed in two steps. Initially, cellulose was dispersed in DMAc in a round bottom flask and activated for 2 h at 120 °C. After two hours, the temperature was lowered to 100 °C and LiCl (6.6 wt%) was added to the flask and the mixture was stirred at 100 °C for another hour. The mixture was then cooled to room temperature and a clear cellulose solution (5 wt%) was obtained. Afterwards, the polymerization reaction of DMAEMA was carried out in a solution of cellulose in DMAc/LiCl. The obtained cellulose solution was weighed into a round bottom flask and heated to 90 °C. After 5 min, a solution of MBA and DMAEMA (50 wt%) in DMAc/LiCl was added to the flask. The weighted amount of initiator Trigonox 21 in DMAc/LiCl (1 wt% towards monomers) was added to the flask 5 min after the addition of monomers. The reaction was carried out for 3 h. After cooling, the synthesized polymers were precipitated in deionized water whereat the spheres were formed for the first phase of research. Hydrogel samples (spheres) were kept in deionized water for 5 days and the deionized water exchanged after 1 and 24 h, two and five days so as to remove all unreacted compounds. Afterwards spheres (samples) in the equilibrium swelling state were frozen and immersed in ethanol, dried in a freeze dryer, and used for further analysis. Weighed spheres (1 g) were placed in a beaker and cooled to −40 °C in a cryostat. The beaker containing the spheres was kept at −40 °C for 30 min and then cold ethanol (35 mL) was poured into the beaker. The beaker was placed in the freezer and the spheres were kept in ethanol at −18 °C for 48 h, with fresh ethanol added after 24 h. After 48 h, the ethanol was decanted and the spheres were dried in vacuum at 50 °C until constant weight. Molar ratio of reactants (monomer (DMAEMA), crosslinking agent (MBA) and cellulose (cel)) in Table 3.

#### 4.2.1. Gamma-Ray Irradiation of Hydrogel

The Cel-g-PDMAEMA hydrogel sample, which was in a rather viscous state, was subjected to a gamma-ray source, with gamma-ray absorbed doses of 10, 30 or 100 kGy [38]. The reaction mixture was placed in flasks, sealed and purged with $N_2$ to achieve an oxygen-free environment, and then subjected to gamma irradiation. Irradiation was performed at the panoramic Co-60 gamma source of the Radiation Chemistry and Dosimetry Laboratory, Ruđer Bošković Institute. Gamma irradiation was used to further increase the conversion of the polymerization reaction, to graft PDMAEMA onto the cellulose and possibly achieve complete crosslinking of the hydrogel. The samples were irradiated for an appropriate duration with a dose rate of 19.8 kGy h$^{-1}$. Dose mapping of the irradiation facility was performed experimentally using ionizing chambers and ECB dosimetric system, and by simulation calculations [78]. The resulting hydrogels were still in a liquid state, though more viscous. Initially, the resultant irradiated (10, 30 or 100 kGy) liquid material from Cel-g-PDMAEMA hydrogel was shaped into spheres, which were used for the study of swelling as well as structure characterization by FTIR. Based on the study of swelling of spheres the reaction mixture for preparation of hydrogel films was irradiated with the dose of 100 kGy. Molar ratio of reactants (monomer (DMAEMA), crosslinking agent (MBA) and cellulose (cel)) and irradiation dose for prepared samples are shown in Table 3.

**Table 3.** Molar ratio of reactants (monomer (DMAEMA), crosslinking agent (MBA) and cellulose (cel)) and irradiation dose for prepared samples.

| Sample | n(cel)/n(DMAEMA) | n(DMAEMA)/n(MBA) | Irradiation Dose (kGy) | AgNO$_3$ |
|---|---|---|---|---|
| 1-1_KS |  |  | 0 | - |
| 1-1_KS + AgNO$_3$ |  |  | 0 | + |
| 1-1_10 | 1:1 | 15:1 | 10 | - |
| 1-1_30 |  |  | 30 | - |
| 1-1_100 |  |  | 100 | - |
| 1-1_100 + AgNO$_3$ |  |  | 100 | + |
| 1-3_KS |  |  | 0 | - |
| 1-3_KS + AgNO$_3$ |  |  | 0 | + |
| 1-3_10 | 1:3 | 15:1 | 10 | - |
| 1-3_30 |  |  | 30 | - |
| 1-3_100 |  |  | 100 | - |
| 1-3_100 + AgNO$_3$ |  |  | 100 | + |
| 1-5_KS |  |  | 0 | - |
| 1-5_KS + AgNO$_3$ |  |  | 0 | + |
| 1-5_10 | 1:5 | 15:1 | 10 | - |
| 1-5_30 |  |  | 30 | - |
| 1-5_100 |  |  | 100 | - |
| 1-5_100 + AgNO$_3$ |  |  | 100 | + |

#### 4.2.2. Preparation of Hydrogel Films

The liquid Cel-g-PDMAEMA hydrogels, originated from chemical synthesis as well as those irradiated (100 kGy), were casted on the glass plate and submersed into a deionized water bath. The deionized water was exchanged after 1 and 24 h, two and five days. The prepared hydrogel films were stored in a refrigerator. Samples were freeze-extracted using same procedure as for spheres.

#### 4.2.3. Decorating Hydrogel Films with Silver Particles

Samples of 10 × 10 × 2 mm$^3$ dimensions were cut from hydrogel films and used for testing antibacterial properties. Additionally, samples of the same dimensions were decorated with silver particles by immersing in 5 mL of silver nitrate solution ($c = 10^{-2}$ mol dm$^{-3}$). After 24 h, the samples were removed from the silver nitrate solution and rinsed with deionized water for 1 h to remove unreacted silver nitrate [79].

### 4.3. Structural Confirmation of Cel-g-PDMAEMA Hydrogels
#### 4.3.1. Fourier Transform Infrared Spectroscopy

Fourier transform infrared spectra (FTIR) were recorded using Perkin-Elmer Spectrum One equipped with ATR module at room temperature in frequency range 650–4000 cm$^{-1}$. The FTIR spectra were processed using the OriginLab software.

#### 4.3.2. Scanning Electron Microscopy Analysis

Scanning electron microscopy (SEM) was used to examine the morphology of the synthesized Cel-g-PDMAEMA hydrogels. The lyophilized hydrogel samples were halved (spheres) or broken (films) and sputter coated by an alloy of gold and palladium (85%/15%) in the argon plasma to enhance their electrical conductivity. Metalized Cel-g-PDMAEMA hydrogel samples were scanned with a TESCAN VEGA 3 Scanning Electron Microscop, with a detector of secondary electrons.

### 4.3.3. Swelling Study

A gravimetric method was used to measure the equilibrium swelling ratio of the Cel-g-PDMAEMA hydrogels. For swelling degree at least five spheres were swollen and the average result is presented. The swelling kinetic of freeze-extracted samples was carried out for 60 days. Dried samples were reswelled in deionized water pH value 5.5 as well as in acidic and basic solution, respectively, to determine the swelling degree. The equilibrium swelling weights were measured at RT for the hydrogel samples in a solution of an appropriate pH value (3.0, 5.5 and 10.0) after wiping excess water from the hydrogel surface with moistened filter paper. The masses of the samples were measured before starting and at defined intervals until equilibrium was reached, i.e., a constant sample mass was achieved. The swelling ratio, $\alpha$, (and the equilibrium swelling ratio, $\alpha_e$,) were calculated according to Equation (1) [80]:

$$\alpha = \frac{m_t - m_0}{m_0} \times 100\% \tag{1}$$

where $m_0$ is mass of dry sample (xerogel) and $m_t$ mass of swollen sample at the time $t$.

### 4.4. X-ray Diffraction of Hydrogel Films Decorated with Silver Particles

Crystalline phase analysis was carried out using X-ray diffraction analysis (XRD) performed on Shimadzu XRD-6000 diffractometer with Cu K$_\alpha$ (1.5406 Å) radiation, operating at 40 kV, with a 2θ range of 30–90°, at a step size of 0.02°. The analyzed sample was ground into fine powder prior to XRD analysis.

### 4.5. Antibacterial Properties

The antimicrobial activity of the prepared films was tested against *Escherichia coli*, *Pseudomonas aeruginosa* and *Bacillus subtilis* by the agar disc diffusion test [81]. The specimens were placed (square shaped with an area of 1 × 1 cm$^2$) on Mueller–Hinton sterilized agar (Mueller Hinton Agar, Biolife), and the agar plate was inoculated uniformly with $10^6$ CFU mL$^{-1}$; lastly, the plates were incubated at 37 °C for 24 h. The volume of the added inoculum was 0.1 mL. The bacteria developed and enveloped the medium, except the resulting sterile area named "inhibition zone", generated by the antimicrobial materials. After 24 h, the inhibition zone diameter was measured. Cellulose and 30 µL of sterile deionized water pipetted on to blank disc were used as a negative control, and Gentamicin (12 µg µL$^{-1}$) as a positive control. All studies were performed in duplicate.

### 4.6. Electrochemical Testing

Two smooth nickel current collectors (1 × 1 cm$^2$) spaced by polymer hydrogel sided with two rGO/CNT sheets were used in order to assemble a supercapacitor. The geometric area of the supercapacitor was 1 cm$^2$. Three polymer hydrogels, 1-1_KS, 1-3_KS and 1-5_KS, were swollen in 0.5 M Na$_2$SO$_4$ for 30 min prior to testing. To determine the performance of the assembled supercapacitor, cyclic voltammetry (CV) and electrochemical impedance spectroscopy (EIS) were used. CV measurements were carried out at a scan rate of 50 mV s$^{-1}$ in the voltage range between 0 and 1.2 V. EIS was performed in the frequency range between $10^5$ and $10^{-3}$ Hz using an AC voltage amplitude of ±5 mV at a DC voltage of 0 V. The measurements were carried out by using PalmSens potentiostat/galvanostat/impedance analyzer and PSTrace 5.8 Software. The resistance of polymer hydrogel electrolyte was calculated based on the EIS response of supercapacitor. Specific capacitance of the cells was calculated by integration of the cyclic voltammogram according to Equation (2) [74]:

$$c_s = \frac{\int_{U_1}^{U_2} I(U) \cdot dU}{2mv \cdot (U_2 - U_1)} \qquad (2)$$

where $c_s$ is specific capacitance (F g$^{-1}$), $I$ current (A), $U$ voltage (V), $U_1$ starting voltage (V), $U_2$ switching voltage (V), $m$ mass of active material of one electrode (g), $v$ scan rate (V s$^{-1}$). All tests were carried out under ambient conditions.

**Supplementary Materials:** The following supporting information can be downloaded at: https://www.mdpi.com/article/10.3390/gels8100636/s1, Figures S1–S3: Pore size distribution of prepared Cel-g-PDMAEMA samples and Figures S4–S6: Testing of antibacterial activity for prepared hydrogel samples.

**Author Contributions:** R.B. and E.V. carried out the experiments and data analysis, D.K.G. performed antibacterial testing and interpretation of the results, M.K.R. performed electrochemical measurements and interpretation of the results, E.V. was responsible for conceptualization and the manuscript writing. All authors have read and agreed to the published version of the manuscript.

**Funding:** This research received no external funding.

**Institutional Review Board Statement:** Not applicable.

**Informed Consent Statement:** Not applicable.

**Acknowledgments:** The authors acknowledge Branka Mihaljević and Katarina Marušić from the Radiation Chemistry and Dosimetry Laboratory, Ruđer Bošković Institute for performing the irradiation procedure at the panoramic Co-60 gamma source.

**Conflicts of Interest:** The authors declare no conflict of interest.

## References

1. Ferreira, N.N.; Ferreira, L.M.B.; Cardoso, V.M.O.; Boni, F.I.; Souza, A.L.R.; Gremião, M.P.D. Recent Advances in Smart Hydrogels for Biomedical Applications: From Self-Assembly to Functional Approaches. *Eur. Polym. J.* **2018**, *99*, 117–133. [CrossRef]
2. Wang, D.; Xu, Y.; Li, Q.; Turng, L.S. Artificial Small-Diameter Blood Vessels: Materials, Fabrication, Surface Modification, Mechanical Properties, and Bioactive Functionalities. *J. Mater. Chem. B* **2020**, *8*, 1801–1822. [CrossRef] [PubMed]
3. Ciolacu, D.E.; Nicu, R.; Ciolacu, F. Natural Polymers in Heart Valve Tissue Engineering: Strategies, Advances and Challenges. *Biomedicines* **2022**, *10*, 1095. [CrossRef]
4. Ruben, M. Soft Contact Lenses: Clinical and Appled Technology. *Arch. Ophthalmol.* **1979**, *97*, 989. [CrossRef]
5. Wichterle, O.; Lim, D. Hydrophilic Gels for Biological Use. *Nature* **1960**, *185*, 117–118. [CrossRef]
6. Liang, Y.; He, J.; Guo, B. Functional Hydrogels as Wound Dressing to Enhance Wound Healing. *ACS Nano* **2021**, *15*, 12687–12722. [CrossRef] [PubMed]
7. Bai, X.; Gao, M.; Syed, S.; Zhuang, J.; Xu, X.; Zhang, X.Q. Bioactive Hydrogels for Bone Regeneration. *Bioact. Mater.* **2018**, *3*, 401–417. [CrossRef]
8. Tran, H.D.N.; Park, K.D.; Ching, Y.C.; Huynh, C.; Nguyen, D.H. A Comprehensive Review on Polymeric Hydrogel and Its Composite: Matrices of Choice for Bone and Cartilage Tissue Engineering. *J. Ind. Eng. Chem.* **2020**, *89*, 58–82. [CrossRef]
9. Mandal, A.; Clegg, J.R.; Anselmo, A.C.; Mitragotri, S. Hydrogels in the Clinic. *Bioeng. Transl. Med.* **2020**, *5*, e10158. [CrossRef]
10. Peppas, N.A.; Van Blarcom, D.S. Hydrogel-Based Biosensors and Sensing Devices for Drug Delivery. *J. Control. Release* **2016**, *240*, 142–150. [CrossRef]
11. Aziz, T.; Ullah, A.; Ali, A.; Shabeer, M.; Shah, M.N.; Haq, F.; Iqbal, M.; Ullah, R.; Khan, F.U. Manufactures of Bio-Degradable and Bio-Based Polymers for Bio-Materials in the Pharmaceutical Field. *J. Appl. Polym. Sci.* **2022**, *139*, e52624. [CrossRef]
12. Gil, C.J.; Li, L.; Hwang, B.; Cadena, M.; Theus, A.S.; Finamore, T.A.; Bauser-Heaton, H.; Mahmoudi, M.; Roeder, R.K.; Serpooshan, V. Tissue Engineered Drug Delivery Vehicles: Methods to Monitor and Regulate the Release Behavior. *J. Control. Release* **2022**, *349*, 143–155. [CrossRef] [PubMed]
13. Nieto, C.; Vega, M.A.; Rodríguez, V.; Esteban, P.P.; Martín del Valle, E.M. Biodegradable Gellan Gum Hydrogels Loaded with Paclitaxel for HER2+ Breast Cancer Local Therapy. *Carbohydr. Polym.* **2022**, *294*, 119732. [CrossRef] [PubMed]
14. Li, W.; Tang, J.; Lee, D.; Tice, T.R.; Schwendeman, S.P.; Prausnitz, M.R. Clinical Translation of Long-Acting Drug Delivery Formulations. *Nat. Rev. Mater.* **2022**, *7*, 406–420. [CrossRef]

15. Durpekova, S.; Filatova, K.; Cisar, J.; Ronzova, A.; Kutalkova, E.; Sedlarik, V. A Novel Hydrogel Based on Renewable Materials for Agricultural Application. *Int. J. Polym. Sci.* **2020**, *2020*, 8363418. [CrossRef]
16. Supare, K.; Mahanwar, P. Starch-Chitosan Hydrogels for the Controlled-Release of Herbicide in Agricultural Applications: A Study on the Effect of the Concentration of Raw Materials and Crosslinkers. *J. Polym. Environ.* **2022**, *30*, 2448–2461. [CrossRef]
17. Abobatta, W. Impact of hydrogel polymer in agricultural sector. *Adv. Agric. Environ. Sci.* **2018**, *1*, 59–64. [CrossRef]
18. Demitri, C.; Scalera, F.; Madaghiele, M.; Sannino, A.; Maffezzoli, A. Potential of Cellulose-Based Superabsorbent Hydrogels as Water Reservoir in Agriculture. *Int. J. Polym. Sci.* **2013**, *2013*, 435073. [CrossRef]
19. Neethu, T.M.; Dubey, P.K.; Kaswala, A.R. Prospects and Applications of Hydrogel Technology in Agriculture. *Int. J. Curr. Microbiol. Appl. Sci.* **2018**, *7*, 3155–3162. [CrossRef]
20. Khan, A.; Othman, M.B.H.; Razak, K.A.; Akil, H.M. Synthesis and Physicochemical Investigation of Chitosan-PMAA-Based Dual-Responsive Hydrogels. *J. Polym. Res.* **2013**, *20*, 273. [CrossRef]
21. Rosiak, J.M.; Yoshii, F. Hydrogels and Their Medical Applications. *Nucl. Instrum. Methods Phys. Res. Sect. B Beam Interact. Mater. Atoms* **1999**, *151*, 56–64. [CrossRef]
22. Yildiz, U.; Kemik, Ö.F.; Hazer, B. The Removal of Heavy Metal Ions from Aqueous Solutions by Novel PH-Sensitive Hydrogels. *J. Hazard. Mater.* **2010**, *183*, 521–532. [CrossRef] [PubMed]
23. Patel, A.M.; Patel, R.G.; Patel, M.P. Nickel and Copper Removal Study from Aqueous Solution Using New Cationic Poly[Acrylamide/N,N-DAMB/N,N-DAPB] Super Absorbent Hydrogel. *J. Appl. Polym. Sci.* **2011**, *119*, 2485–2493. [CrossRef]
24. Zhang, B.; Cui, Y.; Yin, G.; Li, X. Adsorption of Copper (II) and Lead (II) Ions onto Cottonseed Protein-PAA Hydrogel Composite. *Polym.-Plast. Technol. Eng.* **2012**, *51*, 612–619. [CrossRef]
25. Li, J.; Xu, Z.; Wu, W.; Jing, Y.; Dai, H.; Fang, G. Nanocellulose/Poly(2-(dimethylamino)ethyl Methacrylate)Interpenetrating Polymer Network Hydrogels for Removal of Pb(II) and Cu(II) Ions. *Colloids Surf. A Physicochem. Eng. Asp.* **2018**, *538*, 474–480. [CrossRef]
26. Murali Mohan, Y.; Vimala, K.; Varsha, T.; Varaprasad, K.; Sreedhar, B.; Bajpai, S.K.; Mohana Raju, K. Controlling of silver nanoparticles structure by hydrogel networks. *J. Colloid Interface Sci.* **2010**, *342*, 73–82. [CrossRef]
27. Rodríguez Nuñez, Y.A.; Castro, R.I.; Arenas, F.A.; López-Cabaña, Z.E.; Carreño, G.; Carrasco-Sánchez, V.; Marican, A.; Villaseñor, J.; Vargas, E.; Santos, L.S.; et al. Preparation of Hydrogel/Silver Nanohybrids Mediated by Tunable-Size Silver Nanoparticles for Potential Antibacterial Applications. *Polymers* **2019**, *11*, 716. [CrossRef]
28. Ilić-Stojanović, S.; Urošević, M.; Nikolić, L.; Petrović, D.; Stanojević, J.; Najman, S.; Nikolić, V. Intelligent Poly(N-Isopropylmethacrylamide) Hydrogels: Synthesis, Structure Characterization, Stimuli-Responsive Swelling Properties, and Their Radiation Decomposition. *Polymers* **2020**, *12*, 1112. [CrossRef]
29. Chen, Q.; Li, S.; Feng, Z.; Wang, M.; Cai, C.; Wang, J.; Zhang, L. Poly(2-(diethylamino)ethyl Methacrylate)-Based, PH-Responsive, Copolymeric Mixed Micelles for Targeting Anticancer Drug Control Release. *Int. J. Nanomed.* **2017**, *12*, 6857–6870. [CrossRef]
30. Goracci, G.; Arbe, A.; Alegría, A.; García Sakai, V.; Rudić, S.; Schneider, G.J.; Lohstroh, W.; Juranyi, F.; Colmenero, J. Influence of Solvent on Poly(2-(dimethylamino)ethyl Methacrylate) Dynamics in Polymer-Concentrated Mixtures: A Combined Neutron Scattering, Dielectric Spectroscopy, and Calorimetric Study. *Macromolecules* **2015**, *48*, 6724–6735. [CrossRef]
31. Rawlinson, L.B.; Ryan, S.M.; Mantovani, G.; Syrett, J.A.; Haddleton, D.M.; Brayden, D.J. Antibacterial Effects of Poly(2-(dimethylamino ethyl)methacrylate) against Selected Gram-Positive and Gram-Negative Bacteria. *Biomacromolecules* **2010**, *11*, 443–453. [CrossRef] [PubMed]
32. Van de Wetering, P.; Cherng, J.-Y.; Talsma, H.; Crommelin, D.J. Hennink, W. 2-(dimethylamino)ethyl Methacrylate Based (Co)Polymers as Gene Transfer Agents. *J. Control. Release* **1998**, *53*, 145–153. [CrossRef]
33. Van De Wetering, P.; Schuurmans-Nieuwenbroek, N.M.E.; Van Steenbergen, M.J.; Crommelin, D.J.A.; Hennink, W.E. Copolymers of 2-(dimethylamino)ethyl Methacrylate with Ethoxytriethylene Glycol Methacrylate or N-Vinyl-Pyrrolidone as Gene Transfer Agents. *J. Control. Release* **2000**, *64*, 193–203. [CrossRef]
34. De Souza, V.V.; Carretero, G.P.B.; Vitale, P.A.M.; Todeschini, Í.; Kotani, P.O.; Saraiva, G.K.V.; Guzzo, C.R.; Chaimovich, H.; Florenzano, F.H.; Cuccovia, I.M. Stimuli-Responsive Polymersomes of Poly [2-(dimethylamino) ethyl methacrylate]-b-Polystyrene. *Polym. Bull.* **2022**, *79*, 785–805. [CrossRef]
35. Diaz, I.L.; Sierra, C.A.; Jérôme, V.; Freitag, R.; Perez, L.D. Target Grafting of Poly(2-(dimethylamino)ethyl methacrylate) to Biodegradable Block Copolymers. *J. Polym. Sci.* **2020**, *58*, 2168–2180. [CrossRef]
36. Jiang, P.; Chen, S.; Lv, L.; Ji, H.; Li, G.; Jiang, Z.; Wu, Y. Effect of 2-(dimethylamino) ethyl Methacrylate on Nanostructure and Properties of PH and Temperature Sensitive Cellulose-Based Hydrogels. *J. Nanosci. Nanotechnol.* **2020**, *20*, 1799–1806. [CrossRef]
37. Gao, Y.; Zhou, D.; Lyu, J.; Sigen, A.; Xu, Q.; Newland, B.; Matyjaszewski, K.; Tai, H.; Wang, W. Complex Polymer Architectures through Free-Radical Polymerization of Multivinyl Monomers. *Nat. Rev. Chem.* **2020**, *4*, 194–212. [CrossRef]
38. Blažić, R.; Marušić, K.; Vidović, E. Swelling and viscoelastic properties of cellulose based hydrogels prepared by free radical polymerization of dimethylaminoethyl methacrylate in cellulose solution. 2022; *in press*.
39. Chen, W.; He, H.; Zhu, H.; Cheng, M.; Li, Y.; Wang, S. Thermo-Responsive Cellulose-Based Material with Switchable Wettability for Controllable Oil/Water Separation. *Polymers* **2018**, *10*, 592. [CrossRef]
40. Wulandari, W.T.; Rochliadi, A.; Arcana, I.M. Nanocellulose Prepared by Acid Hydrolysis of Isolated Cellulose from Sugarcane Bagasse. *IOP Conf. Ser. Mater. Sci. Eng.* **2016**, *107*, 012045. [CrossRef]

41. Ayu Laksanawati, T.; Novarita Trisanti, P. Sumarno Synthesis and Characterization of Composite Gels Starch-Graftacrylic Acid/Bentonite (St-g-AA/B) Using N'N-methylenebisacrylamide (MBA). *IOP Conf. Ser. Mater. Sci. Eng.* **2019**, *509*, 8–14. [CrossRef]
42. Ilić-Stojanović, S.S.; Nikolić, L.B.; Nikolić, V.D.; Petrović, S.D. Smart Hydrogels for Pharmaceutical Applications. *Polym. Bull.* **2017**, *46*, 278–310.
43. Blažic, R.; Lenac, K.; Vidović, E. Priprava celuloznih hidrogelova modificiranih 2-dimetilaminoetil-metakrilatom i srebrovim nanočesticama. *Kem. Ind.* **2020**, *69*, 269–279. [CrossRef]
44. Angelini, A.; Fodor, C.; Yave, W.; Leva, L.; Car, A.; Meier, W. PH-Triggered Membrane in Pervaporation Process. *ACS Omega* **2018**, *3*, 18950–18957. [CrossRef] [PubMed]
45. Takács, E.; Wojnárovits, L.; Borsa, J.; Földváry, C.; Hargittai, P.; Zöld, O. Effect of $\gamma$-Irradiation on Cotton-Cellulose. *Radiat. Phys. Chem.* **1999**, *55*, 663–666. [CrossRef]
46. Almeida do Nascimento, H.; Didier Pedrosa Amorim, J.; José Galdino da Silva, C., Jr.; D'Lamare Maia de Medeiros, A.; Fernanda de Santana Costa, A.; Carla Napoleão, D.; Maria Vinhas, G.; Asfora Sarubbo, L. Influence of Gamma Irradiation on the Properties of Bacterial Cellulose Produced with Concord Grape and Red Cabbage Extracts. *Curr. Res. Biotechnol.* **2022**, *4*, 119–128. [CrossRef]
47. Driscoll, M.; Stipanovic, A.; Winter, W.; Cheng, K.; Manning, M.; Spiese, J.; Galloway, R.A.; Cleland, M.R. Electron Beam Irradiation of Cellulose. *Radiat. Phys. Chem.* **2009**, *78*, 539–542. [CrossRef]
48. White, R.J.; Budarin, V.; Luque, R.; Clark, J.H.; Macquarrie, D.J. Tuneable Porous Carbonaceous Materials from Renewable Resources. *Chem. Soc. Rev.* **2009**, *38*, 3401–3418. [CrossRef]
49. Henryk, K.K.; Kinga, W.; Sławomir, B. The Effect of Gamma Radiation on the Supramolecular Structure of Pine Wood Cellulose in Situ Revealed by X-ray Diffraction. *Electron. J. Polish Agric. Univ.* **2004**, *7*, Available online: http://www.ejpau.media.pl/volume7/issue1/wood/art-06.html (accessed on 27 July 2022).
50. Salama, A.; Shukry, N.; El-Sakhawy, M. Carboxymethyl Cellulose-g-Poly(2-(dimethylamino) ethyl Methacrylate) Hydrogel as Adsorbent for Dye Removal. *Int. J. Biol. Macromol.* **2015**, *73*, 72–75. [CrossRef]
51. Ibrahim, H.M.M. Green Synthesis and Characterization of Silver Nanoparticles Using Banana Peel Extract and Their Antimicrobial Activity against Representative Microorganisms. *J. Radiat. Res. Appl. Sci.* **2015**, *8*, 265–275. [CrossRef]
52. Lee, S.H.; Jun, B.H. Silver Nanoparticles: Synthesis and Application for Nanomedicine. *Int. J. Mol. Sci.* **2019**, *20*, 865. [CrossRef] [PubMed]
53. Dong, X.Y.; Gao, Z.W.; Yang, K.F.; Zhang, W.Q.; Xu, L.W. Nanosilver as a New Generation of Silver Catalysts in Organic Transformations for Efficient Synthesis of Fine Chemicals. *Catal. Sci. Technol.* **2015**, *5*, 2554–2574. [CrossRef]
54. Haider, A.; Kang, I.K. Preparation of Silver Nanoparticles and Their Industrial and Biomedical Applications: A Comprehensive Review. *Adv. Mater. Sci. Eng.* **2015**, *2015*, 165257. [CrossRef]
55. Mehta, B.K.; Chhajlani, M.; Shrivastava, B.D. Green Synthesis of Silver Nanoparticles and Their Characterization by XRD. *J. Phys. Conf. Ser.* **2017**, *836*, 012050. [CrossRef]
56. Anwar, Y.; Ul-Islam, M.; Mohammed Ali, H.S.H.; Ullah, I.; Khalil, A.; Kamal, T. Silver Impregnated Bacterial Cellulose-Chitosan Composite Hydrogels for Antibacterial and Catalytic Applications. *J. Mater. Res. Technol.* **2022**, *18*, 2037–2047. [CrossRef]
57. Dos Santos, O.A.L.; de Araujo, I.; Dias da Silva, F.; Sales, M.N.; Christofolete, M.A.; Backx, B.P. Surface Modification of Textiles by Green Nanotechnology against Pathogenic Microorganisms. *Curr. Res. Green Sustain. Chem.* **2021**, *4*, 100206. [CrossRef]
58. Gao, A.; Hang, R.; Huang, X.; Zhao, L.; Zhang, X.; Wang, L.; Tang, B.; Ma, S.; Chu, P.K. The Effects of Titania Nanotubes with Embedded Silver Oxide Nanoparticles on Bacteria and Osteoblasts. *Biomaterials* **2014**, *35*, 4223–4235. [CrossRef]
59. Ando, S.; Hioki, T.; Yamada, T.; Watanabe, N.; Higashitani, A. $Ag_2O_3$ Clathrate is a Novel and Effective Antimicrobial Agent. *J. Mater. Sci.* **2012**, *47*, 2928–2931. [CrossRef]
60. Franci, G.; Falanga, A.; Galdiero, S.; Palomba, L.; Rai, M.; Morelli, G.; Galdiero, M. Silver Nanoparticles as Potential Antibacterial Agents. *Molecules* **2015**, *20*, 8856–8874. [CrossRef]
61. Grigor'Eva, A.; Saranina, I.; Tikunova, N.; Safonov, A.; Timoshenko, N.; Rebrov, A.; Ryabchikova, E. Fine Mechanisms of the Interaction of Silver Nanoparticles with the Cells of Salmonella Typhimurium and Staphylococcus Aureus. *BioMetals* **2013**, *26*, 479–488. [CrossRef]
62. Kim, J.S.; Kuk, E.; Yu, K.N.; Kim, J.H.; Park, S.J.; Lee, H.J.; Kim, S.H.; Park, Y.K.; Park, Y.H.; Hwang, C.Y.; et al. Antimicrobial Effects of Silver Nanoparticles. *Nanomed. Nanotechnol. Biol. Med.* **2007**, *3*, 95–101. [CrossRef] [PubMed]
63. Hosny, A.M.S.; Kashef, M.T.; Rasmy, S.A.; Aboul-Magd, D.S.; El-Bazza, Z.E. Antimicrobial Activity of Silver Nanoparticles Synthesized Using Honey and Gamma Radiation against Silver-Resistant Bacteria from Wounds and Burns. *Adv. Nat. Sci. Nanosci. Nanotechnol.* **2017**, *8*, 045009. [CrossRef]
64. Šálek, P.; Trousil, J.; Nováčková, J.; Hromádková, J.; Mahun, A.; Kobera, L. Poly[2-(dimethylamino)ethyl Methacrylate-*co*-Ethylene Dimethacrylate]Nanogel by Dispersion Polymerization for Inhibition of Pathogenic Bacteria. *RSC Adv.* **2021**, *11*, 33461–33470. [CrossRef] [PubMed]
65. Singh, R.; Shedbalkar, U.U.; Wadhwani, S.A.; Chopade, B.A. Bacteriagenic Silver Nanoparticles: Synthesis, Mechanism and Applications. *Appl. Microbiol. Biotechnol.* **2015**, *99*, 4579–4593. [CrossRef] [PubMed]
66. Beer, C.; Foldbjerg, R.; Hayashi, Y.; Sutherland, D.S.; Autrup, H. Toxicity of Silver Nanoparticles-Nanoparticle or Silver Ion? *Toxicol. Lett.* **2012**, *208*, 286–292. [CrossRef] [PubMed]

67. Ruano, G.; Iribarren, J.I.; Pérez-Madrigal, M.M.; Torras, J.; Alemán, C. Electrical and Capacitive Response of Hydrogel Solid-like Electrolytes for Supercapacitors. *Polymers* **2021**, *13*, 1337. [CrossRef] [PubMed]
68. Cao, X.; Jiang, C.; Sun, N.; Tan, D.; Li, Q.; Bi, S.; Song, J. Recent Progress in Multifunctional Hydrogel-Based Supercapacitors. *J. Sci. Adv. Mater. Devices* **2021**, *6*, 338–350. [CrossRef]
69. Gao, H.; Li, J.; Lian, K. Alkaline Quaternary Ammonium Hydroxides and Their Polymer Electrolytes for Electrochemical Capacitors. *RSC Adv.* **2014**, *4*, 21332–21339. [CrossRef]
70. Wu, M.; Yang, C.; Xia, H.; Xu, J. Comparative Analysis of Different Separators for the Electrochemical Performances and Long-Term Stability of High-Power Lithium-Ion Batteries. *Ionics* **2021**, *27*, 1551–1558. [CrossRef]
71. Zhang, J.; Yue, L.; Kong, Q.; Liu, Z.; Zhou, X.; Zhang, C.; Xu, Q.; Zhang, B.; Ding, G.; Qin, B.; et al. Sustainable, Heat-Resistant and Flame-Retardant Cellulose-Based Composite Separator for High-Performance Lithium Ion Battery. *Sci. Rep.* **2014**, *4*, 3935. [CrossRef] [PubMed]
72. Ding, G.; Qin, B.; Liu, Z.; Zhang, J.; Zhang, B.; Hu, P.; Zhang, C.; Xu, G.; Yao, J.; Cui, G. A Polyborate Coated Cellulose Composite Separator for High Performance Lithium Ion Batteries. *J. Electrochem. Soc.* **2015**, *162*, A834–A838. [CrossRef]
73. Zhao, C.; Niu, J.; Xiao, C.; Qin, Z.; Jin, X.; Wang, W.; Zhu, Z. Separator with High Ionic Conductivity and Good Stability Prepared from Keratin Fibers for Supercapacitor Applications. *Chem. Eng. J.* **2022**, *444*, 136537. [CrossRef]
74. Radić, G.; Šajnović, I.; Petrović, Ž.; Roković, M.K. Reduced Graphene Oxide/α-$Fe_2O_3$ Fibres as Active Material for Supercapacitor Application. *Croat. Chem. Acta* **2018**, *91*, 481–490. [CrossRef]
75. Sopčić, S.; Šešelj, N.; Kraljić Roković, M. Influence of Supporting Electrolyte on the Pseudocapacitive Properties of $MnO_2$/Carbon Nanotubes. *J. Solid State Electrochem.* **2019**, *23*, 205–214. [CrossRef]
76. Sopčić, S.; Antonić, D.; Mandić, Z. Effects of the Composition of Active Carbon Electrodes on the Impedance Performance of the AC/AC supercapacitors. *J. Solid State Electrochem.* **2022**, *26*, 591–605. [CrossRef]
77. Ljubek, G.; Roković, M.K. Aktivni materijali koji se upotrebljavaju u superkondenzatorima. *Kem. Ind.* **2019**, *68*, 507–520. [CrossRef]
78. Majer, M.; Roguljić, M.; Knežević, Ž.; Starodumov, A.; Ferenček, D.; Brigljević, V.; Mihaljević, B. Dose Mapping of the Panoramic $^{60}$Co Gamma Irradiation Facility at the Ruđer Bošković Institute–Geant4 Simulation and Measurements. *Appl. Radiat. Isot.* **2019**, *154*, 108824. [CrossRef]
79. Teper, P.; Sotirova, A.; Mitova, V.; Oleszko-Torbus, N.; Utrata-Wesołek, A.; Koseva, N.; Kowalczuk, A.; Mendrek, B. Antimicrobial Activity of Hybrid Nanomaterials Based on Star and Linear Polymers of $N, N'$-Dimethylaminoethyl Methacrylate with in Situ Produced Silver Nanoparticles. *Materials* **2020**, *13*, 3037. [CrossRef] [PubMed]
80. Vidović, E.; Klee, D.; Höcker, H. Evaluation of Water Uptake and Mechanical Properties of Biomedical Polymers. *J. Appl. Polym. Sci.* **2013**, *130*, 3682–3688. [CrossRef]
81. Clinical and Laboratory Standards Institute. *Performance Standards for Antimicrobial Disk Susceptibility Tests, Approved Standard*, 7th ed.; CLSI Document M02-A11; Clinical and Laboratory Standards Institute: Wayne, PA, USA, 2012.

Article

# Use of Carbotrace 480 as a Probe for Cellulose and Hydrogel Formation from Defibrillated Microalgae

Frederik L. Zitzmann *, Ewan Ward and Avtar S. Matharu *

Green Chemistry Centre of Excellence, Department of Chemistry, University of York, York YO10 5DD, UK; ew1057@york.ac.uk
* Correspondence: flz500@york.ac.uk (F.L.Z.); avtar.matharu@york.ac.uk (A.S.M.)

**Abstract:** Carbotrace 480 is a commercially available fluorescent optotracer that specifically binds to cellulose's glycosidic linkages. Herein, the use of Carbotrace 480 is reported as an analytical tool for linking cellulose content to hydrogel formation capability in defibrillated celluloses obtained from proprietary microalgae. Defibrillated celluloses obtained from acid-free hydrothermal microwave processing at low temperature (160 °C) showed poor hydrogel formation attributed to a low cellulose concentration as evidenced through the lack of Carbotrace fluorescence. High temperature (220 °C) processing afforded reasonable gels commensurate with a higher cellulose loading and stronger response to Carbotrace.

**Keywords:** Carbotrace 480; microalgae; defibrillated celluloses; hydrogels

**Citation:** Zitzmann, F.L.; Ward, E.; Matharu, A.S. Use of Carbotrace 480 as a Probe for Cellulose and Hydrogel Formation from Defibrillated Microalgae. *Gels* **2022**, *8*, 383. https://doi.org/10.3390/gels8060383

Academic Editors: Lorenzo Bonetti, Christian Demitri and Laura Riva

Received: 20 May 2022
Accepted: 14 June 2022
Published: 16 June 2022

**Publisher's Note:** MDPI stays neutral with regard to jurisdictional claims in published maps and institutional affiliations.

**Copyright:** © 2022 by the authors. Licensee MDPI, Basel, Switzerland. This article is an open access article distributed under the terms and conditions of the Creative Commons Attribution (CC BY) license (https://creativecommons.org/licenses/by/4.0/).

## 1. Introduction

Microalgae are typically unicellular species that grow in aquatic environments using sunlight and carbon dioxide as sources for photosynthesis to generate their energy [1–3]. They mainly consist of carbohydrates, lipids, proteins and, pigments, the exact split of which varies from species to species [4]. Their photosynthetic ability, coupled with their abundance across oceans, rivers, and lakes, allows microalgae to fix massive amounts of carbon dioxide from the atmosphere with estimates of around 865 mg $CO_2$ $L^{-1}$ $day^{-1}$ [2,5].

Microalgal cell walls comprise a complex three-dimensional structure that varies heavily depending on the species. Generally, microalgal cell walls consist of carbohydrates, lipids, and proteins, with carbohydrates being the most abundant cell-wall building block and proteins only making up around 5–8% of the dry cell-wall mass [6]. There is, however, a lack of scientific research into the exact structural composition, and only very well-known species are well-documented.

Cellulose as a cell-wall component in microalgae plays a lesser role than in terrestrial biomass. In addition to the lesser abundance, the microalgal cell wall is very thick and rigid, making extractions without any pre-treatment very difficult [7–9]. Microwave-assisted extraction (MAE) is a very efficient way to disrupt algal cell walls by creating localized pressure waves formed from the dielectric heating of water within the samples and aiding the production of defibrillated celluloses (D.C.) [10–13]. The microwave process induces hydrolysis of carbohydrates and therefore aids in the weakening and disruption of the cell wall, making extraction easier.

The MAE-derived D.C. can then be further processed to form cellulose-based hydrogels. Cellulose as a gelling agent in combination with water is well-studied [10,11,13]. The abundant hydroxyl groups in cellulose possess the capability to capture and trap water in the wider cellulosic structure through hydrogen bonding to form a three-dimensional hydrogel [14–16]. These hydrogels derived from biomass are becoming increasingly sought after in various fields of application, such as hygiene products, contact lenses, wound healing products, or lubricants [15]. Due to cellulose-based hydrogels being both biodegradable as well as low-cost and highly abundant, they pose a very promising and green candidate for future innovations.

Carbotrace molecules are fluorescent optotracers which specifically bind to the glycosidic linkages in cellulose and are therefore able to visually map cellulose content using confocal laser microscopy [17–19]. They have previously been used to identify cellulose in plant cells for mapping cellulosic nanofibrils in microfluidic devices and anatomical mapping in plant cells [17–19]. However, this paper reports the first use of Carbotrace to analyse defibrillated cellulose obtained from microalgae which were used for the formation of hydrogels.

This paper aims to use Carbotrace 480 for the first time to explore the distribution of cellulose in defibrillated cellulose samples obtained from microalgae prepared according to the method detailed by Zitzmann et al. [10]. The findings of the analysis using Carbotrace 480 will be linked to the hydrogel formation capabilities of the samples. The D.C. samples have been prepared using this acid-free and TEMPO-free green extraction process [10–12]. Carbotrace 480 acts as a non-destructive analytical tool for cellulose content in defibrillated cellulosic matter obtained via acid-free microwave hydrothermal processing of microalgae and their propensity to form hydrogels. The Carbotrace 480 (CT 480) was chosen, amongst many trials, to be the ideal probe molecule as its emission maxima did not interfere with the autofluorescence of the defibrillated celluloses. Therefore, the use of CT 480 allows for a good resolution of the images and a clear designation of cellulose and the other parts of the defibrillated celluloses that do not bind to Carbotrace 480.

## 2. Results and Discussion

### 2.1. Formation of Hydrogels from Defibrillated Cellulose Samples

Hydrogels have been formed from defibrillated cellulose samples prepared from both native microalgae and spent microalgae (industrial processing of ethanol-based alkali extraction of lipids leaving spent biomass) as detailed in Zitzmann et al. [10]. Defibrillated cellulose samples have been prepared according to the method previously described [10]. Results of the hydrogel formation are summarized in Table 1.

**Table 1.** Hydrogel formation results of defibrillated cellulose samples with numbers corresponding to the temperature in °C of the microwave process in the preparation step of the samples.

| Sample Type | Hydrogel Formation |
| --- | --- |
| Native 160 | X |
| Native 180 | X |
| Native 200 | X |
| Native 220 | Reasonably stable gel |
| Spent 160 | X |
| Spent 180 | X |
| Spent 200 | X |
| Spent 220 | Reasonably stable gel |

It was only possible to form hydrogels for the defibrillated cellulose samples that have been treated in the microwave at the highest temperature (Figure 1). All other samples did not yield any stable gel, merely a thick viscous mixture which, upon inversion of the vial, did not show any stability. The hydrogel formed from the Native 220 and Spent 220 samples formed a stable gel upon inversion but began to disintegrate within a couple of minutes after preparation.

With conventional analytical methods such as XRD or solid-state NMR, no clear direct correlation between the ability to form hydrogels and the cellulose content could be established, contrary to other literature findings [11,12] due to the increasingly complex mixture of cell-wall components in microalgae compared to plant cells. Therefore, Carbotrace has been used as a probe to identify the cellulose content in the samples and establish a link to the ability to form hydrogels.

**Figure 1.** Hydrogels formed from defibrillated cellulose samples obtained from native algal biomass and spent algal biomass. Numbers refer to the temperature in °C of the microwave process used for production of the D.C.

## 2.2. Using Carbotrace 480 as a Probe to Image Cellulose Content

To generate reference spectra of both the pure cellulose channels and the autofluorescence of the samples to apply to all the later images for unmixing and clear assignment, as well as to replicate the literature excitation and emission spectra of bound and unbound CT480, CT480 was run on its own in phosphate-buffered saline (PBS) as well as bound to pure cellulose immersed in PBS (see Figure 2).

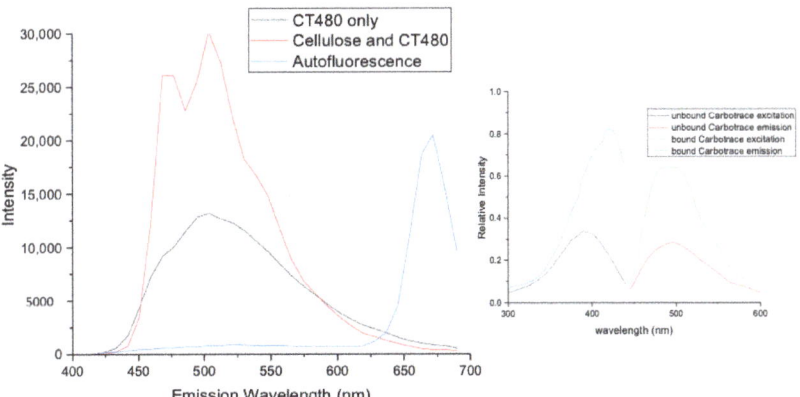

**Figure 2.** Confocal Laser Microscopy emission spectra of unbound CT480, bound CT480 to pure cellulose, defibrillated cellulose autofluorescence, and their comparison to the reference spectra provided by Ebba biotech—right side.

The obtained emission spectra correlate well with the spectra provided by Ebba Biotech with the emission maxima in the region of 480 nm as well as the perceived shift to the left upon Carbotrace binding to cellulose (approx. 20 nm). The intensity in emission also increases by around a factor of two upon binding to cellulose indicating that the intense fluorescence capabilities of the Carbotrace 480 are being switched on due to structural changes in the optotracer backbone upon binding to cellulose.

Also, with the autofluorescence peaking at around 670 nm and no emission appearing in the region where Carbotrace emits, there is excellent spatial separation between the two allowing for confident de-mixing of the channels of the subsequent defibrillated celluloses.

In order to confirm that the CT480 binds to cellulose only and not to other carbohydrate structures that can be found in algae, for example, xylan, both cellulose and xylan were mixed in their pure form with CT480. The images in Figure 3 show that, with the exact instrument settings, the pure cellulose manifests a very bright response to the CT binding, whereas xylan stays almost exclusively black (dark), confirming that the Carbotrace indeed binds to cellulose only.

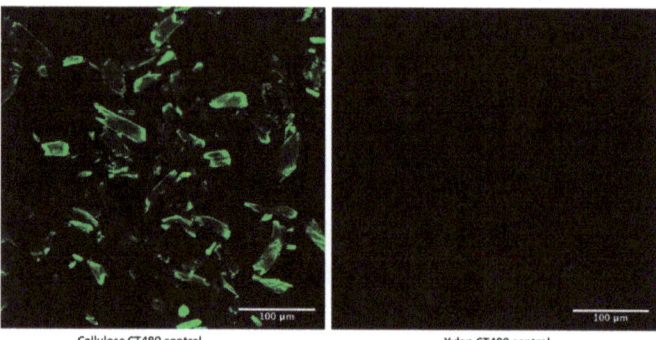

**Figure 3.** Laser Confocal Microscopy images of CT480 mixed with pure cellulose (**left**) and pure xylan (**right**) with the exact same instrument settings.

The morphology and aspect ratio of the D.C.s also clearly reveal themselves. In order to further make sure that the emission reference giving rise to the green colour channel only refers to the CT480 bound to cellulose, the native defibrillated cellulose 220 was run without any Carbotrace (Figure 4) to confirm the autofluorescence as the only component detected.

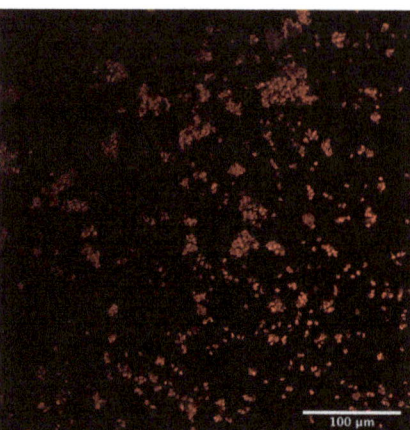

**Figure 4.** Laser Confocal Microscopy Image of unstained defibrillated cellulose 220 from native biomass.

The pure red colour channel obtained (Figure 4) was used to generate the autofluorescence emission spectrum used as a reference for all later defibrillated celluloses to de-mix cellulose bound to Carbotrace and autofluorescence the rest of the sample.

To explore the initial untreated biomass and the differences in cellulose distribution across the microalgae, Figure 5 shows both the initial native and spent biomass mixed with CT480.

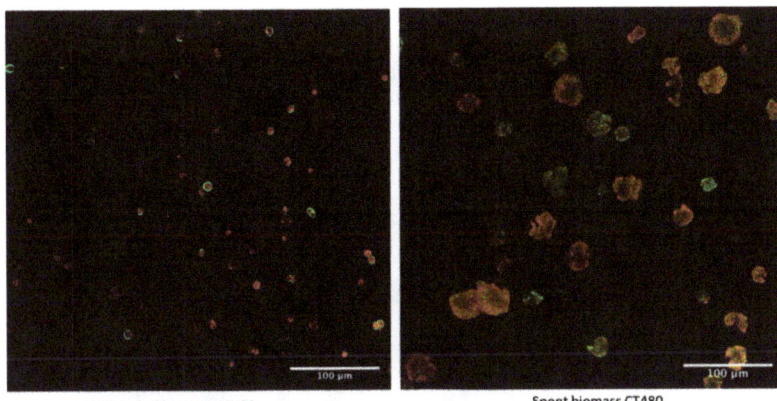

**Figure 5.** Confocal Laser Microscopy image of initial native and spent biomass mixed with Carbotrace 480.

The differences between the initial native and spent biomass can be seen very clearly in these images, with the native biomass showing an array of single microalgal cells, each with a ring of cellulose encapsulating the cells. This is in line with the basic structure of microalgal cells which has a cell wall containing cellulose wrapped around the inner cell. This is displayed more clearly in Figure 6, where only the cellulose channel is shown. Furthermore, the Carbotrace binds well to the cells and shows the cell wall.

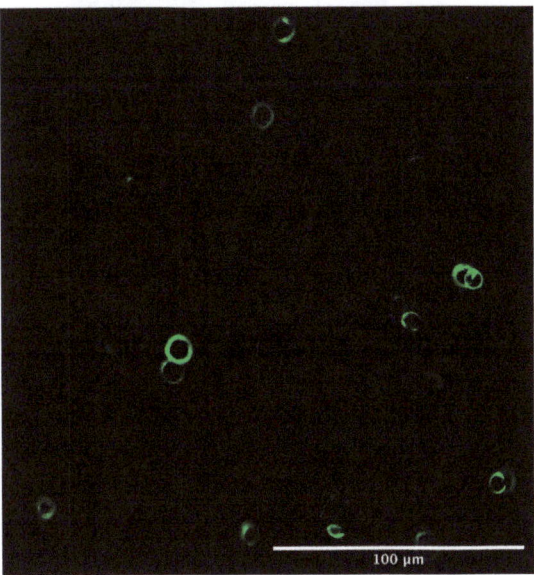

**Figure 6.** Laser Confocal Microscopy Image of the cellulose-only channel with CT480 of initial native biomass.

On the other hand, the spent biomass shows a much more disrupted profile with no individual circular cells able to be made out anymore. Rather an array of smudged and smashed cells with irregular shapes and a more even distribution of cellulose across the whole cell are observed, indicating that the industrial process has indeed destroyed and ruptured the cells. This is in line with previous findings that suggest an easy extraction from spent biomass precisely due to the factors that the image shows of a completely disrupted cell [10].

In order to apply the Carbotrace technology to the defibrillated celluloses and to use a visual tool that can directly identify and spatially show the distribution of the cellulose, all eight defibrillated cellulose samples were mixed with the Carbotrace 480, with the results displayed in Figure 7.

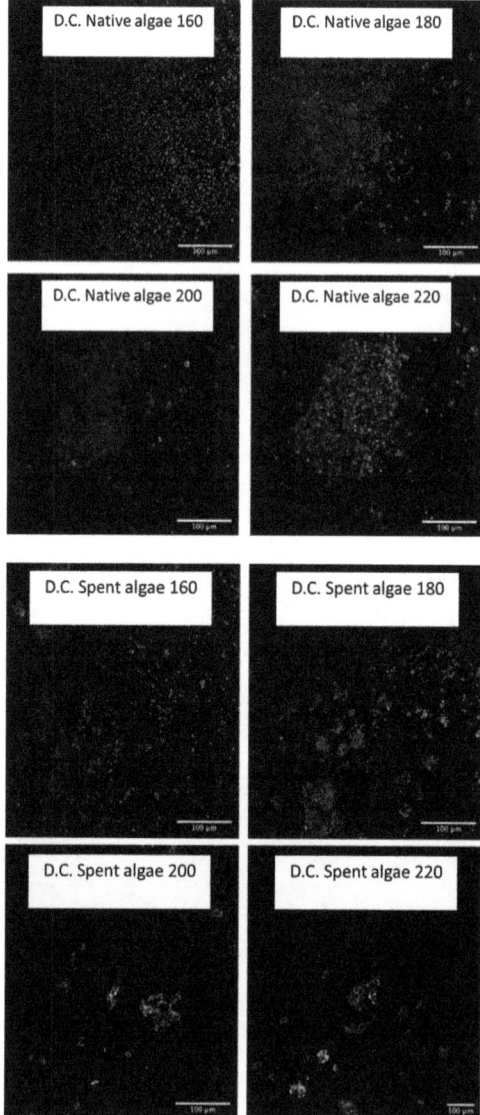

**Figure 7.** Laser Confocal Microscopy Images of all native algal derived and spent derived defibrillated celluloses.

The results show that both types of defibrillated celluloses form very small grains which then lump together into larger aggregates. Both types of defibrillated celluloses clearly show the trend that, with increasing microwave temperature, an increasing amount of cellulose can be observed in the samples, which is in line with previous findings from Zitzmann et al. that suggested that the highest temperatures show the highest correlation with pure cellulose [10].

However, it is interesting that, at the lowest temperature, there is still very little sign of cellulose as the red autofluorescence is much more prominent than any green specks, indicating the presence of cellulose. At the highest temperatures, this ratio is reversed with mostly cellulose being present in the samples. It must be noted that this type of analysis is suitable only for qualitative analysis and cannot be used to quantitatively calculate the exact percentages of cellulose present in the samples. Nevertheless, it still confirms visually what the previous analyses (XRD, NMR) hinted at, namely that cellulose content increases with higher microwave temperatures [10].

Figure 8 shows the nucleus/grain formation of the defibrillated celluloses with the native 200 defibrillated celluloses as a very clear example next to the initial native biomass. The nucleus formation and aggregate formation can be seen very clearly, which bears some resemblance to the initial microalgal cells; however, there is a size difference of 4–5 times that can be made out when comparing them to the initial spray-dried biomass on the left. Also, the shape of the grains is less perfectly circular but rather slightly off-shape. Interestingly, it can be seen that there is again encapsulation of the core by the cellulose which is flagged as green by the Carbotrace. This might be due to the core of the grains being so dense that the Carbotrace molecules are not able to penetrate and therefore form a circular layer around it.

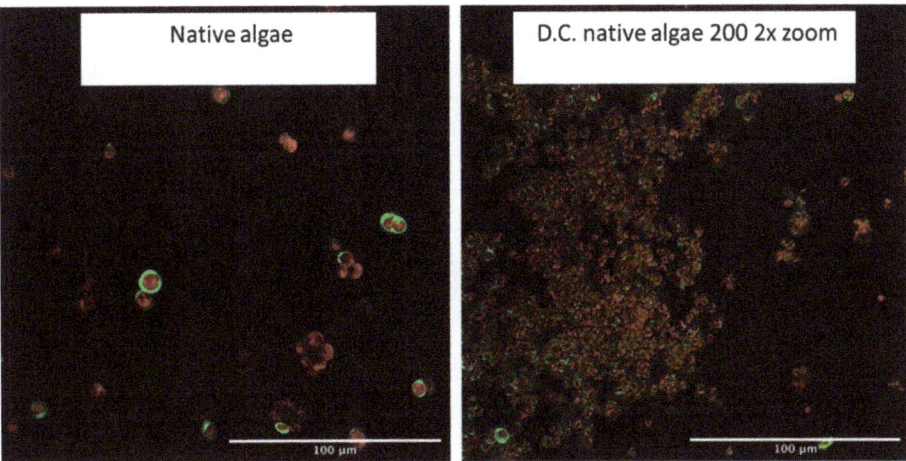

**Figure 8.** Laser Confocal Microscopy Images of initial native algal biomass and defibrillated celluloses 200 from native biomass at 2× zoom.

## 3. Conclusions

The reported findings in this paper, which used Carbotrace 480 as an analytical tool to visualize cellulose content in the defibrillated cellulose samples derived from two types of microalgae, correlate very well with the observed capabilities in forming a hydrogel. The unique fluorescent abilities of the Carbotrace 480 provide a clear link between the cellulose content in the samples, which can be used as a proxy for successful hydrogel formation. It has been shown that, for microalgal samples, a very high percentage of cellulose is required to be able to form any kind of even lightly stable hydrogel.

## 4. Materials and Methods

*Source of Biomass*

Microalgae were obtained from AlgaeCytes, Kent, England, who provided omega-3 enriched biomass from their proprietary microalgal Eustigmatophyceae strain, ALG01. The ALG01 strain was upscaled from Petri dish to 100 L using AlgaeCytes in-house proprietary upstream pyramid process and inoculated into the 1000 L Industrial Plankton seeding tank. Once the culture reached the late exponential phase, it was transferred into AlgaeCytes pilot-plant production module (VariconAqua 12,000 L Phyco-FlowTM). After reaching an appropriate density, it underwent semi-continuous harvesting to provide material for spray drying. On each harvesting day, 1000 L of algal culture was dewatered using an Alfa Laval Clara 20 model disc-stack centrifuge to produce an algal slurry of ~15% ± 5% solids. The algal slurry was subsequently dried using a Büchi mini spray dryer B-290 to produce a dried algal powder of <1% moisture content.

The defibrillated cellulose samples were prepared according to the method detailed by Zitzmann et al. [10] and using a Milestone Synthwave reactor (1500 W, 2.45 GHz).

Hydrogels were produced by mixing the defibrillated celluloses with deionized water (3 wt%) and treating the mixture with a homogenizer for 3 min. The stability of the gel was tested by inverting the vial and recording the time it stably stayed at the top of the vial before descending.

Carbotrace 480 was obtained from Ebba Biotech (Stockholm, Sweden). It was mixed with PBS at a ratio of 1:1000 (pH 7.4) and, subsequently, the defibrillated cellulose samples (0.2 mg) were mixed with aliquots of this stock solution (50 µL) and left to incubate for 30 min at room temperature.

Carbotrace images were captured using a Zeiss LSM980 confocal microscope, AxioObserver Z1 using ZEN 3.4 (blue edition) software and either an EC Plan-Neofluar 10×/0.3 or a Plan Apochromat 20×/0.8 objective. All samples were excited with a 405 nm laser using a 405 nm main beam splitter, and emissions were collected from 411–694 nm in bins of 8.9 nm. The pixel size was 1.657 µm$^2$ or 0.829 um$^2$ for the 10× or 20× objectives, respectively. The pinhole was 1 AU, and the images were taken in 16 bit.

Reference spectra of cellulose stained with Carbotrace 480 and autofluorescence values from unstained sample spectra were collected independently to permit optimal spectral unmixing. Samples were typically averaged ×8 to reduce noise and increase the precision of the spectral unmixing. This process was performed using the in-built application within the Carl Zeiss ZEN 3.4 software (Jena, Germany) on a pixel-to-pixel basis.

The images were unmixed as follows:

SY_temperature samples using SY160 unstained autofluorescence spectra and the cellulose CT480 spectra.

ST_temperature samples using ST160 unstained autofluorescence spectra and the cellulose CT480 spectra.

**Author Contributions:** Conceptualization, A.S.M. and F.L.Z.; methodology, F.L.Z. and E.W.; writing—original draft preparation, F.L.Z.; writing—review and editing, A.S.M. All authors have read and agreed to the published version of the manuscript.

**Funding:** This research received no external funding.

**Institutional Review Board Statement:** Not applicable.

**Informed Consent Statement:** Not applicable.

**Data Availability Statement:** Not applicable.

**Acknowledgments:** We would like to thank Jo Marrison and Grant Calder at the Department of Biology, University of York, UK, for running all the samples on the confocal laser microscope as well as putting a lot of time into editing all the spectra. Furthermore, we thank Donal McGee, Senior Microalgal Scientist at AlgaeCytes, Kent, UK, for kindly providing the proprietary microalgae used in this paper. Finally, thanks to John Bergqvist, Research Engineer at Ebba Biotech, Stockholm, Sweden, for providing the Carbotrace as well as many scientific discussions. We also thank the reviewers for providing valuable feedback to our work and suggesting important experimental evidence for extending the current work in the future.

**Conflicts of Interest:** The authors declare no conflict of interest.

## References

1. Bilanovic, D.; Andargatchew, A.; Kroeger, T.; Shelef, G. Freshwater and Marine Microalgae Sequestering of $CO_2$ at Different C and N Concentrations—Response Surface Methodology Analysis. *Energy Convers. Manag.* **2009**, *50*, 262–267. [CrossRef]
2. Sydney, E.B.; Sturm, W.; de Carvalho, J.C.; Thomaz-Soccol, V.; Larroche, C.; Pandey, A.; Soccol, C.R. Potential Carbon Dioxide Fixation by Industrially Important Microalgae. *Bioresour. Technol.* **2010**, *101*, 5892–5896. [CrossRef] [PubMed]
3. Niccolai, A.; Zittelli, G.C.; Rodolfi, L.; Biondi, N.; Tredici, M.R. Microalgae of Interest as Food Source: Biochemical Composition and Digestibility. *Algal Res.* **2019**, *42*, 101617. [CrossRef]
4. Alhattab, M.; Kermanshahi-Pour, A.; Brooks, M.S.-L. Microalgae Disruption Techniques for Product Recovery: Influence of Cell Wall Composition. *J. Appl. Phycol.* **2019**, *31*, 61–88. [CrossRef]
5. Zhao, B.; Su, Y. Process Effect of Microalgal-Carbon Dioxide Fixation and Biomass Production: A Review. *Renew. Sustain. Energy Rev.* **2014**, *31*, 121–132. [CrossRef]
6. Spain, O.; Plöhn, M.; Funk, C. The Cell Wall of Green Microalgae and Its Role in Heavy Metal Removal. *Physiol. Plant.* **2021**, *173*, 526–535. [CrossRef]
7. Weber, S.; Grande, P.M.; Blank, L.M.; Klose, H. Insights into Cell Wall Disintegration of Chlorella Vulgaris. *PLoS ONE* **2022**, *17*, e0262500. [CrossRef] [PubMed]
8. Colusse, G.A.; Carneiro, J.; Duarte, M.E.R.; de Carvalho, J.C.; Noseda, M.D. Advances in Microalgal Cell Wall Polysaccharides: A Review Focused on Structure, Production, and Biological Application. *Crit. Rev. Biotechnol.* **2021**, *42*, 562–577. [CrossRef] [PubMed]
9. Derenne, S.; Largeau, C.; Hatcher, P.G. Structure of Chlorella Fusca Algaenan: Relationships with Ultralaminae in Lacustrine Kerogens; Species- and Environment-Dependent Variations in the Composition of Fossil Ultralaminae. *Org. Geochem.* **1992**, *18*, 417–422. [CrossRef]
10. Zitzmann, F.L.; Ward, E.; Meng, X.; Matharu, A.S. Microwave-Assisted Defibrillation of Microalgae. *Molecules* **2021**, *26*, 4972. [CrossRef] [PubMed]
11. Gao, Y.; Xia, H.; Sulaeman, A.P.; de Melo, E.M.; Dugmore, T.I.J.; Matharu, A.S. Defibrillated Celluloses via Dual Twin-Screw Extrusion and Microwave Hydrothermal Treatment of Spent Pea Biomass. *ACS Sustain. Chem. Eng.* **2019**, *7*, 11861–11871. [CrossRef]
12. Gao, Y.; Ozel, M.Z.; Dugmore, T.; Sulaeman, A.; Matharu, A.S. A Biorefinery Strategy for Spent Industrial Ginger Waste. *J. Hazard. Mater.* **2021**, *401*, 123400. [CrossRef] [PubMed]
13. de Melo, E.M.; Clark, J.H.; Matharu, A.S. The Hy-MASS Concept: Hydrothermal Microwave Assisted Selective Scissoring of Cellulose for in Situ Production of (Meso)Porous Nanocellulose Fibrils and Crystals. *Green Chem.* **2017**, *19*, 3408–3417. [CrossRef]
14. Chang, C.; Zhang, L. Cellulose-Based Hydrogels: Present Status and Application Prospects. *Carbohydr. Polym.* **2011**, *84*, 40–53. [CrossRef]
15. Sannino, A.; Demitri, C.; Madaghiele, M. Biodegradable Cellulose-Based Hydrogels: Design and Applications. *Materials* **2009**, *2*, 353–373. [CrossRef]
16. Du, H.; Liu, W.; Zhang, M.; Si, C.; Zhang, X.; Li, B. Cellulose Nanocrystals and Cellulose Nanofibrils Based Hydrogels for Biomedical Applications. *Carbohydr. Polym.* **2019**, *209*, 130–144. [CrossRef] [PubMed]
17. Kumar, T.; Soares, R.R.G.; Ali Dholey, L.; Ramachandraiah, H.; Aval, N.A.; Aljadi, Z.; Pettersson, T.; Russom, A. Multi-Layer Assembly of Cellulose Nanofibrils in a Microfluidic Device for the Selective Capture and Release of Viable Tumor Cells from Whole Blood. *Nanoscale* **2020**, *12*, 21788–21797. [CrossRef] [PubMed]
18. Choong, F.X.; Lantz, L.; Shirani, H.; Schulz, A.; Nilsson, K.P.R.; Edlund, U.; Richter-Dahlfors, A. Stereochemical Identification of Glucans by a Donor–Acceptor–Donor Conjugated Pentamer Enables Multi-Carbohydrate Anatomical Mapping in Plant Tissues. *Cellulose* **2019**, *26*, 4253–4264. [CrossRef]
19. Lahchaichi, E. Cellulose Nanofibril-Based Layer-by-Layer System for Immuno-Capture of Circulating Tumor Cells in Microfluidic Devices. 2021. Available online: diva-portal.org (accessed on 5 May 2022).

*Review*

# Advances in Cellulose-Based Hydrogels for Biomedical Engineering: A Review Summary

Pengfei Zou [1,†], Jiaxin Yao [1,†], Ya-Nan Cui [1,†], Te Zhao [1,2], Junwei Che [1,3], Meiyan Yang [1], Zhiping Li [1,*] and Chunsheng Gao [1,*]

1 State Key Laboratory of Toxicology and Medical Countermeasures, Beijing Institute of Pharmacology and Toxicology, Beijing 100850, China; wsygfxj@163.com (P.Z.); yaojiaxin0719@126.com (J.Y.); yanancui518@163.com (Y.-N.C.); zt12080923352021@163.com (T.Z.); chejunwei1019@163.com (J.C.); ymyzi@163.com (M.Y.)
2 School of Chemical and Pharmaceutical Engineering, Hebei University of Science and Technology, Shijiazhuang 050018, China
3 School of Pharmaceutical Sciences, Shandong First Medical University & Shandong Academy of Medical Sciences, Taian 271016, China
* Correspondence: dearwood2010@126.com (Z.L.); gaocs@bmi.ac.cn (C.G.)
† These authors contributed equally to this work.

**Abstract:** In recent years, hydrogel-based research in biomedical engineering has attracted more attention. Cellulose-based hydrogels have become a research hotspot in the field of functional materials because of their outstanding characteristics such as excellent flexibility, stimulus-response, biocompatibility, and degradability. In addition, cellulose-based hydrogel materials exhibit excellent mechanical properties and designable functions through different preparation methods and structure designs, demonstrating huge development potential. In this review, we have systematically summarized sources and types of cellulose and the formation mechanism of the hydrogel. We have reviewed and discussed the recent progress in the development of cellulose-based hydrogels and introduced their applications such as ionic conduction, thermal insulation, and drug delivery. Also, we analyzed and highlighted the trends and opportunities for the further development of cellulose-based hydrogels as emerging materials in the future.

**Keywords:** cellulose; hydrogels; biomedical engineering; application

## 1. Introduction

Growing concerns about environmental issues and the increasing demand for environmentally friendly materials have forced researchers around the world to explore naturally occurring biopolymer or biomimetic materials for their potential applications in various fields [1,2]. Hydrogels are ductile and extremely porous polymers with a three-dimensional network structure, which was first produced by Wicherle and Lim in 1960 [3]. Over time, the research of hydrogels has developed for biomedical applications [4], including wound dressings [5,6], anti-tumor immunotherapy [7,8], anti-central nervous system disorders [9], tissue-engineering [10–14], smart drug-delivery systems [7,15,16], and contraception [15], due to their good biocompatibility, excellent physical and mechanical properties, and long-term implant stability.

So far, hydrogels are divided into physical hydrogels and chemical hydrogels according to the different cross-linking modes [16]. Physical hydrogels are formed by physical forces, such as hydrophobic aggregation, π-π stacking, hydrogen bonding, and electrostatic interaction, which are non-permanent and converted into a solution by heating or other external stimulation. On the contrary, chemical hydrogels formed by chemical cross-linking are permanent and irreversible. In addition, hydrogels are also divided into traditional hydrogels and functional/smart hydrogels according to their response to environmental stimuli [17]. Traditional hydrogels are not sensitive to environmental changes, while functional/smart stimuli-responsive nanocomposite hydrogels [18] produce corresponding changes in physical structure and chemical properties to small changes in the external environment (such as temperature, pH, light, magnetism, etc.) [19,20]. The outstanding feature of these hydrogels is that the swelling behavior changes significantly in response to the environment. They are used as actuators [21], sensors [22], plantable and biodegradable ion batteries [23], thermally insulating materials [24], for tissue transformation [25,26], in controlled-release switches [27,28] or in precise topical administration regimens [29] and programmable and bioinstructive materials systems [30], etc. Therefore, functional/smart stimuli-responsive hydrogels have been one of the most interesting topics for scientific researchers in recent years.

On the other hand, hydrogels can also be divided into synthetic polymer hydrogels and natural polymer hydrogels according to the different synthetic raw materials [31]. The natural polymer has attracted more attention due to its biocompatibility, abundant source, low price and good biomedical application prospects. For instance, cellulose, alginate, chitosan, pectin, and starch are the most important biopolymers used for the fabrication of biopolymer hydrogels. Amongst these natural polymers, research focusing on cellulose-based hydrogels has gained significant attention because of their low cost, strong processability, renewability, biocompatibility, biodegradability and environmental friendliness [32]. However, the poor strength of these natural hydrogels has further limited their application. To address this challenge, synthetic hydrogels and hybrid hydrogels are favored by researchers because of their tunable physical and chemical properties, which include super-adhesion [33], strong toughness, fatigue resistance, self-reinforcement [34], and self-healing [35–38]. In addition, strategies for the design and functionality of aerogels [39,40] and nanofiber-based hydrogels [41] were also considered for intensive attention.

In the present review, the recent advances in cellulose hydrogels are highlighted (Figure 1). Moreover, the sources and types of cellulose, the mechanism of hydrogel formation, the research progress of hybrid cellulose hydrogels, and the different functional types of cellulose hydrogels are mainly discussed. Finally, the application and potential challenges of cellulose-based hydrogels are outlined, and future research directions are considered.

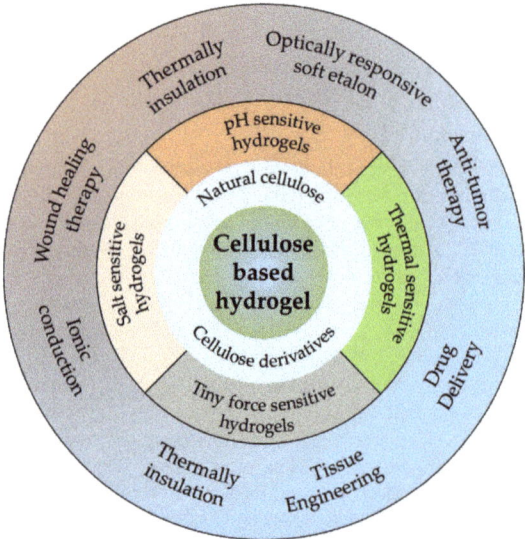

Figure 1. The recent advances of cellulose hydrogels in this review.

## 2. Cellulose

As we know, cellulose was first isolated by French scientist Anselme Payen in 1838 [32]. Subsequently, the polymer form of cellulose was identified by German chemist Staudinger in 1932. Up to now, cellulose has already been extracted from readily available natural resources (such as bacteria, bamboo, jute, algae, biofilm, wood, cotton, hemp, and other plant-based materials), and is the most abundant natural macromolecular compound in the world. Five thousand to fifteen thousand glucose molecules with the molecular formula $(C_6H_{10}O_5)_n$ are covalently bonded together through C1 of the glucose ring and C4 of the adjacent ring (Figure 2a) [10,42], covalently bonded together by the acetal oxygen to form D-glucose with β-1,4 glycosidic bond [43,44]. The structure and size of natural cellulose are different in various sources, and the structural form of cellulose nanomaterials depends on processing technology. In all these multiscale materials, the structure of cellulose is extremely important because it directly affects its mechanical properties [45]. In addition, the polyhydroxy groups in cellulose can produce various forms and different functional properties after specific physical or chemical modification [46]. Therefore, the development of functional cellulose derivatives has great potential, as a means of improving the flexibility and feasibility of cellulose.

### 2.1. Classification of Cellulose

#### 2.1.1. Natural Cellulose

Cellulose is divided into native cellulose and synthetic cellulose according to its source (Table 1) [47]. Natural cellulose is composed of plant and bacterial cellulose (BC) (Figure 2b) [43,48]. Plant cellulose widely exists in cotton, wood, and other plants, such as phloem fiber, seed fiber, and wood fiber, which is the most abundant organic substance in nature [45]. Bacterial fiber refers to the cellulose synthesized by a specific species of microorganisms under different conditions, and it is the finest nano-scale fiber in nature. Multiple microorganisms can synthesize cellulose, such as *Pseudomonas* and *Acetobacter* [49]. Importantly, diverse bacteria produce cellulose with various morphology, structures, characteristics, and functionalities. For example, cellulose was secreted by some fungi and green algae (e.g., *Valonia ventricular*, *Glaucocystis*), and contained the outer cell membrane of some marine ascidians.

A wide range of studies has been conducted on the potential advantages of bacterial and plant cellulose as biomaterials [50]. Compared with plant cellulose, BC has high crystallinity and purity, because it does not produce lignin, hemicellulose, and other accompaniments [51]. It was demonstrated that the superior performance of BC satisfied the essential requirements of indispensable and versatile biomedical materials for all the practical and innovative applications [52], such as tissue engineering and wound repair [49]. The wide biomedical applications of BC are supported by the ease of production, lack of contaminants, and the capability of modulating the material's features during syntheses—such as crystallinity index, aspect ratio, and morphology to perfectly fit the final application requirements [52]. Significantly, natural cellulose-based hydrogels were prepared from pure cellulose solutions by physical cross-linking due to the presence of numerous hydroxyl groups, which can connect the polymer network via hydrogen bonding [28]. Moreover, plant cellulose and bacterial cellulose differ in terms of macromolecular properties. Plant cellulose has a medium water-holding capacity of 60%, and a moderate level of tensile strength and crystallinity. At the same time, BC is chemically pure, hydrophilic, and it has a high water-holding capacity (100%) [52]. However, natural cellulose has multiple shortcomings such as poor solubility, low thermoplasticity, strong hydrophilicity, weak adsorption capacity, and difficult processing, which limits its development and application in the biomedical and pharmaceutical fields [53]. Fortunately, insolubility can be overcome by obtaining cellulose derivatives through various chemical modification procedures, such as esterification, etherification, or oxidation (Figure 2c) [54].

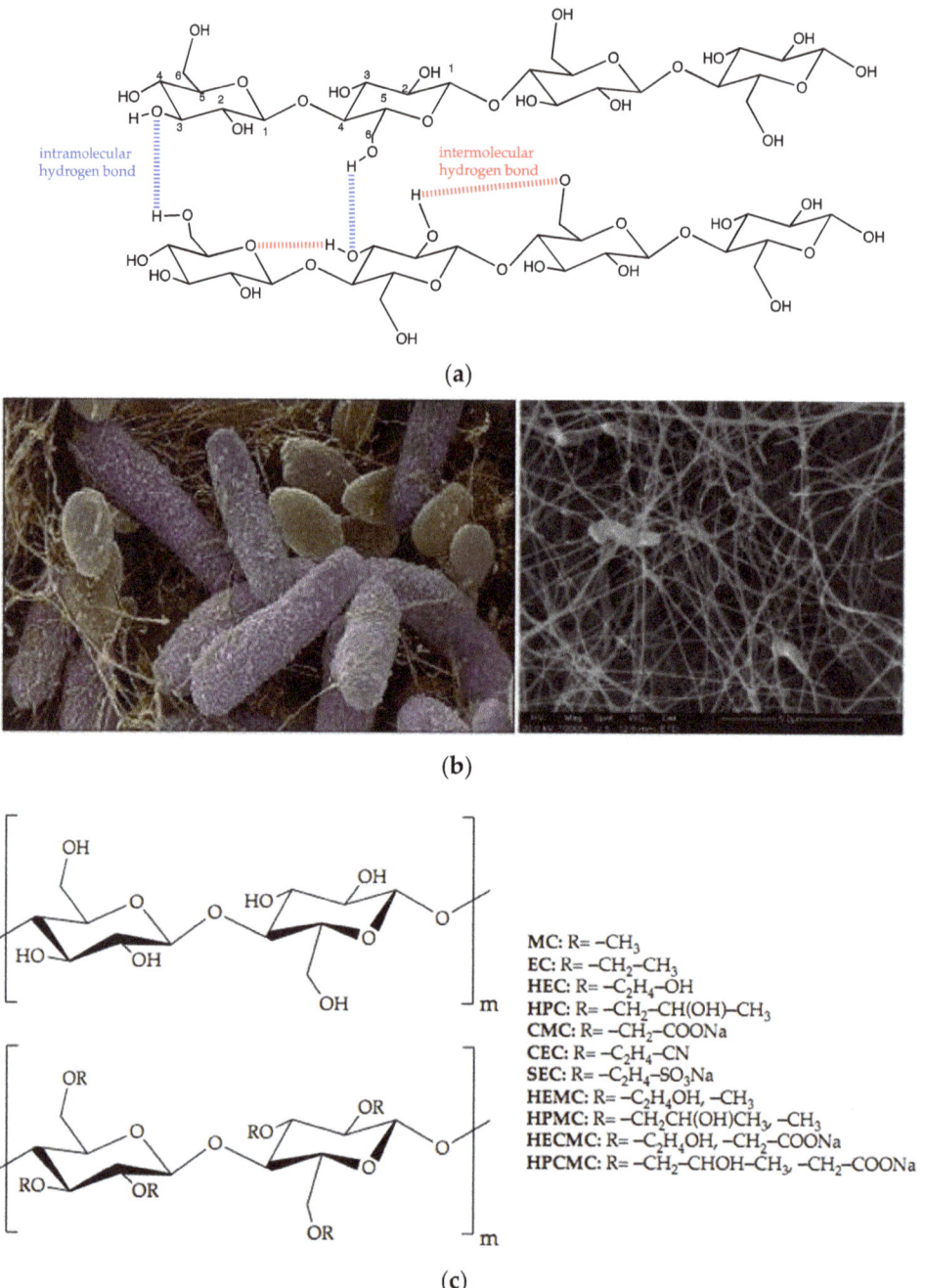

**Figure 2.** Chemical structure of cellulose and its derivatives. (**a**) Chemical structure of cellulose. (**b**) Scanning electron micrograph images of *Acetobacter xylinum* and formation of bacterial cellulose. Adapted with permission from Ref. [43] Copyright 2019, Springer Nature. (**c**) The chemical structure of cellulose and some of its derivatives. Adapted with permission from Ref. [54] Copyright 2016, WILEY-VCH Verlag GmbH & Co. KGaA, Weinheim.

Table 1. Summary of the cellulose classification, property and usage.

| Classification | Name | Property | Usage | Ref. |
|---|---|---|---|---|
| Natural cellulose | Plant cellulose | thermal and mechanical degradation; major components of plant cell walls. | fabrics, ropes, tapes, isolating materials. | [47] |
| Bacterial cellulose | BC | — | dura mater replacement, diagnostic sensors, dental grafting, artificial cornea, wound dressing, drug delivery system, bone tissue engineering. | [49,55] |
| Cellulose derivatives | Phosphate cellulose | — | enrichment agents, ion exchangers. | [56] |
| | Nitrocellulose | flammability, bonding. | coatings, adhesives, cosmetics, food packagings, centrifugation tube materials. | [57] |
| | Cellulose acetate (CA) | biodegradable, renewable, non-corrosive, non-toxic, biocompatible. | nanocomposites for biomedical applications and equipment. | [58] |
| | Methylcellulose (MC) | high water retention, thermogelation, macro-phase separation, syneresis. | slow-release preparation. | [59] |
| | Ethylcellulose (EC) | thermoplastic, water insoluble, nonionic, thermally stable, hydrophobicity. | controlled release formulations, coating agents. | [28,60] |
| | Carboxymethyl cellulose (CMC) | hydrophilic, bioadhesive, non-toxic, pH sensitive, thermally stable. | hindering crystallization or degradation of the drug; enhancing the frequency of drug release. | [28,61] |
| | Hydroxyethyl cellulose (HEC) | suspension, adhesion, emulsification, dispersion, moisture | food stabilizers, thickeners, adhesives, pharmaceutical excipients, stabilizers, film coating agents. | [44] |
| | Hydroxypropyl cellulose (HPC) | biodegradable and biocompatible, self-repairing abilities, shape memory, unique hydrophilic/hydrophobic change. | thermo-responsive hydrogels. | [62] |
| | Hydroxypropyl Methylcellulose (HPMC) | viscous soluble fiber, high viscosity, gelling. | thickener, emulsifier, stabilizer, gelling agents, antioxidants, hypoglycemics. | [63,64] |
| | Hydroxyethyl Methylcellulose (HEMC) | water solubility, thermally stable, gel properties. | hypoglycemics, antioxidant, coatings, medical dressings. | [64,65] |

2.1.2. Cellulose Derivatives

Cellulose derivatives are mainly obtained by two methods including physical and chemical modification (Table 2).

Table 2. Modification method and classification of cellulose derivatives.

| Methods | Classification |
|---|---|
| Physical modification | film cellulose; microcrystalline cellulose; spherical cellulose; nano-cellulose; |
| Chemical modification | degradation reactions: acid-base, oxidative, biodegradation, mechanical processing; hydroxyl derivative reactions: nucleophilic substitution, graft copolymerization, cross-linking reaction, esterification reaction, etherification reaction. |

1. Physical modification

On the one hand, the physical modification is mainly used to obtain new properties and functions by changing the structure and surface properties of cellulose. In short, the physical modification mainly includes mechanical grinding, swelling, recombining, and surface adsorption without changing the chemical composition of cellulose, such as regenerated cellulose (the material obtained after cellulose is dissolved and precipitated) [66], membrane cellulose, microcrystalline cellulose, spherical cellulose [67], and nanocellulose (NC) [68].

Specifically, NC is attracting more attention because it is easy to form a self-assembly structure with stable mechanical energy. They are mainly classified into two categories: cellulose nanocrystals (CNC) and cellulose nanofibers (CNF) [69]. NC has high specific surface area, easy modification, biodegradability, non-toxic biocompatibility, wound-healing characteristics, and antibacterial effects, making it widely used in biomedical fields. For example, Gonzalez reported polyvinyl alcohol (PVA)/CNC composite hydrogels prepared by the freeze-thaw technique. Compared with pure PVA hydrogel, the obtained composite hydrogel maintains transparency, improves thermal stability, and improves mechanical properties [70].

On the other hand, the physical modification is also used to obtain new properties and functions by using inorganic salt solutions such as $CaCl_2$, $ZnCl_2$ or $MgCl_2$. Attributable to the strong binding between metal ions and organic groups, hydrogels are endowed with unique properties such as temperature sensitivity, pH sensitivity and optical properties [71].

2. Chemical modification

The chemical modification includes two types of reactions: degradation of cellulose and derivatization of hydroxyl groups [72]. Among them, degradation reactions include acid/base degradation [73], oxidative degradation [74], biodegradation [75], and mechanical processing degradation. Derivatization, one of the most common methods for modification of cellulose, was used to synthesize derivatives of cellulose. For example, cellulose esters, such as cellulose nitrate, cellulose xanthate, cellulose acetate [58], cellulose acetate phthalate (CAP), and hydroxypropylmethyl cellulose phthalate (HPMCP), are synthesized by esterification of hydroxyl groups with various organic acids in the presence of a strong acid as a catalyst [76]. Cellulose derivatives obtained by esterification have good properties of film-forming, which are suitable for standard coatings. Moreover, it was essential for drug-delivery applications that cellulose esters are not only non-toxic and stable, but are not absorbed from the gastrointestinal tract [28].

Significantly, hydroxyl derivatization (including nucleophilic substitution, graft copolymerization, and cross-linking reactions) is beneficial to improve the water solubility of cellulose. A classic example is cellulose ether, a class of polymer compound with an ether structure—such as methyl cellulose (MC), hydroxypropyl methyl cellulose (HPMC), hydroxyethyl cellulose (HEC), and carboxymethyl cellulose (CMC), in which the hydrogen of the cellulose hydroxyl group is replaced by the hydrocarbon group [46]. Compared with cellulose, cellulose ether has excellent advantages of water-solubility and thermoplasticity, which is significant for the preparation of hydrogels. Therefore, it is expected that the mechanical strength, hydrophobicity, heat resistance, and antibacterial activity of cellulose can be improved by modification, and the application range of cellulose can be expanded [44].

In addition, graft copolymerization is also a hot topic in the research of cellulose modification. Ring-opening grafting of cyclic monomers (e.g., epoxides, lactones, cycloimines, cyclic-thioethers) is one of the commonly used methods for graft modification of cellulose. Different flexible groups can be introduced by grafting modification to plasticize the internal cellulose. This method retaining the inherent properties of cellulose endows it with thermoplasticity [54].

## 2.2. Property of Cellulose Materials

Natural cellulose is neither soluble in water nor in common organic solvents such as ether and acetone at room temperature. To dissolve cellulose, two methods are mainly used in industry: the derivatization method [77] and the direct method [78]. The derivatization method generally swells cellulose in sodium hydroxide and activates it to produce cellulose derivatives which are dissolved in alkali. A special solvent system is used to dissolve cellulose directly without modifying the property of cellulose by the direct method [79] In addition, the strong intramolecular and intermolecular hydrogen bonds of cellulose endue cellulose with many special properties, such as crystallinity, water absorption, self-assembly, and chemical activity, electrical conductivity, thermal conductivity, and optical properties [80].

2.2.1. Mechanical Properties of Cellulose Materials

Some factors, such as crystal structure (Iα, Iβ, II), crystallinity percentage, anisotropy, and performance measurement methods and techniques may affect the measured mechanical properties. The mechanical properties, mainly the elastic properties of several cellulose particles, have been reported previously [42,81]. Here, we have briefly summarized them. In short, there are four basic parameter influenced properties of cellulose-based materials: elastic modulus in the axial direction ($E_A$), the elastic modulus in the transverse direction ($E_T$), tensile strength ($\sigma_f$, tensile testing), and strain to failure ($\varepsilon_f$, tensile testing). The elastic properties have been measured using in situ tensile tests combined with XRD and inelastic X-ray scattering (IXS) to measure strain [42].

2.2.2. Conductive Properties

Hydrogels are composed of polymer networks and more than 90% water, and therefore can be used as ideal ionic conductors when mixed with electrolyte salts. However, the polymerization process is time-consuming and energy-consuming, and the poor adaptability to the extreme environment has seriously hindered its development in the emerging green power field. Therefore, it is still a challenge to prepare ionic conductive hydrogels with high mechanical properties, high electrical conductivity, and good freezing resistance through a simple method.

For example, Wang directly generated ionic conductive cellulose hydrogels (CCHs) with antifreeze properties through a simple one-step chemical cross-linking [82]. Cellulose was dissolved in an aqueous solution of benzyl trimethyl ammonium hydroxide ($BzMe_3NOH$), and CCHs were directly obtained by chemical cross-linking without further treatment. The conductive hydrogel is endowed with sensitivity, high transparency, and elasticity by the solvent, and can still maintain a stable performance at low temperatures. These advantages make the cellulose-based hydrogels show promising application prospects in sensors, energy storage, and wearable devices at sub-zero temperatures.

2.2.3. Thermal Properties

Typically, thermochemical degradation of cellulose occurs between 200–300 °C, depending on the heating rate, particle type, and surface modification type. Iα is composed of single-chain triclinic monomer units, and Iβ is composed of double-chain monoclinic monomer units. The former is mainly cellulose in green algae and bacteria, and the latter is the main crystalline form of cellulose in wood and cotton. This also leads to a certain anisotropy of the thermal conductivity of cellulose [24]. The thermal conductivity of cellulose determines its important position in hydrogel-based thermal insulation materials. For example, Zhang et al. integrated ionic compounds ($ZnCl_2/CaCl_2$) into cellulose hydrogel networks to enhance frost resistance [83]. Their specially designed $ZnCl_2/CaCl_2$ system has excellent freezing resistance and improved the solidification rate of cellulose by extra water or glycerol.

In addition, MC is a typical temperature-responsive water-soluble polymer [84], which can form gels in the presence of salt at 37 °C. Inspired by this criterion, MC has become an attractive candidate for the preparation of thermosensitive physical hydrogels in the field of drug delivery, tissue engineering, etc. [85,86]. For example, a thermo-responsive MC hydrogel was prepared by crosslinking with citric acid for cell sheet engineering, allowing cell detachment from their surface by lowering the temperature below transition temperature [87].

2.2.4. Optical Properties

Anisotropic structures have unique characteristics in many aspects, such as electricity, mechanics and optics, and most biological soft tissues (muscle, skin and cartilage) have multi-layered ordered structures ranging from nano scale to macro scale, which play a crucial role in organisms [88]. As a kind of soft and wet material, hydrogels are widely used in biomedical materials, flexible electrodes, brakes, sensors, etc. Compared with traditional hydrogels, anisotropic hydrogels with oriented structures have greater potential in all aspects [89]. However, the preparation of highly oriented hydrogels remains challenging because the regulation of macromolecular chains is often limited by electrostatic, hydrogen bonding and other interactions.

Based on the regenerated cellulose hydrogels prepared by $ZnCl_2/CaCl_2$ dissolution system in the early stage [83], He Group conducted a series of studies on the preparation of anisotropic cellulose hydrogels induced by $Ca^{2+}$ ions. For example, a simple method for the preparation of multiphase convertible cellulose hydrogels under $Ca^{2+}/Zn^{2+}$ ion exchange was reported (Figure 3a) [71]. The introduction of $Ca^{2+}$ can not only improve the compressive strength of cellulose hydrogel, but also improve the orientation of the hydrogel through compression setting. The ionic hydrogel has good pressure response performance and can effectively monitor the slight bending of fingers and pressure changes. In addition, a new method for producing cellulose hydrogels with high orientation (refraction index: ~$6.4 \times 10^{-3}$) and high-water content (~72%) was proposed (Figure 3b) [90]. This strategy uses $Ca^{2+}$ coordination cycles to break the hydrogen bonds between cellulose chains, and flexibly switches between ion coordination/hydrogen bond dominance, achieving continuous regulation of high-orientation structures. This principle may provide a new way to construct highly oriented structures and prepare a variety of stimulus-response anisotropic materials. Subsequently, they also prepared gradient anisotropic cellulose hydrogels by the $CaCl_2$ solution diffusion method [91]. The orientation of the cellulose chain in the hydrogel shows a characteristic of decreasing along the direction of ionic diffusion. They proved for the first time the concept of sensitive regions in the ordering-disorder transition region of gradient hydrogels. On this basis, a readable strain response colorimetric card that can be used to detect tiny strains was designed (Figure 3c). This strategy has great potential in the fabrication of optical response devices, complex 3-D structures, and bionic structures.

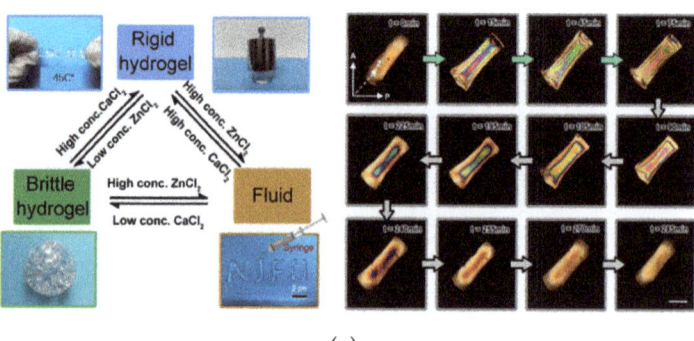

(a)

**Figure 3.** *Cont.*

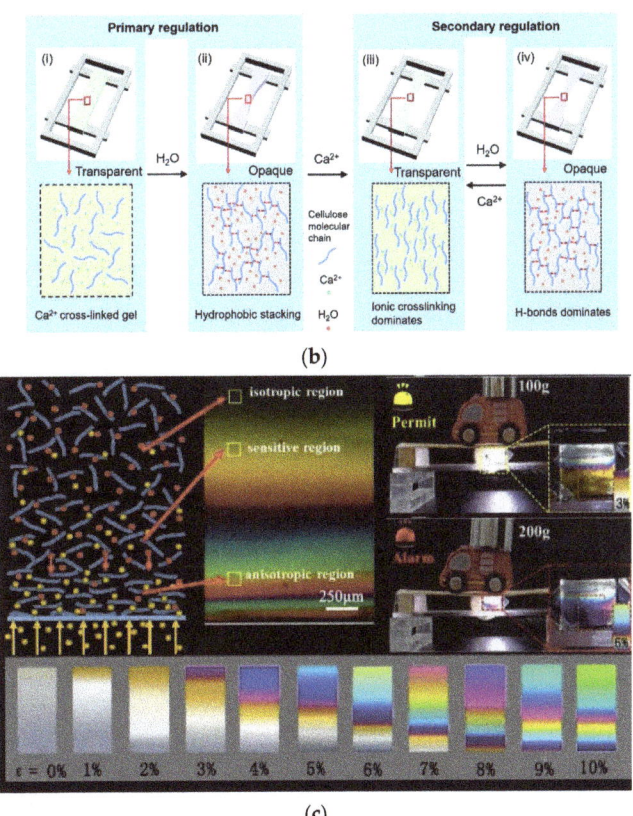

**Figure 3.** A new method for the preparation of anisotropic cellulose hydrogels induced by calcium ions. (**a**) Reversible multiphase transformation of Cellulose hydrogels based on $Ca^{2+}/Zn^{2+}$ exchange and color change of the gel interference under orthogonal polarized light during ion exchange. Adapted with permission from Ref. [71] Copyright 2020, WILEY-VCH Verlag GmbH & Co. KGaA, Weinheim. (**b**) The schematic diagram of the preparation of a highly oriented cellulose hydrogel by $H_2O/Ca^{2+}$ exchange shows that the flexible switch between ion coordination/hydrogen bond dominance is achieved, thus achieving continuous regulation of a high-oriented structure. (i) The initial gel cross-linked by $Ca^{2+}$. (ii) The hydrophobic stacking of cellulose chains triggered by water along the length direction to form an aligned structure. (iii) $Ca^{2+}$ breaks the H-bonds between the cellulose chains and is simultaneously cross-linked with cellulose molecules. (iv) Hydrophobic stacking occurred again when the gel was soaked in water. With the repeated $H_2O$ and $Ca^{2+}$ exchange process, the cellulose molecular chains continued to adjust along the confined direction, resulting in a high orientation structure. Adapted with permission from Ref. [90] Copyright 2020, WILEY-VCH Verlag GmbH & Co. KGaA, Weinheim. (**c**) Mechanism diagram and microstrain test diagram of gradient anisotropic hydrogel prepared by directional diffusion of cellulose sol in $CaCl_2$ solution. Adapted with permission from Ref. [91] Copyright 2021, Elsevier Ltd.

In addition, under the action of external forces, the elastic chemical crosslinking network can ensure the large deformation of hydrogels, while the fracture and rearrangement of the physical crosslinking network can effectively dissipate energy, endowing cellulose hydrogels with good strength, toughness and force-induced optical anisotropy. For example, the Zhang Group prepared a novel cellulose hydrogel with a small amount of epichlorohydrin (EPI) in LiOH/urea solution and subsequent treating with dilute acid by using cellulose cotton short pulp to show sensitive force-induced optical anisotropy properties [92]. It was confirmed that the unique structure endows the hydrogel with excellent mechanical properties and the force-induced optical anisotropy is derived from the force-induced structural orientation. This kind of cellulose hydrogel can be designed as an intelligent soft matter force sensor to sense external forces because of its sensitive force-induced optical anisotropy behavior.

## 3. Preparation Methods of Cellulose Hydrogel

A cellulose-based hydrogel is prepared by physical or chemical cross-linking (Figure 4a,b) [28]. The hydroxyl groups of cellulose and various groups after modification are beneficial to the cross-linking of cellulose to obtain cellulose-based hydrogels (Table 3). The preparation of hydrogels is closely related to the formation of its interconnected network structure. Interactions of physical properties include mechanical chain entanglements, van der Waals interactions, hydrogen bondings, hydrophobic force aggregations or electronic associations. Mastering the preparation principle and method of hydrogel will help us to better design biomedical materials that could meet different needs. In this section, we will briefly introduce the preparation methods of hydrogel, including physical, chemical cross-linking, and interpenetrating networks.

### 3.1. Physical Cross-Linking

Physical hydrogels, also called pseudo hydrogels or thermally reversible hydrogels, are formed through van der Waals forces, hydrogen bonds, ionic bonds, and hydrophobic interactions. The interaction between physical cross-linked cellulose hydrogel molecules is reversible, and the network structure is destroyed with the change in physical conditions [93]. Physical gelation is the self-association of the cellulose chain. Cellulose preferentially binds to cellulose rather than to cellulose solvent, which is often accompanied by a microphase separation. Physical cross-linking methods have the following four mechanisms: (1) interaction between ions: the physical cross-linking network based on PVA and chitosan (CS) was mainly constructed by immersing the $Na_3Cit$ solution as the dynamic bond, and the N-glucosamine unit of the CS chain in the tridentate coordination of the $Cit^{3-}$ anion [94]; (2) crystal cross-linking: Koichi proposed a non-damage enhancement strategy for hydrogels using strain-induced crystallization [34]. For the highly oriented slip-ring hydrogels with polyethylene glycol chains exposed to each other under large deformation, crystallization formed and melted with elongation and contraction, resulting in almost 100% rapid recovery of tensile energy and excellent toughness of 6.6 to 22 megajoules each square meter; (3) hydrogen bond cross-linking: there is a new strategy to dynamically adjust the hydrogen bond cross-linking between PVA and tannic acid (TA) by ethanol, which can be simply coated on the surface of porous substrates by different methods [95]; and (4) hydrophobic association: Sun made the short alkyl side chain modified hydrogel library into a model phase separation hydrogel [96]. With the increase of side chain length, stronger hydrophobic interaction was generated. The longer side chain in the hydrogel promoted a thicker and denser network. Therefore, the layering was faster compared with the short side chain.

### 3.2. Chemical Cross-Linking

Different from physical hydrogels, chemical hydrogels are irreversible with three-dimensional networks connected by covalent bonds. The starting material can be a monomer, polymer or a mixture of monomer and polymer (Figure 4c) [97]. What's more, the preparation process is not spontaneous; on one hand, the reaction is triggered by radiation. On the other hand, the polymer reacts with small molecule cross-linking agents. Chemical cross-linking hydrogels are prepared by functional coupling agents or by cross-linking more than two polymer chains under ultraviolet light. Hydrogels prepared by this method will be more stable and have better swelling performance [98]. Various cross-linking agents and catalysts are used for the chemical cross-linking of cellulose derivatives. The most commonly used cross-linking agents are dialdehydes, acetals, polycarboxylic acids, epichlorohydrin and polyepichlorohydrin. The chemical cross-linking method disrupts the self-association and packing of cellulose leading to swollen transparent coagulated cellulose hydrogels with more uniform morphology and porous structure, lower crystallinity, higher swelling and higher water vapor adsorption affinity [43]. Chemical cross-linking includes radiation polymerization and free radical polymerization (Figure 4d) [99]. For instance, a polyacrylamide/acidified single-walled carbon nanotube composite hydrogel was used to assemble quasi-solid TEC by Chen [100]. The hydrogel was chlorinated with Sn (IV)/Sn (II) chloride ($Sn^{4+}/Sn^{2+}$) as redox pairs, and a simple in-situ free radical polymerization route was used to fabricate a hydrogel with high thermos electro chemical Seebeck Coefficient and excellent thermos electro chemical stability against large mechanical tensile and deformation [101]. In addition, a polyacrylamide-based hydrogel sensor was also designed by Wen, where hydroxyl radicals are caused by the radiolysis of X-ray water molecules [100].

In addition to using the above two methods alone, physical and chemical cross-linking can also be used together. For example, Zhao proposed a strategy using sequential chemical cross-linking and physical cross-linking to form mechanically strong and tough double-crosslinked (DC) cellulose hydrogels from cellulose/alkali hydroxide/urea aqueous solutions [102]. The resulting DC cellulose hydrogels were transparent, foldable and were elastic and featured quick recovery properties. Moreover, the nanostructured network of DC cellulose hydrogels was achieved by covalent cross-linking, hydrogen bonding and chain entanglement between cellulose chains and cellulose II microcrystalline hydrates in DC cellulose hydrogels. The hydrogel effectively improved the mechanical properties of DC cellulose hydrogel, which exceeded the single chemical cross-linking and physical cross-linking cellulose hydrogel. The hydrogel developed in this work represents the first example of a cellulose hydrogel with high strength and toughness, whose excellent mechanical properties stem from the cross-linked structure resulting from a sequential chemical and physical cross-linking strategy. DC cellulose hydrogels are a new class of polysaccharide hydrogels with potential applications in artificial blood vessels and skin, tissue engineering materials, and catalyst carriers.

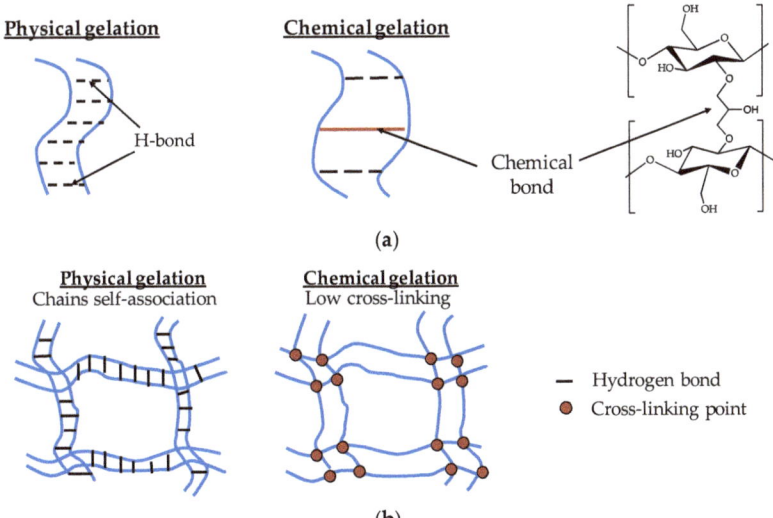

Figure 4. Cont.

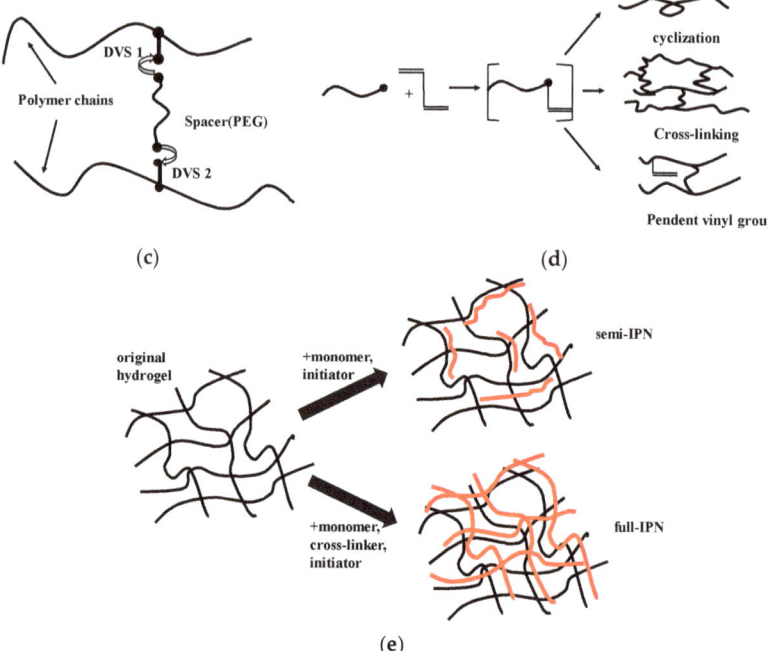

**Figure 4.** Formation mechanism of hydrogel. (**a**) A sketch of network formation in cellulose solutions: physical gelation via self-association of chains and chemical cross-linking; (**b**) a schematic presentation of the structures of physical and chemical cellulose gels; adapted from Ref. [28]. (**c**) Scheme of the crosslinking between celluloses in the presence of spacers. DVS: divinyl sulfone, cross-linked molecules. (**d**) Synthesis route of hydrogel by radical polymerization. Adapted with permission from Ref. [99] copyright 2007, Society of Chemical Industry. (**e**) Formation and structure of semi- and full interpenetrating hydrogels. Adapted with permission from Ref. [103] Copyright 2008, Elsevier Ltd.

### 3.3. Interpenetrating Network

An interpenetrating network polymer (IPN) is an aggregation structure formed by the penetration of two polymers in the form of a network (Figure 4e) [103,104]. One of the polymers is chemically cross-linked and the other penetrates its network of it, so that there is no covalent bond between the two polymers [105]. Cellulose-based hydrogel prepared by the IPN method was a two-component network structure, which has the characteristics of a large swelling ratio and good mechanical properties. It has good application prospects in biomedical tissue engineering, adsorption, and separation [106].

Table 3. Classification of hydrogel crosslinking methods.

| Crosslinking Types | Crosslinking Mechanism | Crosslinking Methods | Reference |
|---|---|---|---|
| Physical crosslinking | Ions interaction | Based on polyvinyl alcohol, a physical cross-linked network was constructed by coordinating the N-glucosamine unit of chitosan chain with the tridentate ligand of $Cit^{3-}$ anion. | [94] |
| | Crystalline crosslinking | Crystallization formed in polyethylene glycol chain slip ring gel and melted with elongation and contraction. | [34] |
| | Hydrogen bonding crosslinking | Between PVA and Tannic Acid (TA). | [95] |
| | Hydrophobic association interaction | Increased side chain length leads to stronger hydrophobic interactions and promotes thicker and denser networks. | [96] |
| Chemical crosslinking | Free radical polymerization | Composite hydrogel formed by in situ radical polymerization of poly-acrylamide/acidified single-walled carbon nanotubes. | [101] |
| | Radiation polymerization | Polymerization of monomers is initiated by high-energy radiation. | [100] |
| | Interpenetrating polymer network | PVA network with high crosslink density (a skeleton restricting the saturated water content of gel) is penetrated by polystyrene sulfonate (PSS) network. | [107] |

The semi-interpenetrating network (Semi-IPN) hydrogel is synthesized by immersing the monomer and initiator into the hydrogel solution, and the full-IPN hydrogel is formed by adding cross-linking agent [103]. Compared with hydrogels prepared by other methods, hydrogels prepared by the interpenetrating network method have good mechanical strength, better flexibility, controllable physical properties, and more effective drug loading. Since the micropores are adjustable, the release behavior of drug conforms to controlled release kinetics [108]. For example, a simple strategy to control the hydration of polymer networks in hydrogels was reported by Zhao, in which highly skeletonized polymer networks can be used to functionalize densely cross-linked polymers as skeletons [109]. The hydration of polymer chains will produce a large number of weakly bound water molecules, thereby promoting water evaporation. A PVA network with a high cross-link density (a skeleton restricting the saturated water content of hydrogel) is penetrated by a polystyrene sulfonate (PSS) network which can be actively hydrated by water molecules through electrostatic and hydrogen bonds. Hydrogels with this interpenetrating polymer network can activate more than 50% of the water into an intermediate state.

## 4. Cellulose-Based Hydrogels

Cellulose-based hydrogels as biomaterials should be biocompatible. In the meantime, it should perform special biochemical, mechanical, and physical properties in order to simulate fundamental aspects in vivo. Biocompatibility signals the likelihood of a material to coexist and interact without harm in the presence of a specific tissue or biological function. Assessing the biocompatibility of a biomaterial requires examining the harm or side effects it may cause to the host. The feature of the cellulose-based hydrogels, such as biocompatibility, biodegradability, adjustable mechanical properties, sensitivity to various stimuli, the ability to encapsulate different therapeutic agents and to control drug release make it an important candidate for biomedical applications.

### 4.1. CNF-Based Hydrogels

CNF-based hydrogels, which have three-dimensional NF networks and unique physical properties, have great applications in elastic hydrogels, ionic conduction, and water purification for emerging materials [41]. CNF is composed of a cellulose chain, which is long and flexible and arranged in the structure controlled by hydrogen bonds. CNF is characterized by a recognized cytocompatibility and a high tolerogenic potential. For example, Hong demonstrated the gelation of carboxylated cellulose nanofibers and the formation of interconnected porous networks by adding divalent or trivalent cations ($Ca^{2+}$, $Zn^{2+}$, $Cu^{2+}$, $Al^{3+}$, and $Fe^{3+}$) to aqueous nanofiber dispersions. Dynamic viscoelastic measurement showed that the hydrogel modulus can be adjusted by appropriate selection of cations. Gelation was induced by screening for repulsive charges on the nanofibers, and gel properties were controlled by ionic crosslinking. We can envision multiple potential applications of CNF-Mn+ hydrogels in biomedicine and other fields, such as drug carriers and the encapsulation of functional molecules.

### 4.2. Cellulose-Based Hybrid Smart Hydrogels

As we know, the poor mechanical properties of traditional biomolecular hydrogels severely limit their application prospects. Traditional hydrogels have been increasingly unable to meet the growing needs of biomedicine. Smart hydrogels (stimuli-responsive hydrogels) have attracted extensive attention in academic and industrial fields due to their high elasticity and adaptability, which can respond to environmental stimuli rapidly and significantly [110]. Therefore, how to prepare smart biocompatible hydrogels with high strength, high toughness and high-water content from biological macromolecules is a hot research topic in hydrogel fields [111,112].

According to the response to external stimuli, intelligent polymer hydrogels can be allocated to pH sensitive hydrogels, thermal sensitive hydrogels, salt sensitive hydrogels, tiny force sensitive hydrogels, and so on [113].

#### 4.2.1. pH Sensitive Hydrogels

pH sensitive hydrogels contain a large number of acid and base groups that are easily hydrolyzed or protonated, such as carboxyl and amino groups. When the external pH value changes, the degree of protonation or deprotonation of these groups changes accordingly, resulting in changes in the electrostatic attraction or repulsion between functional groups, and the swelling degree of hydrogels [114]. At the same time, with the change of pH value, the difference in ion concentration inside and outside the hydrogel also changed, resulting in the change of osmotic pressure inside and outside the hydrogel, and the change in swelling degree of the hydrogel [115]. In addition, the dissociation of these groups will also destroy the corresponding hydrogen bonds in the hydrogel, reduce the cross-linking points of the hydrogel network, cause a change in the hydrogel network structure, and cause hydrogel swelling [116]. For example, polyacrylic acid, polyacrylamide, and chitosan are typical pH sensitive polymer hydrogels. Yin et al. prepared an intelligent pH sensitive hydrogel based on the oxidation of hydroxyethyl cellulose from pineapple peel and the residual carboxymethyl chitosan from Hericium Erinaceus. It was done mainly through the oxidation and alkalization of cellulose hydrogels. Cellulose was modified to oxidized dialdehyde cellulose, and chitin was modified to CMC, which was combined by a Schiff base reaction to form a new hydrogel with excellent characteristics [117].

#### 4.2.2. Thermal Sensitive Hydrogels

In 1978, Tanaka et al. first found the thermal sensitivity of hydrogels when they studied polyacrylamide [118]. The thermal sensitivity of hydrogels refers to the volume mutation that occurs with the change in ambient temperature. The hydrogel has a certain proportion of hydrophobic and hydrophilic groups. The temperature change can affect the hydrophobic effect of these groups and the hydrogen bond between macromolecular chains so that the gel structure changes and the volume changes.

Inspired by this criterion, cellulose derivatives, such as MC, HPC, HPMC, and EHEC, have become an attractive candidate for the preparation of thermosensitive physical hydrogels. For example, a thermoreversible hydrogel was prepared by combining high molecular weight hyaluronic acid with rapidly oxidized nanocellulose, MC and polyethylene glycol to prevent adhesion. Among them, MC ensured the thermal sensitivity of this hydrogel [119].

As thermal responsive hydrogels, HPC-based hydrogels have also been studied in depth. HPC hydrogels respond to temperature changes by volume phase transition. Below the phase transition temperature ($T_t$ = 41 °C), the hydrogel is hydrophilic, swelling in water. The hydrogel becomes hydrophobic, and disintegrates into a small volume when the temperature is above the $T_t$ [120]. Also, HPMC is a polysaccharide derivative with water solubility, pH stability, biodegradability and biocompatibility [121]. Due to the dehydration of the hydrophobic substitution zone of the polymer chain, HPMC can undergo thermal reversible sol-gel phase transition during heating. Although the gel temperature of HPMC was 60 °C, much higher than 37 °C, Wang destroyed the polymer water sheath by adding a high concentration of glycerol, promoting the formation of a hydrophobic region and reducing $T_t$. Furthermore, EHEC has a low critical dissolution temperature, and phase separation occurs above this temperature. With the increase in temperature, EHEC becomes more hydrophobic and induces the formation of large aggregates separated from the water phase. Specifically, ionic surfactant was added to EHEC, which significantly changed this situation and formed low-toxic thermoresponsive hydrogels [122].

In addition, the thermal response of CNF-based hydrogels can also be triggered intelligently. For example, Wei et al. prepared an ionic skin–water solvent network based on high-performance organic hydrogels with superior mechanical response and thermal sensing ability through one-step UV-initiated polymerization, which can be assembled into capacitive sensors for motion monitoring in real life and thermal resistance for dynamic temperature detection. In short, olyacrylamide (PAAm), CNFs, tannic acid (TA), electrolytes (NaCl), and glycerol/water binary solvent are incorporated by UV-initiated free radical polymerization [123]. In addition, based on the temperature-dependent self-association of betaine methacrylate (SBMA), polymer chains and the incorporation of temperature sensitive cellulose/polyaniline nanofibers (CPA NFs) in a glycerol-water binary solvent system, Hao et al. successfully prepared an intelligent temperature-sensing amphoteric hydrogel with superior low-temperature tolerance and conformal adhesion [124].

### 4.2.3. Salt Sensitive Hydrogels

The swelling rate/water absorption rate of salt sensitive hydrogels will vary with the change of salt concentration outside because of its structure. The positive and negative charged groups of these hydrogels are bonded together by covalent bonds, and the addition of small molecular salts can shield and destroy the association of them in the macromolecular chain, resulting in the stretching of the molecular chain. The outstanding advantage of such hydrogels is that the swelling behavior of hydrogels in salt solution presents the anti-polyelectrolyte behavior. In another word, under certain conditions, the swelling ratio of hydrogels in salt does not decrease but increases with the increase of the applied salt concentration. For example, Hai et al. developed a novel fluorescent switch-reporting $ClO^-/SCN^-$ reversible responsive cellulose hydrogel. When $ClO^-$ was added, the hydrogel network of NC hydrogel was destroyed, and the fluorescence was quenched. Hydrogel changes in a completely reversible process by regulating $ClO^-/SCN^-$ [102].

### 4.2.4. Tiny Force Sensitive Hydrogels

Strain sensors can sense external stimulus signals and convert them into recordable electrical signals, which are widely used in human daily motion monitoring, skin perception, and other fields. Ouyang et al. prepared gradient anisotropic carboxymethyl cellulose hydrogel (CMC-$Al^{3+}$) CMC-$Al^{3+}$ by directional diffusion of aluminum chloride solution. The CMC-$Al^{3+}$ was packaged with PVC flame retardant tape, and a strain sensor for detecting micromotion of the human body was made. It can accurately and stably monitor micromotion [125]. Ye et al. found that cellulose was reacted with a small amount of epichlorohydrin (EPI) in LiOH/urea solution, and then treated with dilute acid to prepare tough cellulose hydrogel with deformation-induced anisotropy. This cellulose hydrogel has sensitive mechanical response characteristics and can be used as dynamic light switches and soft sensors to accurately detect small external forces [92].

## 5. Application

During the past decades, cellulose-based hybrid smart hydrogels/aerogels have been developed in an increasing number of applications which significantly contribute to our public healthcare systems [126]. Cellulose is known to be a naturally abundant, mechanically excellent, sustainable technological, and inexpensive material [127]. Among them, naturally derived nanocelluloses possesses unique physicochemical properties and great potential as renewable smart nanomaterials, opening up a large number of new functional materials for multi-sensing applications. Cellulose not only plays a huge role by itself but synergizes with other materials to achieve the manufacture of more advanced and multifunctional hybrid materials. In recent years, many advances have been made in the design of functional materials with excellent mechanical properties using cellulose [45,128]. In this section, we will focus on the advances of cellulose-based hydrogels in biomedical applications.

### 5.1. Ionic Conduction/Battery

With the rapid development of the Internet of Things and the increasing demand for human-machine interfaces, flexible ionic conductors have attracted extensive attention due to their characteristics of high elasticity, transparency, adjustable mechanical properties, and consistent electrical conductivity. Cellulose-based conductive hydrogels for tissue engineering are constantly progressing [10]. Recently, conductive hydrogels showed great prospects for extensive and important applications in the field of sustainable energy such as sensors, batteries, and flexible electronic devices [129] because of their unique characteristics of sufficient flexibility, durability, and versatility. Traditional hydrogels lose their original properties due to freezing at low temperatures, which limits their application. For example, a freeze-resistant ionic conductive cellulose hydrogel, which can be used as a tensile, compression and temperature sensor was prepared [82]. Unlike traditional hydrogels, its frost resistance allows it to work well in sub-zero temperatures. The sensor based on antifreeze conductive hydrogel has stable mechanical property and thermal sensitivity, and fast, reliable, stable and reversible response performance. It is suitable for low temperatures, and is used for soft artificial intelligence devices in complex temperature environments. In recent years, the use of renewable resources and green preparation technologies to prepare smart hybrid materials has attracted the extensive attention of researchers. For example, the most advanced nanocellulose-graphene composites [130] can be used for multifunctional sensing platforms such as mechanical, environmental and human biological signal detection, simulation and field monitoring. Moreover, a number of researchers have also been involved in investigating the exploitation of CFs as load-bearing components for composites. The use of these materials in composites has increased due to their relative cheapness compared to conventional materials such as glass and aramid fibres, their recyclability, and the fact that they compete well in terms of strength per unit weight of material [47].

Recently, a new transparent ionic conductive hydrogel with excellent mechanical properties, high conductivity, high transparency, and freezing resistance based on PVA and CNFs was prepared by the sol-gel transformation method (Figure 5a) [131]. The multi-layer porous structure of CNFs/PVA plays an important role in enhancing the ionic conductivity of organic hydrogels (Figure 5b). Pressure sensors based on conductive hydrogels with high-pressure sensitivity (Figure 5c) are used to detect complex human movements in real time (Figure 5d). In short, this material design demonstrates the synergistic effect of CNFs in improving mechanical properties and ionic conductivity, solving the long-standing dilemma between the strength, toughness, and ionic conductivity of ionic conducting hydrogels, which are widely used in wearable electronic devices.

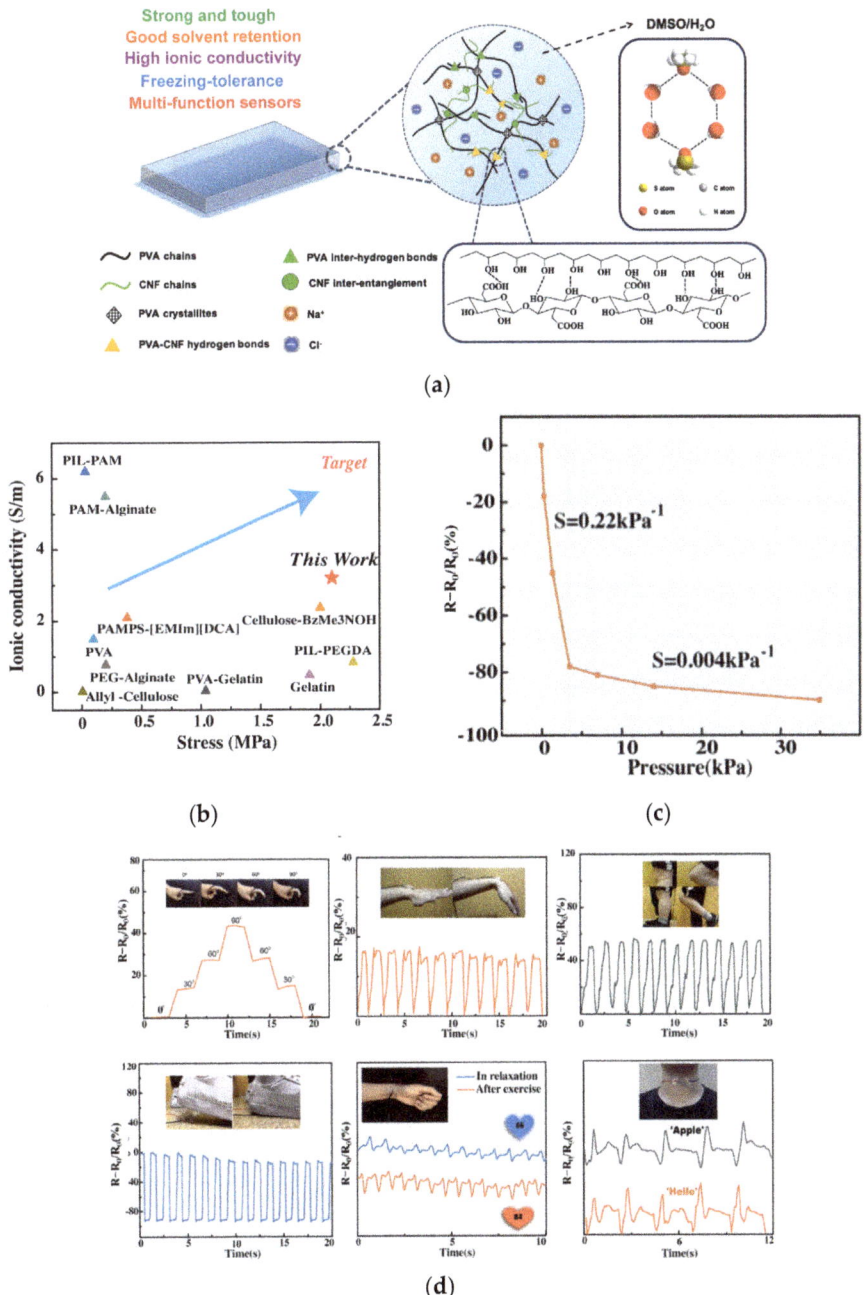

**Figure 5.** Cellulose-based ionic conductive hydrogel for multi-functional sensors. (**a**) Schematic illustration of PVA-CNF organohydrogel. (**b**) Ashby plot of ionic conductivity and tensile stress with other reported ionic conductive (organo) hydrogels. (**c**) Relative resistance changes and pressure sensitivity of PVA-1%CNF organohydrogels-based sensor at varying pressure. (**d**) The relative resistance changes of sensors versus time for real-time monitoring of various human motions. Adapted with permission from Ref. [131], WILEY-VCH Verlag GmbH & Co. KGaA, Weinheim.

Zinc-air batteries are regarded as the ideal power source for the next generation. Limited by the zinc-air battery electrolyte, it has rarely been reported in the improvement of flexibility and wearability. Although there have been many reports of stretchable supercapacitors and stretchable batteries, preparing ultra-long stretchable zinc-air batteries remains a major challenge. Recently, a sodium polyacrylate/cellulose double network hydrogel, which has ~1200% tensile properties in strong alkali as a result of the introduction of cellulose into the sodium polyacrylate, was prepared [132]. Based on sodium polyacrylate/cellulose network hydrogel, the planar electrode was prepared by using a wavy zinc electrode, of which the air electrode has a maximum tensile strength of 800% and the fiber electrode can be stretched up to 500%. The hydrogels with double network structure prepared by them have good alkali resistance, which can also be used in other alkaline electrolyte storage devices.

The transient device is a new electronic device whose main characteristic is that after completing the task, it can completely or partially dissolve or decompose the constituent material through a chemical or physical process, which is considered a new research direction for an implantable device. However, the research on transient devices is still in its infancy, and many challenges need to be overcome, especially since the development of transient energy devices is relatively slow. Recently, a transient zinc ion battery (TZIB) with excellent biocompatibility and complete degradation based on a carefully designed cellulose aerogel-gelatin (CAG) solid electrolyte was reported [23]. The new fully degradable CAG solid electrolyte enables TZIB to achieve controlled degradation and stable electrochemical performance while maintaining excellent mechanical properties. More importantly, TZIB has excellent electrochemical performance while meeting the requirements of controlled degradation. These results demonstrate the potential of TZIB for future clinical applications and provide a new platform for transient electronics. This is the first time that the perfect combination of high flexibility, high mechanical performance, and high biocompatibility has been achieved while maintaining high battery performance. Thus, this work offers new opportunities for future self-powered transient electronic devices or traditional self-powered implantable medical devices, such as implantable cardioverter defibrillators, implantable diagnostic sensors and the rapidly evolving implantable monitoring of diabetes. In addition, a soft stretchable conductive hydrogel composite consisting of alginate, carboxymethyl cellulose, polyacrylamide, and silver flakes was reported [133].

### 5.2. Thermal Insulation

In the next five years, cellulosic polyporous materials with thermal insulation, light weight, and excellent mechanical properties are expected to replace commercial thermal insulation materials (expanded polystyrene, expanded polyurethane and glass wool). Since the thermal conductivity of porous materials is mainly contributed to by the gas and solid phases, the thermal conductivity can be reduced by increasing the Knudsen effect and improving phonon scattering [134]. Cellulosic porous materials with light weight, high mechanical strength, flame retardance, and heat insulation are expected to replace commercial fossil energy materials for indoor insulation in the future. In the long run, due to the large amount of insulation materials, it is necessary to design cellulose-based thermally insulating materials that can be prepared with low energy consumption on a large scale. The research progress of porous materials based on nanocellulose in heat insulation in recent years has been reported previously in some reviews [24].

For example, a mixed aerogel with a high mechanical compression ratio ($\approx$99%) and superhydrophobicity ($\approx$168°) by using BC and methyltrimethoxysilane (MTMS) was reported (Figure 6a) [132]. Improving moisture-proof performance to prevent water penetration is the key premise to ensuring the thermal insulation quality of aerogel. The layered porous structure ensures the ultra-light and thermal insulation properties of aerogel. The aerogel has a strong superhydrophobicity to withstand high humidity due to its fibrous nanostructures, hydrophobic surface parts and stable nanofiber framework. When the relative humidity changed from 30% to 90% (Figure 6b), the thermal conductivity of aerogel remained almost constant. Under experimental conditions (−20 to 150 °C), the thermal insulation performance of hybrid aerogel is comparable to that of down (Figure 6c). Therefore, the aerogel may be a good choice for thermal protection under extreme temperature and humidity. It can be formed in a variety of shapes by freeze-forming and can be scaled up to any desired size for future industrial applications.

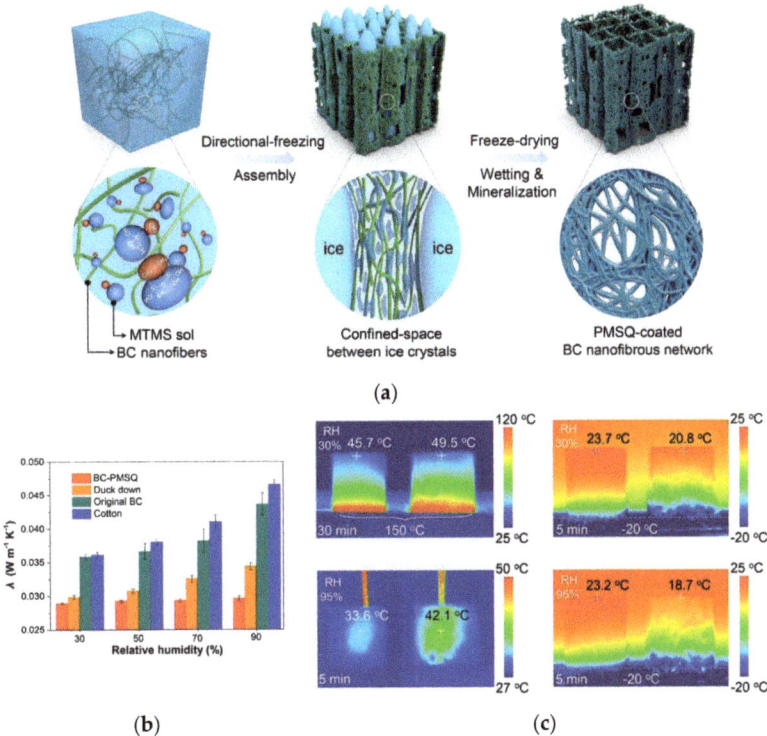

Figure 6. Nanoellulosic hybrid aerogels for thermal insulation. (a) Processing principles and synthesis of the BC–PMSQ hybrid aerogels. (b) Thermal conductivities λ of BC–PMSQ, down feathers, and pure BC membranes. (c) Durable thermal insulation performance was evaluated by optical and infrared images. Adapted with permission from Ref. [132], Copyright 2021, Wiley-VCH GmbH.

## 5.3. Optically Responsive Soft Etalon

Stimuli-responsive optical hydrogel provides a broad platform for the development of smart materials and has been integrated into a variety of optical filters, sensors, indicators and diodes. For example, hydrogels exhibit strong color changes when the periodicity of the structure is stimulated externally. However, it is very difficult to prepare photonic crystals of such high quality. A key advantage is that fine chemical modifications of hydrogel precursors can control the mechanical strength and elasticity of the resultant film. In recent years, since the cellulose itself does not absorb visible light and does not scatter light significantly, cellulose-based optical hydrogels are attracting more attention.

Recently, an optically responsive soft etalon based on a double network cellulose hydrogel was reported [135]. The refractive index and thickness of the hydrogel were changed by the humidity because of optical interference within the metal–insulator–metal (MIM) cavity. Further functionalization of these cellulose hydrogels will facilitate the development of sensors that respond to a variety of external stimuli. In addition, the Zhang group reported cellulose hydrogels with deformation-induced anisotropy that exhibited high toughness by reacting with epichlorohydrin (EPI) in LiOH/urea solution [92]. Force-induced optical anisotropy is derived from force-induced structural orientation. This kind of cellulose hydrogel can be designed as an intelligent soft matter force sensor to sense external forces because of its sensitive force-induced optical anisotropy behavior.

### 5.4. Wound Healing Therapy

Ideal materials for wound healing should have non-cytotoxicity, good biocompatibility, a high moisturizing effect, excellent breathability, and be easy to apply to different types of wounds [136]. To promote wound healing, wound dressings are essential to repair the skin and restore skin function. Cellulose-based hydrogels are an effective treatment material to promote wound repair, and they have attracted significant attention in the field of tissue regeneration because of their unique properties to meet the requirements of wound healing materials [137]. For example, the injectable hydrogel is initially a liquid at room temperature with pre-gelling fluidity which can be applied to any defect or cavity with minimal invasion. After experiencing transformation or responding to pH/temperature changes in a short period of time, the hydrogel will be formed in situ, and can quickly cross-link with the tissues around the wound. He et al. developed a series of injectable pH-responsive self-healing hydrogels based on acryloyl-6-aminohexanoic acid (AA) and AA-N-hydroxysuccinimide (AA-NHS), and further proved its great potential as an endoscopic sprayable bioadhesive material to effectively prevent bleeding and promote wound healing in a pig stomach bleeding/wound model [138].

To further accelerate the wound healing of patients with diabetes, the construction of conductive wound dressings in response to physiological electrical signals and external electric field stimulation at the wound site will help to conduct and distribute electrical signals to the damaged tissue more effectively. Given this, the Lu Group reported a polydopamine-reduced-graphene oxide (PGO)-hybridized cellulose (PGC) bio-nanosheet, and a PGC bio-nanosheet-assembled hydrogel (PGCNSH) with good flexibility, biological stability, electrical conductivity and cell/tissue affinity (Figure 7a) [5]. The effects of hydrogels combined with electric stimulation on cell behavior showed that myoblasts (C2C12 cells) had higher proliferation activity (Figure 7b) and more adhesion spots (Figure 7c) on the PGCNSH hydrogel surface than pure cellulose hydrogel in the absence of electrical stimulation. In addition, the conductive PGCNSH hydrogel can also deliver electrical stimulation in vivo, which can promote the repair of diabetic wounds under the coupling effect of external electrical stimulation (Figure 7d). The results of this study provided an effective synergistic treatment strategy for speeding up diabetic wound repair by coupling electrically conductive cellular-based hydrogel dressings with electrical stimulation. It indicated that conductive PGCNSH hydrogel as an "electronic skin" has the potential to promote chronic wound repair under electrotherapy coupling.

Cook et al. studied a hydrogel dressing which was formed in situ through the reaction between amine-terminated branched poly (ethyleneimine) (PEI) and a bifunctional NHS-activated poly (ethylene glycol) (PEG) cross-linking agent, thiol-ester exchange in the presence of methyl cysteine ( CME ) to leave crosslinking agent and dissolve dressings [139]. This study implements alternative dressings and simple methods for secondary burns.

Human skin is soft and sensitive to environmental changes. Many bionic skin materials have been developed via artificial intelligence, such as wearable sensors and soft robots. These smart devices can convert external stimuli into visual data. However, the complete imitation of human skin's sensory and sensory characteristics will bring great challenges [140]. The Zhao Group proposed a multifunctional electronic skin based on a HPC, PACA and CNT composite conductive cellulose hydrogel which can not only intuitively feedback external stimuli through color changes, but also through quantitative changes in electrical resistance. The dual-signal sensing capability of conductive cellulose nanocomposite hydrogels is expected to open a new chapter in the design and preparation of multifunctional flexible electronic skin [12].

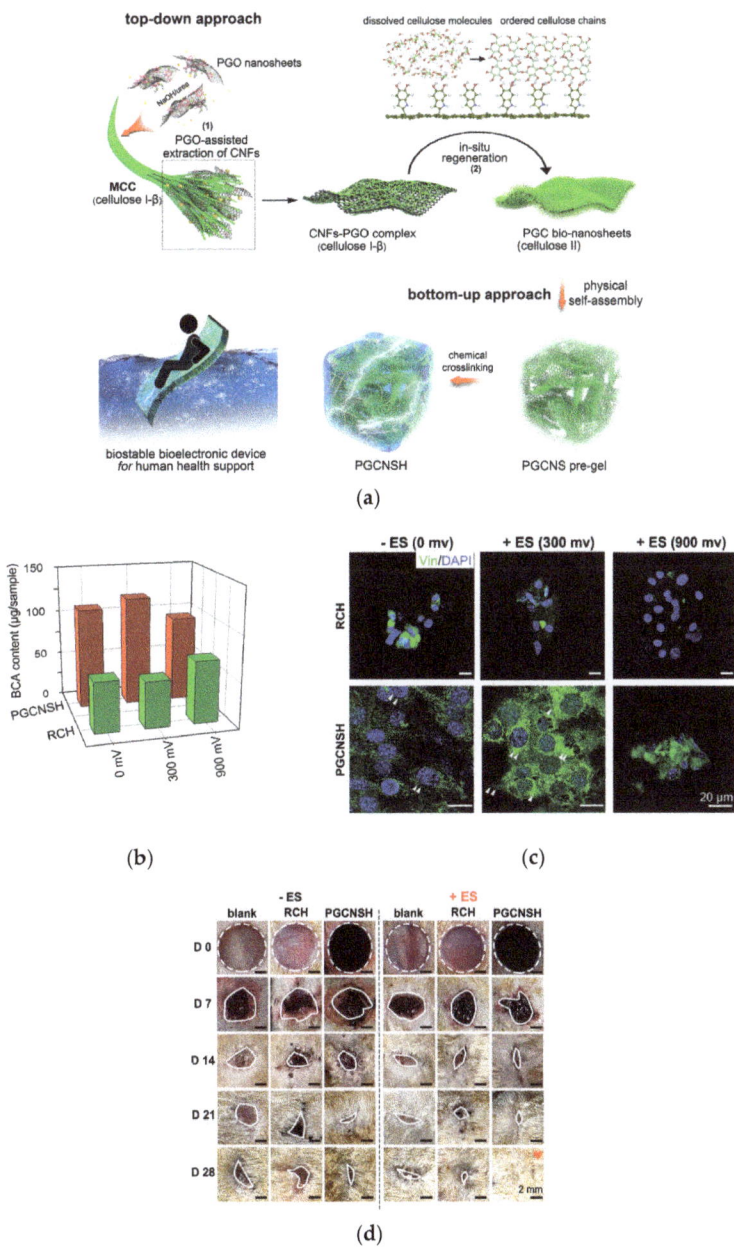

**Figure 7.** Cellulose-based hydrogels for wound healing. (**a**) Design strategy for 2D conductive cellulose nanosheets and their assembly into biostable and conductive 3D bulk hydrogels. (**b**) BCA content of C2C12 cells on the hydrogels after three days of culturing. (**c**) Immunofluorescent staining for focal adhesion formation of C2C12 cells on day three. Vinculin, a focal adhesion protein, was stained with green, and cell nuclei were counterstained with blue. (**d**) Images of hydrogel-treated wounds with and without electrotherapy. Adapted with permission from Ref. [5], Copyright 2021, Wiley-VCH GmbH.

Moreover, Zhang et al. used a sodium alginate/carboxymethyl cellulose blend hydrogel as biological ink for artificial skin, and confirmed that a SC 4:1 blend hydrogel was most suitable for the 3D printing of artificial skin. This study is of great significance for implantable tissue-engineered skin scaffolds and provides the possibility and basis for the repair of large-area skin defects [137].

*5.5. Bacterial Infection Therapy*

A more stringent requirement for biomedical skin wound dressings is to prevent bacterial infection. Infection is a serious complication of chronic wounds. At present, the treatment of chronic wounds depends on dressings. Such dressings often contain silver as a broad-spectrum antibacterial agent, but inappropriate doses may lead to serious side effects. The alkaline environment is related to chronic wound infection. Due to the release of ammonia and polyamines, the bacterial metabolism may lead to the increase in the local pH value, which may damage the healing of wound tissue and lead to necrosis. Cellulose-based hybrid hydrogels seem to be an effective approach for bacterial infection.

For example, a novel MC—based hydrogel was proposed to release silver nanoparticles (AgNPs) locally through an intelligent mechanism activated by pH changes in infected wounds. By optimizing the physicochemical and rheological properties of MC hydrogel, the optimum process conditions of crosslinker (citric acid) concentration, crosslinker time and temperature can be determined. An $^1$H NMR analysis revealed the role of alkali hydrolysis of ester bonds (i.e., cross-linking bonds) in controlling pH response behavior; the swelling and degradation behavior of MC hydrogels depended on pH and temperature, and it was worth noting that pH triggered the release of AgNPs, which was 10 times higher at pH 12 than at pH 4. MC/AgNPs nanocomposite hydrogels were prepared by in-situ synthesis using MC as a capping and reducing agent. TEM and UV–vis measurements assessed the shape, size, and distribution of AgNPs. Finally, inductively coupled plasma and UV Vis measurements supported the quantitative evaluation of the pH triggered release mechanism of AgNPs to develop systems with enhanced antibacterial activity under alkaline conditions.

In addition, For example, Cai Group proposed a multi-modulus component strategy to prepare a high-strength and high-water content double cross-linked cellulose-Go (DCCG) composite hydrogel [141]. The chemically cross-linked DCCG nanosheet region forms a non-covalent interaction and becomes more elastic and flexible. What's more, the photothermal conversion performance of Go nanosheets leads to the excellent photothermal antibacterial performance of composite hydrogels. This study is expected to provide new ideas for the construction of high performance and multifunctional composites from natural polymers. Also, Johnson et al. developed a drug-loaded hydrogel composed of CNF and κ-carrageenan oligosaccharide nanoparticles [142]. They chose two antibiotics as therapeutic agents and prepared hydrogels according to the increase in surfactant concentration. The material has been proved to have antibacterial and anti-inflammatory properties and can be used to treat periodontitis.

*5.6. Drug Delivery*

It is essential to control the release of drugs, as the pharmacological purpose cannot be achieved with a rapid release. An "ideal" drug carrier system should deliver precise amounts of drugs at some pre-planned rates to provide the desired level of drugs for treatment.

Hydrogels based on cellulose derivatives have important applications as drug delivery systems (DDS) and function as external stimuli, such as body temperature and variable pH ranges in different parts of the body, to improve the controlled release of drugs. The double-layer hydrogels have great potential to develop into a novel functional sustained drug delivery system [143]. Cellulose-based hydrogels prepared by physical or chemical methods have different structures and swelling degrees. Compared with the hydrogels formed by physical self-association, chemical crosslinking hydrogels can be loaded with greater amounts of drugs that they release faster.

In recent years, in the treatment of oral diseases, the treatment method of directly releasing drugs into the mouth may be used. There is a large amount of saliva flowing through the oral mucosa in the mouth. Due to the special environment of the mouth, hydrogels with long-term adhesion ability are needed to achieve local medication. HPC hydrogel can be used to prepare a bioadhesive hydrogel system by combining HPC with a polyacrylic acid (PAA) lactose non-adhesive backing layer, which can be used to treat aphthous ulcers by releasing triamcinolone acetonide [144]. Traditional insulin injection is carried out by subcutaneous injection, and long-term repeated injections will lead to reduced patient acceptance. In order to avoid these disadvantages, pH-responsive hydrogels prepared by acrylate grafting of CMC and PAA were developed for oral administration of insulin to increase patient comfort [145]. Another example is that the dual functionalized L-Histidine conjugated chitosan-cellulose nanohybrid hydrogel embedded green zinc oxide nanoparticles were formulated as a sustained drug delivery carrier for the polyphenol drugs–Naringenin, Quercetin, and Curcumin [146].

*5.7. Anti-Tumor Immunotherapy*

Thermosensitive hydrogels help to improve the local and remote effects of cancer immunotherapy. Wang mixed surface-modified nanocellulose with hexadecyl amine as a long chain to construct a cellulose hydrogel network. The hydrogel has been successfully applied to the control and targeted delivery of paclitaxel and has achieved a remarkable anti-tumor effect [147]. The hydrogel system is widely used in humans to block the continuous release of cytotoxic T lymphocyte-associated protein-4 from immune checkpoints by nitric oxide donors and antibodies to achieve efficient and durable anti-tumor immunotherapy. Due to its unique hydrogel formation and degradation characteristics, it can maintain the retention of drugs in tumor tissues, which are triggered and released by the tumor microenvironment, and form in situ micelles suitable for lymphatic absorption.

*5.8. Tissue Engineering*

Cellulose-based hydrogels are immensely important for tissue engineering. When hydrogel is used as a scaffold, it can be used in many aspects of tissue engineering, arthroscopy, vascular stents and skin stents. In tissue engineering, tissue function is affected by cell adhesion, proliferation, differentiation and maturation. Therefore, biocompatibility, bioactivity and biomechanics of materials are critical requirements. Biomaterials need the above three points to support tissue regeneration without eliciting any adverse local or systemic reactions in the eventual host.

For example, Guo et al. applied a small amount of epichlorohydrin to slightly pre-crosslink the cellulose chain to form a permeability network to regulate the rheological properties and to form a loose crosslinking point to regulate the self-assembly of the cellulose chain to obtain excellent mechanical properties. Printed cellulose hydrogel has biomimetic NF topology and remarkable tensile and compressive strength (5.22 and 11.80 MPa), and toughness (1.81 and 2.16 MJ/m$^3$). The original cellulose hydrogel (ALCOGEL) was prepared by Guo et al. using ethanol as an antisolvent. The mechanical properties of biopolymer materials were improved and adjusted by controlling the fiber arrangement, and both of them are expected to be used in tissue engineering [148].

At present, the main method for the treatment of coronary artery disease is to implant vascular stents and shape memory alloys into the dilated artery. Shi et al. constructed a bi-directional shape memory cellulose scaffold which can be adjusted a by mild solution (such as water and alcohol), has excellent biocompatibility, and can support the left coronary artery or left main coronary artery in an open state [149].

## 6. Conclusions and Future Outlook

In the past few decades, the rapid development of biomedical engineering has brought great opportunities and challenges to cellulose-based hydrogels. On the one hand, cellulose is abundant, renewable, green biodegradable and an eco-friendly building block. On the other hand, cellulose-based hydrogels have become a research hotspot in the field of functional materials because of their outstanding characteristics such as excellent flexibility, stimulus-response, biocompatibility, and degradability. With the increasing and deepening of hydrogel research, cellulose-based hydrogel properties are constantly evolving in an attempt to match the multifunctional polymer materials. For example, multi-functional cellulose hydrogels combine self-healing, high strength, adhesion and conductivity. Therefore, the design and construction of multi-functional cellulose-based hydrogel materials to meet the application requirements of different fields is the focus of future research.

At present, most of the research is focused on functional cellulose-based hydrogels as flexible wearable sensors, drug carriers, and wound dressings. However, some disadvantages need to be solved, such as the easy loss of the sensor signal, poor antifreeze performance in low temperatures, the no-matching chemical force between the sensor and the organization, foreign body reaction, and performance loss. At the same time, most functional cellulose hydrogels are still in the experimental stage and have not been put into industrial production because of technical immaturity. Therefore, a lot of research work is needed to popularize functional cellulose-based hydrogels in daily life, realize their commercialization, and give full play to their application potential. Cellulose-based hydrogels have a long way to go in the future.

**Author Contributions:** Conceptualization and original draft writing, P.Z., J.Y. and Y.-N.C.; supervision, conceptualization, writing, and reviewing and editing, P.Z.; resources, validation, reviewing, and editing, P.Z. and J.Y.; resources, reviewing, and editing, P.Z., M.Y., Z.L. and C.G.; proofreading, editing, and budget administration, P.Z., T.Z. and J.C. All authors have read and agreed to the published version of the manuscript.

**Funding:** This research received no external funding.

**Institutional Review Board Statement:** Not applicable.

**Informed Consent Statement:** Not applicable.

**Data Availability Statement:** The data used to support the review summary of this paper are included within the article.

**Conflicts of Interest:** The authors declare that they have no conflict of interest.

## References

1. Ganewatta, M.S.; Wang, Z.; Tang, C. Chemical syntheses of bioinspired and biomimetic polymers toward biobased materials. *Nat. Rev. Chem.* **2021**, *5*, 753–772. [CrossRef]
2. Tang, T.-C.; An, B.; Huang, Y.; Vasikaran, S.; Wang, Y.; Jiang, X.; Lu, T.K.; Zhong, C. Materials design by synthetic biology. *Nat. Rev. Mater.* **2020**, *6*, 332–350. [CrossRef]
3. Wichterle, O.; LÍM, D. Hydrophilic Gels for Biological Use. *Nature* **1960**, *185*, 117–118. [CrossRef]
4. Zhang, K.; Feng, Q.; Fang, Z.; Gu, L.; Bian, L. Structurally Dynamic Hydrogels for Biomedical Applications: Pursuing a Fine Balance between Macroscopic Stability and Microscopic Dynamics. *Chem. Rev.* **2021**, *121*, 11149–11193. [CrossRef]
5. Yan, L.; Zhou, T.; Han, L.; Zhu, M.; Cheng, Z.; Li, D.; Ren, F.; Wang, K.; Lu, X. Conductive Cellulose Bio-Nanosheets Assembled Biostable Hydrogel for Reliable Bioelectronics. *Adv. Funct. Mater.* **2021**, *31*, 2010465. [CrossRef]
6. Ilkar Erdagi, S.; Asabuwa Ngwabebhoh, F.; Yildiz, U. Genipin crosslinked gelatin-diosgenin-nanocellulose hydrogels for potential wound dressing and healing applications. *Int. J. Biol. Macromol.* **2020**, *149*, 651–663. [CrossRef] [PubMed]
7. Liu, M.; Song, X.; Wen, Y.; Zhu, J.L.; Li, J. Injectable Thermoresponsive Hydrogel Formed by Alginate-g-Poly(N-isopropylacrylamide) That Releases Doxorubicin-Encapsulated Micelles as a Smart Drug Delivery System. *ACS Appl. Mater. Interfaces* **2017**, *9*, 35673–35682. [CrossRef]
8. Kim, J.; Francis, D.M.; Sestito, L.F.; Archer, P.A.; Manspeaker, M.P.; O'Melia, M.J.; Thomas, S.N. Thermosensitive hydrogel releasing nitric oxide donor and anti-CTLA-4 micelles for anti-tumor immunotherapy. *Nat. Commun.* **2022**, *13*, 1479. [CrossRef]
9. Li, Q.; Shao, X.; Dai, X.; Guo, Q.; Yuan, B.; Liu, Y.; Jiang, W. Recent trends in the development of hydrogel therapeutics for the treatment of central nervous system disorders. *NPG Asia Mater.* **2022**, *14*, 14. [CrossRef]

10. Gebeyehu, E.K.; Sui, X.; Adamu, B.F.; Beyene, K.A.; Tadesse, M.G. Cellulosic-Based Conductive Hydrogels for Electro-Active Tissues: A Review Summary. *Gels* **2022**, *8*, 140. [CrossRef]
11. Gjorevski, N.; Nikolaev, M.; Brown, T.E.; Mitrofanova, O.; Brandenberg, N.; DelRio, F.W.; Yavitt, F.M.; Liberali, P.; Anseth, K.S.; Lutolf, M.P. Tissue geometry drives deterministic organoid patterning. *Science* **2022**, *375*, eaaw9021. [CrossRef] [PubMed]
12. Zhang, Z.; Chen, Z.; Wang, Y.; Zhao, Y. Bioinspired conductive cellulose liquid-crystal hydrogels as multifunctional electrical skins. *Proc. Natl. Acad. Sci. USA* **2020**, *117*, 18310–18316. [CrossRef]
13. Mredha, M.T.I.; Guo, Y.Z.; Nonoyama, T.; Nakajima, T.; Kurokawa, T.; Gong, J.P. A Facile Method to Fabricate Anisotropic Hydrogels with Perfectly Aligned Hierarchical Fibrous Structures. *Adv. Mater.* **2018**, *30*, 1704937. [CrossRef] [PubMed]
14. Zhao, D.; Zhu, Y.; Cheng, W.; Xu, G.; Wang, Q.; Liu, S.; Li, J.; Chen, C.; Yu, H.; Hu, L. A Dynamic Gel with Reversible and Tunable Topological Networks and Performances. *Matter* **2020**, *2*, 390–403. [CrossRef]
15. Park, J.Y.; Mani, S.; Clair, G.; Olson, H.M.; Paurus, V.L.; Ansong, C.K.; Blundell, C.; Young, R.; Kanter, J.; Gordon, S.; et al. A microphysiological model of human trophoblast invasion during implantation. *Nat. Commun.* **2022**, *13*, 1252. [CrossRef]
16. Sharma, S.; Tiwari, S. A review on biomacromolecular hydrogel classification and its applications. *Int. J. Biol. Macromol.* **2020**, *162*, 737–747. [CrossRef]
17. Sun, X.; Agate, S.; Salem, K.S.; Lucia, L.; Pal, L. Hydrogel-Based Sensor Networks: Compositions, Properties, and Applications-A Review. *ACS Appl. Bio Mater.* **2021**, *4*, 140–162. [CrossRef]
18. Lavrador, P.; Esteves, M.R.; Gaspar, V.M.; Mano, J.F. Stimuli-Responsive Nanocomposite Hydrogels for Biomedical Applications. *Adv. Funct. Mater.* **2020**, *31*, 2005941. [CrossRef]
19. Guo, Y.; Bae, J.; Fang, Z.; Li, P.; Zhao, F.; Yu, G. Hydrogels and Hydrogel-Derived Materials for Energy and Water Sustainability. *Chem. Rev.* **2020**, *120*, 7642–7707. [CrossRef]
20. Nele, V.; Wojciechowski, J.P.; Armstrong, J.P.K.; Stevens, M.M. Tailoring Gelation Mechanisms for Advanced Hydrogel Applications. *Adv. Funct. Mater.* **2020**, *30*, 2002759. [CrossRef]
21. Na, H.; Kang, Y.W.; Park, C.S.; Jung, S.; Kim, H.Y.; Sun, J.Y. Hydrogel-based strong and fast actuators by electroosmotic turgor pressure. *Science* **2022**, *376*, 301–307. [CrossRef] [PubMed]
22. Dobashi, Y.; Yao, D.; Petel, Y.; Nguyen, T.N.; Sarwar, M.S.; Thabet, Y.; Ng, C.L.W.; Scabeni Glitz, E.; Nguyen, G.T.M.; Plesse, C.; et al. Piezoionic mechanoreceptors: Force-induced current generation in hydrogels. *Science* **2022**, *376*, 502–507. [CrossRef] [PubMed]
23. Zhou, J.; Zhang, R.; Xu, R.; Li, Y.; Tian, W.; Gao, M.; Wang, M.; Li, D.; Liang, X.; Xie, L.; et al. Super-Assembled Hierarchical Cellulose Aerogel-Gelatin Solid Electrolyte for Implantable and Biodegradable Zinc Ion Battery. *Adv. Funct. Mater.* **2022**, *32*, 2111406. [CrossRef]
24. Apostolopoulou-Kalkavoura, V.; Munier, P.; Bergstrom, L. Thermally Insulating Nanocellulose-Based Materials. *Adv. Mater.* **2021**, *33*, e2001839. [CrossRef] [PubMed]
25. Choi, S.W.; Guan, W.; Chung, K. Basic principles of hydrogel-based tissue transformation technologies and their applications. *Cell* **2021**, *184*, 4115–4136. [CrossRef] [PubMed]
26. Freedman, B.R.; Kuttler, A.; Beckmann, N.; Nam, S.; Kent, D.; Schuleit, M.; Ramazani, F.; Accart, N.; Rock, A.; Li, J.; et al. Enhanced tendon healing by a tough hydrogel with an adhesive side and high drug-loading capacity. *Nat. Biomed. Eng.* **2022**. [CrossRef] [PubMed]
27. Li, J.; Mooney, D.J. Designing hydrogels for controlled drug delivery. *Nat. Rev. Mater.* **2016**, *1*, 16071. [CrossRef]
28. Ciolacu, D.E.; Nicu, R.; Ciolacu, F. Cellulose-Based Hydrogels as Sustained Drug-Delivery Systems. *Materials* **2020**, *13*, 5270. [CrossRef]
29. Wang, Z.; Ding, B.; Zhao, Y.; Han, Y.; Sheng, Y.; Tao, L.; Shen, X.; Zhou, J.; Jiang, L.; Ding, Y. Tumor-oriented mathematical models in hydrogel regulation for precise topical administration regimens. *J. Control. Release* **2022**, *345*, 610–624. [CrossRef]
30. Hu, Y.; Fan, C. Nanocomposite DNA hydrogels emerging as programmable and bioinstructive materials systems. *Chem* **2022**. [CrossRef]
31. Zhao, S.; Chen, Z.; Dong, Y.; Lu, W.; Zhu, D. The Preparation and Properties of Composite Hydrogels Based on Gelatin and (3-Aminopropyl) Trimethoxysilane Grafted Cellulose Nanocrystals Covalently Linked with Microbial Transglutaminase. *Gels* **2022**, *8*, 146. [CrossRef] [PubMed]
32. Klemm, D.; Heublein, B.; Fink, H.P.; Bohn, A. Cellulose: Fascinating biopolymer and sustainable raw material. *Angew. Chem. Int. Ed. Engl.* **2005**, *44*, 3358–3393. [CrossRef] [PubMed]
33. Li, Y.; Li, L.; Zhang, Z.; Cheng, J.; Fei, Y.; Lu, L. An all-natural strategy for versatile interpenetrating network hydrogels with self-healing, super-adhesion and high sensitivity. *Chem. Eng. J.* **2021**, *420*, 129736. [CrossRef]
34. Liu, C.; Morimoto, N.; Jiang, L.; Kawahara, S.; Noritomi, T.; Yokoyama, H.; Mayumi, K.; Ito, K. Tough hydrogels with rapid self-reinforcement. *Science* **2021**, *372*, 1078–1081. [CrossRef]
35. Hua, M.; Wu, S.; Ma, Y.; Zhao, Y.; Chen, Z.; Frenkel, I.; Strzalka, J.; Zhou, H.; Zhu, X.; He, X. Strong tough hydrogels via the synergy of freeze-casting and salting out. *Nature* **2021**, *590*, 594–599. [CrossRef]
36. Liang, Q.; Xia, X.; Sun, X.; Yu, D.; Huang, X.; Han, G.; Mugo, S.M.; Chen, W.; Zhang, Q. Highly Stretchable Hydrogels as Wearable and Implantable Sensors for Recording Physiological and Brain Neural Signals. *Adv. Sci.* **2022**, *9*, e2201059. [CrossRef]

37. Wang, Z.; Zheng, X.; Ouchi, T.; Kouznetsova, T.B.; Beech, H.K.; Av-Ron, S.; Matsuda, T.; Bowser, B.H.; Wang, S.; Johnson, J.A.; et al. Toughening hydrogels through force-triggered chemical reactions that lengthen polymer strands. *Science* **2021**, *374*, 193–196. [CrossRef]
38. Kim, J.; Zhang, G.; Shi, M.; Suo, Z. Fracture, fatigue, and friction of polymers in which entanglements greatly outnumber cross-links. *Science* **2021**, *374*, 212–216. [CrossRef]
39. Rahmanian, V.; Pirzada, T.; Wang, S.; Khan, S.A. Cellulose-Based Hybrid Aerogels: Strategies toward Design and Functionality. *Adv. Mater.* **2021**, *33*, e2102892. [CrossRef]
40. Chen, Y.; Zhang, L.; Yang, Y.; Pang, B.; Xu, W.; Duan, G.; Jiang, S.; Zhang, K. Recent Progress on Nanocellulose Aerogels: Preparation, Modification, Composite Fabrication, Applications. *Adv. Mater.* **2021**, *33*, e2005569. [CrossRef]
41. Guan, Q.F.; Yang, H.B.; Han, Z.M.; Ling, Z.C.; Yin, C.H.; Yang, K.P.; Zhao, Y.X.; Yu, S.H. Sustainable Cellulose-Nanofiber-Based Hydrogels. *ACS Nano* **2021**, *15*, 7889–7898. [CrossRef] [PubMed]
42. Moon, R.J.; Martini, A.; Nairn, J.; Simonsen, J.; Youngblood, J. Cellulose nanomaterials review: Structure, properties and nanocomposites. *Chem. Soc. Rev.* **2011**, *40*, 3941–3994. [CrossRef] [PubMed]
43. Dutta, S.D.; Patel, D.K.; Lim, K.T. Functional cellulose-based hydrogels as extracellular matrices for tissue engineering. *J. Biol. Eng.* **2019**, *13*, 55. [CrossRef] [PubMed]
44. Sannino, A.; Demitri, C.; Madaghiele, M. Biodegradable Cellulose-based Hydrogels: Design and Applications. *Materials* **2009**, *2*, 353–373. [CrossRef]
45. Ray, U.; Zhu, S.; Pang, Z.; Li, T. Mechanics Design in Cellulose-Enabled High-Performance Functional Materials. *Adv. Mater.* **2021**, *33*, e2002504. [CrossRef]
46. Kabir, S.M.F.; Sikdar, P.P.; Haque, B.; Bhuiyan, M.A.R.; Ali, A.; Islam, M.N. Cellulose-based hydrogel materials: Chemistry, properties and their prospective applications. *Prog. Biomater.* **2018**, *7*, 153–174. [CrossRef]
47. Eichhorn, S.J.; Baillie, C.A.; Zafeiropoulos, N.; Mwaikambo, L.Y.; Ansell, M.P.; Dufresne, A.; Entwistle, K.M.; Herrera-Franco, P.J.; Escamilla, G.C.; Groom, L.; et al. Current international research into cellulosic fibres and composites. *J. Mater. Sci.* **2001**, *36*, 2107–2131. [CrossRef]
48. Moradali, M.F.; Rehm, B.H.A. Bacterial biopolymers: From pathogenesis to advanced materials. *Nat. Rev. Microbiol.* **2020**, *18*, 195–210. [CrossRef]
49. Picheth, G.F.; Pirich, C.L.; Sierakowski, M.R.; Woehl, M.A.; Sakakibara, C.N.; de Souza, C.F.; Martin, A.A.; da Silva, R.; de Freitas, R.A. Bacterial cellulose in biomedical applications: A review. *Int. J. Biol. Macromol.* **2017**, *104*, 97–106. [CrossRef]
50. Ul-Islam, M.; Khan, S.; Ullah, M.W.; Park, J.K. Comparative study of plant and bacterial cellulose pellicles regenerated from dissolved states. *Int. J. Biol. Macromol.* **2019**, *137*, 247–252. [CrossRef]
51. Fang, Q.; Zhou, X.; Deng, W.; Zheng, Z.; Liu, Z. Freestanding bacterial cellulose-graphene oxide composite membranes with high mechanical strength for selective ion permeation. *Sci. Rep.* **2016**, *6*, 33185. [CrossRef] [PubMed]
52. Abeer, M.M.; Mohd Amin, M.C.; Martin, C. A review of bacterial cellulose-based drug delivery systems: Their biochemistry, current approaches and future prospects. *J. Pharm. Pharmacol.* **2014**, *66*, 1047–1061. [CrossRef] [PubMed]
53. Khalil, H.; Jummaat, F.; Yahya, E.B.; Olaiya, N.G.; Adnan, A.S.; Abdat, M.; Nasir, N.A.M.; Halim, A.S.; Kumar, U.S.U.; Bairwan, R.; et al. A Review on Micro- to Nanocellulose Biopolymer Scaffold Forming for Tissue Engineering Applications. *Polymers* **2020**, *12*, 2043. [CrossRef] [PubMed]
54. Kang, H.; Liu, R.; Huang, Y. Cellulose-Based Gels. *Macromol. Chem. Phys.* **2016**, *217*, 1322–1334. [CrossRef]
55. He, W.; Wu, J.; Xu, J.; Mosselhy, D.A.; Zheng, Y.; Yang, S. Bacterial Cellulose: Functional Modification and Wound Healing Applications. *Adv. Wound Care (New Rochelle)* **2021**, *10*, 623–640. [CrossRef]
56. Zhang, Q.; Wang, M.; Mu, G.; Ren, H.; He, C.; Xie, Q.; Liu, Q.; Wang, J.; Cha, R. Adsorptivity of cationic cellulose nanocrystals for phosphate and its application in hyperphosphatemia therapy. *Carbohydr. Polym.* **2021**, *255*, 117335. [CrossRef]
57. Fiume, M.M.; Bergfeld, W.F.; Belsito, D.V.; Hill, R.A.; Klaassen, C.D.; Liebler, D.C.; Marks, J.G., Jr.; Shank, R.C.; Slaga, T.J.; Snyder, P.W.; et al. Safety Assessment of Nitrocellulose and Collodion as Used in Cosmetics. *Int. J. Toxicol.* **2016**, *35*, 50S–59S. [CrossRef]
58. Bifari, E.N.; Bahadar Khan, S.; Alamry, K.A.; Asiri, A.M.; Akhtar, K. Cellulose Acetate Based Nanocomposites for Biomedical Applications: A Review. *Curr. Pharm. Des.* **2016**, *22*, 3007–3019. [CrossRef]
59. Cheng, Y.; Qin, H.; Acevedo, N.C.; Shi, X. Development of methylcellulose-based sustained-release dosage by semisolid extrusion additive manufacturing in drug delivery system. *J. Biomed. Mater. Res. B Appl. Biomater.* **2021**, *109*, 257–268. [CrossRef] [PubMed]
60. Wu, D.; Cheng, J.; Su, X.; Feng, Y. Hydrophilic modification of methylcellulose to obtain thermoviscosifying polymers without macro-phase separation. *Carbohydr. Polym.* **2021**, *260*, 117792. [CrossRef]
61. Javanbakht, S.; Shaabani, A. Carboxymethyl cellul.lose-based oral delivery systems. *Int. J. Biol. Macromol.* **2019**, *133*, 21–29. [CrossRef] [PubMed]
62. Sawant, P.D.; Luu, D.; Ye, R.; Buchta, R. Drug release from hydroethanolic gels. Effect of drug's lipophilicity (logP), polymer-drug interactions and solvent lipophilicity. *Int. J. Pharm.* **2010**, *396*, 45–52. [CrossRef] [PubMed]
63. Burdock, G.A. Safety assessment of hydroxypropyl methylcellulose as a food ingredient. *Food Chem. Toxicol.* **2007**, *45*, 2341–2351. [CrossRef] [PubMed]
64. Ban, S.J.; Rico, C.W.; Um, I.C.; Kang, M.Y. Antihyperglycemic and antioxidative effects of Hydroxyethyl Methylcellulose (HEMC) and Hydroxypropyl Methylcellulose (HPMC) in mice fed with a high fat diet. *Int. J. Mol. Sci.* **2012**, *13*, 3738–3750. [CrossRef] [PubMed]

65. Abd-El Kader, F.H.; Said, G.; El-Naggar, M.M.; Anees, B.A. Characterization and electrical properties of methyl-2-hydroxyethyl cellulose doped with erbium nitrate salt. *J. Appl. Polym. Sci.* **2006**, *102*, 2352–2361. [CrossRef]
66. Liu, X.; Xiao, W.; Ma, X.; Huang, L.; Ni, Y.; Chen, L.; Ouyang, X.; Li, J. Conductive Regenerated Cellulose Film and Its Electronic Devices—A Review. *Carbohydr. Polym.* **2020**, *250*, 116969. [CrossRef]
67. Dong, H.; Ding, Q.; Jiang, Y.; Li, X.; Han, W. Pickering emulsions stabilized by spherical cellulose nanocrystals. *Carbohydr. Polym.* **2021**, *265*, 118101. [CrossRef]
68. Subhedar, A.; Bhadauria, S.; Ahankari, S.; Kargarzadeh, H. Nanocellulose in biomedical and biosensing applications: A review. *Int. J. Biol. Macromol.* **2021**, *166*, 587–600. [CrossRef]
69. Teodoro, K.B.R.; Sanfelice, R.C.; Migliorini, F.L.; Pavinatto, A.; Facure, M.H.M.; Correa, D.S. A Review on the Role and Performance of Cellulose Nanomaterials in Sensors. *ACS Sens.* **2021**, *6*, 2473–2496. [CrossRef]
70. Gonzalez, J.S.; Luduena, L.N.; Ponce, A.; Alvarez, V.A. Poly(vinyl alcohol)/cellulose nanowhiskers nanocomposite hydrogels for potential wound dressings. *Mater. Sci Eng. C Mater. Biol. Appl.* **2014**, *34*, 54–61. [CrossRef]
71. Zhou, S.; Guo, K.; Bukhvalov, D.; Zhang, X.F.; Zhu, W.; Yao, J.; He, M. Cellulose Hydrogels by Reversible Ion-Exchange as Flexible Pressure Sensors. *Adv. Mater. Technol.* **2020**, *5*, 2000358. [CrossRef]
72. Tavakolian, M.; Jafari, S.M.; van de Ven, T.G.M. A Review on Surface-Functionalized Cellulosic Nanostructures as Biocompatible Antibacterial Materials. *Nanomicro Lett.* **2020**, *12*, 73. [CrossRef] [PubMed]
73. Shaheen, T.I.; Emam, H.E. Sono-chemical synthesis of cellulose nanocrystals from wood sawdust using Acid hydrolysis. *Int. J. Biol. Macromol.* **2018**, *107*, 1599–1606. [CrossRef] [PubMed]
74. Chimpibul, W.; Nakaji-Hirabayashi, T.; Yuan, X.; Matsumura, K. Controlling the degradation of cellulose scaffolds with Malaprade oxidation for tissue engineering. *J. Mater. Chem. B* **2020**, *8*, 7904–7913. [CrossRef] [PubMed]
75. Mbarki, K.; Fersi, M.; Louati, I.; Elleuch, B.; Sayari, A. Biodegradation study of PDLA/cellulose microfibres biocomposites by Pseudomonas aeruginosa. *Environ. Technol.* **2021**, *42*, 731–742. [CrossRef] [PubMed]
76. Arca, H.C.; Mosquera-Giraldo, L.I.; Bi, V.; Xu, D.; Taylor, L.S.; Edgar, K.J. Pharmaceutical Applications of Cellulose Ethers and Cellulose Ether Esters. *Biomacromolecules* **2018**, *19*, 2351–2376. [CrossRef] [PubMed]
77. Yang, Y.; Lu, Y.T.; Zeng, K.; Heinze, T.; Groth, T.; Zhang, K. Recent Progress on Cellulose-Based Ionic Compounds for Biomaterials. *Adv. Mater.* **2021**, *33*, e2000717. [CrossRef]
78. Dang, C.; Huang, Z.; Chen, Y.; Zhou, S.; Feng, X.; Chen, G.; Dai, F.; Qi, H. Direct Dissolution of Cellulose in NaOH/Urea/alpha-Lipoic Acid Aqueous Solution to Fabricate All Biomass-Based Nitrogen, Sulfur Dual-Doped Hierarchical Porous Carbon Aerogels for Supercapacitors. *ACS Appl. Mater. Interfaces* **2020**, *12*, 21528–21538. [CrossRef]
79. Van de Vyver, S.; Geboers, J.; Jacobs, P.A.; Sels, B.F. Recent Advances in the Catalytic Conversion of Cellulose. *ChemCatChem* **2011**, *3*, 82–94. [CrossRef]
80. Tashiro, K.; Kobayashi, M. Theoretical evaluation of three-dimensional elastic constants of native and regenerated celluloses: Role of hydrogen bonds. *Polymer* **1991**, *32*, 1516–1526. [CrossRef]
81. Naderi, A. Nanofibrillated cellulose: Properties reinvestigated. *Cellulose* **2017**, *24*, 1933–1945. [CrossRef]
82. Wang, Y.; Zhang, L.; Lu, A. Transparent, Antifreezing, Ionic Conductive Cellulose Hydrogel with Stable Sensitivity at Subzero Temperature. *ACS Appl. Mater. Interfaces* **2019**, *11*, 41710–41716. [CrossRef] [PubMed]
83. Zhang, X.F.; Ma, X.; Hou, T.; Guo, K.; Yin, J.; Wang, Z.; Shu, L.; He, M.; Yao, J. Inorganic Salts Induce Thermally Reversible and Anti-Freezing Cellulose Hydrogels. *Angew. Chem. Int. Ed.* **2019**, *58*, 7366–7370. [CrossRef] [PubMed]
84. Nasatto, P.; Pignon, F.; Silveira, J.; Duarte, M.; Noseda, M.; Rinaudo, M. Methylcellulose, a Cellulose Derivative with Original Physical Properties and Extended Applications. *Polymers* **2015**, *7*, 777–803. [CrossRef]
85. Bonetti, L.; De Nardo, L.; Fare, S. Thermo-Responsive Methylcellulose Hydrogels: From Design to Applications as Smart Biomaterials. *Tissue Eng. Part. B Rev.* **2021**, *27*, 486–513. [CrossRef]
86. Bonetti, L.; Fiorati, A.; D'Agostino, A.; Pelacani, C.M.; Chiesa, R.; Fare, S.; De Nardo, L. Smart Methylcellulose Hydrogels for pH-Triggered Delivery of Silver Nanoparticles. *Gels* **2022**, *8*, 298. [CrossRef] [PubMed]
87. Bonetti, L.; De Nardo, L.; Fare, S. Chemically Crosslinked Methylcellulose Substrates for Cell Sheet Engineering. *Gels* **2021**, *7*, 141. [CrossRef] [PubMed]
88. Gila-Vilchez, C.; Manas-Torres, M.C.; Contreras-Montoya, R.; Alaminos, M.; Duran, J.D.G.; de Cienfuegos, L.A.; Lopez-Lopez, M.T. Anisotropic magnetic hydrogels: Design, structure and mechanical properties. *Philos. Trans. A Math. Phys. Eng. Sci.* **2019**, *377*, 20180217. [CrossRef]
89. Sano, K.; Ishida, Y.; Aida, T. Synthesis of Anisotropic Hydrogels and Their Applications. *Angew. Chem. Int. Ed. Engl.* **2018**, *57*, 2532–2543. [CrossRef]
90. Zhou, S.; Guo, K.; Bukhvalov, D.; Zhu, W.; Wang, J.; Sun, W.; He, M. H-bond/ionic coordination switching for fabrication of highly oriented cellulose hydrogels. *J. Mater. Chem. A* **2021**, *9*, 5533–5541. [CrossRef]
91. Guo, K.; Zhu, W.; Wang, J.; Sun, W.; Zhou, S.; He, M. Fabrication of gradient anisotropic cellulose hydrogels for applications in micro-strain sensing. *Carbohydr. Polym.* **2021**, *258*, 117694. [CrossRef] [PubMed]
92. Ye, D.; Cheng, Q.; Zhang, Q.; Wang, Y.; Chang, C.; Li, L.; Peng, H.; Zhang, L. Deformation Drives Alignment of Nanofibers in Framework for Inducing Anisotropic Cellulose Hydrogels with High Toughness. *ACS Appl. Mater. Interfaces* **2017**, *9*, 43154–43162. [CrossRef] [PubMed]

93. Hu, X.; Vatankhah-Varnoosfaderani, M.; Zhou, J.; Li, Q.; Sheiko, S.S. Weak Hydrogen Bonding Enables Hard, Strong, Tough, and Elastic Hydrogels. *Adv. Mater.* **2015**, *27*, 6899–6905. [CrossRef]
94. Jiang, P.; Lin, P.; Yang, C.; Qin, H.; Wang, X.; Zhou, F. 3D Printing of Dual-Physical Cross-linking Hydrogel with Ultrahigh Strength and Toughness. *Chem. Mater.* **2020**, *32*, 9983–9995. [CrossRef]
95. Bai, Z.; Jia, K.; Liu, C.; Wang, L.; Lin, G.; Huang, Y.; Liu, S.; Liu, X. A Solvent Regulated Hydrogen Bond Crosslinking Strategy to Prepare Robust Hydrogel Paint for Oil/Water Separation. *Adv. Funct. Mater.* **2021**, *31*, 2104701. [CrossRef]
96. Pan, J.; Gao, L.; Sun, W.; Wang, S.; Shi, X. Length Effects of Short Alkyl Side Chains on Phase-Separated Structure and Dynamics of Hydrophobic Association Hydrogels. *Macromolecules* **2021**, *54*, 5962–5973. [CrossRef]
97. Sannino, A.; Maffezzoli, A.; Nicolais, L. Introduction of molecular spacers between the crosslinks of a cellulose-based superabsorbent hydrogel: Effects on the equilibrium sorption properties. *J. Appl. Polym. Sci.* **2003**, *90*, 168–174. [CrossRef]
98. Berthiaume, F.; Maguire, T.J.; Yarmush, M.L. Tissue engineering and regenerative medicine: History, progress, and challenges. *Annu Rev. Chem. Biomol. Eng.* **2011**, *2*, 403–430. [CrossRef] [PubMed]
99. Kopeček, J.; Yang, J. Hydrogels as smart biomaterials. *Polym. Int.* **2007**, *56*, 1078–1098. [CrossRef]
100. Jiang, L.; Li, W.; Nie, J.; Wang, R.; Chen, X.; Fan, W.; Hu, L. Fluorescent Nanogel Sensors for X-ray Dosimetry. *ACS Sens.* **2021**, *6*, 1643–1648. [CrossRef]
101. Liang, L.; Lv, H.; Shi, X.-L.; Liu, Z.; Chen, G.; Chen, Z.-G.; Sun, G. A flexible quasi-solid-state thermoelectrochemical cell with high stretchability as an energy-autonomous strain sensor. *Mater. Horiz.* **2021**, *8*, 275–276. [CrossRef] [PubMed]
102. Hai, J.; Zeng, X.; Zhu, Y.; Wang, B. Anions reversibly responsive luminescent nanocellulose hydrogels for cancer spheroids culture and release. *Biomaterials* **2019**, *194*, 161–170. [CrossRef] [PubMed]
103. Hoare, T.R.; Kohane, D.S. Hydrogels in drug delivery: Progress and challenges. *Polymer* **2008**, *49*, 1993–2007. [CrossRef]
104. Yue, Z.; Wen, F.; Gao, S.; Ang, M.Y.; Pallathadka, P.K.; Liu, L.; Yu, H. Preparation of three-dimensional interconnected macroporous cellulosic hydrogels for soft tissue engineering. *Biomaterials* **2010**, *31*, 8141–8152. [CrossRef]
105. Wang, M.; Fang, Y.; Hu, D. Preparation and properties of chitosan-poly(N-isopropylacrylamide) full-IPN hydrogels. *React. Funct. Polym.* **2001**, *48*, 215–221. [CrossRef]
106. Zhang, X.Z.; Wu, D.Q.; Chu, C.C. Synthesis, characterization and controlled drug release of thermosensitive IPN-PNIPAAm hydrogels. *Biomaterials* **2004**, *25*, 3793–3805. [CrossRef]
107. Ruan, Z.; Li, C.; Li, J.-R.; Qin, J.; Li, Z. A relay strategy for the mercury (II) chemodosimeter with ultra-sensitivity as test strips. *Sci. Rep.* **2015**, *5*, 15987. [CrossRef]
108. Muniz, E.C.; Geuskens, G. Polyacrylamide hydrogels and semi-interpenetrating networks (IPNs) with poly(N-isopropylacrylamide): Mechanical properties by measure of compressive elastic modulus. *J. Mater. Sci Mater. Med.* **2001**, *12*, 879–881. [CrossRef]
109. Zhou, X.; Guo, Y.; Zhao, F.; Shi, W.; Yu, G. Topology-Controlled Hydration of Polymer Network in Hydrogels for Solar-Driven Wastewater Treatment. *Adv. Mater.* **2020**, *32*, e2007012. [CrossRef]
110. Xia, L.W.; Xie, R.; Ju, X.J.; Wang, W.; Chen, Q.; Chu, L.Y. Nano-structured smart hydrogels with rapid response and high elasticity. *Nat. Commun.* **2013**, *4*, 2226. [CrossRef]
111. Tu, H.; Zhu, M.; Duan, B.; Zhang, L. Recent Progress in High-Strength and Robust Regenerated Cellulose Materials. *Adv. Mater.* **2020**, *33*, 2000682. [CrossRef] [PubMed]
112. Zhao, D.; Huang, J.; Zhong, Y.; Li, K.; Zhang, L.; Cai, J. High-Strength and High-Toughness Double-Cross-Linked Cellulose Hydrogels: A New Strategy Using Sequential Chemical and Physical Cross-Linking. *Adv. Funct. Mater.* **2016**, *26*, 6279–6287. [CrossRef]
113. Bae, S.W.; Kim, J.; Kwon, S. Recent Advances in Polymer Additive Engineering for Diagnostic and Therapeutic Hydrogels. *Int. J. Mol. Sci.* **2022**, *23*, 2955. [CrossRef]
114. Kurkuri, M.D.; Aminabhavi, T.M. Poly(vinyl alcohol) and poly(acrylic acid) sequential interpenetrating network pH-sensitive microspheres for the delivery of diclofenac sodium to the intestine. *J. Control. Release* **2004**, *96*, 9–20. [CrossRef]
115. Bazban-Shotorbani, S.; Hasani-Sadrabadi, M.M.; Karkhaneh, A.; Serpooshan, V.; Jacob, K.I.; Moshaverinia, A.; Mahmoudi, M. Revisiting structure-property relationship of pH-responsive polymers for drug delivery applications. *J. Control. Release* **2017**, *253*, 46–63. [CrossRef]
116. Qiu, Y.; Park, K. Environment-sensitive hydrogels for drug delivery. *Adv. Drug Deliv. Rev.* **2012**, *64*, 49–60. [CrossRef]
117. Yin, H.; Song, P.; Chen, X.; Xiao, M.; Tang, L.; Huang, H. Smart pH-Sensitive Hydrogel Based on the Pineapple Peel-Oxidized Hydroxyethyl Cellulose and the Hericium erinaceus Residue Carboxymethyl Chitosan for Use in Drug Delivery. *Biomacromolecules* **2022**, *23*, 253–264. [CrossRef] [PubMed]
118. Tanaka, F.; Koga, T.; Kaneda, I.; Winnik, F.M. Hydration, phase separation and nonlinear rheology of temperature-sensitive water-soluble polymers. *J. Phys. Condens. Matter.* **2011**, *23*, 284105. [CrossRef] [PubMed]
119. Sultana, T.; Gwon, J.G.; Lee, B.T. Thermal stimuli-responsive hyaluronic acid loaded cellulose based physical hydrogel for post-surgical de novo peritoneal adhesion prevention. *Mater. Sci. Eng. C Mater. Biol. Appl.* **2020**, *110*, 110661. [CrossRef]
120. Weng, H.; Zhou, J.; Tang, L.; Hu, Z. Tissue responses to thermally-responsive hydrogel nanoparticles. *J. Biomater. Sci. Polym. Ed.* **2004**, *15*, 1167–1180. [CrossRef]
121. Wang, T.; Chen, L.; Shen, T.; Wu, D. Preparation and properties of a novel thermo-sensitive hydrogel based on chitosan/hydroxypropyl methylcellulose/glycerol. *Int. J. Biol. Macromol.* **2016**, *93*, 775–782. [CrossRef] [PubMed]

122. Calejo, M.T.; Kjoniksen, A.L.; Pinazo, A.; Perez, L.; Cardoso, A.M.; de Lima, M.C.; Jurado, A.S.; Sande, S.A.; Nystrom, B. Thermoresponsive hydrogels with low toxicity from mixtures of ethyl(hydroxyethyl) cellulose and arginine-based surfactants. *Int. J. Pharm.* **2012**, *436*, 454–462. [CrossRef]
123. Wei, Y.; Xiang, L.; Zhu, P.; Qian, Y.; Zhao, B.; Chen, G. Multifunctional Organohydrogel-Based Ionic Skin for Capacitance and Temperature Sensing toward Intelligent Skin-like Devices. *Chem. Mater.* **2021**, *33*, 8623–8634. [CrossRef]
124. Hao, S.; Meng, L.; Fu, Q.; Xu, F.; Yang, J. Low-Temperature tolerance and conformal adhesion zwitterionic hydrogels as electronic skin for strain and temperature responsiveness. *Chem. Eng. J.* **2022**, *431*, 133782. [CrossRef]
125. Ouyang, K.; Zhuang, J.; Chen, C.; Wang, X.; Xu, M.; Xu, Z. Gradient Diffusion Anisotropic Carboxymethyl Cellulose Hydrogels for Strain Sensors. *Biomacromolecules* **2021**, *22*, 5033–5041. [CrossRef]
126. Fu, L.H.; Qi, C.; Ma, M.G.; Wan, P. Multifunctional cellulose-based hydrogels for biomedical applications. *J. Mater. Chem. B* **2019**, *7*, 1541–1562. [CrossRef] [PubMed]
127. Li, T.; Chen, C.; Brozena, A.H.; Zhu, J.Y.; Xu, L.; Driemeier, C.; Dai, J.; Rojas, O.J.; Isogai, A.; Wagberg, L.; et al. Developing fibrillated cellulose as a sustainable technological material. *Nature* **2021**, *590*, 47–56. [CrossRef]
128. Wong, L.C.; Leh, C.P.; Goh, C.F. Designing cellulose hydrogels from non-woody biomass. *Carbohydr. Polym.* **2021**, *264*, 118036. [CrossRef]
129. Keplinger, C.; Sun, J.Y.; Foo, C.C.; Rothemund, P.; Whitesides, G.M.; Suo, Z. Stretchable, transparent, ionic conductors. *Science* **2013**, *341*, 984–987. [CrossRef]
130. Brakat, A.; Zhu, H. Nanocellulose-Graphene Hybrids: Advanced Functional Materials as Multifunctional Sensing Platform. *Nanomicro. Lett.* **2021**, *13*, 94. [CrossRef]
131. Zhang, J.; Cheng, Y.; Xu, C.; Gao, M.; Zhu, M.; Jiang, L. Hierarchical Interface Engineering for Advanced Nanocellulosic Hybrid Aerogels with High Compressibility and Multifunctionality. *Adv. Funct. Mater.* **2021**, *31*, 2009349. [CrossRef]
132. Ma, L.; Chen, S.; Wang, D.; Yang, Q.; Mo, F.; Liang, G.; Li, N.; Zhang, H.; Zapien, J.A.; Zhi, C. Super-Stretchable Zinc–Air Batteries Based on an Alkaline-Tolerant Dual-Network Hydrogel Electrolyte. *Adv. Energy Mater.* **2019**, *9*, 1803046. [CrossRef]
133. Park, K.; Choi, H.; Kang, K.; Shin, M.; Son, D. Soft Stretchable Conductive Carboxymethylcellulose Hydrogels for Wearable Sensors. *Gels* **2022**, *8*, 92. [CrossRef]
134. Zhao, S.; Siqueira, G.; Drdova, S.; Norris, D.; Ubert, C.; Bonnin, A.; Galmarini, S.; Ganobjak, M.; Pan, Z.; Brunner, S.; et al. Additive manufacturing of silica aerogels. *Nature* **2020**, *584*, 387–392. [CrossRef] [PubMed]
135. Dong, Y.; Akinoglu, E.M.; Zhang, H.; Maasoumi, F.; Zhou, J.; Mulvaney, P. An Optically Responsive Soft Etalon Based on Ultrathin Cellulose Hydrogels. *Adv. Funct. Mater.* **2019**, *29*, 1904290. [CrossRef]
136. Guo, B.; Dong, R.; Liang, Y.; Li, M. Haemostatic materials for wound healing applications. *Nat. Rev. Chem.* **2021**, *5*, 773–791. [CrossRef]
137. Zhang, K.; Wang, Y.; Wei, Q.; Li, X.; Guo, Y.; Zhang, S. Design and Fabrication of Sodium Alginate/Carboxymethyl Cellulose Sodium Blend Hydrogel for Artificial Skin. *Gels* **2021**, *7*, 115. [CrossRef]
138. He, J.; Zhang, Z.; Yang, Y.; Ren, F.; Li, J.; Zhu, S.; Ma, F.; Wu, R.; Lv, Y.; He, G.; et al. Injectable Self-Healing Adhesive pH-Responsive Hydrogels Accelerate Gastric Hemostasis and Wound Healing. *Nanomicro. Lett.* **2021**, *13*, 80. [CrossRef]
139. Cook, K.A.; Naguib, N.; Kirsch, J.; Hohl, K.; Colby, A.H.; Sheridan, R.; Rodriguez, E.K.; Nazarian, A.; Grinstaff, M.W. In situ gelling and dissolvable hydrogels for use as on-demand wound dressings for burns. *Biomater. Sci.* **2021**, *9*, 6842–6850. [CrossRef]
140. Yuk, H.; Lu, B.; Zhao, X. Hydrogel bioelectronics. *Chem. Soc. Rev.* **2019**, *48*, 1642–1667. [CrossRef]
141. Wei, P.; Wang, L.; Xie, F.; Cai, J. Strong and tough cellulose–graphene oxide composite hydrogels by multi-modulus components strategy as photothermal antibacterial platform. *Chem. Eng. J.* **2022**, *431*, 133964. [CrossRef]
142. Johnson, A.; Kong, F.; Miao, S.; Lin, H.V.; Thomas, S.; Huang, Y.C.; Kong, Z.L. Therapeutic effects of antibiotics loaded cellulose nanofiber and kappa-carrageenan oligosaccharide composite hydrogels for periodontitis treatment. *Sci. Rep.* **2020**, *10*, 18037. [CrossRef] [PubMed]
143. Hu, Y.; Hu, S.; Zhang, S.; Dong, S.; Hu, J.; Kang, L.; Yang, X. A double-layer hydrogel based on alginate-carboxymethyl cellulose and synthetic polymer as sustained drug delivery system. *Sci. Rep.* **2021**, *11*, 9142. [CrossRef] [PubMed]
144. Peppas, N. Hydrogels in pharmaceutical formulations. *Eur. J. Pharm. Biopharm.* **2000**, *50*, 27–46. [CrossRef]
145. Gao, X.; Cao, Y.; Song, X.; Zhang, Z.; Zhuang, X.; He, C.; Chen, X. Biodegradable, pH-responsive carboxymethyl cellulose/poly(acrylic acid) hydrogels for oral insulin delivery. *Macromol. Biosci.* **2014**, *14*, 565–575. [CrossRef] [PubMed]
146. George, D.; Maheswari, P.U.; Begum, K. Chitosan-cellulose hydrogel conjugated with L-histidine and zinc oxide nanoparticles for sustained drug delivery: Kinetics and in-vitro biological studies. *Carbohydr. Polym.* **2020**, *236*, 116101. [CrossRef]
147. Ning, L.; You, C.; Zhang, Y.; Li, X.; Wang, F. Synthesis and biological evaluation of surface-modified nanocellulose hydrogel loaded with paclitaxel. *Life Sci.* **2020**, *241*, 117137. [CrossRef]
148. Guo, Y.Z.; Nakajima, T.; Mredha, M.T.I.; Guo, H.L.; Cui, K.; Zheng, Y.; Cui, W.; Kurokawa, T.; Gong, J.P. Facile preparation of cellulose hydrogel with Achilles tendon-like super strength through aligning hierarchical fibrous structure. *Chem. Eng. J.* **2022**, *428*, 132040. [CrossRef]
149. Shi, S.; Cui, M.; Sun, F.; Zhu, K.; Iqbal, M.I.; Chen, X.; Fei, B.; Li, R.K.Y.; Xia, Q.; Hu, J. An Innovative Solvent-Responsive Coiling-Expanding Stent. *Adv. Mater.* **2021**, *33*, e2101005. [CrossRef]

*Article*

# Electrospun Core (HPMC–Acetaminophen)–Shell (PVP–Sucralose) Nanohybrids for Rapid Drug Delivery

Xinkuan Liu [1,†], Mingxin Zhang [1,†], Wenliang Song [1], Yu Zhang [2], Deng-Guang Yu [1,*] and Yanbo Liu [3,*]

1. School of Materials & Chemistry, University of Shanghai for Science and Technology, Shanghai 200093, China; xinkuanliu@usst.edu.cn (X.L.); 203613006@st.usst.edu.cn (M.Z.); wenliang.song@usst.edu.cn (W.S.)
2. School of Pharmacy, Shanghai University of Medicine & Health Sciences, Shanghai 201318, China; zhangy_21@sumhs.edu.cn
3. School of Textile Science and Engineering, Wuhan Textile University, Wuhan 430200, China
* Correspondence: ydg017@usst.edu.cn (D.-G.Y.); yanboliu@gmail.com (Y.L.)
† These authors contributed equally to this work.

**Abstract:** The gels of cellulose and its derivatives have a broad and deep application in pharmaceutics; however, limited attention has been paid to the influences of other additives on the gelation processes and their functional performances. In this study, a new type of electrospun core–shell nanohybrid was fabricated using modified, coaxial electrospinning which contained composites of hydroxypropyl methyl cellulose (HPMC) and acetaminophen (AAP) in the core sections and composites of PVP and sucralose in the shell sections. A series of characterizations demonstrated that the core–shell hybrids had linear morphology with clear core–shell nanostructures, and AAP and sucralose distributed in the core and shell section in an amorphous state separately due to favorable secondary interactions such as hydrogen bonding. Compared with the electrospun HPMC–AAP nanocomposites from single-fluid electrospinning of the core fluid, the core–shell nanohybrids were able to promote the water absorbance and HMPC gelation formation processes, which, in turn, ensured a faster release of AAP for potential orodispersible drug delivery applications. The mechanisms of the drug released from these nanofibers were demonstrated to be a combination of erosion and diffusion mechanisms. The presented protocols pave a way to adjust the properties of electrospun, cellulose-based, fibrous gels for better functional applications.

**Keywords:** HPMC; coaxial electrospinning; core–shell nanohybrids; orodispersible drug delivery; fast dissolution; poorly water-soluble drug

## 1. Introduction

Cellulose, as one of the most abundant natural resources, has been exploited in the field of drug delivery for many years [1–6]. Particularly, it has a wide variety of derivatives, which have different chemical and physical properties for different drug controlled-release performances [7–11]. For example, acetate cellulose, an insoluble derivative, is frequently utilized to provide a sustained drug release profile [12–14]. In sharp contrast, HPMC, as a water-soluble drug, is a popular drug carrier for ensuring the fast dissolution of poorly water-soluble drugs [15–18]. Meanwhile, these derivatives can be combined with other synthetic polymers to offer multiple-phase drug controlled-release profiles, such as the typical double-phase release that includes an initial, immediate release for easing patient symptoms and a later, sustained release for the sake of reduced administration time [19].

However, cellulose's functional applications have not yet been fully expanded [20–26]. On the one hand, advanced techniques in science and engineering have been able to improve the functional performances of cellulose and its derivatives [27–33]. On the other hand, most cellulose derivatives are insert materials, where the active ingredients and other additives need to be encapsulated into the cellulose matrices to gain advantages such as biocompatibility, stability and ease of use [34,35].

Electrospinning, as an electrohydrodynamic technique, is broadly exploited to treat cellulose and its derivatives for biomedical and other applications [36–38]. There are numerous research publications and many review articles about this topic [39,40]. The popularity of this interdisciplinary field is due to the usefulness of cellulose derivatives and the ease of the preparation of nanofibers using electrospinning [41–46]. A single-step and straightforward operation of electrospinning on a co-dissolving solution of a cellulose derivative and a drug can bring out medicated nanoproducts with unique properties for targeted applications, which have been demonstrated in many investigations [47–51]. What is more, the fast developments of electrospinning in the following three directions have further produced more and more cellulose-based, medicated products in the forms of composites or hybrids. These three directions are: (1) the double-fluid coaxial, side-by-side and tri-fluid complex processes, by which core–shell, Janus and their combined nanostructures of cellulose can be easily prepared directly and robustly [52–55]; (2) approaching nanofibers on a large scale using the non-needle processes or multiple-needle processes [56,57]; (3) the reasonable combinations of electrospinning and other traditional pharmaceutical methods and, also, the combinations of cellulose derivatives and other excipients [58–61].

In this study, with a common cellulose derivative, i.e., hydroxypropyl methyl cellulose (HPMC) as the filament-forming matrix and also the drug carrier, its homogeneous electrospun nanocomposite (containing APP from a single-fluid blending process) and its heterogeneous core–shell nanohybrids (having an additional PVP–sucralose shell from a modified, coaxial process) were fabricated and characterized in parallel. A well-known poorly water-soluble drug, acetaminophen (AAP), was explored as the model's active ingredient [62,63]. The results show that an additional PVP–sucralose shell effectively promotes the fast gelation of HMPC and also the dissolution of AAP, as well as masking the poor taste of AAP for potential oral administration.

## 2. Results and Discussion

### 2.1. The Two Different Working Processes from the Same Apparatus

In an electrospinning system, the spinneret is the most vital component [64–66]. This is not only because the electrohydrodynamic process happens around the working fluids and high voltage convergent here, but also because its macrostructures directly determine the nanostructures of the final products and, often, the categories of the electrospinning processes. Figure 1a is a diagram of the elements of a typical, coaxial electrospinning apparatus. The power supply is utilized to provide the electrical energy, the two pumps are exploited to drive the shell and core working fluids quantitatively, the collector is used to collect nanofibers and the spinneret is used to guide the two working fluids in the electrical field in an organized manner. For safety, all the elements must be grounded.

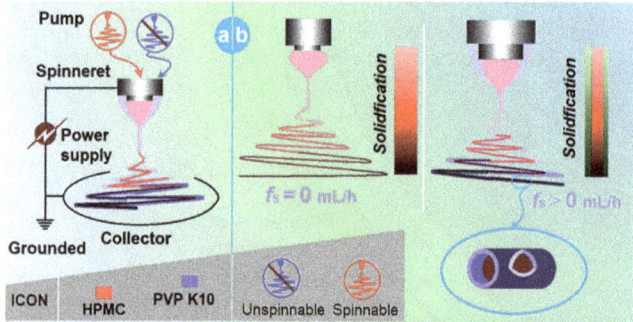

**Figure 1.** A diagram showing the coaxial apparatus that was explored for both a single-fluid blending process and a modified, coaxial process: (**a**) the parts of the electrospinning apparatus; (**b**) the two processes with different sheath fluid flow rates ($f_s$).

In the traditional sense, the concentric spinneret is utilized to prepare the core–shell nanofibers [67,68]. However, in the coaxial system, when the shell fluid flow rate ($f_s$) is adjusted to 0 mL/h, the process is degraded to a single-fluid blending process of the core fluid, and the final products are, correspondingly, homogeneous nanocomposites. In contrast, when $f_s$ is larger than 0 mL/h, the final products are core–shell nanohybrids (Figure 1b). The real implementations of the electrospinning processes are presented in Figure 2. Figure 2a is a whole view of the coaxial electrospinning apparatus which was used for preparing the nanohybrids F2. The collected film showed a slightly pink color. In contrast, the film in the upper-right corner showed a purplish red. Figure 2b shows that the spinneret was easy to connect with two pumps, one by a syringe containing the shell fluid and the other by a highly elastic silicon tubing. An alligator clip was used to transfer the electrical energy to the working fluids. Figure 2c–f shows digital photos of the typical working processes. Figure 2c,d shows the processes of single-fluid and coaxial blending, respectively. The Taylor cone was observed using a camera under a magnification of 12×. For the single-fluid electrospinning of the core fluid, the Taylor cone, which showed a purplish red color, was as shown in Figure 2e. The typical compound Taylor cone when the shell fluid was pumped was as shown in Figure 2f; the purplish core fluid was surrounded by transparent shell fluid. Within the compound Taylor cone, the shell and core fluids were clearly separated into their own colors. However, the collected films of core–shell nanofibers still showed a pink color, as in Figure 2a. Although this color was significantly paler than that of nanofibers F1 from the single-fluid of core liquid shown in the upper-right inset of Figure 2a, it gives a hint that some basic fuchsin in the core solutions might escape to the shell section during the fast drying processes of electrospinning. In the present experiments, both the solutes in the core and shell layers dissolved in the other layers. This means that very small interfacial tensions existed between the shell and core working fluids, which benefit a stable and robust continuous preparation of core–shell nanofibers, although a little diffusion may occur during the bending and whipping processes.

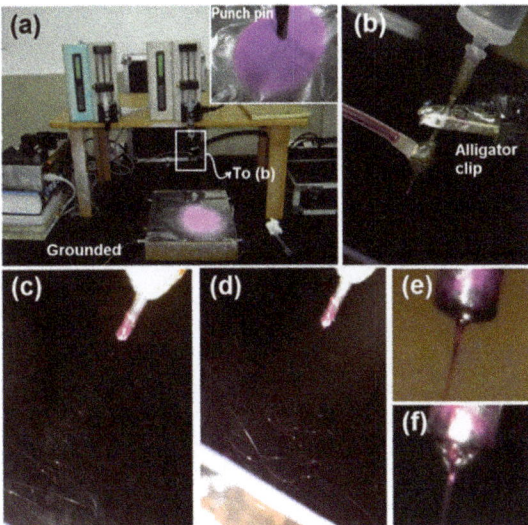

**Figure 2.** Digital photos exhibiting the preparation processes: (**a**) a whole image of the electrospinning apparatus, the upper-right inset showing the collected nanofibers films prepared by the single-fluid blending process of the core fluid; (**b**) the connection of the spinneret with the two working fluids and the power supply; (**c**) a typical process for creating HPMC-based nanocomposites solely from the core fluid; (**d**) a typical process for producing the core–shell hybrids from the modified coaxial processes; (**e**,**f**) Taylor cones for preparing monolithic composites and core–shell nanohybrids, respectively.

## 2.2. The Morphologies and Inner Structure of the Nanofibers

The SEM images of the electrospun nanocomposites F1 and the core–shell nanohybrids are included in Figure 3. Figure 3a,b shows the surface and cross-section morphology of the nanocomposites F1 from the single-fluid electrospinning, respectively. It is obvious that these nanofibers were linear without any discerned beads or spindles, suggesting the fine electrospinnability of the core HPMC working fluid [69,70]. These nanofibers had an average diameter of $680 \pm 120$ nm, which was estimated by calculating the mean values of over 50 points in the SEM images. The upper-right inset of Figure 3b shows an enlarged image of the cross-section of the homogeneous composites F1. No phase separations were observed for either the cross-sections or the surfaces of the nanofibers, giving a hint that the drug AAP presented in the composites F1 in a homogeneous manner with HPMC, most probably on a molecular level.

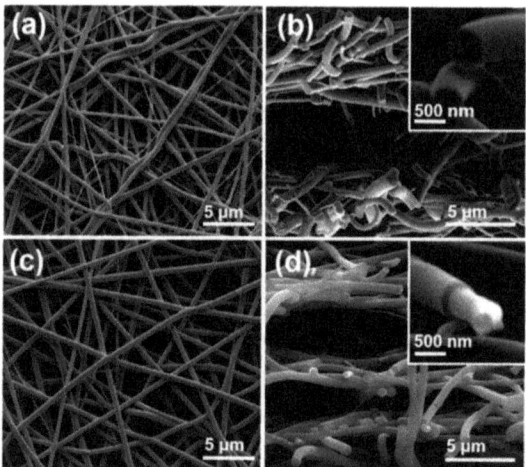

**Figure 3.** SEM images of the generated nanofibers: (**a,b**) the surface and cross-section morphology of the nanocomposites from the single-fluid electrospinning, respectively, the upper-right inset showing an enlarged image for the homogeneous composites; (**c,d**) the surface and cross-section morphology of the core–shell nanohybrids from the coaxial electrospinning, respectively, the upper-right inset showing an enlarged image for the heterogeneous hybrids.

Figure 3c,d are the SEM images of the surface and cross-section of the core–shell nanohybrids F2 from the coaxial electrospinning, respectively. Although the shell fluid PVP solution had no electrospinnability, straight linear nanofibers could still be robustly prepared. They had an average diameter of $730 \pm 80$ nm by estimation. The upper-right inset of the figure shows an enlarged image of the heterogeneous hybrids, in which a core–shell structure was observed. This is closely related to the different mechanical properties of the shell and core sections. There were no spindles or beads on the surface of the hybrids or their cross-sections.

The inner structures of the nanocomposites F1 and the nanohybrids F2 were detected using TEM. The images are exhibited in Figure 4. The nanocomposites F1 showed a gradually reduced gray level from the center of the nanofiber to its boundaries (Figure 4a). It was the thicknesses that made this difference, giving a hint that the different parts of nanofibers F1 had no differences in their elements and density and also no solid phase separation, demonstrating that it was a homogeneous nanocomposite. In sharp contrast, the core–shell hybrids F2 had a significant difference between the core section and the shell section, as shown by the two parallel nanofibers in Figure 4b. By estimation, the core section had a diameter of $540 \pm 110$ nm and a shell thickness of about 90 nm.

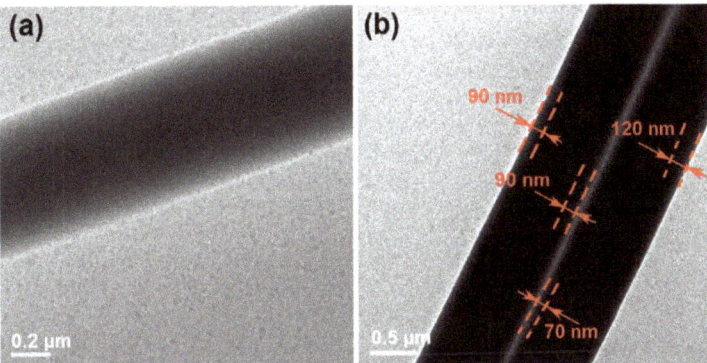

**Figure 4.** TEM images of the generated nanofibers: (**a**) the nanocomposites from the single-fluid electrospinning; (**b**) the core–shell nanohybrids from the coaxial electrospinning process.

### 2.3. The Physical Forms of Components and Their Compatibility

The measured XRD patterns of the starting raw materials, HPMC, ATP, PVP and sucralose, and their electrospun nanocomposites F1 and core–shell nanohybrids are shown in Figure 5. Just as anticipated, the drug AAP and sucralose had many sharp peaks in their patterns, suggesting that they were crystalline materials originally. HPMC and PVP had no sharp peaks but halos, indicating that they were amorphous polymers. Both the nanocomposites F1 and core–shell nanohybrids had no sharp peaks of the AAP and sucralose, demonstrating that they were converted into amorphous, polymer-based composites. The nanofibers F1 were homogeneous composites containing HMPC and AAP. The core and shell sections of the core–shell nanofibers F2 were composed of homogeneous core composites of HMPC and AAP and a homogeneous shell composite of PVP and sucralose, in general, a hybrid of the core section and shell section.

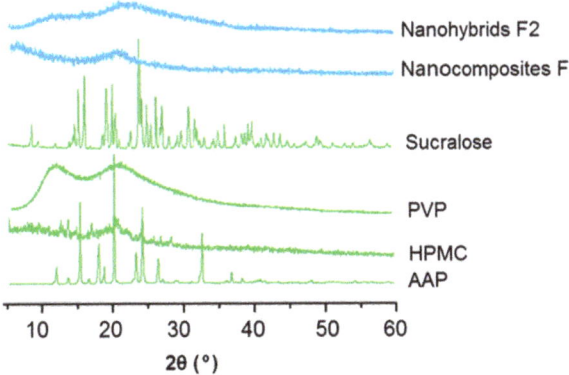

**Figure 5.** XRD measured results of the starting raw materials (HPMC, ATP, PVP and sucralose) and their different electrospun products.

The ATR-FTIR spectra of the starting raw materials (HPMC, ATP, PVP and sucralose) and their different electrospun products are included in Figure 6. The comparisons of sucralose and AAP spectra with their composites F1 and nanohybrids F2 showed that almost all the sharp peaks in the spectra of the raw materials greatly decreased their intensities or totally disappeared from the spectra of the electrospun products. For example, the characteristic peaks at 1655 cm$^{-1}$ for –C=O and at 1611, 1565 and 1507 cm$^{-1}$ for the benzene ring in the spectra of AAP could not be discerned in the spectra of either the nanocomposites F1 or nanohybrids F2. This showed the close relationship between the secondary interactions of the host polymeric carrier and the guest active ingredient. As shown by their molecular formulae, a sucralose molecule has five –OH groups, whereas, a PVP molecule has many –C=O groups; thus, hydrogen bonding is easy between them and favors the stability of the composites in the shell sections. Similarly, an HPMC molecule has many –OH group, whereas an AAP molecule has a –C=O group, suggesting that the possible hydrogen bonding between them could benefit the stability of nanocomposites F1 and the core section of the core–shell nanohybrids. However, the spectra of nanohybrids F2 had sharp peaks of 1655 cm$^{-1}$ and 957 cm$^{-1}$, which are characteristic peaks of PVP and HPMC. This phenomenon clearly suggests that PVP and HPMC were organized in the core–shell nanofibers in a separate manner, with each having their own region, i.e., a typical hybrid material.

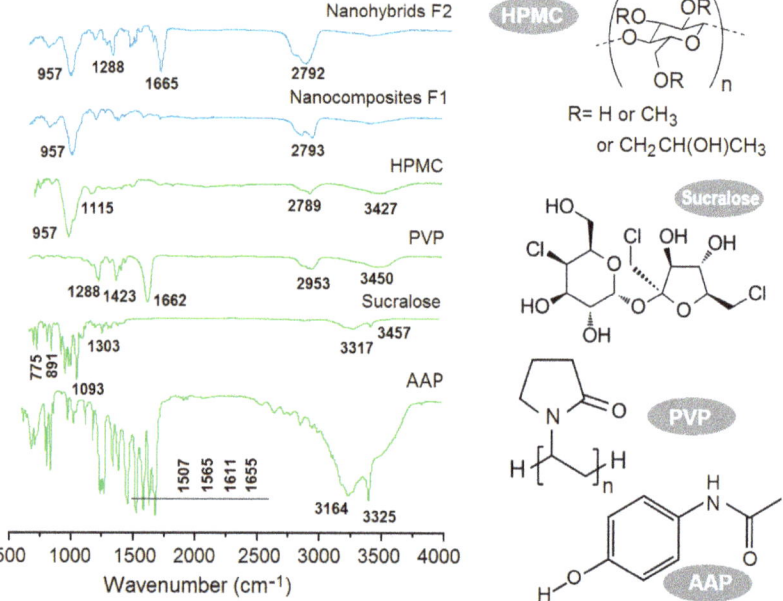

**Figure 6.** ATR-FTIR spectra of the starting raw materials (HPMC, ATP, PVP and sucralose), their different electrospun products and their molecular formulae.

### 2.4. The Hydrophilic Properties

Both HPMC and PVP are water-soluble, polymeric excipients broadly utilized in the pharmaceutical industry [71,72]. However, their dissolution behaviors have differences. PVP is very hygroscopic and can be dissolved in water all at once and, thus, is reported to promote the dissolution of nearly 200 poorly water-soluble drugs [73,74]. In contrast, HPMC is a hydrophilic polysaccharide, which contains partly O-(2-hydroxypropylated) and partly O-methylated cellulose. HPMC shows adjustable solubility based on the degree of substitution [75,76]. In this study, the HPMC was able to dissolve in water by forming a viscous, colloidal solution.

Shown in Figure 7 are the performances of nanocomposites F1 and core–shell nanofibers F2. When a drop of water (3 μL) was placed on their surfaces, the recession processes showed significant differences. In 1 s, the water contact angles for F1 (Figure 7a) and F2 (Figure 7b) were 11 and 6 degrees, respectively. After 3 s, the water droplet totally disappeared from the surface of F2, but the angle for the HPMC–AAP nanocomposites still remained at 8 degrees. The PVP–sucralose shell layer increased the hydrophilicity of the fibrous films.

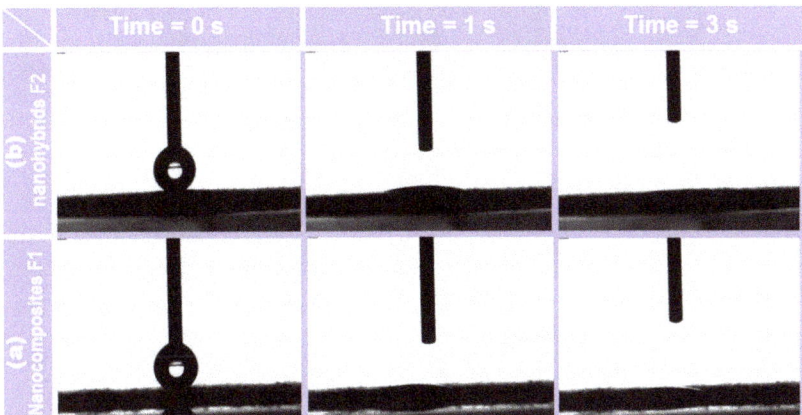

**Figure 7.** Water contact angle experimental results: (**a**) electrospun core–shell nanohybrids; (**b**) electrospun nanocomposites.

A punch pin with an inner hole of 10 mm, as shown in the upper-right inset of Figure 2a, was utilized to cut circles from the electrospun films. These films were placed on the surface of wet paper (mimicking the tongue). The behaviors of the cut circles caused by the electrospun nanocomposites F1 and core–shell nanohybrids F2 are shown in Figure 8. The nanocomposites F1 showed typical water absorbance and gradual gelation processes. The purplish color was deepened, but the circle showed no significant enlargement. Meanwhile, the purplish color was always within the residues of circle, suggesting that the HPMC gels were able to hold the basic fuchsin well.

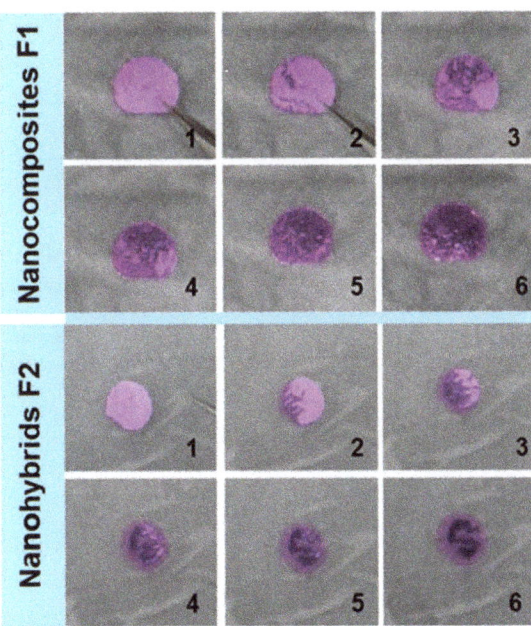

**Figure 8.** Fast disintegrating experimental results on the wet paper: the electrospun HPMC-based nanocomposites and the electrospun core–shell nanohybrids.

The core–shell nanohybrids F2 exhibited different behaviors to nanocomposites F1 in the following aspects: (1) the circle was slightly enlarged, but the pink color was light, which was attributed to the diffusion of basic fuchsin dissolved from the PVP–sucralose shell sections; (2) the colors at different regions were different—some showed a slightly pink color and some showed a deep purplish color, which was closely related to the core–shell nanostructures; (3) the deep purplish color section self-assembled into a strange shape, which was attributed to the movement and aggregation of the core HPMC molecules during the gelation processes and further indicated that the HMPC molecules had a strong capability of holding the basic fuchsin.

## 2.5. The In Vitro Drug Release Profiles and the Mechanisms

The in vitro drug release profiles of the electrospun nanocomposites F1 and the core–shell nanohybrids F2 are included in Figure 9. In general, the core–shell nanohybrids F2 were able to provide a faster release effect than the nanocomposites F1. This judgement was made based on the following aspects: (1) after five minutes, nanofibers F1 and F2 released 28.3 ± 4.3% and 34.8 ± 3.5% of the loaded AAP, respectively; and (2) 35.6 and 23.9 min were needed to release 90% of the loaded AAP for nanocomposites F1 and nanohybrids, respectively.

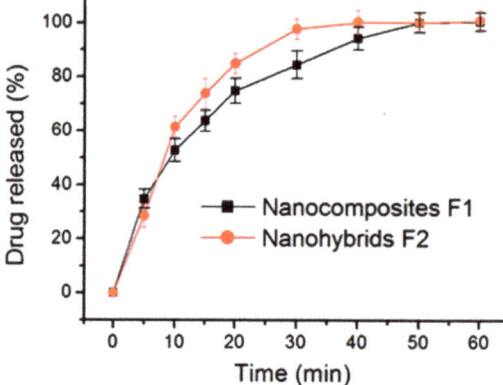

**Figure 9.** The in vitro dissolution tests of nanocomposites F1 and core–shell nanohybrids.

HPMC is frequently utilized both in the pharmaceutical industry and in scientific research as a film coating agent, thickening agent, drug release modifier, drug stabilizer, table binder and suspending ingredient in some liquid dosage forms for oral administration [77,78]. It is well known that the drug release mechanisms of HPMC-based drug delivery systems are often complicated. Various dynamic processes are active during the course of the gelation, diffusion and dissolution processes. These processes are often closely related to the viscosity of HPMC. Higher HPMC often means less erosion and corresponding, longer time period of sustained release. In this study, the drug release data were treated using the famous Peppas equation (Equation (1)) [79]:

$$P = Q_t/Q_0 = k \times t^n \qquad (1)$$

in which $Q_t$ and $Q_0$ represent the drug released into the dissolution media from its carriers at time point (t), k and n are two constants and P represents accumulative drug release percentage. The drug release mechanisms can be judged by the value of n. It is common knowledge that an n value smaller than 0.45 indicates a diffusion mechanism, a value larger than 0.90 indicates an erosion mechanism and a value between 0.45 and 0.90 represents a complex mechanism involving both diffusion and erosion.

The regressed equation for the electrospun nanofibers is included in Figure 10. For the electrospun nanocomposites F1, the equation is:

$$\log(P) = 1.23 + 0.48 \log(t) \text{ or } P = 16.98\, t^{0.48} (R = 0.9925) \quad (2)$$

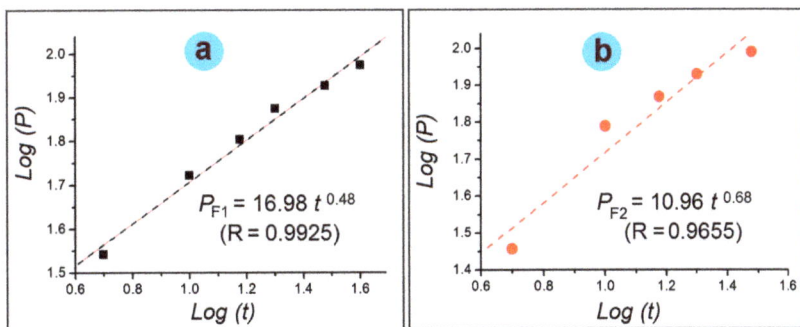

**Figure 10.** The drug release mechanism judgement: (**a**) the electrospun nanocomposites F1; (**b**) the electrospun core–shell nanohybrids F2.

For the electrospun core–shell nanohybrids F2, the equation is:

$$\log(P) = 1.04 + 0.68 \log(t) \text{ or } P = 10.96\, t^{0.68} (R = 0.9655) \quad (3)$$

Thus, it is clear that AAP from the electrospun nanocomposites F1 and the core–shell nanohybrids are all a combination of erosion and diffusion mechanisms because of an n value between 0.45 and 0.90.

### 2.6. The Mechanism of the Influence of Shell PVP on the Gelatin of Core-Medicated HPMC

Although the drug release behaviors from the two types of electrospun nanoproduct involved both diffusion and erosion mechanisms, it is obvious that the core–shell nanohybrids F2 were more closed to the erosion mechanism due to a value of 0.68 compared to 0.48 for the nanocomposites F1. A schematic showing the different behaviors of the monolithic nanocomposites F1 and core–shell nanohybrids F2 is given in Figure 11. The replacement of a surface PVP coating in the core–shell nanohybrids F2 adds the benefit of faster initial absorbance of water than monolithic HPMC–AAP nanofibers because of the highly solubility and strong hygroscopicity [73,74]. This case further highlights the easier swelling, gelation and dissolution of HPMC molecules from the core sections. Certainly, the relatively small diameter of the core section in the core–shell nanohybrids F2 compared to the nanocomposites F1 played a role in promoting the faster release of AAP from the nanofibers F2. For potential orodispersible drug delivery, the faster release of the drug, the better it is for the patients. Thus, a shell coating of PVP and sucralose makes the electrospun core–shell hybrids a more welcome product than the electrospun nanocomposites F1. Incidentally, HPMC and PVP are both tasteless and odorless excipients, and, thus, they have broad applications in traditional compressed tablets and medicated films. However, the drug AAP has a bitter taste, which reduces the patients' compliance. The addition of sucralose with PVP in the shell section endows the core–shell nanohybrids with a favorable taste, and, in turn, could improve the patients' compliance when exploited as oral, disintegrating films. Using the concept demonstrated here, many new, medicated materials can be further developed through the combination of cellulose-based gels and other types of polymer, e.g., biodegradable PLGA [80]. Particularly, those soluble polymers with a natural source will be able to play a more and more important role in developing new orodispersible drug delivery systems. This is because these polymers often have fine biocompatibility, are non-toxic and have easy processability, and some of them are currently popular in traditional orodispersible tablets [81,82]. Certainly, the coaxial electrospinning, side-by-side electrospinning and also the single-fluid blending processes can be combined with traditional methods (such as casting films) for treating cellulose-based gels to offer sustained release and multiple-phase release profiles [83–85].

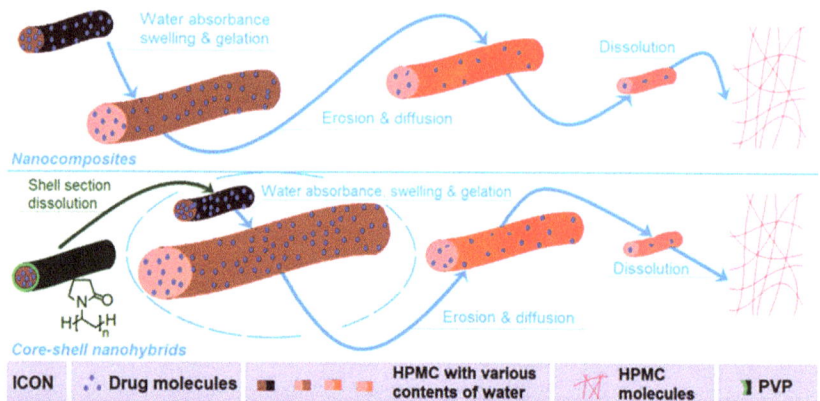

Figure 11. Schematics showing the different behaviors of the monolithic nanocomposites F1 and core–shell nanohybrids F2.

## 3. Conclusions

With HPMC as a key filament-forming polymeric matrix, both single-fluid blending electrospinning and modified, coaxial electrospinning were implemented to preparing HPMC–AAP monolithic nanocomposites F1 and core (HPMC–AAP)–shell (PVP–sucralose) heterogeneous nanohybrids F2. Both F1 and F2 had linear morphologies without any discerned beads of spindles on them, as demonstrated by the SEM images. TEM images verified that F1 were homogeneous nanocomposites, and F2 were core–shell nanohybrids containing two layers of composites. XRD results suggested that AAP presented in F1 and F2 in an amorphous state. This was attributed to the favorable interactions between APP and HPMC, which were demonstrated by ATR-FTIR measurements. Water contact angle experiments and tongue-mimicking tests clearly demonstrated the differences between F1 and F2 in their hydrophilicity and gelation processes. In vitro dissolution tests demonstrated that the core–shell hybrids F2 were able to offer a faster release of the loaded AAP. The drug release was demonstrated to be controlled by a complex combination mechanism involving both diffusion and erosion. The present study shows a new way of adjusting the properties of electrospun, cellulose-based nanofibers for better drug delivery applications.

## 4. Materials and Methods

### 4.1. Materials

Acetaminophen AAP (white powders, purity 99.8%) was obtained from Hua-Shi Big Pharmacy (Shanghai, China). Hydroxypropyl methyl cellulose (HPMC, white powders, 2910, 5 cps, $M_n$ = 428,000 g/mol, methoxy content 28.0–30.0%, hydroxypropoxy content 7.5–12.0%) was purchased from Shandong Fine Chemical Co., Ltd. (Jinan, China). Sucralose, polyvinylpyrrolidone (PVP K10, $M_w$ = 8000) dichloromethane (DCM), anhydrous ethanol and basic fuchsin were bought from Shanghai Chemical Regents Co., Ltd. (Shanghai, China). All other chemicals were analytical reagents. Water was doubly distilled before usage.

### 4.2. Preparations

The electrospinning instrument was homemade and comprised four typical parts: one power supply, two fluid drivers, one collector and one spinneret. After some pre-experiments, the working fluids and experimental conditions were fixed as follows. The core fluid contained 8.0 g HPMC and 2.0 g AAP in a mixture of DCM and anhydrous ethanol with a volume ratio of 1:1. A $10^{-3}$ mg/mL amount of basic fuchsin was added into the core fluids for optimizing the working processes and also exhibiting the disintegrating measurement. The shell fluid contained 10.0 g PVP K30 and 2.0 g sucralose in anhydrous ethanol.

Two types of electrospun nanofiber were fabricated. One was the monolithic, fibrous nanocomposite (denoted as F1) from single-fluid electrospinning of the core liquid; the other was the core–shell fibrous nanohybrid (denoted as F2) from modified, coaxial electrospinning. The applied voltages were adjusted to ensure continuous spinning and that, meanwhile, no droplets were dropped during the working processes; the values were between 11 and 14 kV. The core fluid flow rate ($f_c$) was fixed at 2.0 mL/h. The sheath fluid flow rate ($f_s$) was 0.0 mL/h and 0.5 mL/h for generating F1 and F2, respectively. The ambient conditions were a temperature of $21 \pm 4\,°C$ and a relatively humidity of $52 \pm 5\%$.

*4.3. Characterizations*

4.3.1. Morphologies and Inner Structures

The morphologies of electrospun nanofibers were assessed using scanning electron microscope (SEM, Quanta FEG450, FEI Corporation, Hillsboro, OR, USA). The cross-sections of fibers were prepared by manually breaking the fibrous strip after immersion in liquid nitrogen for about 20 min. The inner structures were evaluated using transmission electron microscope (TEM, JEM2200F, JEOL, Tokyo, Japan) under an applied electron voltage of 300 kV.

4.3.2. Physical Forms and Compatibility

All the raw materials (HPMC, PVP, sucralose and AAP) and their electrospun products F1 and F2 experienced X-ray diffraction (XRD, Bruker-AXS, Karlsruhe, Germany) tests. Attenuated total reflectance Fourier-transform infrared (ATR-FTIR, PerkinElmer, Billerica, MA, USA) was exploited to investigate the compatibility between the polymeric carriers and active ingredients.

4.3.3. Properties

Water contact angle measurements and a homemade experiment on wet paper were conducted to evaluate the fast disintegrating properties of the electrospun nanocomposites F1 and core–shell nanohybrids F2, and the gelation processes were recorded using a camera (PowerShot SX50 HS, Canon, Tokyo, Japan).

4.3.4. Functional Performances

In vitro dissolution tests were carried out using paddle methods according to the Chinese Pharmacopoeia (Ed. 2020). An RCZ-8A dissolution apparatus (Tianjin University Radio Factory, Tianjin, China) was used. Samples equivalent to 20 mg AAP were placed into the dissolution vessels in which 600 mL physiological saline (PS) was kept at a temperature of $37 \pm 1\,°C$. The paddle rotation rate was 50 rpm. A UV–vis spectrophotometer (UV-2102PC, Unico Instrument Co., Ltd., Shanghai, China) was used to measure the AAP concentration at a $\lambda_{max} = 243$ nm. All the measurements were repeated 6 times.

**Author Contributions:** Conceptualization, X.L., Y.L. and D.-G.Y.; methodology, D.-G.Y. and M.Z.; validation, D.-G.Y., W.S. and Y.Z.; formal analysis, M.Z.; investigation, X.L. and M.Z.; resources, D.-G.Y.; data curation, M.Z., W.S. and Y.Z.; writing—original draft preparation, X.L.; writing—review and editing, D.-G.Y. and Y.L.; supervision, D.-G.Y.; project administration, D.-G.Y.; funding acquisition, D.-G.Y. and Y.L. All authors have read and agreed to the published version of the manuscript.

**Funding:** The Natural Science Foundation of Shanghai (20ZR1439000), the National Natural Science Foundation of China (51973168) and the Innovation project of USST students (nos. SH2022225 & 2022229).

**Institutional Review Board Statement:** Not applicable.

**Informed Consent Statement:** Not applicable.

**Data Availability Statement:** The data supporting the findings of this manuscript are available from the corresponding authors upon reasonable request.

**Conflicts of Interest:** The authors declare no conflict of interest.

## References

1. Madaghiele, M.; Demitri, C.; Surano, I.; Silvestri, A.; Vitale, M.; Panteca, E.; Zohar, Y.; Rescigno, M.; Sannino, A. Biomimetic Cellulose-Based Superabsorbent Hydrogels for Treating Obesity. *Sci. Rep.* **2021**, *11*, 21394. [CrossRef]
2. Bonetti, L.; De Nardo, L.; Farè, S. Chemically Crosslinked Methylcellulose Substrates for Cell Sheet Engineering. *Gels* **2021**, *7*, 141. [CrossRef] [PubMed]
3. Gallo Stampino, P.; Riva, L.; Punta, C.; Elegir, G.; Bussini, D.; Dotelli, G. Comparative Life Cycle Assessment of Cellulose Nanofibres Production Routes from Virgin and Recycled Raw Materials. *Molecules* **2021**, *26*, 2558. [CrossRef] [PubMed]
4. Yang, J.; Liu, D.; Song, X.; Zhao, Y.; Wang, Y.; Rao, L.; Fu, L.; Wang, Z.; Yang, X.; Li, Y.; et al. Recent Progress of Cellulose-Based Hydrogel Photocatalysts and Their Applications. *Gels* **2022**, *8*, 270. [CrossRef]
5. Gebeyehu, E.K.; Sui, X.; Adamu, B.F.; Beyene, K.A.; Tadesse, M.G. Cellulosic-Based Conductive Hydrogels for Electro-Active Tissues: A Review Summary. *Gels* **2022**, *8*, 140. [CrossRef]
6. Riva, L.; Fiorati, A.; Punta, C. Synthesis and Application of Cellulose-Polyethyleneimine Composites and Nanocomposites: A Concise Review. *Materials* **2021**, *14*, 473. [CrossRef] [PubMed]
7. Bonetti, L.; Fiorati, A.; D'Agostino, A.; Pelacani, C.M.; Chiesa, R.; Farè, S.; De Nardo, L. Smart Methylcellulose Hydrogels for PH-Triggered Delivery of Silver Nanoparticles. *Gels* **2022**, *8*, 298. [CrossRef]
8. Jo, C.; Kim, S.S.; Rukmanikrishnan, B.; Ramalingam, S.; Prabakaran, D.S.; Lee, J. Properties of Cellulose Pulp and Polyurethane Composite Films Fabricated with Curcumin by Using NMMO Ionic Liquid. *Gels* **2022**, *8*, 248. [CrossRef] [PubMed]
9. Feng, X.; Hao, J. Identifying New Pathways and Targets for Wound Healing and Therapeutics from Natural Sources. *Curr. Drug Deliv.* **2021**, *18*, 1064–1084. [CrossRef]
10. Park, K.; Choi, H.; Kang, K.; Shin, M.; Son, D. Soft Stretchable Conductive Carboxymethylcellulose Hydrogels for Wearable Sensors. *Gels* **2022**, *8*, 92. [CrossRef]
11. Barba, A.A.; d'Amore, M.; Chirico, S.; Lamberti, G.; Titomanlio, G. Swelling of cellulose derivative (HPMC) matrix systems for drug delivery. *Carbohydr. Polym.* **2009**, *78*, 469–474. [CrossRef]
12. Wang, M.; Hou, J.; Yu, D.-G.; Li, S.; Zhu, J.; Chen, Z. Electrospun Tri-Layer Nanodepots for Sustained Release of Acyclovir. *J. Alloys Compd.* **2020**, *846*, 156471. [CrossRef]
13. Huang, C.; Dong, J.; Zhang, Y.; Chai, S.; Wang, X.; Kang, S.; Yu, D.; Wang, P.; Jiang, Q. Gold Nanoparticles-Loaded Polyvinylpyrrolidone/Ethylcellulose Coaxial Electrospun Nanofibers with Enhanced Osteogenic Capability for Bone Tissue Regeneration. *Mater. Des.* **2021**, *212*, 110240. [CrossRef]
14. Ghosal, K.; Augustine, R.; Zaszczynska, A.; Barman, M.; Jain, A.; Hasan, A.; Kalarikkal, N.; Sajkiewicz, P.; Thomas, S. Novel Drug Delivery Systems Based on Triaxial Electrospinning Based Nanofibers. *React. Funct. Polym.* **2021**, *163*, 104895. [CrossRef]
15. Huang, C.; Dong, H.; Zhang, Z.; Bian, H.; Yong, Q. Procuring the Nano-Scale Lignin in Prehydrolyzate as Ingredient to Prepare Cellulose Nanofibril Composite Film with Multiple Functions. *Cellulose* **2020**, *27*, 9355–9370. [CrossRef]
16. Araujo, R.; Soares, M.; Mazzei, J.L.; Ramos, M.; Siani, A.C. A Comparative Study of Hard Gelatin and Hypromellose Capsules Containing a Dry Extract of Senna (Cassia Angustifolia) under Controlled Temperature and Relative Humidity. *Indian J. Pharm. Sci.* **2021**, *82*, 718–723. [CrossRef]
17. Khalid, G.M.; Selmin, F.; Musazzi, U.M.; Gennari, C.G.M.; Minghetti, P.; Cilurzo, F. Trends in the Characterization Methods of Orodispersible Films. *Curr. Drug Deliv.* **2021**, *18*, 941–952. [CrossRef]
18. Ortega, C.A.; Favier, L.S.; Cianchino, V.A.; Cifuente, D.A. New Orodispersible Mini Tablets of Enalapril Maleate by Direct Compression for Pediatric Patients. *Curr. Drug Deliv.* **2020**, *17*, 505–510. [CrossRef]
19. He, H.; Wu, M.; Zhu, J.; Yang, Y.; Ge, R.; Yu, D.-G. Engineered Spindles of Little Molecules Around Electrospun Nanofibers for Biphasic Drug Release. *Adv. Fiber Mater.* **2022**, *4*, 305–317. [CrossRef]
20. Yu, D.-G. Preface-Bettering Drug Delivery Knowledge from Pharmaceutical Techniques and Excipients. *Curr. Drug Deliv.* **2021**, *18*, 2–3. [CrossRef]
21. Song, X.; Jiang, Y.; Zhang, W.; Elfawal, G.; Wang, K.; Jiang, D.; Hong, H.; Wu, J.; He, C.; Mo, X.; et al. Transcutaneous Tumor Vaccination Combined with Anti-Programmed Death-1 Monoclonal Antibody Treatment Produces a Synergistic Antitumor Effect. *Acta Biomater.* **2022**, *140*, 247–260. [CrossRef]
22. Zhang, Y.; Li, S.; Xu, Y.; Shi, X.; Zhang, M.; Huang, Y.; Liang, Y.; Chen, Y.; Ji, W.; Kim, J.R.; et al. Engineering of Hollow Polymeric Nanosphere-Supported Imidazolium-Based Ionic Liquids with Enhanced Antimicrobial Activities. *Nano Res.* **2022**, *15*, 5556–5568. [CrossRef]
23. Kose, M.D.; Ungun, N.; Bayraktar, O. Eggshell Membranebased Turmeric Extract Loaded Orally Disintegrating Films. *Curr. Drug Deliv.* **2021**, *18*, 547–559. [CrossRef]
24. Verma, N.; Tiwari, A.; Bajpai, J.; Bajpai, A.K. Swelling Triggered Release of Cisplatin from Gelatin Coated Gold Nanoparticles. *Inorg. Nano-Met. Chem.* **2022**, *52*, 1–13. [CrossRef]

25. Sultana, M.; Sultana, S.; Hussain, K.; Saeed, T.; Butt, M.A.; Raza, S.A.; Mahmood, R.; Hassan, S.; Anwer, U.U.; Bukhari, N.I. Enhanced Mefenamic Acid Release from Poloxamer-Silicon Dioxide Gel Filled in Hard Gelatin Capsules—An Application of Liquid Semisolid Matrix Technology for Insoluble. *Curr. Drug Deliv.* **2021**, *18*, 801–811. [CrossRef]
26. Yang, H.; Lan, X.; Xiong, Y. In Situ Growth of Zeolitic Imidazolate Framework-L in Macroporous PVA/CMC/PEG Composite Hydrogels with Synergistic Antibacterial and Rapid Hemostatic Functions for Wound Dressing. *Gels* **2022**, *8*, 279. [CrossRef]
27. Esim, O.; Hascicek, C. Lipid-Coated Nanosized Drug Delivery Systems for an Effective Cancer Therapy. *Curr. Drug Deliv.* **2021**, *18*, 147–161. [CrossRef] [PubMed]
28. Ejeta, F.; Gabriel, T.; Joseph, N.M.; Belete, A. Formulation, Optimization and In Vitro Evaluation of Fast Disintegrating Tablets of Salbutamol Sulphate Using a Combination of Superdisintegrant and Subliming Agent. *Curr. Drug Deliv.* **2021**, *19*, 129–141. [CrossRef]
29. Bigogno, E.R.; Soares, L.; Mews, M.H.R.; Zétola, M.; Bazzo, G.C.; Stulzer, H.K.; Pezzini, B.R. It Is Possible to Achieve Tablets with Good Tabletability from Solid Dispersions-The Case of the High Dose Drug Gemfibrozil. *Curr. Drug Deliv.* **2021**, *18*, 460–470. [CrossRef] [PubMed]
30. Li, J.K.; Guan, S.M.; Su, J.J.; Liang, J.H.; Cui, L.L.; Zhang, K. The Development of Hyaluronic Acids Used for Skin Tissue Regeneration. *Curr. Drug Deliv.* **2021**, *18*, 836–846. [CrossRef]
31. Bakadia, B.M.; Zhong, A.; Li, X.; Boni, B.O.O.; Ahmed, A.A.Q.; Souho, T.; Zheng, R.; Shi, D.; Lamboni, L.; Yang, G. Biodegradable and Injectable Poly(Vinyl Alcohol) Microspheres in Silk Sericin Based Hydrogel for the Controlled Release of Antimicrobials: Application to Deep Full Thickness Burn Wound Healing. *Adv. Compos. Hybrid Mater.* **2022**, *5*, 1–26. [CrossRef]
32. Xu, L.; Liu, Y.; Zhou, W.; Yu, D.-G. Electrospun Medical Sutures for Wound Healing: A Review. *Polymers* **2022**, *14*, 1637. [CrossRef]
33. Yue, Y.; Liu, X.; Pang, L.; Liu, Y.; Lin, Y.; Xiang, T.; Li, J.; Liao, S.; Jiang, Y. Astragalus Polysaccharides/PVA Nanofiber Membranes Containing Astragaloside IV-Loaded Liposomes and Their Potential Use for Wound Healing. *Evid.-Based Complement. Alternat. Med.* **2022**, *2022*, 9716271. [CrossRef] [PubMed]
34. Liu, Y.; Chen, X.; Yu, D.-G.; Liu, H.; Liu, Y.; Liu, P. Electrospun PVP-Core/PHBV-Shell Fibers to Eliminate Tailing Off for an Improved Sustained Release of Curcumin. *Mol. Pharm.* **2021**, *18*, 4170–4178. [CrossRef] [PubMed]
35. Murugesan, R.; Raman, S. Recent Trends in Carbon Nanotubes Based Prostate Cancer Therapy: A Biomedical Hybrid for Diagnosis and Treatment. *Curr. Drug Deliv.* **2022**, *19*, 229–237. [CrossRef]
36. Liu, H.; Wang, H.; Lu, X.; Murugadoss, V.; Huang, M.; Yang, H.; Wan, F.; Yu, D.G.; Guo, Z. Electrospun Structural Nanohybrids Combining Three Composites for Fast Helicide Delivery. *Adv. Compos. Hybrid Mater.* **2022**, *5*. [CrossRef]
37. Riva, L.; Lotito, A.D.; Punta, C.; Sacchetti, A. Zinc- and Copper-Loaded Nanosponges from Cellulose Nanofibers Hydrogels: New Heterogeneous Catalysts for the Synthesis of Aromatic Acetals. *Gels* **2022**, *8*, 54. [CrossRef]
38. Zhao, K.; Lu, Z.-H.; Zhao, P.; Kang, S.-X.; Yang, Y.-Y.; Yu, D.-G. Modified Tri–Axial Electrospun Functional Core-Shell Nanofibrous Membranes for Natural Photodegradation of Antibiotics. *Chem. Eng. J.* **2021**, *425*, 131455. [CrossRef]
39. Liu, R.; Hou, L.; Yue, G.; Li, H.; Zhang, J.; Liu, J.; Miao, B.; Wang, N.; Bai, J.; Cui, Z.; et al. Progress of Fabrication and Applications of Electrospun Hierarchically Porous Nanofibers. *Adv. Fiber Mater.* **2022**, *4*, 1–27. [CrossRef]
40. Zhou, Y.; Liu, Y.; Zhang, M.; Feng, Z.; Yu, D.-G.; Wang, K. Electrospun Nanofiber Membranes for Air Filtration: A Review. *Nanomaterials* **2022**, *12*, 1077. [CrossRef]
41. Bukhary, H.; Williams, G.R.; Orlu, M. Fabrication of Electrospun Levodopa-Carbidopa Fixed-Dose Combinations. *Adv. Fiber Mater.* **2020**, *2*, 194–203. [CrossRef]
42. Wang, M.; Tan, Y.; Li, D.; Xu, G.; Yin, D.; Xiao, Y.; Xu, T.; Chen, X.; Zhu, X.; Shi, X. Negative Isolation of Circulating Tumor Cells Using a Microfluidic Platform Integrated with Streptavidin-Functionalized PLGA Nanofibers. *Adv. Fiber Mater.* **2021**, *3*, 192–202. [CrossRef]
43. Chen, M.; Le, T.; Zhou, Y.; Kang, F.; Yang, Y. Ultra-High Rate and Ultra-Stable Lithium Storage Enabled by Pore and Defect Engineered Carbon Nanofibers. *Electrochim. Acta* **2022**, *420*, 140429. [CrossRef]
44. El-Shanshory, A.A.; Agwa, M.M.; Abd-Elhamid, A.I.; Soliman, H.M.A.; Mo, X.; Kenawy, E.-R. Metronidazole Topically Immobilized Electrospun Nanofibrous Scaffold: Novel Secondary Intention Wound Healing Accelerator. *Polymers* **2022**, *14*, 454. [CrossRef] [PubMed]
45. Sivan, M.; Madheswaran, D.; Valtera, J.; Kostakova, E.K.; Lukas, D. Alternating Current Electrospinning: The Impacts of Various High-Voltage Signal Shapes and Frequencies on the Spinnability and Productivity of Polycaprolactone Nanofifibers. *Mater. Des.* **2022**, *213*, 110308. [CrossRef]
46. Brimo, N.; Serdaroglu, D.C.; Uysal, B. Comparing Antibiotic Pastes with Electrospun Nanofibers as Modern Drug Delivery Systems for Regenerative Endodontics. *Curr. Drug Deliv.* **2021**, *18*, 7. [CrossRef]
47. Ziyadi, H.; Baghali, M.; Bagherianfar, M.; Mehrali, F.; Faridi-Majidi, R. An Investigation of Factors Affecting the Electrospinning of Poly(Vinyl Alcohol)/Kefiran Composite Nanofibers. *Adv. Compos. Hybrid Mater.* **2021**, *4*, 768–779. [CrossRef]
48. Song, Y.; Huang, H.; He, D.; Yang, M.; Wang, H.; Zhang, H.; Li, J.; Li, Y.; Wang, C. Gallic Acid/2-Hydroxypropyl-B-Cyclodextrin Inclusion Complexes Electrospun Nanofibrous Webs: Fast Dissolution, Improved Aqueous Solubility and Antioxidant Property of Gallic Acid. *Chem. Res. Chin. Univ.* **2021**, *37*, 450–455. [CrossRef]
49. Wang, K.; Wang, X.; Jiang, D.; Pei, Y.; Wang, Z.; Zhou, X.; Wu, J.; Mo, X.; Wang, H. Delivery of Mrna Vaccines and Anti-PDL1 Sirna through Non-Invasive Transcutaneous Route Effectively Inhibits Tumor Growth. *Compos. Part B Eng.* **2022**, *233*, 109648. [CrossRef]

50. Zhang, C.; Sun, J.; Lyu, S.; Lu, Z.; Li, T.; Yang, Y.; Li, B.; Han, H.; Wu, B.; Sun, H.; et al. Poly(Lactic Acid)/Artificially Cultured Diatom Frustules Nanofibrous Membranes with Fast and Controllable Degradation Rates for Air Filtration. *Adv. Compos. Hybrid Mater.* **2022**, *5*. [CrossRef]
51. Garcia Cervantes, M.Y.; Han, L.; Kim, J.; Chitara, B.; Wymer, N.; Yan, F. N-Halamine-Decorated Electrospun Polyacrylonitrile Nanofibrous Membranes: Characterization and Antimicrobial Properties. *React. Funct. Polym.* **2021**, *168*, 105058. [CrossRef]
52. Yang, X.; Li, L.; Yang, D.; Nie, J.; Ma, G. Electrospun Core-Shell Fibrous 2D Scaffold with Biocompatible Poly (Glycerol Sebacate) and Poly-L-Lactic Acid for Wound Healing. *Adv. Fiber Mater.* **2020**, *2*, 105–117. [CrossRef]
53. Xu, X.; Zhang, M.; Lv, H.; Zhou, Y.; Yang, Y.; Yu, D.-G. Electrospun Polyacrylonitrile-Based Lace Nanostructures and Their Cu(II) Adsorption. *Sep. Purif. Technol.* **2022**, *288*, 120643. [CrossRef]
54. Ghazalian, M.; Afshar, S.; Rostami, A.; Rashedi, S.; Bahrami, S.H. Fabrication and Characterization of Chitosan-Polycaprolactone Core-Shell Nanofibers Containing Tetracycline Hydrochloride. *Colloids Surf. A* **2022**, *636*, 128163. [CrossRef]
55. Xu, H.; Zhang, F.; Wang, M.; Lv, H.; Yu, D.-G.; Liu, X.; Shen, H. Electrospun Hierarchical Structural Films for Effective Wound Healing. *Biomater. Adv.* **2022**, *136*, 212795. [CrossRef]
56. Kang, S.; Zhao, K.; Yu, D.-G.; Zheng, X.; Huang, C. Advances in Biosensing and Environmental Monitoring Based on Electrospun Nanofibers. *Adv. Fiber Mater.* **2022**, *4*, 404–435. [CrossRef]
57. Jiang, S.; Schmalz, H.G.; Agarwal, S.; Greiner, A. Electrospinning of ABS Nanofibers and Their High Filtration Performance. *Adv. Fiber Mater.* **2020**, *2*, 34–43. [CrossRef]
58. Zhang, M.; Song, W.; Tang, Y.; Xu, X.; Huang, Y.; Yu, D.-G. Polymer-Based Nanofiber-Nanoparticle Hybrids and Their Medical Applications. *Polymers* **2022**, *14*, 351. [CrossRef] [PubMed]
59. Zhan, L.; Deng, J.; Ke, Q.; Li, X.; Ouyang, Y.; Huang, C.; Liu, X.; Qian, Y. Grooved Fibers: Preparation Principles through Electrospinning and Potential Applications. *Adv. Fiber Mater.* **2021**, *3*, 203–213. [CrossRef]
60. Zhang, X.; Guo, S.; Qin, Y.; Li, C. Functional Electrospun Nanocomposites for Efficient Oxygen Reduction Reaction. *Chem. Res. Chin. Univ.* **2021**, *37*, 379–393. [CrossRef]
61. Yu, D.-G.; Lv, H. Preface-Striding into Nano Drug Delivery. *Curr. Drug Deliv.* **2022**, *19*, 1–3. [CrossRef]
62. Diaz-Guerrero, A.M.; Castillo-Miranda, C.A.; Peraza-Vazquez, H.; Morales-Cepeda, A.B.; Pena-Delgado, A.F.; Rivera-Armenta, J.L.; Castro-Guerrero, C.F. Modelling of Acetaminophen Release from Hydroxyethylcellulose/Polyacrylamide Hydrogel. *Mater. Res. Express* **2021**, *8*, 015310. [CrossRef]
63. Gaurkhede, S.G.; Osipitan, O.O.; Dromgoole, G.; Spencer, S.A.; Di Pasqua, A.J.; Denga, J. 3D Printing and Dissolution Testing of Novel Capsule Shells for Use in Delivering Acetaminophen. *J. Pharm. Sci.* **2021**, *110*, 3829–3837. [CrossRef]
64. Ning, T.; Zhou, Y.; Xu, H.; Guo, S.; Wang, K.; Yu, D.-G. Orodispersible Membranes from a Modified Coaxial Electrospinning for Fast Dissolution of Diclofenac Sodium. *Membranes* **2021**, *11*, 802. [CrossRef]
65. Lv, H.; Guo, S.; Zhang, G.; He, W.; Wu, Y.; Yu, D.-G. Electrospun Structural Hybrids of Acyclovir-Polyacrylonitrile at Acyclovir for Modifying Drug Release. *Polymers* **2021**, *13*, 4286. [CrossRef] [PubMed]
66. Yu, D.-G.; Wang, M.; Ge, R. Strategies for Sustained Drug Release from Electrospun Multi-Layer Nanostructures. *WIRES Nanomed. Nanobi.* **2022**, *14*, e1772. [CrossRef] [PubMed]
67. Liu, Y.; Chen, X.; Liu, Y.; Gao, Y.; Liu, P. Electrospun Coaxial Fibers to Optimize the Release of Poorly Water-Soluble Drug. *Polymers* **2022**, *14*, 469. [CrossRef]
68. Chen, J.; Zhang, G.; Zhao, Y.; Zhou, M.; Zhong, A.; Sun, J. Promotion of Skin Regeneration through Co-Axial Electrospun Fibers Loaded with Basic Fibroblast Growth Factor. *Adv. Compos. Hybrid Mater.* **2022**, *5*, 1–15. [CrossRef]
69. Silva, P.M.; Prieto, C.; Andrade, C.C.P.; Lagaron, J.M.; Pastrana, L.M.; Coimbra, M.A.; Vicente, A.A.; Cerqueira, M.A. Hydroxypropyl methylcellulose-based micro- and nanostructures for encapsulation of melanoidins: Effect of electrohydrodynamic processing variables on morphological and physicochemical properties. *Int. J. Biol. Macromol.* **2022**, *202*, 453–467. [CrossRef] [PubMed]
70. Jozo, M.; Simon, N.; Yi, L.; Moczo, J.; Pukanszky, B. Improved Release of a Drug with Poor Water Solubility by Using Electrospun Water-Soluble Polymers as Carriers. *Pharmaceutics* **2022**, *14*, 34. [CrossRef]
71. Guo, S.; Jiang, W.; Shen, L.; Zhang, G.; Gao, Y.; Yang, Y.Y.; Yu, D.-G. Electrospun Hybrid Films for Fast and Convenient Delivery of Active Herbs Extracts. *Membranes* **2022**, *12*, 398. [CrossRef]
72. Kang, S.; Hou, S.; Chen, X.; Yu, D.-G.; Wang, L.; Li, X.; Williams, G.R. Energy-Saving Electrospinning with a Concentric Teflon-Core Rod Spinneret to Create Medicated Nanofibers. *Polymers* **2020**, *12*, 2421. [CrossRef] [PubMed]
73. Balusamy, B.; Celebioglu, A.; Senthamizhan, A.; Uyar, T. Progress in the design and development of "fast-dissolving" electrospun nanofibers based drug delivery systems—A systematic review. *J. Control. Release* **2020**, *326*, 482–509. [CrossRef]
74. Anderson, B.D. Predicting Solubility/Miscibility in Amorphous Dispersions: It Is Time to Move Beyond Regular Solution Theories. *J. Pharm. Sci.* **2018**, *107*, 24–33. [CrossRef]
75. Vovko, A.D.; Hodzic, B.; Brec, T.; Hudovornik, G.; Vrecer, F. Influence of Formulation Factors, Process Parameters, and Selected Quality Attributes on Carvedilol Release from Roller-Compacted Hypromellose-Based Matrix Tablets. *Pharmaceutics* **2022**, *14*, 876. [CrossRef] [PubMed]
76. Maskova, E.; Kubova, K.; Raimi-Abraham, B.T.; Vllasaliu, D.; Vohlidalova, E.; Turanek, J.; Masek, J. Hypromellose—A traditional pharmaceutical excipient with modern applications in oral and oromucosal drug delivery. *J. Control. Release* **2020**, *324*, 695–727. [CrossRef]

77. Latif, M.S.; Azad, A.K.; Nawaz, A.; Rashid, S.A.; Rahman, M.H.; Al Omar, S.Y.; Bungau, S.G.; Aleya, L.; Abdel-Daim, M.M. Ethyl Cellulose and Hydroxypropyl Methyl Cellulose Blended Methotrexate-Loaded Transdermal Patches: In Vitro and Ex Vivo. *Polymers* **2021**, *13*, 3455. [CrossRef]
78. Kurniawansyah, I.S.; Rusdiana, T.; Sopyan, I.; Ramoko, H.; Wahab, H.A.; Subarnas, A. In Situ Ophthalmic Gel Forming Systems of Poloxamer 407 and Hydroxypropyl Methyl Cellulose Mixtures for Sustained Ocular Delivery of Chloramphenicole: Optimization Study by Factorial Design. *Heliyon* **2020**, *6*, e05365. [CrossRef]
79. Peppas, N. Analysis of Fickian and Non-Fickian Drug Release from Polymers. *Pharm. Acta Helv.* **1985**, *60*, 110–111.
80. Zhang, Y.; Song, W.; Lu, Y.; Xu, Y.; Wang, C.; Yu, D.-G.; Kim, I. Recent Advances in Poly(A-L-Glutamic Acid)-Based Nanomaterials for Drug Delivery. *Biomolecules* **2022**, *12*, 636. [CrossRef]
81. Morath, B.; Sauer, S.; Zaradzki, M.; Wagner, A.H. Orodispersible films—Recent developments and new applications in drug delivery and therapy. *Biochem. Pharmacol.* **2022**, *200*, 115036. [CrossRef] [PubMed]
82. Wiedey, R.; Kokott, M.; Breitkreutz, J. Orodispersible tablets for pediatric drug delivery: Current challenges and recent advances. *Expert Opin. Drug Deliv.* **2021**, *18*, 1873–1890. [CrossRef] [PubMed]
83. Liu, H.; Jiang, W.; Yang, Z.; Chen, X.; Yu, D.-G.; Shao, J. Hybrid Films Prepared from a Combination of Electrospinning and Casting for Offering a Dual-Phase Drug Release. *Polymers* **2022**, *14*, 2132. [CrossRef]
84. Wang, M.; Yu, D.-G.; Williams, G.R.; Bligh, S.W.A. Co-Loading of Inorganic Nanoparticles and Natural Oil in the Electrospun Janus Nanofibers for a Synergetic Antibacterial Effect. *Pharmaceutics* **2022**, *14*, 1208. [CrossRef]
85. Chen, W.; Zhao, P.; Yang, Y.; Yu, D.G. Electrospun Beads-on-the-string Nanoproducts: Preparation and Drug Delivery Application. *Curr. Drug Deliv.* **2022**, *19*. [CrossRef] [PubMed]

Article

# Tissue Adhesive, Conductive, and Injectable Cellulose Hydrogel Ink for On-Skin Direct Writing of Electronics

Subin Jin [1], Yewon Kim [2], Donghee Son [2,3,4,*] and Mikyung Shin [1,4,5,*]

1 Department of Intelligent Precision Healthcare Convergence, Sungkyunkwan University (SKKU), Suwon 16419, Korea; subinjin@g.skku.edu
2 Department of Electrical and Computer Engineering, Sungkyunkwan University (SKKU), Suwon 16419, Korea; ywkim0726@gmail.com
3 Department of Superintelligence Engineering, Sungkyunkwan University (SKKU), Suwon 16419, Korea
4 Center for Neuroscience Imaging Research, Institute for Basic Science (IBS), Suwon 16419, Korea
5 Department of Biomedical Engineering, Sungkyunkwan University (SKKU), Suwon 16419, Korea
* Correspondence: daniel3600@g.skku.edu (D.S.); mikyungshin@g.skku.edu (M.S.)

**Abstract:** Flexible and soft bioelectronics used on skin tissue have attracted attention for the monitoring of human health. In addition to typical metal-based rigid electronics, soft polymeric materials, particularly conductive hydrogels, have been actively developed to fabricate biocompatible electrical circuits with a mechanical modulus similar to biological tissues. Although such conductive hydrogels can be wearable or implantable in vivo without any tissue damage, there are still challenges to directly writing complex circuits on the skin due to its low tissue adhesion and heterogeneous mechanical properties. Herein, we report cellulose-based conductive hydrogel inks exhibiting strong tissue adhesion and injectability for further on-skin direct printing. The hydrogels consisting of carboxymethyl cellulose, tannic acid, and metal ions (e.g., $HAuCl_4$) were crosslinked via multiple hydrogen bonds between the cellulose backbone and tannic acid and metal-phenol coordinate network. Owing to this reversible non-covalent crosslinking, the hydrogels showed self-healing properties and reversible conductivity under cyclic strain from 0 to 400%, as well as printability on the skin tissue. In particular, the on-skin electronic circuit printed using the hydrogel ink maintained a continuous electrical flow under skin deformation, such as bending and twisting, and at high relative humidity of 90%. These printable and conductive hydrogels are promising for implementing structurally complicated bioelectronics and wearable textiles.

**Keywords:** carboxymethylcellulose; tannic acid; conductive hydrogel; injectable hydrogel; adhesive hydrogel; 3D printing; direct printing

## 1. Introduction

Flexible and soft electronics, such as wearable monitoring devices [1–3], sensors [4–7], electronic skin [8,9], and human–machine interfaces [10], have recently attracted attention in the biomedical field [11–13]. Owing to their deformable characteristics, flexible electronics can be used as next-generation electronic devices that contact the human skin instead of conventional hard electronic devices [14,15]. In recent studies, a strategy for printing inexpensive and flexible electronic materials called electronic tattoos or epidermal electronics to directly write on the skin was evaluated [16,17]. Printed electronics on the skin can be used in various fields, such as light-emitting diodes (LEDs) [18], sensors [19,20], and transistors [21]. To achieve a high conductivity efficiency in printed electronics, most studies have focused on the low resistance, stable mechanical properties, and printability of conductive ink [22–24]. However, few studies have been conducted on maintaining its stability on elastic skin tissues, and synthetic polymers have been used to maintain electronic circuits and variable physical properties.

The application of direct writing to sensitive biological tissues requires high levels of flexibility, repetitive deformation due to movement, and low toxicity [25,26]. Hydrogels are useful for satisfying these conditions because of their tissue-like mechanical properties, conductivity, and biocompatibility [27–29]. In particular, cellulose, which is found in many plants, has been used for flexible/transparent substrates, separators, electron and ion conductors, electrolytes, and electrodes owing to its flexibility, ease of fabrication, mechanical strength, and biodegradability [30–33]. Cellulose can also be used as an ink for direct writing because of its stable mechanical properties for deformation, injectability, and biocompatibility [34–36]. However, natural polymer-based hydrogels are unsuitable for printing on tissue systems because of their low tissue adhesion strength and injectability due to covalent crosslinking [37,38]. To resolve these issues, polyphenol is a good candidate for providing tissue adhesion capability and injectability to natural polymer-based hydrogel inks. Polyphenols, which have a large number of phenol groups derived from various plants, may exhibit self-healing and injectable properties through stable structure maintenance of the hydrogel and reversible crosslinking through many hydrogen bonds with the polymer backbone, as well as π–π stacking of the polyphenol itself [39,40]. In addition, because polyphenols are known to have strong adhesion to tissues, they are often used to manufacture adhesive ink [38,41]. In particular, tannic acid (TA) is used to produce a conductive hydrogel with metal ions by forming a metal–phenolic network with variable metal ions and may exhibit strong adhesion to a structure [42,43].

In this study, the fabrication of conductive and adhesive cellulose (CAC) hydrogel ink was evaluated by mixing TA with three types of metal ions—gold chloride ($HAuCl_4$), silver nitrate ($AgNO_3$), and ferric chloride ($FeCl_3$)—and carboxymethyl cellulose (CMC) (Figure 1a). The carboxylate group of CMC and the hydroxy group of TA form strong hydrogen bonds. The adhesive cellulose (AC) hydrogel has tissue adhesive properties owing to TA, self-healing properties, and injectability through non-covalent hydrogen bonding (Figure 1b, top). In addition, in CAC hydrogels, metal–carboxyl coordination and metal–phenolic networks are formed when metal ions of Au, Ag, and Fe are mixed together in the AC hydrogel (Figure 1b, bottom) [43–45]. The CAC hydrogel prepared as described above has injectability due to non-covalent crosslinking and can be used as an ink for direct printing on tissues through tissue adhesion. It also has conductivity because it contains metal ions (Figure 1c). To demonstrate and optimize the mechanical and electrical properties of the CAC hydrogel, the rheological behavior, adhesiveness, and resistance to various metal ions and concentrations in the hydrogel were evaluated. Based on the investigated characteristics, we successfully demonstrated that the LED maintained stable light emission without electrical malfunction even under various deformations of the electronic circuit printed directly on the tissue.

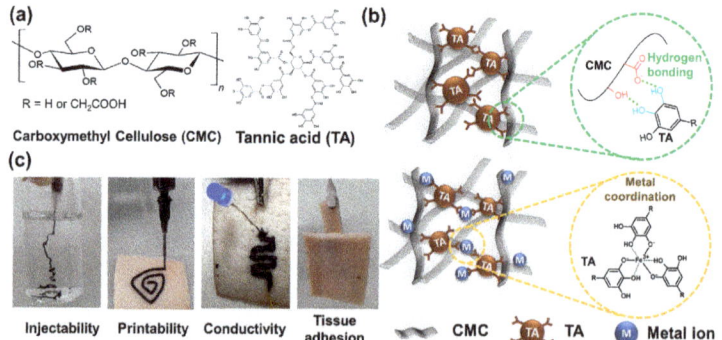

**Figure 1.** Schematic of the conductive adhesive CMC hydrogel. (**a**) Chemical structures of CMC and TA. (**b**) Fabrication mechanisms for the AC and CAC hydrogels. (**c**) Digital images of various properties of the CAC hydrogel.

## 2. Results and Discussion

### 2.1. Fabrication and Rheological Characterization of AC Hydrogels

A common approach for the fabrication of AC hydrogels is the mixing of CMC and TA. High-viscosity hydrogels are manufactured because the carboxyl and hydroxyl groups of CMC form strong hydrogen bonds with the phenol groups of TA. Therefore, the number of functional groups capable of hydrogen bonding depends on the change in the concentration of CMC and TA, and the mechanical properties of the hydrogel are determined. Therefore, hydrogels were prepared with different concentrations of CMC and TA and named $CMC_{0.5}$, AC-1, AC-2, and AC-3 (Table 1).

**Table 1.** Precursor compositions of different AC hydrogels.

|  | CMC (mg) | TA (mg) | DDW (mL) |
|---|---|---|---|
| $CMC_{0.5}$ | 25 | 0 | 1 |
| AC-1 | 25 | 500 | 1 |
| AC-2 | 33.2 | 333 | 1 |
| AC-3 | 37.5 | 250 | 1 |

The rheological properties of the AC hydrogels were evaluated in CMC and TA solutions at different volume ratios (Figure 2a,b). As indicated by the graphs of the storage (G′) and loss (G″) moduli of $CMC_{0.5}$ and AC-1 measured at frequencies ranging from 0 to 10 Hz, the storage modulus increased over all frequency ranges, in addition to TA, indicating that the interaction between TA and CMC created a solid hydrogel (Figure 2a). In addition, measuring the storage modulus and tan(δ) at different volume ratios (AC-1, AC-2, and AC-3) of CMC and TA indicated that the storage modulus increased and tan(δ) decreased as the CMC concentration increased. Therefore, the mechanical properties of the AC hydrogel depended on the concentrations of CMC and TA, and the most stable hydrogel was AC-3. Reversible crosslinking due to hydrogen bonding between TA and CMC resulted in AC hydrogels with self-healing properties, which make them stable during injection, printing, and deformation after printing. To demonstrate this, we investigated the changes in the mechanical modulus of the AC hydrogels upon treatment with urea—which has a role of disrupting multiple hydrogen bonds (Figure S1). If the hydrogen bonds between CMC and TA are major driving forces for the gelation, the mechanical modulus would decrease upon adding urea into the polymeric network. As expected, a decrease in both the G′ and G″ values of the hydrogels was monitored as urea concentration became high, clearly indicating that the hydrogen bonds between CMC and TA are involved in the gelation. In addition, to evaluate the self-healing properties of the AC hydrogels, the storage and loss moduli were measured (Figure 2c). When AC-3 hydrogels were measured at a frequency of 1 Hz with strains of 0.5% and 1000% at 3-min intervals, recovery of the storage modulus to the 0.5% strain level after 1000% strain was observed.

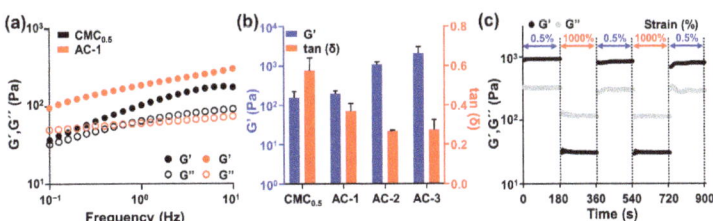

**Figure 2.** Rheological characterization of the AC hydrogel with different concentrations of CMC. (a) Oscillation frequency sweep measurements of $CMC_{0.5}$ (black) and AC-1 (red) hydrogels. The filled circles represent the storage modulus (G′), and the empty circles represent the loss modulus (G″). (b) Comparison of the storage modulus and tan(δ) values among $CMC_{0.5}$, AC-1, AC-2, and AC-3. (c) Disruption and recovery of the storage (G′) and loss (G″) moduli of AC-3 hydrogels under alternating strains of 0.5% and 1000%.

## 2.2. Fabrication and Rheological Characterization of CAC Hydrogel

Various studies have been conducted to form metal–phenolic networks by reacting TA with metal ions to fabricate conductive hydrogels [44,46,47]. In this study, metal–carboxylate coordination and metal–phenolic networks were constructed using Au, Ag, and Fe ions to fabricate conductive hydrogels. When 20 μL of metal ions was mixed with 0.3 mL of CMC, the metal ions and carboxylate of CMC changed to a weak gel through metal coordination bonding (Figure 3a, top). When 0.8 mL of TA was added to the metal ion and CMC mixture, the hydrogel became sticky (Figure 3a, bottom). The color and mechanical properties of the CAC hydrogel, along with the labeling of the samples, depended on the type of metal ion (Table 2). In addition, the rheological properties of the CAC hydrogel depended on the type of metal ion (Figure 3b). There were no significant differences in the storage modulus: 476.9 ± 10.5 Pa for $CAC_{Au}$, 319.1 ± 63.8 Pa for $CAC_{Ag}$, and 242.2 ± 12.6 Pa for $CAC_{Fe}$. However, differences were observed for tan(δ): $CAC_{Au}$ (0.31 ± 0.03), $CAC_{Ag}$ (0.43 ± 0.05), and $CAC_{Fe}$ (0.63 ± 0.01); $CAC_{Au}$ was the toughest, and $CAC_{Fe}$ was the softest. These results indicate that the degree to which TA–TA interactions and CMC–TA interactions interfere depends on the type of metal ion.

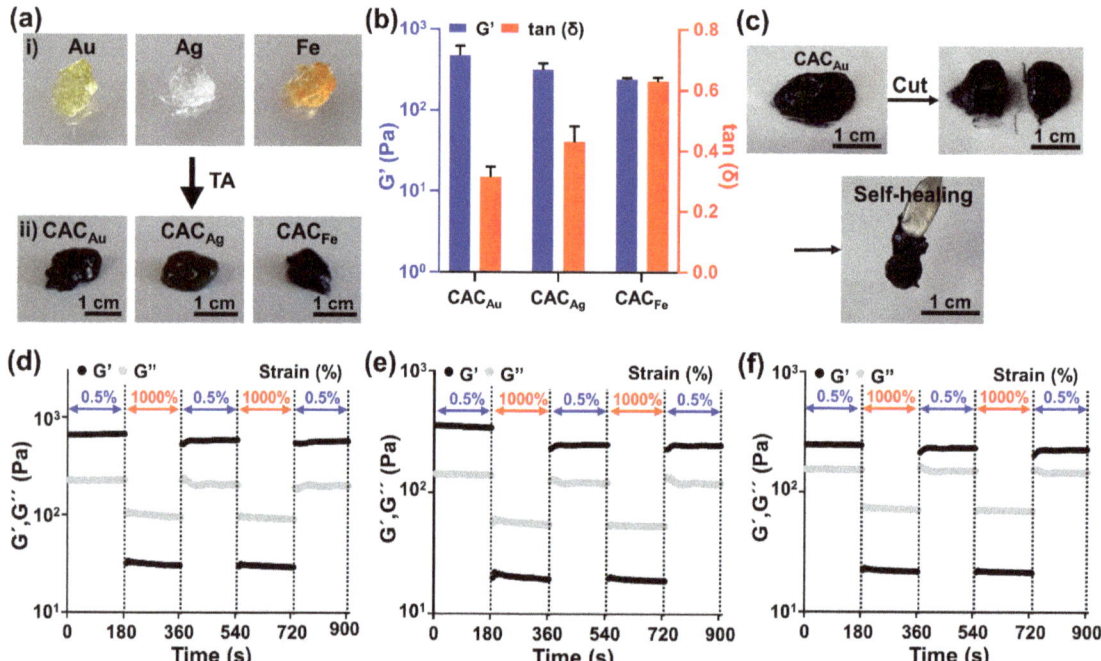

**Figure 3.** Fabrication and mechanical characterization of CAC hydrogels. (**a**) Digital images of (i) the CMC and metal ion mixture and (ii) $CAC_{Au}$, $CAC_{Ag}$, and $CAC_{Fe}$. (**b**) Comparison of the storage modulus (G′, blue) and tan(δ) (red) values among $CAC_{Au}$, $CAC_{Ag}$, and $CAC_{Fe}$. (**c**) Macroscopic self-healing images of $CAC_{Au}$. (**d**–**f**) Evaluation of the disruption and recovery of the storage (G′) and loss (G″) moduli under alternating strains of 0.5% and 1000%: (**d**) $CAC_{Au}$; (**e**) $CAC_{Ag}$; (**f**) $CAC_{Fe}$.

**Table 2.** Precursor compositions of different CAC hydrogels.

| | CMC (mg) | TA (mg) | Metal Ion (mg) | DDW (mL) |
|---|---|---|---|---|
| $CAC_{Au}$ | 37.5 | 250 | Au/5 | 1 |
| $CAC_{Ag}$ | 37.5 | 200 | Ag/5 | 1 |
| $CAC_{Fe}$ | 37.5 | 250 | Fe/1 | 1 |
| $CAC_{Au}@5$ | 37.5 | 250 | Au/1.25 | 1 |
| $CAC_{Au}@10$ | 37.5 | 250 | Au/2.5 | 1 |

The self-healing property of the CAC hydrogel, which is caused by metal ions and phenol groups, suppresses the breakage of the crosslinking by shear stress during printing and the damage to the printed electronics caused by surface deformation after printing. Rheological property analysis, cutting, and contact experiments were conducted to demonstrate the self-healing properties of the CAC hydrogel. When the bulk $CAC_{Au}$ hydrogel was cut in half and brought into contact again, $CAC_{Au}$ was immediately reconstructed (Figure 3c). In addition, in the rheological behavior analysis, the storage modulus, which decreased from the 1000% strain, recovered from the 0.5% strain (Figure 3d). This rheological behavior was also observed for $CAC_{Ag}$ and $CAC_{Fe}$ hydrogels (Figure 3e,f). These results indicate that CAC hydrogels can self-heal with the fast recombination of reversible metal coordination bonds between CMC/TA, and metal ions and hydrogen bonds between CMC and TA.

### 2.3. Tissue Adhesion Capability of CAC Hydrogel

In the direct printing of electronics on the skin, the tissue adhesion capability of ink is important for the electronics to remain stable on the skin. The strong tissue adhesion of TA in CAC hydrogels allows the direct printing of hydrogel ink. To evaluate the tissue adhesion capability of the CAC hydrogel, shear adhesion tests were conducted on skin tissues using a universal testing machine (UTM) (Figure 4a). The adhesion strength of $CMC_{0.5}$ was 2.3 ± 0.3 kPa, indicating that $CMC_{0.5}$ has non-tissue adhesion properties (Figure 4b). The adhesion strength of AC-1 was 14.1 ± 4.3 kPa, which was higher than that of AC-3 (8.9 ± 3.8 kPa). This was caused by an increase in the tissue adhesion strength due to the larger amount of TA. The adhesion strengths of $CAC_{Au}$ and $CAC_{Ag}$ were 13.0 ± 0.8 and 12.2 ± 2.5 kPa, respectively (similar to that of AC-1), and $CAC_{Fe}$ exhibited the highest adhesion strength among the samples (18.2 ± 1.8 kPa). In addition, the amount of gold ions used in the gel formation was adjusted to 5, 10, and 20 µL, and the resulting adhesion was evaluated (Figure 4c). The tissue adhesion strength increased with the amount of gold ions (from 6.7 ± 0.7 kPa for $CAC_{Au}@5$ to 8.5 ± 1.7 kPa for $CAC_{Au}@10$). The results for the tissue adhesion capability of the CAC hydrogel indicated that the addition of metal ions to the AC hydrogel enhanced the stability of the hydrogel structure through the formation of a metal–phenolic network, which increased the cohesive strength and thus the adhesive strength. Furthermore, to demonstrate that the tissue adhesion capability of CAC hydrogels supports the direct printing of electronics on tissue to maintain a stable contact, a qualitative evaluation was conducted involving the deformation of skin tissue in various forms after direct printing of a CAC hydrogel with a 23-gauge needle on pig skin (Figure 4d). At the maximum tensile, bending, and twist deformations of the pig skin, $CAC_{Au}$ remained stable, without separation from the tissue. In addition, when the pig skin was immersed in a 1× phosphate-buffered saline (PBS) solution for 30 min, the initial form was maintained, without excessive swelling.

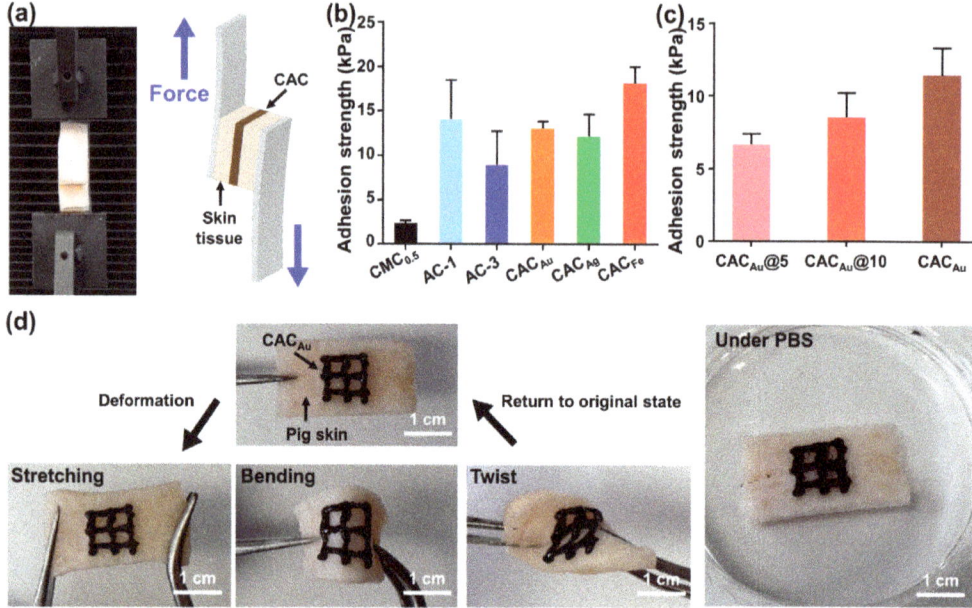

**Figure 4.** Tissue adhesion capabilities of CAC hydrogels. (**a**) Digital images and schematic of the shear adhesion test method using a UTM. (**b**) Evaluation of the shear adhesion strength of CMC0.5, AC-1, AC-3, $CAC_{Au}$, $CAC_{Ag}$, $CAC_{Fe}$. (**c**) Evaluation of the shear adhesion strength of $CAC_{Au}$ depending on the amount of gold ions. (**d**) Macroscopic images of tissue adhesion in direct writing of $CAC_{Au}$ on tissue with various deformations of pig skin (**left**) and the stability in PBS (**right**).

## 2.4. Electrical Properties of CAC Hydrogel

For long-term use, electronic devices printed on the skin must resist excessive increases in electrical resistance due to the daily movement of the skin. To evaluate the change in resistance per unit tensile strain of the CAC hydrogel, we followed the experimental method used in a previous study (Figure 5a) [48]. Regardless of the type of metal ion, the initial resistance of the CAC hydrogel was maintained at approximately 2 kΩ (Figure 5b–d). According to the results of a tensile strain test, the resistances of $CAC_{Au}$ (Figure 5b) and $CAC_{Ag}$ (Figure 5c) at 300% strain increased to 8351 and 9155 Ω, respectively, while the resistance of $CAC_{Fe}$ (Figure 5d) increased rapidly with 300% strain to 12,617 Ω. In addition, in a 10-min cyclic stretching test (0–200% strain), $CAC_{Au}$ maintained a resistance difference of ≤3 kΩ for the initial resistance, and $CAC_{Ag}$ maintained a resistance difference of ≤4 kΩ (Figure 5e,f). $CAC_{Fe}$ also maintained a constant change in resistance, but the resistance increased to 10 kΩ (Figure 5g). Compared with the CAC hydrogels containing other metal ions, a larger conductivity loss occurred during tensile deformation owing to the strong interaction between TA and Fe ions. Therefore, $CAC_{Au}$ is the best candidate for use in direct tissue writing electronics, and these results provide an appropriate electrical demonstration of the electrical properties of each of the different CAC hydrogels.

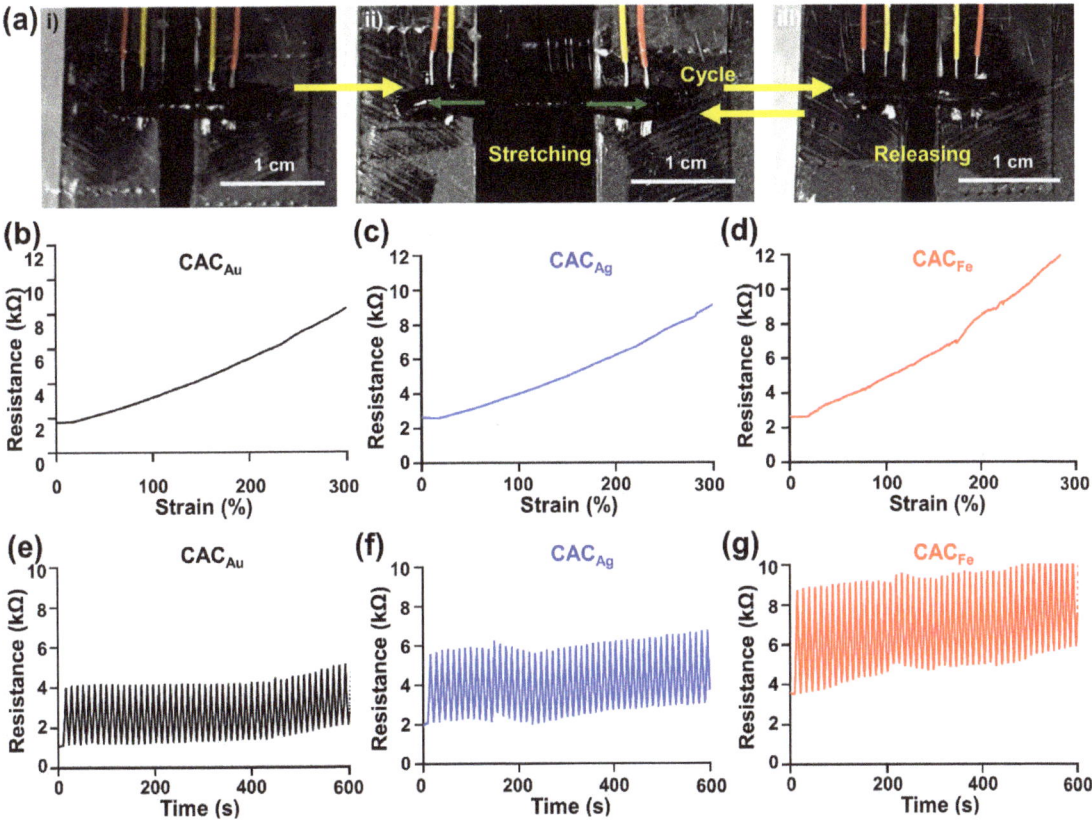

**Figure 5.** Electrical characterization of CAC hydrogels. (**a**) Macroscopic images: (i) pristine; (ii) stretching; (iii) release. (**b**–**d**) Continuous tensile strain of the samples at a speed of 3 mm/min: (**b**) $CAC_{Au}$; (**c**) $CAC_{Ag}$; (**d**) $CAC_{Fe}$. (**e**,**f**) Repetitive cyclic stretching test with strains ranging from 0% to 200% at a speed of 1 mm/s for 10 min: (**e**) $CAC_{Au}$; (**f**) $CAC_{Ag}$; (**g**) $CAC_{Fe}$.

## 2.5. Electrical Properties of the Filaments Printed Using CAC Hydrogel Inks

These rheological, adhesive, and electrical properties of the CAC hydrogel inks enable their direct writing on biological tissue. Considering the further biomedical potential of the inks, we confirmed their cytocompatibility at 24 and 48 h after the treatment of the $CAC_{Au}$ ink (Figure S2). As a result, ~90% of the cells were alive upon treatment of the sample below 1 mg/mL. Furthermore, the concentration of TA included in the $CAC_{Au}$ ink was 250 mg/mL, which is less than the TA concentration previously used in the fabrication of tissue adhesive materials implantable in vivo [49,50]. That is, the inks might not cause severe toxicity and skin irritation after direct on-tissue printing.

To evaluate the electrical properties and conductivity stability of CAC electronics printed directly on skin tissues before and after tissue deformation, $CAC_{Au}$ hydrogels were printed on porcine skin using a three-dimensional (3D) printer (Figure 6a). In addition, $CAC_{Au}$ printed in the form of concentric circles, grids, and circuits on the skin tissue had a space for wire and LEDs to contact. The printed electric circuit consisting of $CAC_{Au}$ had a high resolution and a uniform filament structure, and no slippage from the skin or breakage of ink was observed during the printing process (Figure 6b). Furthermore, in the evaluation of the circuit functionality using LEDs, light emission from both sides was observed when zigzag and concentric circuits were directly printed and connected to LEDs (Figure 6c). In addition, despite deformations such as stretching, bending, and twisting of the skin tissue, the conductivity was maintained, and light emission of the LEDs was observed (Figure 6d). Moreover, when half of the skin tissue with printed electronics was immersed in PBS, the LED was emitted without electrical malfunction. These results indicate that CAC hydrogels are effective for the on-skin direct writing of electronics.

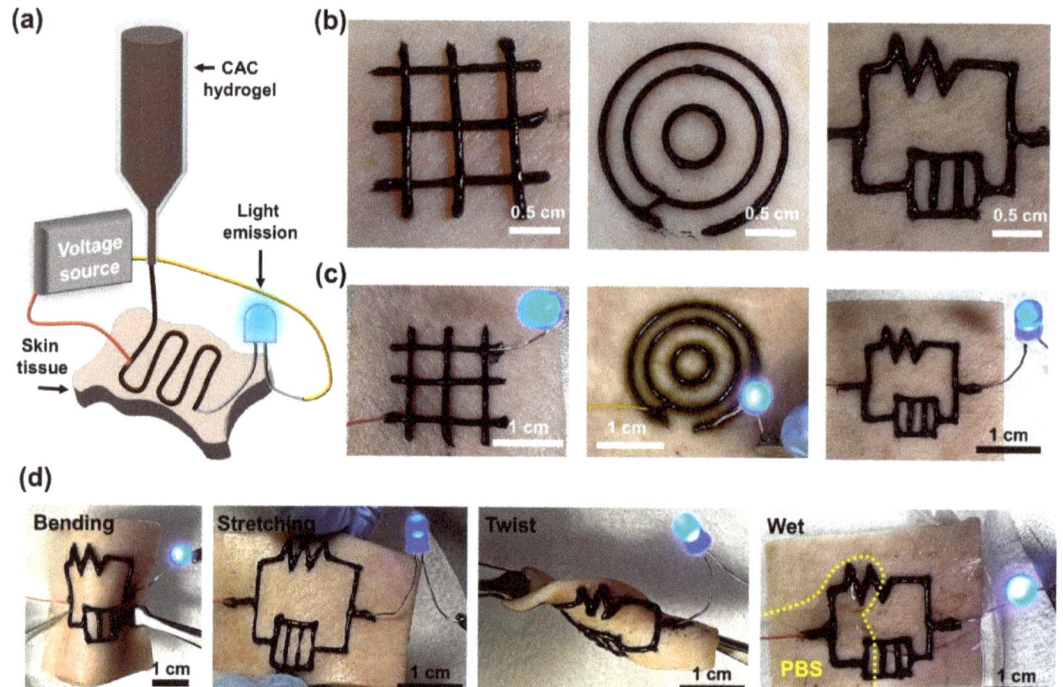

**Figure 6.** Evaluation of electronics directly printed on tissue with the CAC hydrogel. (**a**) Schematic of the direct-printing process. (**b**) Macroscopic images of concentric (**left**), grid (**middle**), and circuit-shaped (**right**) electronics printed using the $CAC_{Au}$ hydrogel. (**c**) Photographs of LED emission with printed electronics. (**d**) LED light emission with a stable conductivity under stretching (**left**) and bending (**right**) deformation.

## 2.6. Characterization of CAC Hydrogels under High Humidity Condition and Repeated Tissue Deformation

The hydrogel inks for on-skin electronics should maintain their properties even under continuous exposure to high relative humidity (e.g., sweat) and repeated mechanical deformation by daily movement. Thus, we evaluated the stability of self-healing, adhesiveness, and electrical resistance of the $CAC_{Au}$ hydrogels in high relative humidity (80 or 90%) (Figure 7a–d). They still showed self-healing behavior even after a 2 h incubation in a humid chamber (~90%) (Figure 7a), and the G' value recovered up to 70% of its initial value (Figure 7b). Regarding their tissue adhesion, although the adhesive strength slightly decreased down to ~5 kPa under humidified conditions when compared to 9.4 ± 1.2 kPa of their initial status (e.g., before incubation in humid chamber), the value was retained for 8 h (Figure 7c). There was also no significant difference in their ionic conductivity (~6 kΩ) before/after exposure to high relative humidity (80 and 90%) (Figure 7d). Moreover, such electrical resistance was maintained even under daily deformation (e.g., repeated bending and releasing every 12 h) (Figure 7e,f). During deformation of the skin substrates with a bending–releasing cycle of 5 min every 12 h, the resistance value of the $CAC_{Au}$ hydrogel was at a level of 6–10 kΩ, and there was no significant increase even after 48 h. All results indicate that the physical and electrical properties of the CAC hydrogels are stable enough against high humidity and continuous mechanical stress to be directly printable on the tissue.

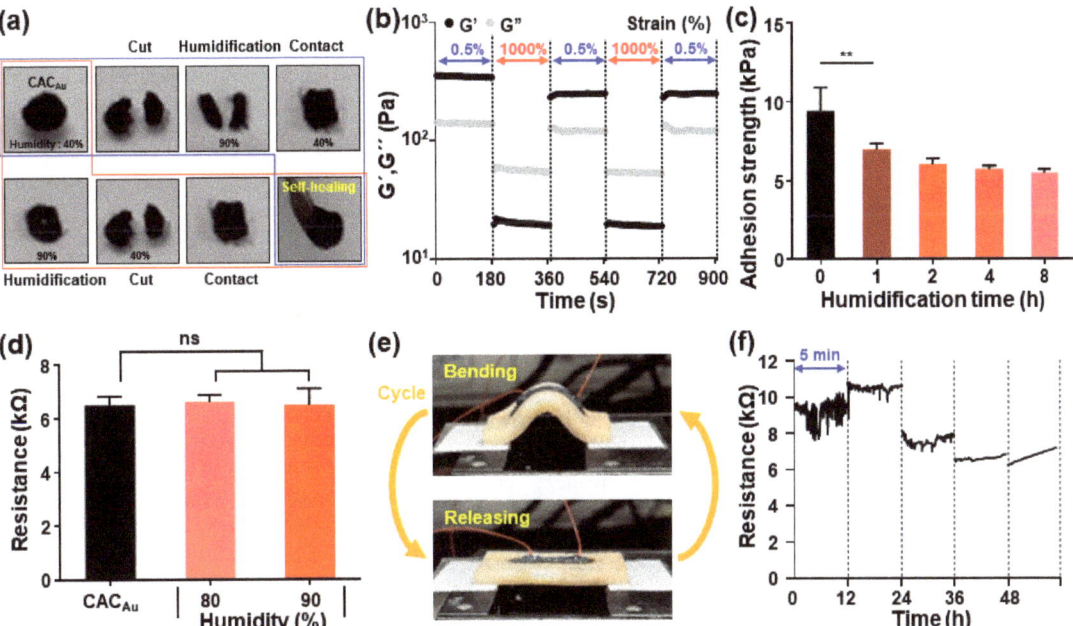

Figure 7. Stability of self-healing property, adhesiveness, and electrical resistance of CAC hydrogels. (a) Photos to show self-healing behavior of $CAC_{Au}$ hydrogels under humidified condition (90%), and (b) their G' and G'' values recovered under alternating strains of 0.5% and 1000% after 2-h incubation of the hydrogels in 90% humidity chamber. (c) Shear adhesion strength of the $CAC_{Au}$ hydrogels as a function of incubation time in 90% humidity chamber. ** $p < 0.01$, one-way ANOVA. (d) Electrical resistance of the hydrogels under high relative humidity (80 or 90%). ns for not significant, one-way ANOVA. (e) Experimental settings to measure electrical resistance of the $CAC_{Au}$ filament directly printed on porcine skin tissue during bending/releasing cycle. (f) Changes in the resistance of the printed filament during the cyclic deformation every 12 h (total 48 h).

3. Conclusions

We synthesized and evaluated CAC hydrogel ink for the direct printing of electronics on tissues using CMC, TA, and metal ions (to increase the conductivity). The CAC hydrogel ink exhibited excellent self-healing properties, regardless of the type of metal ion. An evaluation of the tissue adhesion of the CAC hydrogel using a UTM indicated that a sufficient tissue adhesion strength for direct printing was achieved regardless of the type of metal ion, and it was verified that the tissue adhesion was determined by the amount of metal ions. In addition, an analysis of the resistance changes under deformation from tensile strain and cyclic strain indicated that $CAC_{Au}$ had the lowest resistance at 300% strain (8 kΩ), and the difference in the resistance due to deformation was the smallest (3 kΩ). In contrast, $CAC_{Fe}$ had a high resistance of ≥12 kΩ at 300% strain, and a change of ≥7 kΩ in the electrical resistance occurred in the 200% strain cycle. Owing to its mechanical properties, adhesion, and electrical conductivity, the $CAC_{Au}$ hydrogel was selected as the most suitable for direct printing. Finally, the electrical performance of electronics printed directly on tissue using the CAC hydrogel was evaluated. Electronics with various patterns printed directly on the skin tissue maintained a high resolution and uniform filament thickness and stable conductive properties under deformation due to tension, bending, and twisting. Owing to its various advantages, we expect that the CAC hydrogel will be used reliably as an ink material for the direct printing of electronics on tissue.

## 4. Materials and Methods

### 4.1. Preparation of AC and CAC Hydrogels

Sodium carboxymethyl cellulose (CMC, 700 kDa), TA, ferric chloride ($FeCl_3$), gold trihydrate chloride ($HAuCl_4$), and silver nitrate ($AgNO_3$) were purchased from Sigma-Aldrich (Burlington, MA, USA).

An AC hydrogel was fabricated by mixing CMC and TA. The CMC solution was prepared by adding CMC to deionized distilled water (DDW) at a concentration of 5%, followed by vortexing. The AC hydrogel was prepared by mixing a TA solution (1 g/mL) with CMC solutions of different volume ratios. Each completed sample was labeled according to the volume ratios of CMC and TA.

The CAC hydrogel was prepared by mixing CMC, metal ions, and TA. The metal ion solution was prepared by dissolving $HAuCl_4$ (10 wt.%), $AgNO_3$ (10 wt.%), and $FeCl_3$ (2.5 wt.%) in DDW. To synthesize the CAC hydrogel, 20 μL of the metal solution was added to 0.3 mL of the CMC solution, followed by mixing. Then, 80 μL of the TA solution was added, and the mixture was stirred vigorously for 5 min.

### 4.2. Rheological Characterization of AC and CAC Hydrogels

The oscillation frequency sweep test and self-healing measurements through the step strain test of the rheological properties of the AC and CAC hydrogels were conducted using a TA Instruments Discovery Hybrid Rheometer 2 (TA Instruments, (New Castle, DE, USA)). All rheological measurements were conducted using a 20-mm parallel-plate geometry with a gap size of 300 μm. The storage ($G'$) and loss ($G''$) moduli were measured at room temperature under oscillation frequency sweeps (0.1–10 Hz, at 1% strain). To demonstrate the self-healing property, $G'$ and $G''$ were measured under repeated application strains of 0.5% and 1000% for 180 s, respectively, at an osculation frequency of 1 Hz. In addition, the rheological properties of the AC-3 hydrogels (0.4 mL) after addition of urea (1, 5, and 10 M, 0.1 mL) were investigated by same oscillation frequency sweep test.

### 4.3. Tissue Adhesion Capabilities of AC and CAC Hydrogels

To evaluate the stability of CAC printing on skin tissue, the adhesion of CAC to tissue was investigated using a UTM (34SC-1, Instron, Norwood, MA, USA). First, pig skin was cut into 2 × 1 $cm^2$ pieces. The CMC solution, AC, and CAC were spread between the skins, and the terminals of the substrates were pulled at a speed of 1 mm/s. The adhesion strength (kPa) was calculated by dividing the maximum load (N) by the attached area ($m^2$).

### 4.4. Resistance per Strain

First, the electrical wires were fixed on an automatic one-axis stretcher (SMC-100, Jaeil Optical System (Incheon, Korea)) using double-sided tape after peeling off 0.5 cm of the sheath at one end of the wires. The hydrogel was loaded on the stretcher by approximately 2 cm using a 3-mL syringe. Finally, electrical wires and a digital multimeter (Keithley 2450 source meter, Tektronics (Seoul, Korea) were connected to the samples to measure their electrical resistances. The samples were measured in the stretched state at a rate of 3 mm/min, and for the cyclic test, the rate was 20 mm/min at strains of 0% and 200% for 100 cycles. In all cases, the initial length before the measurement was 3 mm. In addition, liquid metal (eutectic gallium indium, Alfa Aesar (Haverhill, MA, USA)) was used to maintain a stable connection between the wires and the hydrogel during measurement.

## 4.5. Direct Writing of CAC on Skin Tissue

To confirm that direct writing of electronics using CAC hydrogel is possible, CAC ink was printed on pig skin tissue using a 3D printer (Dr. INVIVO 4D, Rokit Healthcare (Seoul, Korea)). For design and printing, the "New creator K v1.57.71" program was used. The printing was conducted using a 0.6-mm nozzle and an output speed of 5 mm/s. The availability of CAC hydrogels printed in various forms, such as circles, grids, and circuits, on pig skin tissues was demonstrated using an LED test. The CAC hydrogel transformed the printed skin tissue by stretching, bending, twisting, and wetting in PBS, and the LED emission was confirmed.

## 4.6. In Vitro Cytotoxicity Test

To evaluate cytotoxicity of the CAC hydrogels, mouse fibroblast cells (L929) were pre-cultured in Dulbecco Modified Eagle Medium (DMEM supplemented with 10% fetal bovine serum and 1% penicillin/streptomycin). To collect the eluates of the $CAC_{Au}$ inks, 40 mg of the inks was incubated in cell culture media (10 mL) for 24 h. The cells were seeded in a 48-well plate ($2 \times 10^4$ cells per well) and cultured in a 5% humidified $CO_2$ incubator at 37 °C. After culturing overnight and washing with Dulbecco's Phosphate-Buffered Saline (DPBS), the media containing the sample eluates was supplemented in the well. The cell viability at 24 and 48 h was evaluated using Live/Dead assay kit (Thermo Fisher Sci. (Seoul, Korea)) The live cells were stained with Calcein AM solution (2 µM) and dead cells were stained with Ethidium homodimer-1 solution (4 µM) (0.2 mL of total working solution), and then incubated for 1 h at 37 °C. Finally, the live/dead cells were observed using fluorescence microscopy (DMi8, Leica (Wetzlar, Germany)). The number of either green dots for live cells or red dots for dead cells was counted using Image J software, and cell viability (%) was calculated as the ratio of the number of live cells to the total number of cells.

## 4.7. The Stability Characterization of CAC Hydrogel in High Relative Humidity Conditions

Self-healing property, tissue adhesiveness, and ionic conductivity of CAC hydrogels were evaluated in high humidity condition (80–90%). The humid chamber equipped with a hygrometer was manually prepared using acrylic plates. First, the self-healing property of $CAC_{Au}$ hydrogel was confirmed after cutting and 2-h incubation of the hydrogel pieces in the chamber. In addition, G' and G'' values of the hydrogels after 24-h incubation in the chamber were monitored using oscillatory rheometer (at 1 Hz, each step for 180 s under alternating strain between 0.5% and 1000%). Second, tissue adhesiveness of the hydrogels as a function of incubation time in the humid chamber was measured in a same manner with Section 4.3. Finally, the ionic conductivity of the hydrogels was also measured without strain in the same manner as in Section 4.4 as a function of incubation time under humidity conditions of 80% and 90%.

## 4.8. The Long-Term Stability of Electrical Resistance of CAC Hydrogels

To evaluate the stability of electrical resistance of the CAC hydrogel against long-term repeated tissue deformation, the single filament (length = 2 cm, 18-gauge needle) was printed using $CAC_{Au}$ hydrogel on the porcine skin. The substrate tissue was fixed on the one-axis stretcher using double-sided tape. Electrical wires and a digital multimeter were connected to the samples to measure their electrical resistance in between two terminals of the printed filament. For the cyclic test, the sample was bent at a rate of 20 mm/min with 1 cm for 5 cycles and the measurement was conducted every 12 h.

**Supplementary Materials:** The following supporting information can be downloaded at: https://www.mdpi.com/article/10.3390/gels8060336/s1, Figure S1. Evaluation and comparison of storage modulus (G') and loss modulus (G'') due to hy-drogen bond collapse between CMC and TA due to disruption of hydrogen bond by various con-centration of urea; Figure S2. In vitro cytocompatibility of CACAu. (a) The fluorescent images of L929 cells at 24 and 48 hours after treatment of the CACAu relesates as a function of concentration (0, 0.5, 1, and 2 mg/mL). (b) Quantitative analysis of the cell viability. All data are expressed as mean ± s.d. One-way ANOVA, **** $p < 0.0001$, and ns for not significant; Movie S1. The stability of the on-tissue printed filaments upon soaking in PBS and against physi-cal deformation of porcine skin.

**Author Contributions:** Conceptualization, M.S.; methodology, M.S. and S.J.; software, S.J. and Y.K.; validation, S.J.; formal analysis, M.S. and S.J.; investigation, S.J. and Y.K.; resources, M.S. and S.J.; data curation, D.S. and S.J.; writing—original draft preparation, S.J. and Y.K.; writing—review and editing, D.S. and M.S.; visualization, S.J. and Y.K.; supervision, D.S. and M.S.; project administration, D.S. and M.S.; funding acquisition, D.S. and M.S. All authors have read and agreed to the published version of the manuscript.

**Funding:** This research was supported by the National Research Foundation of Korea (NRF) grant funded by the Korean government (MSIT) (Nos. 2020R1C1C1003903 (M.S.) and 2020R1C1C1005567 (D.S.)). This research was also funded by a Korea Medical Device Development Fund grant funded by the Korean government (the Ministry of Science and ICT, the Ministry of Trade, Industry and Energy, the Ministry of Health & Welfare, and the Ministry of Food and Drug Safety) (No. 202012D28), Institute of Information & communications Technology Planning & Evaluation (IITP) grant funded by the Korea government (MSIT) (No. 2020-0-00261, Development of low power/low delay/self-power suppliable RF simultaneous information and power transfer system and stretchable electronic epineurium for wireless nerve bypass implementation), and MSIT (Ministry of Science and ICT), Korea, under the ICT Creative Consilience program (IITP-2020-0-01821) supervised by the IITP (Institute for Information & Communications Technology Planning & Evaluation). This research was also supported by Institute for Basic Science (IBS-R015-D1).

**Institutional Review Board Statement:** Not applicable.

**Informed Consent Statement:** Not applicable.

**Data Availability Statement:** The data presented in this study are available in the article.

**Conflicts of Interest:** The funders had no role in the design of the study; in the collection, analyses, or interpretation of data; in the writing of the manuscript, or in the decision to publish the results.

# References

1. Son, D.; Lee, J.; Qiao, S.; Ghaffari, R.; Kim, J.; Lee, J.E.; Song, C.; Kim, S.J.; Lee, D.J.; Jun, S.W.; et al. Multifunctional wearable devices for diagnosis and therapy of movement disorders. *Nat. Nanotechnol.* **2014**, *9*, 397–404. [CrossRef] [PubMed]
2. Wang, Y.; Wang, L.; Yang, T.; Li, X.; Zang, X.; Zhu, M.; Wang, K.; Wu, D.; Zhu, H. Wearable and highly sensitive graphene strain sensors for human motion monitoring. *Adv. Funct. Mater.* **2014**, *24*, 4666–4670. [CrossRef]
3. Lee, H.; Song, C.; Hong, Y.S.; Kim, M.S.; Cho, H.R.; Kang, T.; Shin, K.; Choi, S.H.; Hyeon, T.; Kim, D.-H. Wearable/disposable sweat-based glucose monitoring device with multistage transdermal drug delivery module. *Sci. Adv.* **2017**, *3*, e1601314. [CrossRef] [PubMed]
4. Su, M.; Li, F.; Chen, S.; Huang, Z.; Qin, M.; Li, W.; Zhang, X.; Song, Y. Nanoparticle based curve arrays for multirecognition flexible electronics. *Adv. Mater.* **2016**, *28*, 1369–1374. [CrossRef]
5. Xu, K.; Lu, Y.; Takei, K. Multifunctional skin-inspired flexible sensor systems for wearable electronics. *Adv. Mater. Technol.* **2019**, *4*, 1800628. [CrossRef]
6. Zhao, S.; Li, J.; Cao, D.; Zhang, G.; Li, J.; Li, K.; Yang, Y.; Wang, W.; Jin, Y.; Sun, R.; et al. Recent advancements in flexible and stretchable electrodes for electromechanical sensors: Strategies, materials, and features. *ACS Appl. Mater. Interfaces* **2017**, *9*, 12147–12164. [CrossRef]
7. Yang, T.; Jiang, X.; Zhong, Y.; Zhao, X.; Lin, S.; Li, J.; Li, X.; Xu, J.; Li, Z.; Zhu, H. A wearable and highly sensitive graphene strain sensor for precise home-based pulse wave monitoring. *ACS Sens.* **2017**, *2*, 967–974. [CrossRef]
8. Tee, B.C.-K.; Chortos, A.; Berndt, A.; Nguyen, A.K.; Tom, A.; McGuire, A.; Lin, Z.C.; Tien, K.; Bae, W.-G.; Wang, H.; et al. A skin-inspired organic digital mechanoreceptor. *Science* **2015**, *350*, 313–316. [CrossRef]
9. Wang, X.; Dong, L.; Zhang, H.; Yu, R.; Pan, C.; Wang, Z.L. Recent progress in electronic skin. *Adv. Sci.* **2015**, *2*, 1500169. [CrossRef]
10. Wang, H.; Ma, X.; Hao, Y. Electronic devices for human-machine interfaces. *Adv. Mater. Interfaces* **2017**, *4*, 1600709. [CrossRef]
11. Kang, J.; Tok, J.B.-H.; Bao, Z. Self-healing soft electronics. *Nat. Electron.* **2019**, *2*, 144–150. [CrossRef]
12. Rogers, J.A. Electronics for the human body. *JAMA* **2015**, *313*, 561–562. [CrossRef] [PubMed]
13. Choi, C.; Choi, M.K.; Hyeon, T.; Kim, D.H. Nanomaterial-based soft electronics for healthcare applications. *ChemNanoMat* **2016**, *2*, 1006–1017. [CrossRef]
14. Llerena Zambrano, B.; Renz, A.F.; Ruff, T.; Lienemann, S.; Tybrandt, K.; Vörös, J.; Lee, J. Soft electronics based on stretchable and conductive nanocomposites for biomedical applications. *Adv. Healthc. Mater.* **2021**, *10*, 2001397. [CrossRef] [PubMed]

15. Liu, S.; Shah, D.S.; Kramer-Bottiglio, R. Highly stretchable multilayer electronic circuits using biphasic gallium-indium. *Nat. Mater.* **2021**, *20*, 851–858. [CrossRef]
16. Saadi, M.; Maguire, A.; Pottackal, N.; Thakur, M.S.H.; Ikram, M.M.; Hart, A.J.; Ajayan, P.M.; Rahman, M.M. Direct Ink Writing: A 3D Printing Technology for Diverse Materials. *Adv. Mater.* **2022**, 2108855. [CrossRef]
17. Ershad, F.; Thukral, A.; Yue, J.; Comeaux, P.; Lu, Y.; Shim, H.; Sim, K.; Kim, N.-I.; Rao, Z.; Guevara, R.; et al. Ultra-conformal drawn-on-skin electronics for multifunctional motion artifact-free sensing and point-of-care treatment. *Nat. Commun.* **2020**, *11*, 3823. [CrossRef]
18. Guo, R.; Sun, X.; Yao, S.; Duan, M.; Wang, H.; Liu, J.; Deng, Z. Semi-Liquid-Metal-(Ni-EGaIn)-Based Ultraconformable Electronic Tattoo. *Adv. Mater. Technol.* **2019**, *4*, 1900183. [CrossRef]
19. Kabiri Ameri, S.; Ho, R.; Jang, H.; Tao, L.; Wang, Y.; Wang, L.; Schnyer, D.M.; Akinwande, D.; Lu, N. Graphene electronic tattoo sensors. *ACS Nano* **2017**, *11*, 7634–7641. [CrossRef]
20. Wang, Y.; Qiu, Y.; Ameri, S.K.; Jang, H.; Dai, Z.; Huang, Y.; Lu, N. Low-cost, µm-thick, tape-free electronic tattoo sensors with minimized motion and sweat artifacts. *NPJ Flex. Electron.* **2018**, *2*, 6. [CrossRef]
21. Lai, S.; Zucca, A.; Cosseddu, P.; Greco, F.; Mattoli, V.; Bonfiglio, A. Ultra-conformable Organic Field-Effect Transistors and circuits for epidermal electronic applications. *Org. Electron.* **2017**, *46*, 60–67. [CrossRef]
22. Williams, N.X.; Noyce, S.; Cardenas, J.A.; Catenacci, M.; Wiley, B.J.; Franklin, A.D. Silver nanowire inks for direct-write electronic tattoo applications. *Nanoscale* **2019**, *11*, 14294–14302. [CrossRef] [PubMed]
23. Gao, Y.; Li, H.; Liu, J. Direct writing of flexible electronics through room temperature liquid metal ink. *PLoS ONE* **2012**, *7*, e45485. [CrossRef] [PubMed]
24. Boley, J.W.; White, E.L.; Chiu, G.T.C.; Kramer, R.K. Direct writing of gallium-indium alloy for stretchable electronics. *Adv. Funct. Mater.* **2014**, *24*, 3501–3507. [CrossRef]
25. Zhao, Y.; Kim, A.; Wan, G.; Tee, B.C. Design and applications of stretchable and self-healable conductors for soft electronics. *Nano Converg.* **2019**, *6*, 25. [CrossRef]
26. Chen, Z.; Gao, N.; Chu, Y.; He, Y.; Wang, Y. Ionic network based on dynamic ionic liquids for electronic tattoo application. *ACS Appl. Mater. Interfaces* **2021**, *13*, 33557–33565. [CrossRef]
27. Wang, C.; Yokota, T.; Someya, T. Natural biopolymer-based biocompatible conductors for stretchable bioelectronics. *Chem. Rev.* **2021**, *121*, 2109–2146. [CrossRef] [PubMed]
28. Liu, H.; Li, M.; Ouyang, C.; Lu, T.J.; Li, F.; Xu, F. Biofriendly, stretchable, and reusable hydrogel electronics as wearable force sensors. *Small* **2018**, *14*, 1801711. [CrossRef]
29. Xie, C.; Wang, X.; He, H.; Ding, Y.; Lu, X. Mussel-inspired hydrogels for self-adhesive bioelectronics. *Adv. Funct. Mater.* **2020**, *30*, 1909954. [CrossRef]
30. Zhao, D.; Zhu, Y.; Cheng, W.; Chen, W.; Wu, Y.; Yu, H. Cellulose-based flexible functional materials for emerging intelligent electronics. *Adv. Mater.* **2021**, *33*, 2000619. [CrossRef]
31. Chen, C.; Hu, L. Nanocellulose toward advanced energy storage devices: Structure and electrochemistry. *Acc. Chem. Res.* **2018**, *51*, 3154–3165. [CrossRef] [PubMed]
32. Russo, A.; Ahn, B.Y.; Adams, J.J.; Duoss, E.B.; Bernhard, J.T.; Lewis, J.A. Pen-on-paper flexible electronics. *Adv. Mater.* **2011**, *23*, 3426–3430. [CrossRef] [PubMed]
33. Zhao, D.; Zhu, Y.; Cheng, W.; Xu, G.; Wang, Q.; Liu, S.; Li, J.; Chen, C.; Yu, H.; Hu, L. A dynamic gel with reversible and tunable topological networks and performances. *Matter* **2020**, *2*, 390–403. [CrossRef]
34. Wang, Q.; Sun, J.; Yao, Q.; Ji, C.; Liu, J.; Zhu, Q. 3D printing with cellulose materials. *Cellulose* **2018**, *25*, 4275–4301. [CrossRef]
35. Shi, R.; Zhang, J.; Yang, J.; Xu, Y.; Li, C.; Chen, S.; Xu, F. Direct-Ink-Write Printing and Electrospinning of Cellulose Derivatives for Conductive Composite Materials. *Materials* **2022**, *15*, 2840. [CrossRef]
36. Du, H.; Liu, W.; Zhang, M.; Si, C.; Zhang, X.; Li, B. Cellulose nanocrystals and cellulose nanofibrils based hydrogels for biomedical applications. *Carbohydr. Polym.* **2019**, *209*, 130–144. [CrossRef]
37. Shin, M.; Song, K.H.; Burrell, J.C.; Cullen, D.K.; Burdick, J.A. Injectable and conductive granular hydrogels for 3D printing and electroactive tissue support. *Adv. Sci.* **2019**, *6*, 1901229. [CrossRef]
38. Shin, M.; Galarraga, J.H.; Kwon, M.Y.; Lee, H.; Burdick, J.A. Gallol-derived ECM-mimetic adhesive bioinks exhibiting temporal shear-thinning and stabilization behavior. *Acta Biomater.* **2019**, *95*, 165–175. [CrossRef]
39. Kim, K.; Shin, M.; Koh, M.Y.; Ryu, J.H.; Lee, M.S.; Hong, S.; Lee, H. TAPE: A medical adhesive inspired by a ubiquitous compound in plants. *Adv. Funct. Mater.* **2015**, *25*, 2402–2410. [CrossRef]
40. Shin, M.; Park, E.; Lee, H. Plant-inspired pyrogallol-containing functional materials. *Adv. Funct. Mater.* **2019**, *29*, 1903022. [CrossRef]
41. Rahim, M.A.; Centurion, F.; Han, J.; Abbasi, R.; Mayyas, M.; Sun, J.; Christoe, M.J.; Esrafilzadeh, D.; Allioux, F.M.; Ghasemian, M.B.; et al. Polyphenol-Induced Adhesive Liquid Metal Inks for Substrate-Independent Direct Pen Writing. *Adv. Funct. Mater.* **2021**, *31*, 2007336. [CrossRef]
42. Chen, Y.; Zhang, Y.; Mensaha, A.; Li, D.; Wang, Q.; Wei, Q. A plant-inspired long-lasting adhesive bilayer nanocomposite hydrogel based on redox-active Ag/Tannic acid-Cellulose nanofibers. *Carbohydr. Polym.* **2021**, *255*, 117508. [CrossRef] [PubMed]

43. Hao, S.; Shao, C.; Meng, L.; Cui, C.; Xu, F.; Yang, J. Tannic acid–silver dual catalysis induced rapid polymerization of conductive hydrogel sensors with excellent stretchability, self-adhesion, and strain-sensitivity properties. *ACS Appl. Mater. Interfaces* **2020**, *12*, 56509–56521. [CrossRef]
44. Guo, Z.; Xie, W.; Lu, J.; Guo, X.; Xu, J.; Xu, W.; Chi, Y.; Takuya, N.; Wu, H.; Zhao, L. Tannic acid-based metal phenolic networks for bio-applications: A review. *J. Mater. Chem. B* **2021**, *9*, 4098–4110. [CrossRef]
45. Fan, H.; Ma, X.; Zhou, S.; Huang, J.; Liu, Y.; Liu, Y. Highly efficient removal of heavy metal ions by carboxymethyl cellulose-immobilized Fe3O4 nanoparticles prepared via high-gravity technology. *Carbohydr. Polym.* **2019**, *213*, 39–49. [CrossRef] [PubMed]
46. Fan, H.; Wang, J.; Zhang, Q.; Jin, Z. Tannic acid-based multifunctional hydrogels with facile adjustable adhesion and cohesion contributed by polyphenol supramolecular chemistry. *ACS Omega* **2017**, *2*, 6668–6676. [CrossRef]
47. Shao, C.; Wang, M.; Meng, L.; Chang, H.; Wang, B.; Xu, F.; Yang, J.; Wan, P. Mussel-inspired cellulose nanocomposite tough hydrogels with synergistic self-healing, adhesive, and strain-sensitive properties. *Chem. Mater.* **2018**, *30*, 3110–3121. [CrossRef]
48. Choi, Y.; Park, K.; Choi, H.; Son, D.; Shin, M. Self-healing, stretchable, biocompatible, and conductive alginate hydrogels through dynamic covalent bonds for implantable electronics. *Polymers* **2021**, *13*, 1133. [CrossRef]
49. Shin, M.; Kim, K.; Shim, W.; Yang, J.W.; Lee, H. Tannic acid as a degradable mucoadhesive compound. *ACS Biomater. Sci. Eng.* **2016**, *2*, 687–696. [CrossRef]
50. Jin, S.; Kim, S.; Kim, D.S.; Son, D.; Shin, M. Optically Anisotropic Topical Hemostatic Coacervate for Naked-Eye Identification of Blood Coagulation. *Adv. Funct. Mater.* **2021**, *32*, 2110320. [CrossRef]

*Article*

# Development of Bigels Based on Date Palm-Derived Cellulose Nanocrystal-Reinforced Guar Gum Hydrogel and Sesame Oil/Candelilla Wax Oleogel as Delivery Vehicles for Moxifloxacin

Hamid M. Shaikh [1,*], Arfat Anis [1], Anesh Manjaly Poulose [1], Niyaz Ahamad Madhar [2] and Saeed M. Al-Zahrani [1]

[1] SABIC Polymer Research Centre, Department of Chemical Engineering, King Saud University, P.O. Box 800, Riyadh 11421, Saudi Arabia; aarfat@ksu.edu.sa (A.A.); apoulose@ksu.edu.sa (A.M.P.); szahrani@ksu.edu.sa (S.M.A.-Z.)
[2] Department of Physics and Astronomy, College of Sciences, King Saud University, P.O. Box 2455, Riyadh 11451, Saudi Arabia; nmadhar@ksu.edu.sa
\* Correspondence: hamshaikh@ksu.edu.sa

**Abstract:** Bigels are biphasic semisolid systems that have been explored as delivery vehicles in the food and pharmaceutical industries. These formulations are highly stable and have a longer shelf-life than emulsions. Similarly, cellulose-based hydrogels are considered to be ideal for these formulations due to their biocompatibility and flexibility to mold into various shapes. Accordingly, in the present study, the properties of an optimized guar gum hydrogel and sesame oil/candelilla wax oleogel-based bigel were tailored using date palm-derived cellulose nanocrystals (dp-CNC). These bigels were then explored as carriers for the bioactive molecule moxifloxacin hydrochloride (MH). The preparation of the bigels was achieved by mixing guar gum hydrogel and sesame oil/candelilla wax oleogel. Polarizing microscopy suggested the formation of the hydrogel-in-oleogel type of bigels. An alteration in the dp-CNC content affected the size distribution of the hydrogel phase within the oleogel phase. The colorimetry studies revealed the yellowish-white color of the samples. There were no significant changes in the FTIR functional group positions even after the addition of dp-CNC. In general, the incorporation of dp-CNC resulted in a decrease in the impedance values, except BG3 that had 15 mg dp-CNC in 20 g bigel. The BG3 formulation showed the highest firmness and fluidity. The release of MH from the bigels was quasi-Fickian diffusion mediated. BG3 showed the highest release of the drug. In summary, dp-CNC can be used as a novel reinforcing agent for bigels.

**Keywords:** date palm-derived cellulose nanocrystal; guar gum hydrogel; sesame oil; candelilla wax; oleogel; drug delivery

**Citation:** Shaikh, H.M.; Anis, A.; Poulose, A.M.; Madhar, N.A.; Al-Zahrani, S.M. Development of Bigels Based on Date Palm-Derived Cellulose Nanocrystal-Reinforced Guar Gum Hydrogel and Sesame Oil/Candelilla Wax Oleogel as Delivery Vehicles for Moxifloxacin. *Gels* **2022**, *8*, 330. https://doi.org/10.3390/gels8060330

Academic Editors: Lorenzo Bonetti, Christian Demitri and Laura Riva

Received: 28 April 2022
Accepted: 23 May 2022
Published: 24 May 2022

**Publisher's Note:** MDPI stays neutral with regard to jurisdictional claims in published maps and institutional affiliations.

**Copyright:** © 2022 by the authors. Licensee MDPI, Basel, Switzerland. This article is an open access article distributed under the terms and conditions of the Creative Commons Attribution (CC BY) license (https://creativecommons.org/licenses/by/4.0/).

## 1. Introduction

In recent years, bigel-based delivery systems have been proposed as novel biphasic systems [1]. Such systems are developed by mixing two types of gelled systems of different polarities, namely hydrogels and oleogels, under a controlled temperature [2]. Hydrogels are gelled systems that are hydrophilic systems of aqueous solvent [3], while oleogels are hydrophobic gelled systems of edible oils [4]. The mixing of the aforesaid gelled systems leads to the formation of biphasic systems, which are structurally similar to emulsions [5]. However, unlike emulsions, bigels are semisolid in nature [6]. Depending on the distribution of the gelled systems, bigels are categorized either as oleogel-in-hydrogel or hydrogel-in-oleogel [2]. Moreover, some authors have reported the formation of bi-continuous bigel [2,6]. This type of bigel does not have either a clearly dispersed or continuum phase. The main advantage of bigels is their improved stability and shelf-life [7]. The composition of the constituting oleogels and hydrogels alters the properties of the bigels. Also, the proportion of the oleogel and the hydrogel can govern their properties [8,9]. Since the bigels consist of both hydrophobic (oleogels) and hydrophilic (hydrogels) components, they are a good candidate for delivering both hydrophobic and hydrophilic bioactive agents, either individually or simultaneously [10–12]. Due to the versatile properties of bigels, they have been used for pharmaceutical, food, and cosmetic applications.

Oleogels are gel-based systems of lipids and oils [13,14]. These gels are hydrophobic, unlike hydrophilic hydrogels. In recent times, the use of oleogels in food and pharmaceutical applications has received much attention as a solid fat replacer/alternative [15]. The use of oleogels helps reduce the saturated fatty acid content and consequently increases the unsaturated fatty acid content in food products [16–18]. The oleogels entrap oils within a network structure of the gelators, which are hydrophobic. One of the common types of gelators are vegetable waxes (e.g., mango butter, cocoa butter, sunflower wax, and candelilla wax) [18–20]. Vegetable wax-based oleogels are prepared by the direct dispersion method, which is one of the most common and easiest methods of oleogel preparation [21]. Since vegetable waxes are rich in fat molecules, fat crystals are formed within the gelator network of oleogels during the synthesis process [19,20].

In the present study, candelilla wax (CW) and sesame oil (SO)-based oleogel will be used as the model oleogel. CW is extracted from the leaves of candelilla shrubs (*Euphorbia cerifera* and *Euphorbia antisyphilitica*; family: Euphorbiaceae) [22]. The shrub is mainly found in the region of northern Mexico and the southwestern United States [23,24]. CW is yellowish-brown in color and has a melting point of 62–70 °C [25]. Due to its vegan origin, it has been proposed for food applications. Oleogels of CW have been explored in recent times as saturated fat replacers [26]. SO is extracted from sesame seeds (*Sesamum indicum*; family: Pedaliaceae) [27]. The oil has a high linoleic acid and oleic acid content, which combined constitute nearly 80% of the total fatty acid content [28,29]. It is one of the most widely used cooking oils across the globe and has also been used to develop various food products.

Hydrogels are polymeric networks that entrap water molecules. In recent years, polysaccharides have been explored for developing hydrogels. Polysaccharides are naturally-occurring polymers and are inherently biocompatible. Among various polysaccharides (e.g., guar gum, alginic acid, chitosan, gum tragacanth), guar gum (GG) is one of the most widely used polysaccharides. GG is obtained from the seed of the guar plant (*Cyamopsis tetragonolobus*; family: Leguminosae) [30,31]. The polysaccharide is composed of D-mannopyranose (M) monomer units [32,33]. This natural polymer has been used in the food, pharmaceutical, and cosmetics industries for a long time. Researchers have proposed altering the properties of GG-based polymeric architectures with cellulose nanocrystals (CNCs) [34]. The inclusion of CNC into GG-based formulations help to tailor the properties of the formulations.

Similarly, nanocellulose has a remarkable skeletal structure, due to its numerous hydrophilic functional groups and nano size effect, which allows it to maintain the hydrogel's three dimensional structure to a large extent while maintaining the moisture content [35]. Also, a high degree of polymerization and a large surface area to volume ratio result in increased drug loading and binding capacity for drug release. However, a variety of factors, such as cellulose source, isolation strategy, size, and shape determine the optimum performance of nanocellulose [36]. Cellulose nanocrystals can be obtained from agricultural waste and used in a range of applications. In earlier work, we isolated cellulose nanocrystals from date palm (dp-CNC) tree residues. Furthermore, the reinforcing influence of the dp-CNCs on the bigels has not been studied yet. Therefore, it is reasonable to investigate the potential of this nanocellulose to tailor the properties of bigels.

Maharana, V. et al. [37] have proposed altering the properties of the filled hydrogels by reinforcing the dispersed phase. In the study, the authors developed gelatin-tamarind gum-filled hydrogels (also known as bigels), wherein the tamarind gum was reinforced with carbon nanotubes. The authors demonstrated that the reinforcement of the dispersed phase significantly affected the properties of the filled hydrogels. These carbon nanotubes were found to improve associative interaction among hydrogel components and to maintain the architecture of the hydrogels. Accordingly, in this study, we developed a GG hydrogel-in-CW/SO oleogel bigel. The GG hydrogel was reinforced with varying amounts of date palm-derived CNC (dp-CNC). The synthesis of the bigels was performed by the facile mixing approach. The properties of the bigels were then analyzed by microscopic, FTIR spectroscopic, XRD, impedance spectroscopic, and texture analysis methods. The developed bigels were also explored as carriers for Moxifloxacin (MH).

Moxifloxacin (MH) is categorized under the quinolone group of antibiotics. The drug is a broad-spectrum antibiotic and is used to treat a wide variety of bacterial infections. Accordingly, it has been used to treat several bacteria-induced diseases. Some of the diseases where MH is used as a drug of choice include respiratory tract infections, conjunctivitis, tuberculosis, endocarditis, and pneumonia. Though the drug's elimination half-life is ~12 h, most of the drug is excreted through feces or urine. Accordingly, there is a need to design controlled drug delivery systems that can prolong the release of the drug and hence improve bioavailability. In this regard, bigels have been proposed as a novel drug delivery system by many researchers. The knowledge gathered through this study could allow the scientific community to understand how the reinforcement of the internal phase of bigels with CNC affects the properties of the bigels.

## 2. Results and Discussion

### 2.1. Microscopic Evaluation

The analysis of the optical micrographs of bigels can provide information about the dispersion of the dispersed phase (Figure 1). Observation of the micrographs suggest the presence of globular phases within a continuum phase. As the proportion of the hydrogel phase was only 25%, it can be expected that the dispersed phase would be the hydrogel [38]. Consequently, the continuum phase would be the oleogel phase. It can be seen that BG0 (Figure 1a) showed a wide distribution in the size of the globular phases. A large number of bigger globular structures could be observed in BG0 (Figure 1a). As dp-CNC was incorporated in BG1 (Figure 1b), there was a drastic reduction in the size of the globular structures. However, some bigger size droplets could be seen. A corresponding increase in the dp-CNC content in BG2 and BG3 (Figure 1c,d), resulted in smaller globular structures, respectively. Apart from the formation of smaller globular structures, there was increased homogeneity. It can be observed that the globular size distribution was highly homogenous in BG3 (Figure 1c). A further increase in the dp-CNC content in BG4 (Figure 1e) caused a slight increase in the size of the globular structures and increased size distribution. Additionally, it can also be observed that some of the globular structures in BG0, BG1, and BG4 (Figure 1a,b,e) were apparently deformed. Nevertheless, the globular structures in BG2 and BG3 (Figure 1c,d) were relatively non-deformed. The variation in the mean droplet size (Figure 1f) suggested that the addition of dp-CNC reduced the droplet size of the internal phase. The difference in the droplet size of the internal phase among the dp-CNC-containing formulations was not significant. The droplet size was calculated manually using ImageJ software.

The polarized light micrographs of the bigels are presented in Figure 2. It can be observed from the micrographs that the dispersed phase was darker than the continuum phase, samples BG0-BG4 (Figure 2a–e). It is well-established that fat crystals appear brighter under polarizing conditions [39]. This has been attributed to the ability of fat crystals to diffract light. The diffracted light can be captured by polarizing microscopy. On the other hand, hydrogels are amorphous structures and appear as dark objects under polarizing microscopy, i.e., samples BG4 (Figure 2e). Analyzing the microarchitecture confirms that the dispersed phase was the hydrogel phase, while the continuum phase is composed of the oleogels. Further, the polarizing micrographs also corroborated the observations from the optical micrographs. Therefore, it can be concluded that the addition of dp-CNC within the inner phase of the hydrogel-in-oleogel bigels can help to tailor the size and distribution of the dispersed phase. It could be further seen that dp-CNC helped form an un-deformed dispersed phase of the hydrogels in bigels. The alteration in the properties of the dispersed phase occurs in a concentration-dependent manner.

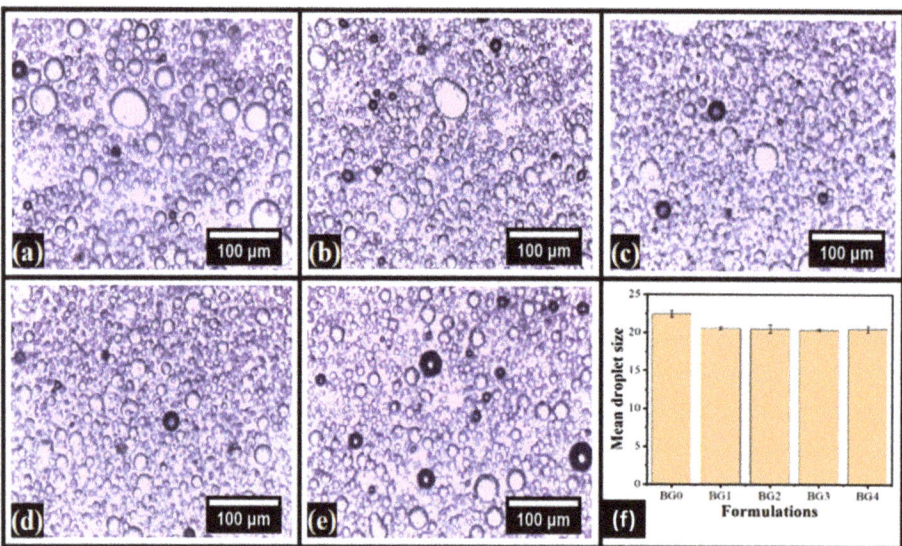

**Figure 1.** Bright-field micrographs of the bigels. (**a**) BG0, (**b**) BG1, (**c**) BG2, (**d**) BG3, and (**e**) BG4. (**f**) Variation in the mean droplet size.

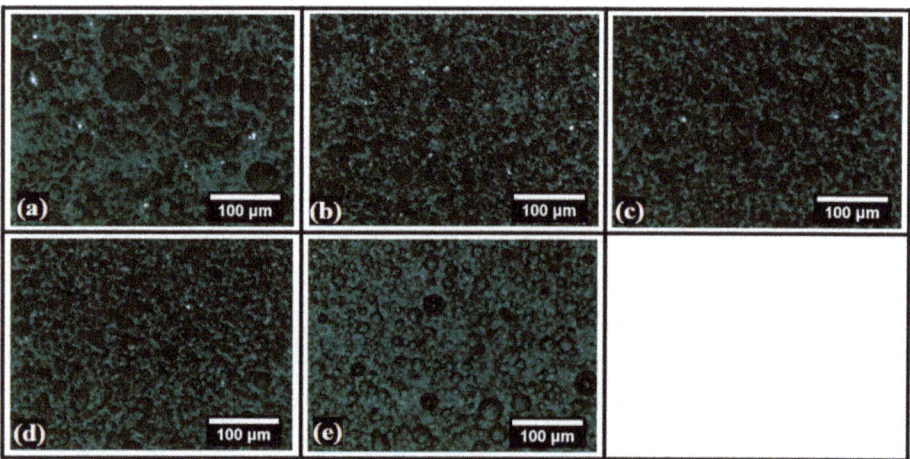

**Figure 2.** Polarized light micrographs of the bigels. (**a**) BG0, (**b**) BG1, (**c**) BG2, (**d**) BG3, and (**e**) BG4.

### 2.2. Colorimetry

The results of the colorimetric analysis have been compiled in Figure 3 and Table 1. The color analysis was carried out in the CIELab color plane, wherein L *, a *, and b * values are obtained [40]. The main advantage of this color model is that it encompasses the entire range of human visual color perception [41]. The "L *" parameter is independent of chromaticity information. It basically represents the perpetual lightness of the samples [42]. From the results, it can be observed that the average "L *" value was in the range of 97.76 and 99.07. However, the "L *" values were not statistically significant from each other. On the other hand, "a *" and "b *" values are chromic parameters, wherein "a *" and "b *" parameters represent green-red and blue-yellow opponent colors, respectively [43,44]. The negative values of "a *" and "b *" represent green and blue colors. The positive values of "a *" and "b *" represent red and yellow colors. Analysis of the "a *" and "b *" values suggest the presence of green and yellow hues within the system. The variation in the values of the parameters was not statistically significant. Apart from the aforesaid primary color parameters, derived parameters, namely whiteness index (WI) and yellowness index (YI), were also obtained from the instrument [45]. The WI was in the range of 75% and 80%, while the YI was in the range of 27% and 35%. There was no significant difference in the WI and YI values of the formulations. In summary, the formulations were highly reflective in nature with predominantly green and yellow hues. The addition of dp-CNC did not alter the color of the formulations.

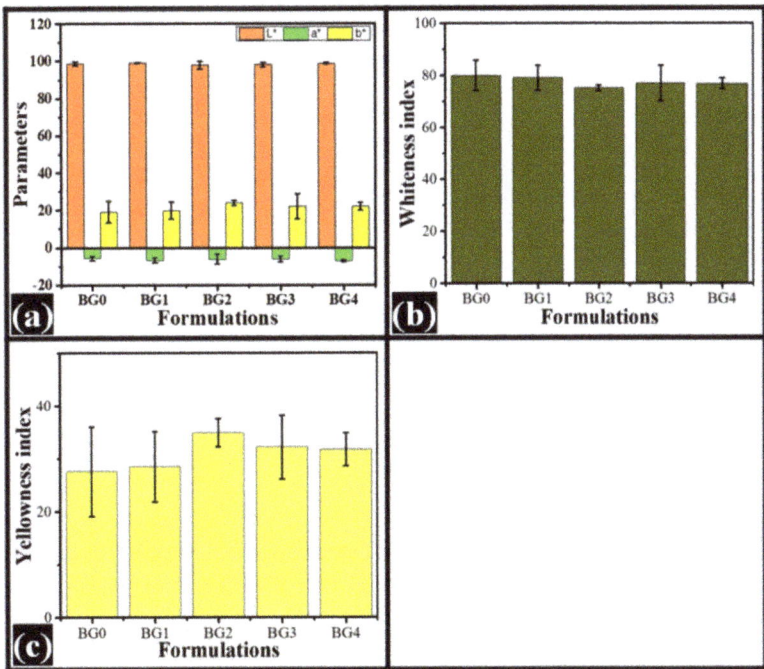

**Figure 3.** Colorimeter profile of bigels. (**a**) CIELab color parameters, (**b**) Whiteness index, (**c**) Yellowness index.

**Table 1.** Color parameters of bigels.

| Sample | L * | a * | b * | WI | YI |
| --- | --- | --- | --- | --- | --- |
| BG0 | 98.67 ± 0.96 | −5.74 ± 1.16 | 19.06 ± 5.70 | 80.02 ± 5.77 | 27.60 ± 8.49 |
| BG1 | 99.07 ± 0.19 | −6.64 ± 1.40 | 19.80 ± 4.57 | 79.10 ± 4.78 | 28.55 ± 6.64 |
| BG2 | 97.76 ± 2.03 | −6.16 ± 2.67 | 23.92 ± 1.38 | 75.20 ± 1.09 | 34.95 ± 2.68 |
| BG3 | 98.19 ± 1.17 | −6.12 ± 1.45 | 22.13 ± 6.74 | 76.97 ± 6.82 | 32.20 ± 6.06 |
| BG4 | 98.81 ± 0.37 | −7.05 ± 0.44 | 21.97 ± 2.11 | 76.90 ± 2.12 | 31.77 ± 3.15 |

### 2.3. FTIR Spectroscopy

The FTIR spectrum provides useful information about the presence of potential functional groups and interactions within them in bigel formulations. The FTIR spectra of the bigels have been provided in Figure 4. The analysis of the functional groups is mainly done in the functional group region. The FTIR spectra of the bigels showed that there was a sharp peak at 1740 cm$^{-1}$. This peak is associated with the -C=O stretching vibrations of the esters that are present in lipids and fatty acids [46].

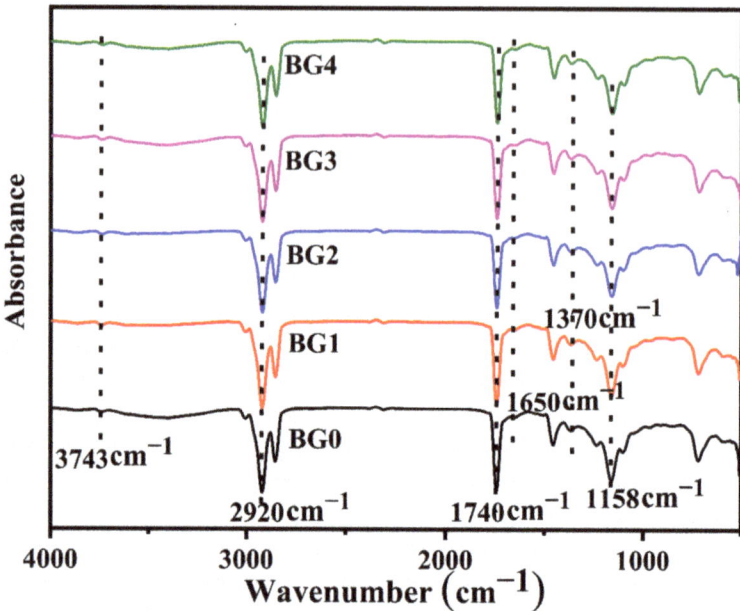

**Figure 4.** FTIR spectra of the bigels.

The presence of dual peaks at 2854 cm$^{-1}$ and 2920 cm$^{-1}$ is due to the sp3 hybridized carbon atom C-H stretching, which was in abundance in SO, CW, GG, and dp-CNC [47]. The broad peak at 3425 cm$^{-1}$ was due to the O-H stretching vibrations due to hydrogen-bonded hydroxyl groups [48]. Further, another peak was observed at 3743 cm$^{-1}$ that can be explained by the O-H stretching of the free hydroxyl groups [49]. The hydroxyl functional groups were in abundance in SO, CW, GG, and dp-CNC. Apart from these major peaks in the functional group region, a couple of minor peaks were also observed. The minor peak at 1650 cm$^{-1}$ was due to the N-H bending of primary amines that are present in GG. Apart from the peaks mentioned in the functional group region, some additional peaks at 1370.7 cm$^{-1}$, 1234.6 cm$^{-1}$, 1158.4 cm$^{-1}$, and 1102.7 cm$^{-1}$ were observed. These peaks are considered to be in the fingerprint region. These peaks can be associated with different vibration modes of methyl (-CH$_3$) and methylene (-CH$_2$) groups in SO, CW, GG, and dp-CNC [50]. In summary, all the formulations (control and dp-CNC-containing bigels) showed peaks precisely at the exact location. This indicates the presence of similar types of interactions among all the formulations, including the control sample. However, there were some variations in the intensity levels.

## 2.4. Analysis of Impedance

The impedance profiles of the bigels are provided in Figure 5. The impedance profiles showed typical capacitive behavior; wherein there is a higher impedance in the low-frequency range that reduces to a basal level in the high-frequency range [51]. The analysis of the impedance profiles suggested that the control bigel (BG0) was relatively higher than the dp-CNC-containing bigels, except BG3. This indicated that the addition of dp-CNC in the bigels reduced the impedance of the bigels in general. Among the dp-CNC-containing bigels, BG1 had the lowest impedance. In other words, the electrical conductivity of BG1 was the highest. The impedance value was correspondingly increased in BG2 and BG3, respectively, with increase in dp-CNC content. In fact, the impedance of BG3 was the highest. Thereafter, a corresponding fall in the impedance value was observed in BG4, which contained the highest amount of dp-CNC.

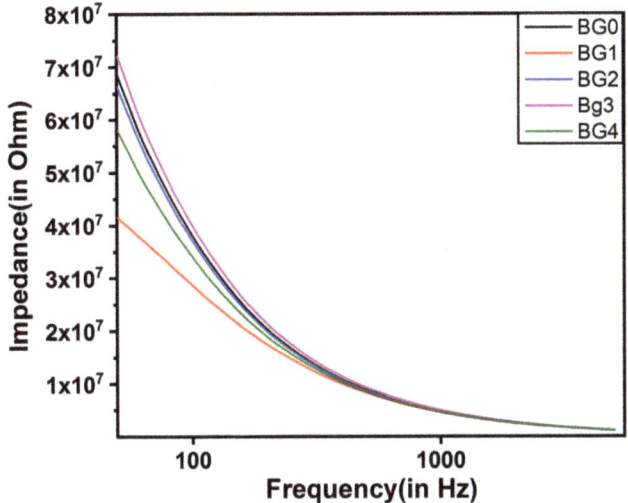

**Figure 5.** Impedance profiles of the bigels in 50 and 5000 Hz frequency range.

The observed impedance can be related to the microstructure of the bigels. From the micrographs, it was observed that BG3 showed smaller and homogenous droplets. As the amount of the hydrogel was the same in all the bigels, the smaller droplets resulted in the presence of more droplets. These droplets acted as numerous capacitive elements, thereby resulting in the highest impedance of BG3. The impedance of the other bigel formulations can also be related to droplet size and homogeneity.

### 2.5. Stress Relaxation Studies

The stress relaxation (SR) profiles and their parameter values are presented in Figure 6 and Table 2. The $F_0$ value, maximum force attained in the SR profile, of BG0 was significantly lower than the dp-CNC containing formulations ($p < 0.05$). This could be associated with the reinforcing properties of CNC. Among the dp-CNC containing formulations, there was an increase in the $F_0$ values from BG1 to BG3. Subsequently, an increase in the dp-CNC content decreased the $F_0$ value in BG4. However, the differences in the $F_0$ values among BG1 and BG2, BG2 and BG4, and BG3 and BG4 were statistically insignificant ($p > 0.05$). The $F_{60}$ values, the minimum force value at the end of the relaxation process, followed a trend similar to $F_0$ values.

The %SR value provides information about the fluidity (or elasticity) of formulations [52,53]. A higher %SR value indicated higher fluidic nature of the formulations. The analysis of the %SR values suggested that BG0 was highly fluidic in nature and had the highest %SR value. However, the %SR value of BG0 was similarly valued with that of BG3 and BG4 ($p > 0.05$). The %SR value of BG1 and BG2 was the lowest ($p > 0.05$). This suggested that the elastic component of BG1 and BG2 was higher than all the other formulations ($p < 0.05$). Interestingly, an increase in the dp-CNC content in BG3 and BG4 significantly increased the fluidic component within the bigels that were similarly to each other to control (BG0) ($p > 0.05$). In brief, BG3 had the highest firmness and, at the same time, improved fluidity.

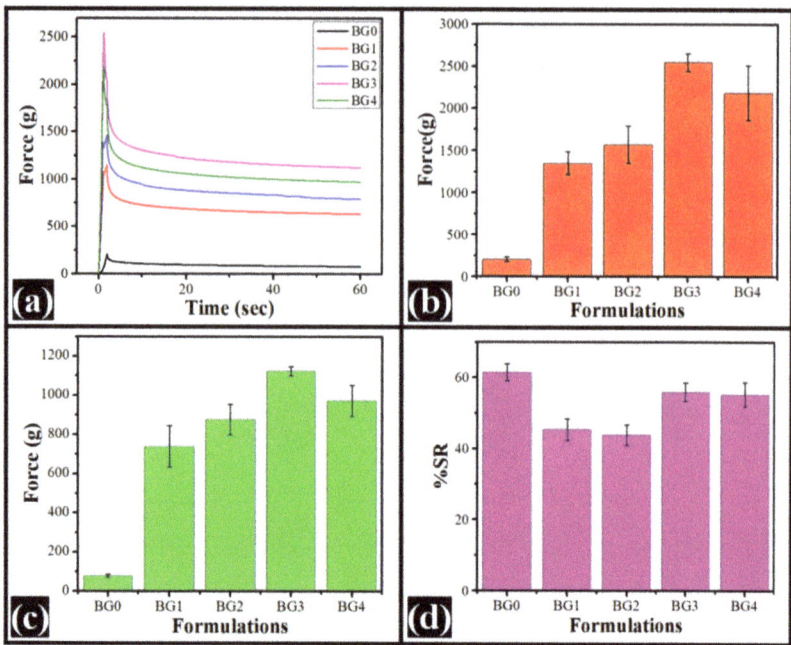

**Figure 6.** Stress relaxation study of the bigels. (**a**) SR profile, (**b**) $F_0$ profile (average ± standard deviation), (**c**) $F_{60}$ profile (average ± standard deviation), and (**d**) % SR profiles (average ± standard deviation).

**Table 2.** Stress relaxation parameters of bigel formulations.

| Formulations | F0 | F60 | %SR |
| --- | --- | --- | --- |
| BG0 | 205.37 ± 26.95 | 78.85 ± 8.09 | 61.60 ± 2.43 |
| BG1 | 1349.64 ± 133.67 | 739.42 ± 105.43 | 45.21 ± 3.00 |
| BG2 | 1569.40 ± 219.49 | 877.00 ± 77.08 | 44.18 ± 2.94 |
| BG3 | 2548.71 ± 106.68 | 1123.40 ± 25.01 | 55.92 ± 2.60 |
| BG4 | 2186.90 ± 328.30 | 971.89 ± 79.98 | 55.55 ± 3.45 |

*2.6. Drug Release Study*

The Moxifloxacin HCl is a new fluoroquinolone antibacterial agent that works against both gram-positive and gram-negative bacteria. It has better activity against anaerobes and gram-positive bacteria (such as streptococci, staphylococci, and enterococci) than ciprofloxacin [54]. Moxifloxacin HCl is used to treat bacterial infections of the respiratory tract, including community-acquired pneumonia, sinusitis, and acute exacerbations of chronic bronchitis [55]. The FDA authorized moxifloxacin HCl ophthalmic solution in April 2003 for the treatment of bacterial conjunctivitis caused by susceptible species [56]. An improved antibacterial activity was reported from chitosan/β-glycerophosphate in situ-forming thermo-sensitive hydrogel loaded with moxifloxacin HCl (0.25% $w/v$) compared to moxifloxacin HCl solution (0.5% $w/v$) [57]. Accordingly, formulations of 0.25% of moxifloxacin HCl were developed in this study.

The release profiles of the drug moxifloxacin HCl from the formulations are provided in Figure 7. It can be seen from the drug release profiles that an increase in the dp-CNC content correspondingly increased the CPDR (cumulative percent drug release) till BG3D. However, there was a subsequent decrease in the CPDR values from BG4D. Among the dp-CNC-containing formulations, the CPDR values were in the same order as that of the impedance values in the low-frequency region. Statistically, the CPDR values of the dp-CNC containing formulations were significantly higher than the control ($p < 0.05$). This is suggestive of the fact that the addition of dp-CNC promoted the release of the drug. Further, the differences in the CPDR values at the end of the experiment of BG1D and BG2D, BG2D and BG4D, and BG3D and BG4D were statistically similar to each other ($p > 0.05$). Overall, the enhanced drug release, due to the addition of dp-CNC, could be explained by faster diffusion of the drug molecules within the hydrogel matrices, which improved water absorption capacity, due to dp-CNC, may have promoted.

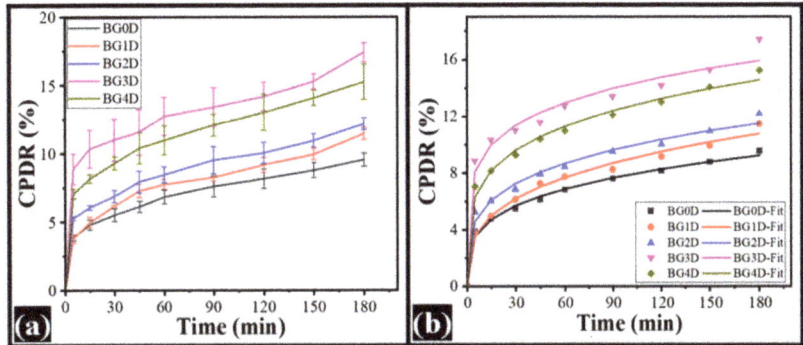

**Figure 7.** Drug release profile of bigel formulations. (a) CPDR profile and (b) KP model of CPDR.

The CPDR profiles were then fitted to the Korsmeyer-Peppas (KP) model (Equation (1)) [58]. The parameters of the KP model are formulated in Table 3. It can be observed that the diffusion factor (K) of BG0D, with respect to BG1D and BG3D, was statistically insignificant ($p > 0.05$). The incorporation of dp-CNC resulted in increase in the 'K' value in a concentration-dependent manner. However, the 'K' values of BG1D and BG3D, BG2D and BG3D, and BG3D-BG4D were statistically insignificant ($p > 0.05$). The diffusion exponent (n) values were less than 0.45, suggesting quasi-Fickian diffusion of the drug molecules during the drug release process:

$$CPDR = K \times t^n \tag{1}$$

where CPDR is cumulative percent drug release, K is the diffusion factor, t is the time of sampling, and n is the diffusion exponent.

**Table 3.** Korsmeyer-Peppas model drug release parameters.

| Formulations | Model Parameters | | |
|---|---|---|---|
| | K | n | $R^2$ |
| BG0D | $2.294 \pm 0.037$ | $0.267 \pm 0.016$ | 0.995 |
| BG1D | $2.186 \pm 0.188$ | $0.308 \pm 0.017$ | 0.993 |
| BG2D | $3.045 \pm 0.227$ | $0.256 \pm 0.007$ | 0.988 |
| BG3D | $6.023 \pm 1.726$ | $0.193 \pm 0.050$ | 0.987 |
| BG4D | $4.362 \pm 0.394$ | $0.232 \pm 0.020$ | 0.992 |

## 3. Conclusions

GG hydrogel and SO/CW oleogel-containing bigels were developed using an easy and facile method in the present study. Dp-CNC was then incorporated within the bigel in varied amounts to alter the physical and biochemical properties of the bigel. It was found that the developed bigels were smooth and stable. The bigels were yellowish-white in color. The formation of biphasic formulations was confirmed by bright field and polarizing microscopy. The IR spectra suggested that there was no change in the interactions of the bigel components after the addition of dp-CNC. The incorporation of dp-CNC improved the firmness of the formulations in a composition-dependent manner. This was associated with the reinforcing property exerted by CNCs. At lower dp-CNC content, the bigels contained more elastic components, while the fluidic component was more predominant when the dp-CNC content was on the higher side. The release of the drug MH was diffusion-mediated and followed quasi-Fickian release kinetics. In closing, it was observed that dp-CNC could be explored as a novel reinforcing and drug release agent, which could be used to develop delivery systems for antimicrobial agents. Finally, the summary of the key parameters is tabulated in Table 4.

Table 4. Summary of the key parameters.

| Formulations | Mean Droplet Size | %SR | Diffusion Factor (K) |
| --- | --- | --- | --- |
| B0 | 22.42 ± 0.42 | 61.60 ± 2.43 | 2.294 ± 0.037 |
| B1 | 20.53 ± 0.16 | 45.21 ± 3.00 | 2.186 ± 0.188 |
| B2 | 20.45 ± 0.53 | 44.18 ± 2.94 | 3.045 ± 0.227 |
| B3 | 20.30 ± 0.14 | 55.92 ± 2.60 | 6.023 ± 1.726 |
| B4 | 20.45 ± 0.41 | 55.55 ± 3.45 | 4.362 ± 0.394 |

## 4. Materials and Methods

### 4.1. Materials

Candelilla wax (CW, BiOrigins, Hampshire, UK), and Sesame oil (SO, Massy Cedex, France) were procured from the local hypermarket. Guar gum was used from Scharlab, Barcelona, Spain. Disodium hydrogen phosphate, and potassium dihydrogen phosphate were procured from Merck, Darmstadt, Germany. Date palm-derived CNCs were synthesized in our laboratory as per the method described earlier [36]. This has nanoparticles with sizes ranging from 26 nm to 61 nm, a negative zeta potential of −35 mV, and 89% crystallinity. Also, double distilled water was used throughout the study.

### 4.2. Preparation of the Formulations

#### 4.2.1. Preparation of the Oleogel

The critical gelation concentration of CW for inducing gelation of SO was initially determined. For this purpose, a specified amount of CW was added to SO. The mixture was then heated at 65 °C for 30 min to induce the dissolution of CW in SO. Then, the hot mixture was kept at 25 °C for 120 min to allow the gelation process to complete. The amount of CW was varied to determine the CGC. The CGC of CW for SO was found to be 7%. The oleogel prepared at CGC was used for further studies.

#### 4.2.2. Preparation of the GG Hydrogels

GG hydrogel was prepared by dispersing 1 g of GG in 99 g of water that was kept stirring at 800 RPM. The dp-CNC loaded GG hydrogels were prepared by dispersing a sufficient amount of dp-CNC in water, followed by the addition of GG in water. The amount of dp-CNC was added so as to maintain the dp-CNC content in the formulations, as mentioned in Table 5.

Table 5. Composition of the bigels.

| Code | Hydrogel (g) | dp-CNC (mg) | Oleogel (g) | Moxifloxacin HCl (% w/w) |
|---|---|---|---|---|
| BG0 | 25 | - | 75 | - |
| BG1 | 25 | 5 | 75 | - |
| BG2 | 25 | 10 | 75 | - |
| BG3 | 25 | 15 | 75 | - |
| BG4 | 25 | 20 | 75 | - |
| BG0D | 25 | - | 75 | 0.25 |
| BG1D | 25 | 5 | 75 | 0.25 |
| BG2D | 25 | 10 | 75 | 0.25 |
| BG3D | 25 | 15 | 75 | 0.25 |
| BG4D | 25 | 20 | 75 | 0.25 |

### 4.2.3. Preparation of the Bigels

Specified amounts of GG hydrogels (60 °C) were slowly added to the oleogel (60 °C), which was kept on homogenization at 800 RPM. The homogenization was further continued for 20 min. Then, the mixture was kept at room temperature (25 °C) for 1 h. The reduction of the temperature of the mixture induced gelation to form bigels. The drug-containing bigels were prepared by incorporating the drug (moxifloxacin hydrochloride; moxifloxacin HCl) within the GG hydrogel phase. The composition of the bigels is presented in Table 5.

### 4.3. Characterization

The microstructures of the bigels were initially visualized using an optical bright field microscope. Subsequently, polarized microscopy was performed using the same microscope, which was attached with an in-house built polarizer and an analyzer. The microscope was fitted with an ICC50-HD camera for imaging. The prepared bigels were then subjected to colorimetric analysis using a reflective colorimeter (X-Rite Ci7600 spectrophotometer, Grand Rapids, MI, USA).

The functional group identification and their interactions were studied using an FTIR (Fourier Transform infra-red spectroscope Nicolet iN10, Thermo Scientific, Winsford, UK) working in the ATR (Attenuated total reflectance) mode. The analysis was performed in the wavenumber range of 3500 cm$^{-1}$ to 500 cm$^{-1}$. An average of 32 scans was used for the final spectra. The resolution of the instrument was 4 cm$^{-1}$. The impedance analyses of the bigels were carried out using an impedance analyzer (Digilant, Pullman, WA, USA). The analysis was carried out through the parallel plate capacitive method. Briefly, the stainless electrode system, consisting of two circular parallel plates (diameter: 1 cm, distance between plates: 1 cm), was inserted into the sample. Then, the impedance was measured in the frequency range of 50 and 5000 Hz.

The viscoelastic properties of the bigels were analyzed by the stress relaxation (SR) study. The SR study was carried out using a mechanical tester (TA HD-plus, Stable Micro Systems, Haskmere, UK). The study was conducted using a 30 mm flat probe. The probe was allowed to penetrate the sample by 1 mm after sensing a force of 5 g. The initial force value at this position was regarded as $F_0$. Thereafter, the probe remained at the same location for 1 min. The force at the end of this time was regarded as $F_{60}$. During this time, the variation in the force profiles was recorded and consequently analyzed. The percentage SR (%SR) was calculated using Equation (1):

$$\%SR = \frac{F_0 - F_{60}}{F_0} \times 100 \qquad (2)$$

where %SR stands for percentage stress relaxation, $F_0$ stands for peak force at 1 mm distance, and $F_{60}$ stands for the force at the end of the 1 min relaxation process.

The drug release from the bigels was studied in a dissolution apparatus (DS-8000, LabIndia analytical Instruments Pvt. Ltd., Mumbai, India). The dissolution apparatus was connected with a basket type (Type-I) sample holder attachment. The dissolution flask was filled with 400 mL of phosphate buffer (pH 6.8, 37 °C). The phosphate buffer solution was prepared by dissolving 28.20 g of disodium hydrogen phosphate and 11.45 g of potassium dihydrogen phosphate in 1000 mL double distilled water. Then, 1 g of the bigels was inserted into the basket. The basket was then rotated at a speed of 100 RPM. Next, at a specified time period (5, 15, 30, 45, 60, 90, 120, 150, and 180 min), 5 mL of the dissolution media was withdrawn for further analysis. The withdrawn dissolution media was replaced with fresh dissolution media. Thereafter, at the end of the study, the withdrawn dissolution media was analyzed in a UV-visible spectrometer (Shimadzu 3600 UV-VIS-NIR, Kyoto, Japan). The analysis was conducted at a wavelength of 290 nm to determine the concentration of the drug released into the dissolution media. The UV-vis spectra of the standard moxifloxacin HCl solution is compiled in Figure 8.

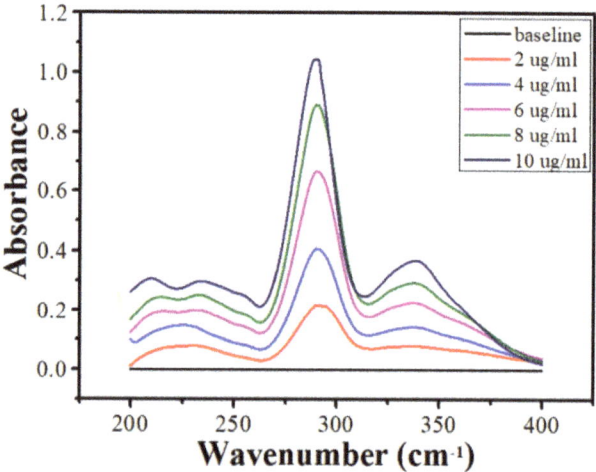

**Figure 8.** UV-vis spectra of the moxifloxacin HCl standard solutions.

**Author Contributions:** Conceptualization, H.M.S. and A.A.; methodology, H.M.S. and A.A.; investigation, H.M.S., A.M.P., and N.A.M.; resources, S.M.A.-Z. and H.M.S., writing—original draft preparation, H.M.S.; writing—review and editing, H.M.S. and A.A.; supervision, S.M.A.-Z.; funding acquisition, H.M.S. All authors have read and agreed to the published version of the manuscript.

**Funding:** This project was funded by the National Plan for Science, Technology, and Innovation (MAARIFAH), King Abdulaziz City for Science and Technology, Kingdom of Saudi Arabia, Award Number (13-NAN1120-02).

**Institutional Review Board Statement:** Not applicable.

**Informed Consent Statement:** Not applicable.

**Data Availability Statement:** Data are contained within the article.

**Conflicts of Interest:** The authors declare no conflict of interest.

## References

1. Martín-Illana, A.; Notario-Pérez, F.; Cazorla-Luna, R.; Ruiz-Caro, R.; Bonferoni, M.C.; Tamayo, A.; Veiga, M.D. Bigels as drug delivery systems: From their components to their applications. *Drug Discov. Today* **2021**, *27*, 1008–1026. [CrossRef] [PubMed]
2. Zhu, Q.; Gao, J.; Han, L.; Han, K.; Wei, W.; Wu, T.; Li, J.; Zhang, M. Development and characterization of novel bigels based on monoglyceride-beeswax oleogel and high acyl gellan gum hydrogel for lycopene delivery. *Food Chem.* **2021**, *365*, 130419. [CrossRef] [PubMed]
3. Gao, Y.; Han, X.; Chen, J.; Pan, Y.; Yang, M.; Lu, L.; Yang, J.; Suo, Z.; Lu, T. Hydrogel–mesh composite for wound closure. *Proc. Natl. Acad. Sci. USA* **2021**, *118*, e2103457118. [CrossRef] [PubMed]
4. Pan, J.; Tang, L.; Dong, Q.; Li, Y.; Zhang, H. Effect of oleogelation on physical properties and oxidative stability of camellia oil-based oleogels and oleogel emulsions. *Food Res. Int.* **2021**, *140*, 110057. [CrossRef]
5. Samui, T.; Goldenisky, D.; Rosen-Kligvasser, J.; Davidovich-Pinhas, M. The development and characterization of novel in-situ bigel formulation. *Food Hydrocoll.* **2021**, *113*, 106416. [CrossRef]
6. Soni, K.; Gour, V.; Agrawal, P.; Haider, T.; Kanwar, I.L.; Bakshi, A.; Soni, V. Carbopol-olive oil-based bigel drug delivery system of doxycycline hyclate for the treatment of acne. *Drug Dev. Ind. Pharm.* **2021**, *47*, 954–962. [CrossRef]
7. Zhuang, X.; Clark, S.; Acevedo, N. Bigels—Oleocolloid matrices—As probiotic protective systems in yogurt. *J. Food Sci.* **2021**, *86*, 4892–4900. [CrossRef]
8. Singh, V.K.; Banerjee, I.; Agarwal, T.; Pramanik, K.; Bhattacharya, M.K.; Pal, K. Guar gum and sesame oil based novel bigels for controlled drug delivery. *Colloids Surf. B Biointerfaces* **2014**, *123*, 582–592. [CrossRef]
9. Singh, V.K.; Anis, A.; Banerjee, I.; Pramanik, K.; Bhattacharya, M.K.; Pal, K. Preparation and characterization of novel carbopol based bigels for topical delivery of metronidazole for the treatment of bacterial vaginosis. *Mater. Sci. Eng. C* **2014**, *44*, 151–158. [CrossRef]
10. Sagiri, S.S.; Singh, V.K.; Kulanthaivel, S.; Banerjee, I.; Basak, P.; Battachrya, M.; Pal, K. Stearate organogel–gelatin hydrogel based bigels: Physicochemical, thermal, mechanical characterizations and in vitro drug delivery applications. *J. Mech. Behav. Biomed. Mater.* **2015**, *43*, 1–17. [CrossRef]
11. Kodela, S.P.; Pandey, P.M.; Nayak, S.K.; Uvanesh, K.; Anis, A.; Pal, K. Novel agar–stearyl alcohol oleogel-based bigels as structured delivery vehicles. *Int. J. Polym. Mater. Polym. Biomater.* **2017**, *66*, 669–678. [CrossRef]
12. Satapathy, S.; Singh, V.K.; Sagiri, S.S.; Agarwal, T.; Banerjee, I.; Bhattacharya, M.K.; Kumar, N.; Pal, K. Development and characterization of gelatin-based hydrogels, emulsion hydrogels, and bigels: A comparative study. *J. Appl. Polym. Sci.* **2015**, *132*, 41502. [CrossRef]
13. Hasda, A.M.; Vuppaladadium, S.S.R.; Qureshi, D.; Prasad, G.; Mohanty, B.; Banerjee, I.; Shaikh, H.; Anis, A.; Sarkar, P.; Pal, K. Graphene oxide reinforced nanocomposite oleogels improves corneal permeation of drugs. *J. Drug Deliv. Sci. Technol.* **2020**, *60*, 102024. [CrossRef]
14. Dhal, S.; Pal, K.; Banerjee, I.; Giri, S. Upconversion nanoparticle incorporated oleogel as probable skin tissue imaging agent. *Chem. Eng. J.* **2020**, *379*, 122272. [CrossRef]
15. Lee, S. Utilization of foam structured hydroxypropyl methylcellulose for oleogels and their application as a solid fat replacer in muffins. *Food Hydrocoll.* **2018**, *77*, 796–802. [CrossRef]
16. Chen, C.; Zhang, C.; Zhang, Q.; Ju, X.; Wang, Z.; He, R. Study of monoglycerides enriched with unsaturated fatty acids at sn-2 position as oleogelators for oleogel preparation. *Food Chem.* **2021**, *354*, 129534. [CrossRef]
17. Sahu, D.; Bharti, D.; Kim, D.; Sarkar, P.; Pal, K. Variations in Microstructural and Physicochemical Properties of Candelilla Wax/Rice Bran Oil–Derived Oleogels Using Sunflower Lecithin and Soya Lecithin. *Gels* **2021**, *7*, 226. [CrossRef]
18. Bharti, D.; Kim, D.; Cerqueira, M.A.; Mohanty, B.; Habibullah, S.; Banerjee, I.; Pal, K. Effect of Biodegradable Hydrophilic and Hydrophobic Emulsifiers on the Oleogels Containing Sunflower Wax and Sunflower Oil. *Gels* **2021**, *7*, 133. [CrossRef]
19. Qureshi, D.; Nadikoppula, A.; Mohanty, B.; Anis, A.; Cerqueira, M.; Varshney, M.; Pal, K. Effect of carboxylated carbon nanotubes on physicochemical and drug release properties of oleogels. *Colloids Surf. A Physicochem. Eng. Asp.* **2021**, *610*, 125695. [CrossRef]
20. Dhal, S.; Qureshi, D.; Mohanty, B.; Maji, S.; Anis, A.; Kim, D.; Sarkar, P.; Pal, K. Kokum butter and rice bran oil-based oleogels as novel ocular drug delivery systems. In *Advances and Challenges in Pharmaceutical Technology*; Elsevier: Amsterdam, The Netherlands, 2021; pp. 147–179.
21. Pinto, T.C.; Martins, A.J.; Pastrana, L.; Pereira, M.C.; Cerqueira, M.A. Oleogel-based systems for the delivery of bioactive compounds in foods. *Gels* **2021**, *7*, 86. [CrossRef]
22. Martinez, R.; Rosado, C.; Velasco, M.; Lannes, S.; Baby, A. Main features and applications of organogels in cosmetics. *Int. J. Cosmet. Sci.* **2019**, *41*, 109–117. [CrossRef] [PubMed]
23. Aranda-Ledesma, N.E.; Bautista-Hernández, I.; Rojas, R.; Aguilar-Zárate, P.; del Pilar Medina-Herrera, N.; Castro-López, C.; Martínez-Ávila, G.C.G. Candelilla wax: Prospective suitable applications within the food field. *LWT* **2022**, *159*, 113170. [CrossRef]
24. Navarro-Guajardo, N.; García-Carrillo, E.M.; Espinoza-González, C.; Téllez-Zablah, R.; Dávila-Hernández, F.; Romero-García, J.; Ledezma-Pérez, A.; Mercado-Silva, J.A.; Torres, C.A.P.; Pariona, N. Candelilla wax as natural slow-release matrix for fertilizers encapsulated by spray chilling. *J. Renew. Mater.* **2018**, *6*, 226–236. [CrossRef]
25. Redondas, C.E.; Baümler, E.R.; Carelli, A.A. Sunflower wax recovered from oil tank settlings: Revaluation of a waste product from the oilseed industry. *J. Sci. Food Agric.* **2020**, *100*, 201–211. [CrossRef]

26. Temkov, M.; Mureșan, V. Tailoring the Structure of Lipids, Oleogels and Fat Replacers by Different Approaches for Solving the Trans-Fat Issue—A Review. *Foods* **2021**, *10*, 1376. [CrossRef]
27. Kumar, Y.; Singh, S.; Patel, K.K. Effect of sowing dates on severity of the pathogen Myrothecium leaf spot of sesame (*Sesamum indicum* L.). *Pharma Innov. J.* **2022**, *11*, 504–506.
28. Feng, W.; Qin, C.; Chu, Y.; Berton, M.; Lee, J.B.; Zgair, A.; Bettonte, S.; Stocks, M.J.; Constantinescu, C.S.; Barrett, D.A.; et al. Natural sesame oil is superior to pre-digested lipid formulations and purified triglycerides in promoting the intestinal lymphatic transport and systemic bioavailability of cannabidiol. *Eur. J. Pharm. Biopharm.* **2021**, *162*, 43–49. [CrossRef]
29. Choi, J.-Y.; Moon, K.-D. Non-destructive discrimination of sesame oils via hyperspectral image analysis. *J. Food Compos. Anal.* **2020**, *90*, 103505. [CrossRef]
30. Sharma, G.; Sharma, S.; Kumar, A.; Ala'a, H.; Naushad, M.; Ghfar, A.A.; Mola, G.T.; Stadler, F.J. Guar gum and its composites as potential materials for diverse applications: A review. *Carbohydr. Polym.* **2018**, *199*, 534–545. [CrossRef]
31. George, A.; Shah, P.A.; Shrivastav, P.S. Guar gum: Versatile natural polymer for drug delivery applications. *Eur. Polym. J.* **2019**, *112*, 722–735. [CrossRef]
32. Abdel-raouf, M.E.-S.; Sayed, A.; Mostafa, M. Application of Guar Gum and Its Derivatives in Agriculture. In *Gums, Resins and Latexes of Plant Origin: Chemistry, Biological Activities and Uses*; Murthy, H.N., Ed.; Springer International Publishing: Cham, Switzerland, 2021; pp. 1–17.
33. Palumbo, G.; Berent, K.; Proniewicz, E.; Banaś, J. Guar Gum as an Eco-Friendly Corrosion Inhibitor for Pure Aluminium in 1-M HCl Solution. *Materials* **2019**, *12*, 2620. [CrossRef] [PubMed]
34. Yagoub, H.; Zhu, L.; Shibraen, M.H.; Xu, X.; Babiker, D.M.; Xu, J.; Yang, S. Complex membrane of cellulose and chitin nanocrystals with cationic guar gum for oil/water separation. *J. Appl. Polym. Sci.* **2019**, *136*, 47947. [CrossRef]
35. Wang, C.; Bai, J.; Tian, P.; Xie, R.; Duan, Z.; Lv, Q.; Tao, Y. The Application Status of Nanoscale Cellulose-Based Hydrogels in Tissue Engineering and Regenerative Biomedicine. *Front. Bioeng. Biotechnol.* **2021**, *9*, 939. [CrossRef] [PubMed]
36. Shaikh, H.M.; Anis, A.; Poulose, A.M.; Al-Zahrani, S.M.; Madhar, N.A.; Alhamidi, A.; Alam, M.A. Isolation and Characterization of Alpha and Nanocrystalline Cellulose from Date Palm (*Phoenix dactylifera* L.) Trunk Mesh. *Polymers* **2021**, *13*, 1893. [CrossRef] [PubMed]
37. Maharana, V.; Gaur, D.; Nayak, S.K.; Singh, V.K.; Chakraborty, S.; Banerjee, I.; Ray, S.S.; Anis, A.; Pal, K. Reinforcing the inner phase of the filled hydrogels with CNTs alters drug release properties and human keratinocyte morphology: A study on the gelatin- tamarind gum filled hydrogels. *J. Mech. Behav. Biomed. Mater.* **2017**, *75*, 538–548. [CrossRef]
38. Paul, S.R.; Qureshi, D.; Yogalakshmi, Y.; Nayak, S.K.; Singh, V.K.; Syed, I.; Sarkar, P.; Pal, K. Development of bigels based on stearic acid-rice bran oil oleogels and tamarind gum hydrogels for controlled delivery applications. *J. Surfactants Deterg.* **2018**, *21*, 17–29. [CrossRef]
39. Da Pieve, S.; Calligaris, S.; Co, E.; Nicoli, M.C.; Marangoni, A.G. Shear Nanostructuring of Monoglyceride Organogels. *Food Biophys.* **2010**, *5*, 211–217. [CrossRef]
40. Masaoka, K.; Jiang, F.; Fairchild, M.D.; Heckaman, R.L. Analysis of color volume of multi-chromatic displays using gamut rings. *J. Soc. Inf. Disp.* **2020**, *28*, 273–286. [CrossRef]
41. Salueña, B.H.; Gamasa, C.S.; Rubial, J.M.D.; Odriozola, C.A. CIELAB color paths during meat shelf life. *Meat Sci.* **2019**, *157*, 107889.
42. Anwarul, S. An Efficient Minimum Spanning Tree-Based Color Image Segmentation Approach. In *International Advanced Computing Conference*; Springer: Cham, Switzerland, 2021; pp. 588–598.
43. Palugan, L.; Spoldi, M.; Rizzuto, F.; Guerra, N.; Uboldi, M.; Cerea, M.; Moutaharrik, S.; Melocchi, A.; Gazzaniga, A.; Zema, L. What's next in the use of opacifiers for cosmetic coatings of solid dosage forms? Insights on current titanium dioxide alternatives. *Int. J. Pharm.* **2022**, *616*, 121550. [CrossRef]
44. Ware, C.; Turton, T.L.; Bujack, R.; Samsel, F.; Shrivastava, P.; Rogers, D.H. Measuring and modeling the feature detection threshold functions of colormaps. *IEEE Trans. Vis. Comput. Graph.* **2018**, *25*, 2777–2790. [CrossRef] [PubMed]
45. Kumar, A.; Srivastav, P.P.; Pravitha, M.; Hasan, M.; Mangaraj, S.; Prithviraj, V.; Verma, D.K. Comparative study on the optimization and characterization of soybean aqueous extract based composite film using response surface methodology (RSM) and artificial neural network (ANN). *Food Packag. Shelf Life* **2022**, *31*, 100778.
46. Tenyang, N.; Tiencheu, B.; Tonfack Djikeng, F.; Morfor, A.T.; Womeni, H.M. Alteration of the lipid of red carp (*Cyprinus carpio*) during frozen storage. *Food Sci. Nutr.* **2019**, *7*, 1371–1378. [CrossRef]
47. Alam, M.S.; Mukherjee, N.; Ahmed, S.F. Optical properties of diamond like carbon nanocomposite thin films. In *AIP Conference Proceedings*; AIP Publishing LLC: Melville, NY, USA, 2018; p. 090018.
48. Makarem, M.; Lee, C.M.; Sawada, D.; O'Neill, H.M.; Kim, S.H. Distinguishing surface versus bulk hydroxyl groups of cellulose nanocrystals using vibrational sum frequency generation spectroscopy. *J. Phys. Chem. Lett.* **2018**, *9*, 70–75. [CrossRef] [PubMed]
49. Gaweł, B.A.; Ulvensøen, A.; Łukaszuk, K.; Arstad, B.; Muggerud, A.M.F.; Erbe, A. Structural evolution of water and hydroxyl groups during thermal, mechanical and chemical treatment of high purity natural quartz. *RSC Adv.* **2020**, *10*, 29018–29030. [CrossRef] [PubMed]
50. Calabrò, E.; Magazù, S. Methyl and methylene vibrations response in amino acids of typical proteins in water solution under high-frequency electromagnetic field. *Electromagn. Biol. Med.* **2019**, *38*, 271–278. [CrossRef] [PubMed]
51. Mech-Dorosz, A.; Khan, M.S.; Mateiu, R.V.; Hélix-Nielsen, C.; Emnéus, J.; Heiskanen, A. Impedance characterization of biocompatible hydrogel suitable for biomimetic lipid membrane applications. *Electrochim. Acta* **2021**, *373*, 137917. [CrossRef]

52. Wei, C.; Wu, M. An Eulerian nonlinear elastic model for compressible and fluidic tissue with radially symmetric growth. *arXiv* **2021**, arXiv:2103.09427.
53. Joshi, Y.M. Thixotropy, nonmonotonic stress relaxation, and the second law of thermodynamics. *J. Rheol.* **2022**, *66*, 111–123. [CrossRef]
54. Balfour, J.A.B.; Wiseman, L.R. Moxifloxacin. *Drugs* **1999**, *57*, 363–373. [CrossRef]
55. Muijsers, R.B.R.; Jarvis, B. Moxifloxacin in uncomplicated skin and skin structure infections. *Drugs* **2002**, *62*, 967–973; discussion 974–965. [CrossRef] [PubMed]
56. Al Omari, M.M.; Jaafari, D.S.; Al-Sou'od, K.A.; Badwan, A.A. Moxifloxacin hydrochloride. *Profiles Drug Subst. Excip. Relat. Methodol.* **2014**, *39*, 299–431. [CrossRef] [PubMed]
57. Asfour, M.H.; Abd El-Alim, S.H.; Awad, G.E.A.; Kassem, A.A. Chitosan/β-glycerophosphate in situ forming thermo-sensitive hydrogel for improved ocular delivery of moxifloxacin hydrochloride. *Eur. J. Pharm. Sci.* **2021**, *167*, 106041. [CrossRef] [PubMed]
58. Ge, M.; Li, Y.; Zhu, C.; Liang, G.; SM, J.A.; Hu, G.; Gui, Y. Preparation of organic-modified magadiite–magnetic nanocomposite particles as an effective nanohybrid drug carrier material for cancer treatment and its properties of sustained release mechanism by Korsmeyer–Peppas kinetic model. *J. Mater. Sci.* **2021**, *56*, 14270–14286. [CrossRef]

Article

# Preparation and Enzyme Degradability of Spherical and Water-Absorbent Gels from Sodium Carboxymethyl Cellulose

Sayaka Fujita [1,*], Toshiaki Tazawa [2] and Hiroyuki Kono [1,*]

1. Division of Applied Chemistry and Biochemistry, National Institute of Technology, Tomakomai College, Tomakomai 059-1275, Japan
2. R&D Center, S.T. Corporation, Shinjuku-ku, Tokyo 161-0033, Japan; t-tazawa@st-c.co.jp
* Correspondence: fujita@tomakomai-ct.ac.jp (S.F.); kono@tomakomai-ct.ac.jp (H.K.); Tel.: +81-144-67-8038 (S.F.); +81-144-67-8036 (H.K.)

**Abstract:** To synthesize a biodegradable alternative to spherical polyacrylic acid absorbent resin, spherical hydrogel particles were prepared from carboxymethyl cellulose (CMC) dissolved in an aqueous solution, using ethylene glycol diglycidyl ether (EGDE) as a crosslinking agent. The effect of varying the initial CMC concentration and feed amount of EGDE on the shape, water absorbency, water-holding capacity, and enzyme degradability of the resultant CMC hydrogels was determined. The reaction solution was poured into fluid paraffin, and spherical hydrogel particles were obtained via the shear force from stirring. The shape and diameter of the spherical hydrogel particles in the swollen state depended on the CMC concentration. The spherical hydrogel particles obtained by increasing the amount of EGDE resulted in a decrease in absorbency. Additionally, all the spherical hydrogel particles were degraded by cellulase. Thus, spherical biodegradable hydrogel particles were prepared from CMC, and the particle size and water absorption of the hydrogel could be controlled in the range of 5–18 mm and 30–90 g·g$^{-1}$ in the swollen state, respectively. As an alternative to conventional superabsorbent polymers, the spherical CMC hydrogels are likely to be useful in industrial and agricultural applications.

**Keywords:** carboxymethyl cellulose; superabsorbent gels; enzymatic biodegradability; spherical control; molding processability

## 1. Introduction

Superabsorbent polymers (SAPs) can absorb and retain extremely large amounts of water [1]. The most popular SAP is crosslinked sodium polyacrylate (SPA), which is industrially produced by copolymerizing acrylic acid and acrylic acid sodium salt [2]. SAPs are widely used in various applications, such as baby diapers, feminine sanitary products, adult incontinence pads, carriers for air fresheners, and agricultural water retention materials. One of the biggest drawbacks of industrial SAPs, including SPA, is that they are non-biodegradable, despite their widespread use in disposable goods [3]. As society moves toward more sustainable and fossil fuel-free commodities and processes, developing biodegradable SAPs is imperative to prevent microplastic accumulation associated with synthetic polymers that causes environmental pollution.

The primary candidates of raw materials for biodegradable SAPs are naturally occurring and inedible polysaccharides, such as cellulose [4], chitin [5], chitosan [6], guar gum [7], starch [8], and carrageenan [9]. These polysaccharides exhibit biocompatible and biodegradable properties; thus, they are of considerable interest for the development of environmentally friendly and biocompatible materials. Structurally, these polysaccharides are polymers consisting of neutral and/or glycosidically bonded amino sugars; hence, they exhibit no water-absorbing properties. Therefore, various methods have been reported to impart the same water-absorbing performance as that of SAPs onto these polysaccharides, while maintaining their biodegradability [4–9]. Most of these methods are based on the substitution of the hydroxyl and/or amino groups of the polysaccharide chain with sodium carboxylate groups and intermolecular crosslinking between the polysaccharide chains. For example, carboxymethyl cellulose sodium salt (CMC), a water-soluble cellulose derivative, can be converted into a biodegradable superabsorbent polymer by intermolecular crosslinking using diepoxy crosslinking agents, such as ethylene glycol diglycidyl ether (EGDE). Here, the sodium carboxylate groups of the CMC and the intermolecular crosslinking between the CMC chains are responsible for the absorption and retention of water molecules inside the structure, respectively [10–12]. The same method was used to develop water-absorbent polysaccharides. The biodegradability of these polysaccharide-based water-absorbent polymers strongly depends on the crosslinking agent used for their preparation and the crosslinking density of the resultant polymer; an increase in the crosslinking density generally leads to a decrease in biodegradability. Therefore, when designing a polysaccharide-based SAP, it is important to consider not only its water-absorbing properties but also the crosslinking agent and crosslinking density.

Difficulty in molding processability is another challenge to overcome during the synthesis of polysaccharide-based SAPs. SAPs with different shapes are utilized for different purposes. The powder form of SAP is used for baby diapers and feminine sanitary products, while the spherical form is used as a carrier for air fresheners and horticultural water retention materials. Most polysaccharide-based SAPs reported to date are obtained as solid lumps, which are then crushed into a powder [13,14]. CMC-based SAPs are no exception. The majority of the CMC-based SAPs are obtained in irregular shapes. This is because polysaccharide-based SAPs, including CMC-based SAPs, are not suitable for injection molding or extrusion molding, which are both molding processes for resins, due to their poor thermal plasticity and solubility in organic solvents. In the case of SPA, the spherical shape of the resin can be controlled by emulsifying droplets of the starting material, consisting of a mixture containing acrylic acid, acrylic acid sodium salt, crosslinking agent, etc., in a poor solvent. In addition, the spherical resin size can be regulated by controlling the emulsified micelle size through the selection of surfactants [15,16]. Therefore, spherical resin particles of various sizes are available for application, while particles with a diameter of 5–15 mm in the swollen state are mainly used for disposable applications. On the other hand, research on the synthesis of polysaccharide-based spherical particles has primarily focused on nanoscale particles for biological applications, such as drug delivery carriers [17,18]. This method forms micelles from a polysaccharide aqueous solution in a poor solvent using a surfactant, followed by producing nanomicelles by ultrasound irradiation [19–21]. This makes it difficult to synthesize polysaccharide-based SAPs for general applications. Therefore, to develop polysaccharide-based SAPs as alternatives to SPA, it is necessary to develop synthesis technologies that allow for the precise control of their shape and size in general and disposable applications.

In this study, a polysaccharide-based SAP with a water absorption capacity and shape comparable to that of the existing spherical SAP was prepared using CMC and EGDE as the crosslinking agent. An aqueous alkaline solution, comprising CMC and a crosslinking agent, was added to liquid paraffin, and the shear force of stirring the liquid paraffin and the surface tension of the water–oil interface produced spherical particles of the alkaline solution (Figure 1). The initial CMC ratio and the feed amount of EGDE were varied to determine their effects on the shape, water absorbency, water-holding capacity, and cellulase degradability of the resultant CMC hydrogel (CMCG). Commercial SPA particles were used as a comparative reference. This study addressed the challenge of molding processability of polysaccharide-based SAPs and contributed to the development of a sustainable society through the use of biodegradable carriers for air fresheners and horticultural water retention materials.

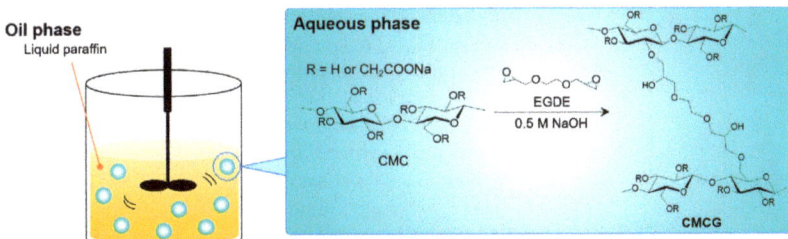

**Figure 1.** Preparation of the spherical carboxymethyl cellulose sodium salt hydrogel (CMCG) by the formation of a water droplet in liquid paraffin.

## 2. Results and Discussion

### 2.1. Preparation and Characterization of the Spherical CMCG

#### 2.1.1. Optimization of the Initial CMC Concentration

To obtain spherical CMCG particles, the reaction mixture containing CMC and EGDE dissolved in the alkaline solution was dropped into liquid paraffin and stirred to facilitate the crosslinking reaction (Figure 1). After being placed in the liquid paraffin, the CMC mixture precipitated at the bottom of the beaker at a stirring speed of 200 rpm. Thus, these samples were lumpy, with no spherical particles formed. Because the surfacing of the reaction mixture prevented adhesion to the stirring blade and the bottom of the beaker, spherical particles were obtained at stirring speeds exceeding 300 rpm. The particle size was affected by the stirring speed, and the particle size dropped as the stirring speed increased (Figure S1). The shear force increased as the stirring speed increased. The increased stirring speed was expected to increase shear force and slice the CMC mixture into smaller droplets [22,23], reducing the CMCG particle size. While maintaining a constant stirring speed of 300 rpm, a series of seven CMCG samples were prepared by varying the CMC concentration and EGDE feed amount (Table 1) to investigate their effects on the shape, water absorption properties, and biodegradability of the CMCGs obtained.

**Table 1.** Initial feed amount of carboxymethyl cellulose sodium salt (CMC) and ethylene glycol diglycidyl ether (EGDE) and yields from the preparation of the spherical CMC hydrogel (CMCG) samples.

| Sample [a] | Initial Feed Amount | | CMC Concentration (wt%) [c] | Feed Mass Ratio of EGDE/CMC | Yields [d] |
|---|---|---|---|---|---|
| | CMC (AGU [b]) | EGDE | | | |
| CMCG$_{2.5,0.4}$ | 1.0 g (4.6 mmol) | 0.4 g (2.3 mmol) | 2.5 | 0.4 | - |
| CMCG$_{5,0.4}$ | 2.0 g (9.2 mmol) | 0.8 g (4.6 mmol) | 5 | 0.4 | - |
| CMCG$_{7.5,0.4}$ | 3.0 g (13.8 mmol) | 1.2 g (6.9 mmol) | 7.5 | 0.4 | 2.5 g (58%) |
| CMCG$_{10,0.4}$ | 4.0 g (18.3 mmol) | 1.6 g (9.2 mmol) | 10 | 0.4 | 5.3 g (90%) |
| CMCG$_{15,0.4}$ | 6.0 g (27.5 mmol) | 2.4 g (13.8 mmol) | 15 | 0.4 | 5.6 g (65%) |
| CMCG$_{10,0.2}$ | 4.0 g (18.3 mmol) | 0.8 g (4.6 mmol) | 10 | 0.2 | 4.3 g (86%) |
| CMCG$_{10,0.6}$ | 4.0 g (18.3 mmol) | 2.4 g (13.8 mmol) | 10 | 0.6 | 4.0 g (60%) |

[a] X and Y in the abbreviation CMCG$_{X,Y}$ refer to the CMC concentration (wt%) and mass ratio of EGDE/CMC, respectively. [b] AGU is the mole of anhydroglucose unit of CMC. [c] CMC dissolved in 40 mL of 0.5 M NaOH. [d] Yields (%) were determined using the following equation: (mass of CMCG/g) × 100/(sum of mass of CMC and EGDE/g).

To investigate the effect of the CMC concentration on the formation of spherical shapes, **CMCG$_{2.5,0.4}$**, **CMCG$_{5,0.4}$**, **CMCG$_{7.5,0.4}$**, **CMCG$_{10,0.4}$**, and **CMCG$_{15,0.4}$** were prepared by setting the initial CMC concentrations to 2.5, 5.0, 7.5, 10, and 15 wt%, respectively (Table 1). The feed mass ratio of EGDE to CMC (EGDE/CMC) was fixed at 0.4. The yields of the spherical CMCG are summarized in Table 1. For **CMCG$_{2.5,0.4}$** and **CMCG$_{5,0.4}$**, when the reaction mixture was placed in the liquid paraffin, it immediately precipitated at the bottom of the beaker. Thus, these samples were lumpy, and no spherical particles were formed. For **CMCG$_{7.5,0.4}$**, although spherical particles were formed by floating in paraffin, a portion of the aqueous solution precipitated at the bottom, resulting in a relatively low yield (58%). For **CMCG$_{10,0.4}$**, most of the reaction mixture was converted into spherical particles because of the increase in CMC concentration. Thus, the yield reached 90%, which was almost in agreement with the yield of the CMCG samples (80–91%) prepared without liquid paraffin [10]. However, when the CMC concentration was increased to 15 wt% (**CMCG$_{15,0.4}$**), the aqueous solution adhered to the surface of the stirring blade, owing to its high viscosity. Furthermore, increasing the CMC concentration to 15 wt% decreased the formation of spherical particles with a relatively low yield of 65%. The yield data suggest that the concentration of CMC, which is related to the specific gravity and viscosity of the reaction solution, was critical for the formation of the spherical particles and that 10 wt% of the initial CMC concentration was optimal.

Then, the absorbency of **CMCG$_{7.5,0.4}$**, **CMCG$_{5,0.4}$**, and **CMCG$_{15,0.4}$** was investigated using phosphate-buffered saline (PBS) as the absorbing solution. Figure 2 compares the time-dependency of the PBS absorbency for the CMCG samples and SPA. The absorbency increased with time and reached equilibrium after 24 h for **CMCG$_{10,0.4}$**, **CMCG$_{15,0.4}$**, and SPA, and after 48 h for **CMCG$_{7.5,0.4}$**. The maximum absorbency at equilibrium increased with a decrease in the initial CMC concentration. This phenomenon can be attributed to the molecular structure of CMCG, including its degree of crosslinking. It has been reported that in a CMC hydrogel using epichlorohydrin, an epoxide crosslinking agent [24], the low concentration of CMC in the reaction mixture led to a decrease in the crosslinking density of the CMC chains. The Fourier transform infrared (FTIR) spectra of CMCGs suggest that the degree of crosslinking decreases with decreasing initial CMC concentration (Figure S2 and Table S1). The less crosslinked structure permits the expansion of the CMC molecules in the hydrogels, resulting in higher absorption.

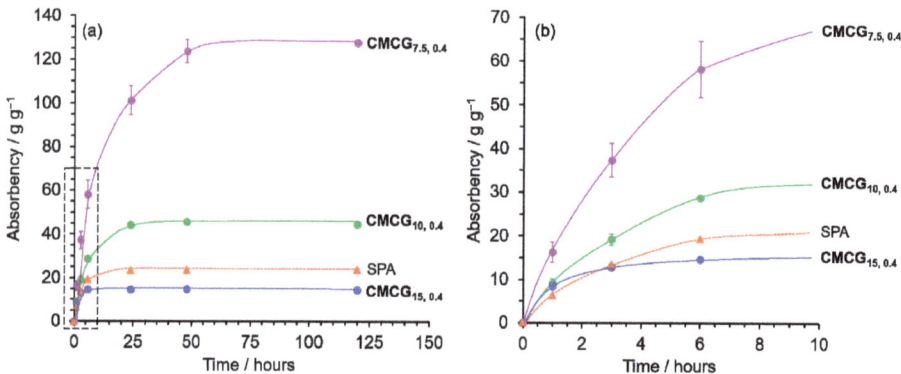

**Figure 2.** Time-dependence of the absorbency of the CMCGs and sodium polyacrylate (SPA) (**a**) toward phosphate-buffered saline (PBS) and (**b**) initial stage of the absorbency.

The photographs of **CMCG$_{7.5,0.4}$**, **CMCG$_{5,0.4}$**, and **CMCG$_{15,0.4}$** in the dried and swollen states are shown in Figure 3, whereas Figure S4 shows photographs of SPA for comparison. The dried CMCGs were white or yellowish-white particles, which changed to more yellowish as the initial CMC concentration increased. Because EGDE is a slightly yellow-colored liquid and CMC is a white powder, this finding suggests that as the CMC concentration was increased, the crosslinking degree of EGDE also increased; this supports the absorbency results. The dried CMCG transformed into transparent spherical particles after swelling in PBS (Figure 3b). Furthermore, none of the CMCG samples deformed under their weight and retained their shape, except for **CMCG$_{7.5,0.4}$**. This was because **CMCG$_{7.5,0.4}$** had a low CMC concentration, which led to decreased crosslinking between the CMC chains. Because some of the CMCG particles were flattened, the major and minor diameters were measured from 50 random particles. Figure 4 shows the size distribution of the major and minor diameters of CMCG swollen with PBS. The average values of the major diameters were 14.0 ± 2.1 (**CMCG$_{7.5,0.4}$**), 10.7 ± 2.4 (**CMCG$_{10,0.4}$**), and 8.8 ± 1.9 mm (**CMCG$_{15,0.4}$**). The average values of the minor diameters were 12.6 ± 2.2 (**CMCG$_{7.5,0.4}$**), 10.0 ± 1.8 (**CMCG$_{10,0.4}$**), and 6.7 ± 0.8 mm (**CMCG$_{15,0.4}$**). The major and minor diameters decreased with increasing CMC concentrations. When the CMC solution was poured into the liquid paraffin, the CMC solution was separated, owing to the difference in surface tension between the water and liquid paraffin. The CMC solution was sliced by the shear force of the propeller blade and formed sphere droplets to reduce the surface area. However, the CMC solution was not only divided into droplets by the shear force but also coalesced into bigger droplets. As the reaction progressed, an intermolecular crosslinked structure was formed, and the CMC droplets were stably dispersed in the liquid paraffin without coalescing. A high CMC concentration is expected to increase the chance of contact between the unsubstituted hydroxy groups of CMC and EGDE, which results in the rapid formation of an intermolecular crosslinked structure. Thus, it is likely that an intermolecular crosslinked structure was formed in the early stages of the reaction, which stabilized the droplet shape. Furthermore, the size of the CMCG decreased as coalescence was suppressed. In contrast, at low CMC concentrations, the reaction progressed slower than at high CMC concentrations. Because it takes time to stabilize the droplet shape by intermolecular crosslink formation, the CMCG size increased, owing to the coalescing droplets. Furthermore, the high surface tension of the CMC solution led to a decrease in the droplet size. The surface tension increased with increasing solution viscosity, which increased with increasing CMC concentration. That is, the surface tension increased with increasing CMC concentration, which caused a decrease in the droplet size.

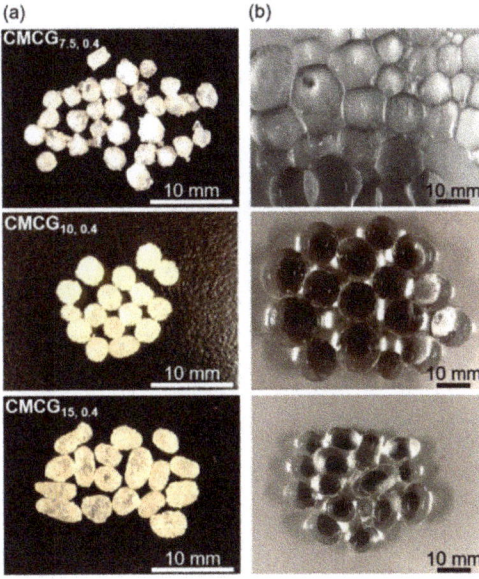

**Figure 3.** Photographs showing the appearance of the CMCGs: (**a**) dried and (**b**) swollen with PBS.

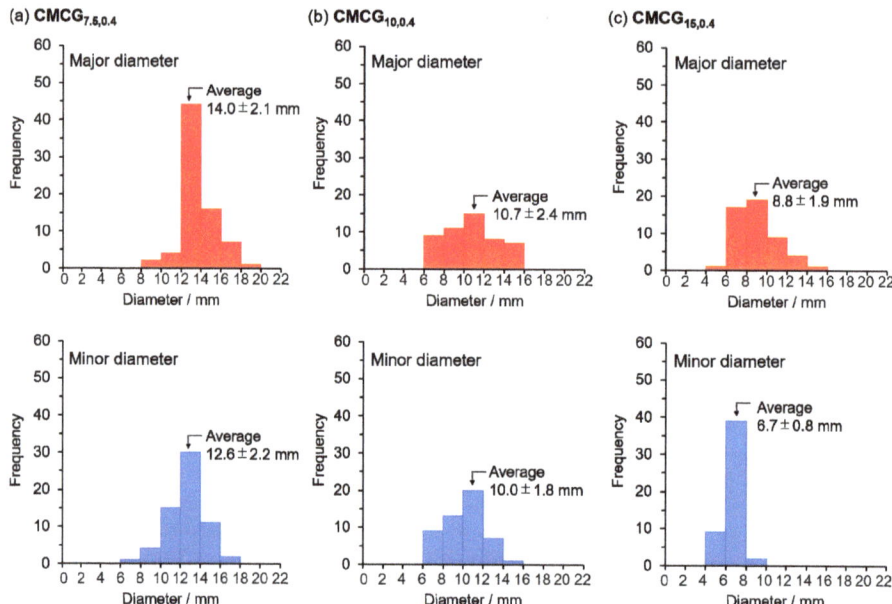

**Figure 4.** Size distributions of the major and minor diameters of swollen in PBS.

A ratio of the major to minor diameters close to 1 (i.e., the smaller the difference between the major and minor diameters) indicates that the particles are spherical. To provide a benchmark, the SPA had an average value of the major diameter of $10.6 \pm 0.4$ mm and a minor diameter of $10.5 \pm 0.3$ mm, resulting in a ratio of 1.00, indicating that the particles were spherical. CMCG, on the other hand, had ratios of 1.11 (**CMCG$_{7.5,0.4}$**), 1.07 (**CMCG$_{10,0.4}$**), and 1.31 (**CMCG$_{15,0.4}$**). The ratio of **CMCG$_{10,0.4}$** was closest to 1, indicating that the particles were almost spherical. **CMCG$_{15,0.4}$** was obtained as non-spherical particles; the oval particles are apparent in the photographs (Figure 3b). The high viscosity of the CMC solution seems to have caused the miniaturization of the CMC solution while stretching and forming oval droplets by shear force. These results confirm that a CMC concentration of 10 wt% is appropriate for obtaining spherical CMCG and was used for the subsequent experiments.

2.1.2. Effect of the Feed Amount of the Crosslinking Agent

The effect of the EGDE concentration (0.2, 0.4, and 0.6 g per 1 g of CMC; CMC concentration was kept constant at the optimal 10 wt%) on the CMCGs (**CMCG$_{10,0.2}$**, **CMCG$_{10,0.4}$**, and **CMCG$_{10,0.6}$**) was determined. Figure 5 shows the time-dependency of the PBS absorbency with the CMCGs and that with SPA as a reference (dotted line). The absorbency gradually increased over time and reached equilibrium at its maximum after 24 h for **CMCG$_{10,0.4}$**, **CMCG$_{10,0.6}$**, and SPA, and after 48 h for **CMCG$_{10,0.2}$**. The maximum absorbency of all the CMCGs at equilibrium was higher than that of SPA. These results suggest that the maximum absorbency at equilibrium decreases with an increasing EGDE/CMC ratio. In our previous study [10,25], the degree of crosslinking increased with an increase in the feed ratio of the crosslinking agent to CMC, during hydrogel preparation. According to the FTIR spectra of CMCGs, it is suggested that the degree of crosslinking of CMCG also increases as the amount of EGDE increases (Figure S3 and Table S1). The highly crosslinked structure suppressed the expansion of the CMC molecules in the hydrogels, resulting in lower absorption.

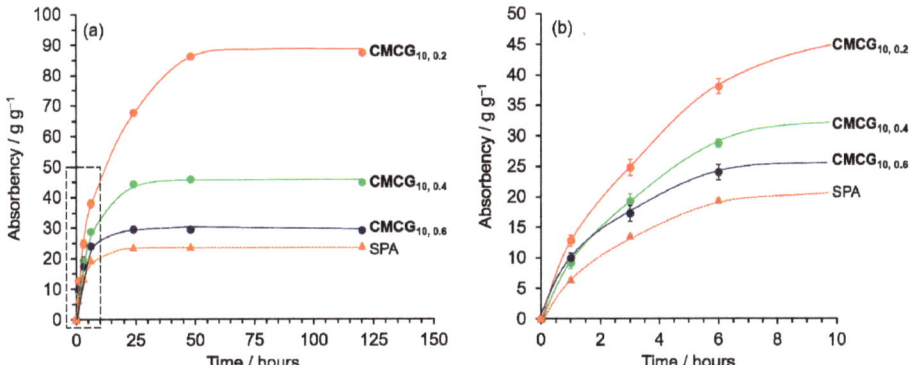

**Figure 5.** Time-dependence of the absorbency of the CMCGs and SPA (**a**) toward PBS and (**b**) initial stage of the absorbency.

Figure 6 shows the absorbencies of CMCG in PBS and pure water after 48 h. Both CMCG and SPA exhibited an absorbency toward pure water that was more than twice as high as that of PBS. In general, the absorbency of ionic gels depends on the difference in the cation concentration between the inner and outer liquids of the gels [26–28]. Ionic gels have functional groups (e.g., sodium carboxylates) that dissociate in water. The free cation, such as $Na^+$, produced as a result, causes a concentration difference between the inner and outer solutions, and water permeates the gel by osmotic pressure. The difference in the ionic concentration between the inside of the gel and the outer solution decreases in electrolyte solutions, such as PBS. Thus, the absorbency toward PBS decreased because the water penetration was suppressed. The absorbency of SPA toward pure water was 109 g g$^{-1}$, whereas that for PBS was approximately 20% of the water absorbency. In contrast, the absorbency of CMCG toward PBS was approximately 40–48% of that toward pure water. These values were higher than those of SPA. The CMC backbone contains several hydrophilic hydroxy groups. The unsubstituted hydroxy groups on the CMC backbone increased the affinity for water as hydrophilic groups, resulting in a higher absorbency toward PBS compared to SPA.

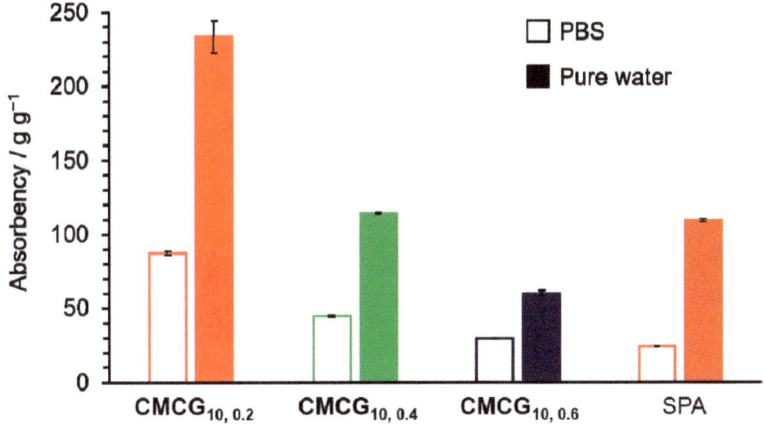

**Figure 6.** Absorbency of the CMCGs and SPA toward PBS (open column) and pure water (closed column) after 48 h.

The photographs of CMCG before and after absorption in PBS and pure water are shown in Figure 7. The dried CMCG became a transparent spherical hydrogel after swelling in PBS. None of the CMCGs deformed under their weight and were able to retain their shape. Although some CMCG particles were flattened, nearly spherical particles were observed, indicating that the amount of EGDE feed did not affect the shape of CMCG. Following that, the major and minor diameters of each CMCG were measured, and the average value was used as the CMCGs' diameter. The size distribution histograms of the CMCGs were obtained by measuring the diameters of 50 randomly selected CMCGs (Figure 8). The mean value and standard deviation were obtained from the histograms and are shown in Figure 9. The average diameter of the swollen CMCG decreased with an increasing EGDE feed amount. Similarly, the absorbency decreased, indicating that the increase in crosslinking density resulted in the construction of a strong network structure, which suppressed the expansion of the network and decreased the size of the swollen particles. The absorbency and size of the CMCG could be controlled by the feed amount of EGDE. So far, the results reveal that the CMC concentration mostly determined the shape of CMCG, whereas the feed amount of EGDE controlled its absorbency.

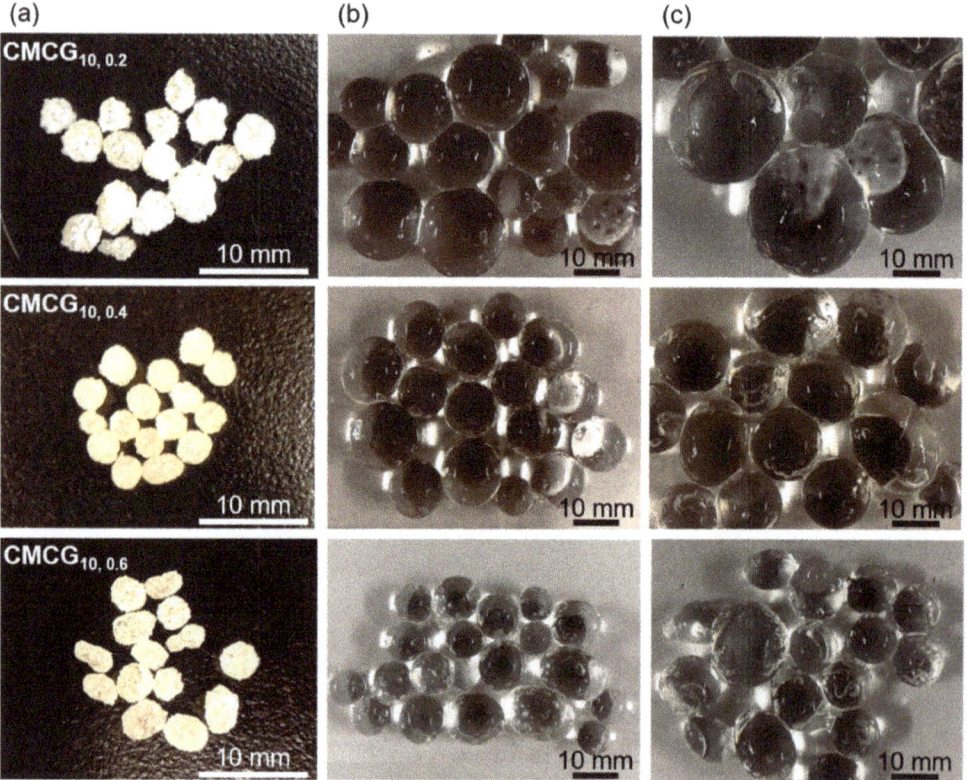

**Figure 7.** Photographs showing the appearance of the CMCGs: (**a**) dried, swollen with (**b**) PBS, and (**c**) pure water, after 48 h.

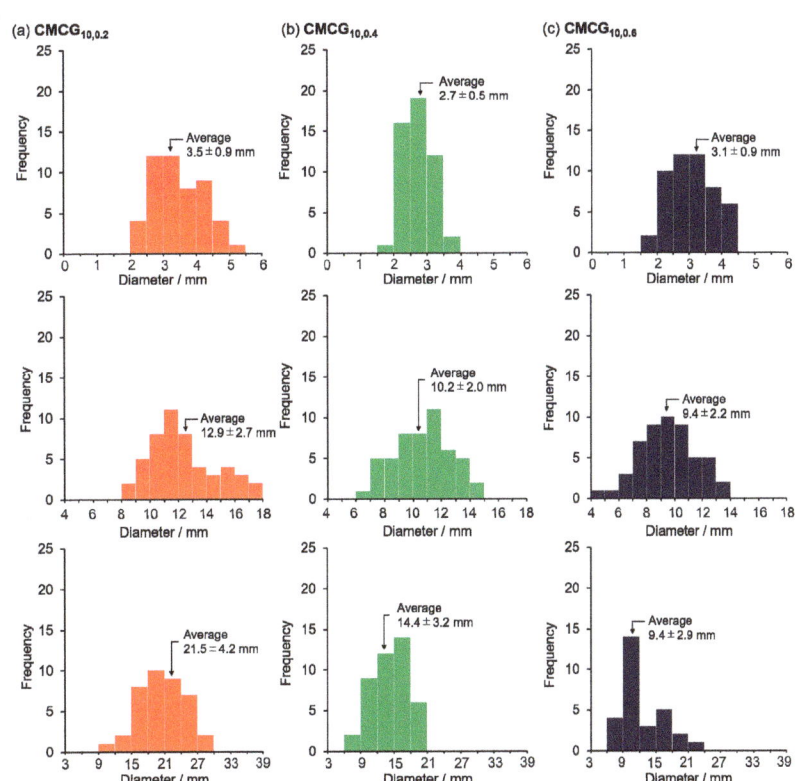

**Figure 8.** Size distributions of the CMCGs: dried (top) and swollen with PBS (middle) and pure water (bottom).

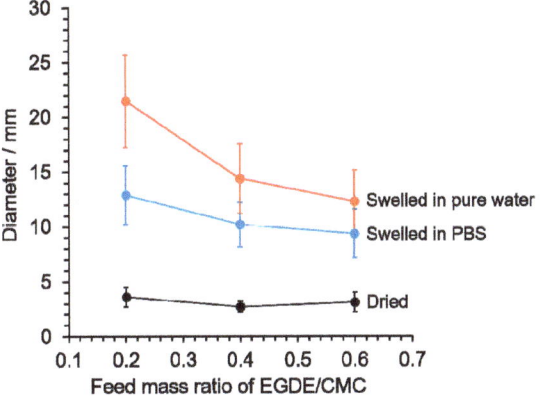

**Figure 9.** Effect of the mass ratio of ethylene glycol diglycidyl ether (EGDE)/cellulose sodium salt (CMC) on the dried CMCGs and on those swollen in PBS and pure water.

Because the absorbency of water was higher than that of PBS, the particle size of CMCG that swelled toward pure water was larger than that of PBS. The intermolecular network structure expanded because of the large amount of water absorbed, and the CMCG was brittle and exhibited cracks. The higher the mass ratio of EGDE/CMC, the more cracks were observed in the CMCG. The degree of crosslinking increased with an increase in the mass ratio of EGDE/CMC, and the flexibility between the molecular chains decreased. The cracks were caused by the inclusion of a large amount of water between the molecular chains with reduced flexibility, owing to the high degree of crosslinking.

The absorbency in PBS of the CMC hydrogel crosslinked with polyethylene glycol diglycidyl ether (PEGDE) as a crosslinking agent was 90–230 $g \cdot g^{-1}$ [25], which is higher than that of CMCG (30–90 $g \cdot g^{-1}$). The absorbency of a CMC hydrogel crosslinked with PEGDE is similarly affected by the PEGDE feed quantity, which is 0.04–0.23 equivalents per CMC anhydroglucose unit. In contrast, CMCGs were prepared by adding 0.25–0.75 equivalents of EGDE per anhydroglucose unit of CMC, which was an excess of PEGDE. By reducing the amount of EGDE added, it is envisaged that CMCG will be able to reproduce the same level of absorption as the CMC hydrogel with PEGDE. However, as observed in the images of CMCG swollen with pure water, the spherical shape may not be maintained, due to cracks induced by the significant amount of absorption (Figure 7).

## 2.2. Water-Holding Capacity

After testing for absorbency, the water-holding capacity of the CMCG swollen with PBS was determined by placing it in a chamber at a temperature of 298 K and a humidity of 55%. The time-dependence of the water released from the CMCG is shown in Figure 10. The water-holding capacity of the SPA is presented as a dotted line in the figure for comparison. All the hydrogels, including SPA, released water over time. On day 1 of the test, the water-holding capacity for all the hydrogels was almost the same. However, that of SPA rapidly decreased after 3 days and was 14.8% after 14 days. The water-holding capacities of the CMCGs also gradually decreased with time, but were 80% (**CMCG$_{10,0.2}$**), 58% (**CMCG$_{10,0.4}$**), and 30% (**CMCG$_{10,0.6}$**) on day 14, which were higher than those of SPA. In the CMCGs, the water release rate was inhibited, suggesting that the unsubstituted hydroxy groups of cellulose increased its affinity toward water. However, the release rate of water accelerated with an increase in the feed amount of EGDE to CMC. The water-holding capacity was almost 0% after 21 days for **CMCG$_{10,0.6}$**, and 28 days for **CMCG$_{10,0.4}$**. In contrast, **CMCG$_{10,0.2}$** showed a high water-holding capacity of 42% even after 49 days. This was because the degree of crosslinking increased, potentially leading to a decrease in the number of unsubstituted hydroxy groups as the mass ratio of EGDE/CMC increased. It is suggested that a decrease in the number of hydroxy groups decreases the affinity toward water and causes an increase in the rate of water release.

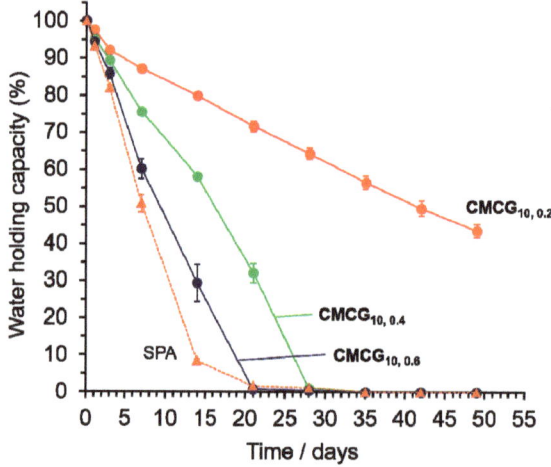

**Figure 10.** Time-dependence of the water-holding capacities of the CMCGs and SPA at a temperature of 298 K and a humidity of 55%.

## 2.3. Cellulase Degradability

CMCGs with different feed mass ratios of EGDE/CMC were enzymatically degraded using cellulase in a nylon teabag. The degradation was conducted using cellulase derived from *Trichoderma*, which is commonly found in soil [29]. The cellulase degraded the CMCG into low molecular weight fractions, which were eluted from the nylon teabag. Cellulase degradability was calculated from the difference between the weight of the CMCG remaining in the teabag and its initial weight. The time-dependence of degradation is shown in Figure 11. The dotted line in this figure represents the weight change in the SPA, which is almost unchanged over time, indicating that SPA was not degraded by cellulase because it is a polyacrylic acid. In contrast, the cellulase degradability of all the CMCG samples was excellent and increased with time. After 7 days, all the CMCGs were completely degraded. **$CMCG_{10,0.2}$**, exhibited 89% degradation in 1 day, which confirmed that CMCG was almost completely degraded after 2 days. **$CMCG_{10,0.6}$** showed degradation of 40% in 1 day and required 5 days to be completely degraded. That is, the degradation rate decreased with an increase in the mass ratio of EGDE/CMC. The degradation of cellulose-based hydrogels is not dependent on the water absorbency and water-holding capacity, but decreases because the interior or surface cellulose chains are not easily accessible to cellulase with an increasing degree of crosslinking [10,25,30,31]. This suggests that the contact between cellulase and the cellulose chains of CMCG was inhibited as the mass ratio of EGDE/CMC increased, which caused a decrease in degradation in the initial stages. However, as the degradation of CMCG progressed, cellulase was able to access the exposed cellulose chains, leading to the complete degradation of CMCG. Furthermore, the cellulase degradability of CMC hydrogels crosslinked with EGDE and PEGDE was a maximum of 70%/5 days and 62%/3 days, depending on the feed amount of crosslinking agent [10,25]. This degradability was calculated based on the amount of the reducing sugars released by the cellulase degradation, indicating complete degradation of CMC to glucose. The degradability of CMCG was determined in this study based on the change in mass of the CMCG remaining in the teabag before and after cellulase degradation, and the low molecular weight fractions eluted from the teabag are also included in the degradability. Therefore, it is expected that CMCG will take longer to completely degrade to glucose than the day it exhibited 100% degradability. Although more research into the microbial degradation of CMCG in the natural environment, such as soil, is required, these findings indicate that CMCG can be degraded by microorganisms.

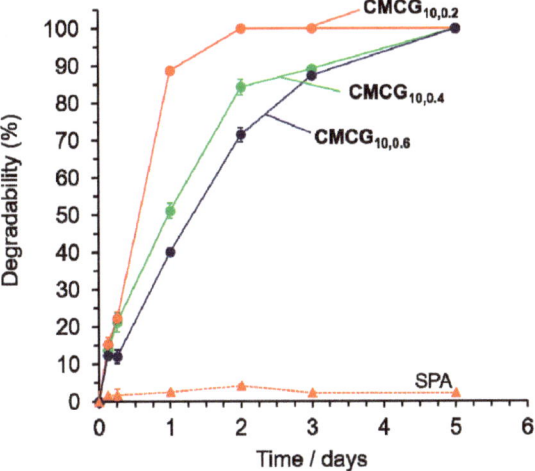

**Figure 11.** Time-dependency of the degradability of the CMCGs and SPA by cellulase in the 100 mM sodium acetate buffer (pH 5.0) at 313 K.

## 3. Conclusions

The crosslinking reaction of CMC with EGDE was conducted in an aqueous droplet of liquid paraffin, formed by the shear force from stirring and the surface tension of the water–oil interface, yielding spherical hydrogel particles. The particle size decreased as the CMC concentration increased, and the shape turned non-spherical. A CMC concentration of 10 wt% was optimal to obtain spherical CMCG. Furthermore, increasing the mass ratio of EGDE/CMC suppressed the electrostatic repulsion between the CMC molecules in the gel structure, decreasing the absorbency. These findings suggest that particle shape and absorbency can be controlled by CMC concentration and EGDE feed quantity, respectively. The spherical CMC hydrogel particles were completely degraded by cellulase. A strong dependence of the degradation rate on the feed amount of EGDE was observed, where the degradation rate increased with a decrease in the feed amount of EGDE. Our results indicate that the spherical hydrogel particles prepared in this study have the potential for a variety of uses, such as air fresheners and horticultural water-holding agents, as alternatives to current synthetic SAPs. In addition, since the spherical CMCG has numerous carboxy groups in its structure, further functionalization can be expected by utilizing them as cross-linking points [32,33]. Nonetheless, the raw material cost of CMCG is higher than the price of existing synthetic SAP, and further research, including cost reduction, is required to establish its utility as an SAP alternative.

## 4. Materials and Methods

### 4.1. Materials

CMC (molecular weight = $8$–$10 \times 10^4$ as specified by manufacturer) with a 0.70 degree of substitution was obtained from Daicel Co., (Osaka, Japan). EGDE was purchased from Fujifilm Wako Pure Chemical Co., Ltd. (Osaka, Japan). Liquid paraffin (Moresco white P-100, 19.04 $mm^2/s$) was purchased from Moresco Co., (Kobe, Japan). Sodium hydroxide was purchased from Kanto Chemical Co., Inc. (Tokyo, Japan). Methanol was purchased from Godo Co., Ltd. (Tokyo, Japan). The SPA was supplied by S. T. Co., (Tokyo, Japan). *Trichoderma viride* cellulase ONOZUKA R-10 was purchased from Yakult Pharmaceutical Co., Ltd. (Tokyo, Japan). The PBS and sodium acetate buffer used were analytical grade chemicals, i.e., sodium chloride, potassium chloride, sodium dihydrogen phosphate, potassium dihydrogen phosphate, sodium acetate, and acetic acid, purchased from Kanto Chemical Co., Inc. (Tokyo, Japan).

### 4.2. Preparation of CMCG

First, CMC (1.0 g) was completely dissolved in 0.5 mol $L^{-1}$ aqueous NaOH (40 mL), followed by adding EGDE (0.40 g) with stirring at 250 rpm for 20 min using a Teflon impeller. Then, the reaction mixture was injected into 200 mL of liquid paraffin in a 300 mL beaker while stirring at 300 rpm, using stainless steel propeller blades at 328 K for 3 h to allow the crosslinking reaction of CMC to occur. Once the reaction was complete, the CMCG was washed with deionized water containing 1 wt% natural detergent, until the liquid paraffin on the surface of the CMCG was completely removed. Subsequently, the CMCG was immersed in methanol until pH 7.0 was reached to remove any unreacted materials and NaOH. Finally, it was dried at 318 K under reduced pressure ($CMCG_{2.5,0.4}$). The dried CMCG samples larger than 2 mm were collected using a sieve with a 2 mm mesh opening, which were used in various tests. The CMCG samples with varying CMC and EGDE concentrations were prepared using a similar method (Table 1).

### 4.3. Water Absorbency

The absorbency of CMCG in PBS was determined using the following method: dried CMCG (250 mg) was immersed in a PBS solution at 298 K. After 1, 3, 6, 24, 48, and 120 h, CMCG was removed from the PBS solution, and any excess solution was drained onto a mesh for 5 min. The weight of the hydrogels was measured and the absorbency was calculated using Equation (1), which is as follows:

$$\text{Absorbency} = \frac{W_s - W_d}{W_d}, \qquad (1)$$

where $W_d$ and $W_s$ are the weights of the dried and swollen CMCG, respectively, at a certain time. The absorbency measurements were performed three times. The absorbency of CMCG in pure water instead of PBS was also similarly investigated. For comparison, the same experiments were performed using the SPA instead of CMCG.

*4.4. Determination of Particle Size*

Photographs of the dried and swollen CMCG were taken, and the digitalized images were analyzed using the *Image J* (version 1.53 m) image analysis program developed by the U.S. National Institutes of Health. Because the particles were flattened shapes, as observed in the photographs, the major and minor diameters were measured using dried and swollen CMCG. The size distribution was determined with the diameters calculated using 50 randomly selected CMCG in the photographs. Using data obtained from 50 CMCGs, the mean diameters and standard deviations were calculated.

*4.5. Water-Holding Capacity*

The CMCG (250 mg) immersed in PBS solution at 298 K for 5 days was placed in a chamber maintained at a temperature and humidity of 298 K and 55%, respectively. After 1, 3, 7, 14, 21, 28, 35, 42, and 49 days, the weight of the CMCG was measured. The water-holding capacity was determined using Equation (2), which is as follows:

$$\text{Water holding capacity} = \frac{W_p - W_d}{W_i - W_d} \times 100 \qquad (2)$$

where $W_i$ is the weight of the swollen CMCG at its initial state and $W_p$ is the weight of the swollen hydrogel at a certain time. $W_d$ is the initial weight of the dried CMCG. The water-holding capacity was measured thrice. For comparison, the same experiments were performed using SPA.

*4.6. Cellulase Degradation*

Cellulase degradability was determined based on the change in the CMCG weight before and after the enzymatic reaction. A teabag (50 mm × 50 mm) was prepared from a nylon sheet with a pore size of 255 mesh using a heat sealer. Then, CMCG (200 mg) was placed in the teabag, which was immersed in 29 mL of 100 mM sodium acetate buffer (pH 5.0) for 1 day. Cellulase dissolved in 1 mL of the same buffer solution was added to the solution containing CMCG to give a cellulase concentration of $5 \times 10^{-5}$ wt%. The teabag immersed in the solution was incubated at 313 K with shaking at 100 rpm. After 3, 6 h, and 1, 2, 3, 5 days, the teabag was removed from the solution and left in a boiling bath for 10 min to inactivate the cellulase. Thereafter, the teabag was dried in an oven at 378 K, and the weight of the teabag ($W_a$) was measured. Cellulase degradability was calculated using Equation (3), which is as follows:

$$\text{Degradability} = \frac{W_d - (W_a - W_t)}{W_d} \times 100 \qquad (3)$$

where $W_d$ is the initial weight of the dried CMCG and $W_t$ is the weight of the empty teabag. Cellulase degradation was performed in triplicate. For comparison, the same experiments were performed using SPA.

**Supplementary Materials:** The following supporting information can be downloaded at: https://www.mdpi.com/xxx/s1, Figure S1: Photographs showing the appearance of the CMCGs swollen with PBS after 48 h. CMCG was prepared at stirring speeds of (a) 300 rpm and (b) 400 rpm; Figure S2: FTIR spectra of CMCGs with different CMC concentrations and CMC as a reference. These spectra were normalized by the peak intensity of the carboxylate groups at 1593 cm$^{-1}$; Figure S3: FTIR spectra of CMCGs with different feed ratios of EGDE/CMC and CMC as a reference. These spectra were normalized by the peak intensity of the carboxylate groups at 1593 cm$^{-1}$; Table S1: Peak area ratio of methylene group to carboxylate group ($A_{CH2}/A_{COONa}$) from FTIR spectra; Figure S4: Photographs showing the appearance of SPA: (a) dried, swollen with (b) PBS and (c) pure water, after 48 h.

**Author Contributions:** Conceptualization, H.K.; methodology, S.F., T.T. and H.K.; validation, S.F., T.T. and H.K.; formal analysis, S.F.; investigation, S.F. and T.T.; data curation, S.F.; writing—original draft preparation, S.F. and H.K.; writing—review and editing, H.K.; visualization, S.F.; supervision, H.K.; project administration, T.T. and H.K.; funding acquisition, S.F., T.T. and H.K. All authors have read and agreed to the published version of the manuscript.

**Funding:** This research was partially supported by the Japan Society for the Promotion of Science (JSPS) (grant no. JP21K14664 (S.F.) and JP21K05175 (H.K.)).

**Institutional Review Board Statement:** Not applicable.

**Informed Consent Statement:** Not applicable.

**Data Availability Statement:** The data presented in this study, supporting the results, are available in the main text. Additional data are available upon request from the corresponding authors.

**Conflicts of Interest:** The authors declare no conflict of interest.

## References

1. Omidian, H.; Hashemi, S.A.; Sammes, P.G.; Meldrum, I. A model for the swelling of superabsorbent polymers. *Polymer* **1998**, *39*, 6697–6704. [CrossRef]
2. Zhang, J.; Liu, R.; Li, A.; Wang, A. Preparation, swelling behaviors, and slow-release properties of a poly(acrylic acid-co-acrylamide)/sodium humate superabsorbent composite. *Ind. Eng. Chem. Res.* **2006**, *45*, 48–53. [CrossRef]
3. Chen, J.; Wu, J.; Raffa, P.; Picchioni, F.; Koning, C.E. Superabsorbent polymers: From long-established, microplastics generating systems, to sustainable, biodegradable and future proof alternatives. *Prog. Polym. Sci.* **2022**, *125*, 101475. [CrossRef]
4. Kono, H.; Fujita, S.; Oeda, I. Comparative study of homogeneous solvents for the esterification crosslinking of cellulose with 1,2,3,4-butanetetracarboxylic dianhydride and water absorbency of the reaction products. *J. Appl. Polym. Sci.* **2012**, *5*, 478–486. [CrossRef]
5. Liu, T.G.; Wang, Y.T.; Li, B.; Deng, H.B.; Huang, Z.L.; Qian, L.W.; Wang, X. Urea free synthesis of chitin-based acrylate superabsorbent polymers under homogeneous conditions: Effects of the degree of deacetylation and the molecular weight. *Carbohydr. Polym.* **2017**, *174*, 464–473. [CrossRef]
6. Kono, H.; Oeda, I.; Nakamura, T. The preparation, swelling characteristics, and albumin adsorption and release behaviors of a novel chitosan-based polyampholyte hydrogel. *React. Funct. Polym.* **2013**, *73*, 97–107. [CrossRef]
7. Sami, A.J.; Khalid, M.; Jamil, T.; Aftab, S.; Mangat, S.A.; Shakoori, A.R.; Iqbal, S. Formulation of novel chitosan guargum based hydrogels for sustained drug release of paracetamol. *Int. J. Biol. Macromol.* **2018**, *108*, 324–332. [CrossRef]
8. Ghobashy, M.M.; El-Wahab, H.A.; Ismail, M.A.; Naser, A.M.; Abdelhai, F.; El-Damhougy, B.K.; Nady, N.; Meganid, A.S.; Alkhursani, S.A. Characterization of starch-based three components of gamma-ray cross-linked hydrogels to be used as a soil conditioner. *Mater. Sci. Eng. B* **2020**, *260*, 114645. [CrossRef]
9. Sharma, S.; Sharma, G.; Kumar, A.; AlGani, T.S.; Naushad, M.; Alothman, Z.A.; Stadler, F. Adsorption of cationic dyes onto carrageenan and itaconic acid-based superabsorbent hydrogel: Synthesis, characterization and isotherm analysis. *J. Hazard. Mater.* **2022**, *421*, 126729. [CrossRef]
10. Kono, H. Carboxymethyl cellulose-based hydrogels. In *Cellulose and Cellulose Derivatives: Synthesis, Modification and Applications*, 1st ed.; Mondal, M.I.H., Ed.; Nova Science Publishers Inc.: New York, NY, USA, 2015; pp. 243–258, ISBN 9781634831277.
11. Jeoung, D.; Joo, S.W.; Hu, Y.; Shinde, V.V.; Cho, E.; Jung, S. Carboxymethyl cellulose-based superabsorbent hydrogels containing carboxymehtyl β-cyclodextrin for enhanced mechanical strength and effective drug delivery. *Eur. Polym. J.* **2018**, *105*, 17–25. [CrossRef]
12. Chang, C.; Duan, B.; Cai, J.; Zhang, L. Superabsorbent hydrogels based on cellulose for smart swelling and controllable delivery. *Eur. Polym. J.* **2010**, *46*, 92–100. [CrossRef]
13. Cheng, S.; Liu, X.; Zhen, J.; Lei, Z. Preparation of superabsorbent resin with fast water absorption rate based on hydroxymethyl cellulose sodium and its application. *Carbohydr. Polym.* **2019**, *225*, 115214. [CrossRef] [PubMed]

14. Wu, F.; Zhang, Y.; Liu, L.; Yao, J. Synthesis and characterization of a novel cellulose-g-poly(acrylic acid-co-acrylamide) superabsorbent composite based on flax yarn waste. *Carbohydr. Polym.* **2012**, *87*, 2519–2525. [CrossRef]
15. Lim, D.-W.; Song, K.-G.; Yoon, K.-J.; Ko, S.-W. Synthesis of acrylic acid-based superabsorbent interpenetrated with sodium PVA sulfate using inverse-emulsion polymerization. *Eur. Polym. J.* **2002**, *38*, 579–586. [CrossRef]
16. Benda, D.; Šňupárek, J.; Čermák, V. Inverse emulsion polymerization of acrylamide and salts of acrylic acid. *Eur. Polym. J.* **1997**, *33*, 1345–1352. [CrossRef]
17. Calvo, P.; Remon-Lopez, C.; Vila-Jato, J.L.; Alonso, M.J. Novel hydrophilic chitosan-polyethylene oxide nanoparticles as protein carriers. *J. Appl. Polym. Sci.* **1997**, *63*, 125–132. [CrossRef]
18. Afshar, M.; Dini, G.; Vaezifar, S.; Mehdikhani, M.; Movahedi, B. Preparation and characterization of sodium alginate/polyvinyl alcohol hydrogel containing drug-loaded chitosan nanoparticles as a drug delivery system. *J. Drug Deliv. Sci. Technol.* **2020**, *56*, 101530. [CrossRef]
19. Mousaviasl, S.; Saleh, T.; Shojaosadati, S.A.; Boddohi, S. Synthesis and characterization of schizophyllan nanogels via inverse emulsion using biobased materials. *Int. J. Biol. Macromol.* **2018**, *120*, 468–474. [CrossRef]
20. Milašinović, N.; Čalijac, B.; Vidović, B.; Sakač, M.C.; Vujić, Z.; Knežević-Jugović, Z. Sustained release of α-lipoic acid from chitosan microbeads synthetized by inverse emulsion method. *J. Taiwan Inst. Chem. Eng.* **2016**, *60*, 106–112. [CrossRef]
21. Riegger, B.R.; Bäurer, B.; Mirzayeva, A.; Tovar, G.E.M.; Bach, M. A systematic approach of chitosan nanoparticle preparation via emulsion crosslinking as potential adsorbent in wastewater treatment. *Carbohydr. Polym.* **2018**, *180*, 46–54. [CrossRef]
22. Yaghoobi, M.; Zaheri, P.; Mousavi, S.H.; Ardehali, B.A.; Yousefi, T. Evaluation of mean diameter and drop size distribution of an emulsion liquid membrane system in a horizontal mixer-settler. *Chem. Eng. Res. Des.* **2021**, *167*, 231–241. [CrossRef]
23. Bąk, A.; Podgórska, W. Investigation of drop breakage and coalescence in the liquid–liquid system with nonionic surfactants Tween 20 and Tween 80. *Chem. Eng. Sci.* **2012**, *74*, 181–191. [CrossRef]
24. Zhang, Z.; Qiao, X. Influences of cation valence on water absorbency of crosslinked carboxymethyl cellulose. *Int. J. Biol. Macromol.* **2021**, *177*, 149–156. [CrossRef] [PubMed]
25. Kono, H. Characterization and properties of carboxymethyl cellulose hydrogels crosslinked by polyethylene glycol. *Carbohydr. Polym.* **2014**, *106*, 84–93. [CrossRef]
26. Zhao, Y.; Su, H.; Fang, L.; Tan, T. Superabsorbent hydrogels from poly(aspartic acid) with salt-, temperature- and pH-responsiveness properties. *Polymer* **2005**, *46*, 5368–5376. [CrossRef]
27. Huang, Y.; Zeng, M.; Ren, J.; Wang, J.; Fan, L.; Xu, Q. Preparation and swelling properties of graphene oxide/poly(acrylic acid-co-acrylamide) super-absorbent hydrogel nanocomposites. *Colloids Surf. A Physicochem. Eng. Asp.* **2012**, *401*, 97–106. [CrossRef]
28. Zhang, M.; Hou, S.; Li, Y.; Hu, S.; Yang, P. Swelling characterization of ionic responsive superabsorbent resin containing carboxylate sodium groups. *React. Funct. Polym.* **2022**, *170*, 105144. [CrossRef]
29. Asad, S.A.; Tabassum, A.; Hameed, A.; Hassan, F.U.; Afzal, A.; Khan, S.A.; Ahmed, R.; Shahzad, M. Determination of lytic enzyme activities of indigenous *Trichoderma* isolates from Pakistan. *Braz. J. Microbiol.* **2015**, *46*, 1053–1064. [CrossRef]
30. Kono, H.; Hara, H.; Hashimoto, H.; Shimizu, Y. Nonionic gelation agents prepared from hydroxypropyl guar gum. *Carbohydr. Polym.* **2015**, *117*, 636–643. [CrossRef]
31. Kono, H.; Fujita, S. Biodegradable superabsorbent hydrogels derived from cellulose by esterification crosslinking with 1,2,3,4-butanetetracarboxylic dianhydride. *Carbohydr. Polym.* **2012**, *87*, 2582–2588. [CrossRef]
32. Kono, H.; Onishi, K.; Nakamura, T. Characterization and bisphenol A adsorption capacity of β-cyclodextrin-carboxymethylcellulose-based hydrogels. *Carbohydr. Polym.* **2013**, *98*, 784–792. [CrossRef] [PubMed]
33. Ogata, M.; Yamanaka, T.; Koizumi, A.; Sakamoto, M.; Aita, R.; Endo, H.; Yachi, Y.; Yamauchi, N.; Otsubo, T.; Ikeda, K.; et al. Application of novel sialoglyco particulates enhances the detection sensitivity of the equine influenza virus by real-time reverse transcriptase polymerase chain reaction. *ACS Appl. Bio Mater.* **2019**, *2*, 1255–1261. [CrossRef] [PubMed]

Article

# Smart Methylcellulose Hydrogels for pH-Triggered Delivery of Silver Nanoparticles

Lorenzo Bonetti [1,*], Andrea Fiorati [1,2], Agnese D'Agostino [1,2], Carlo Maria Pelacani [1], Roberto Chiesa [1,2], Silvia Farè [1,2] and Luigi De Nardo [1,2]

[1] Department of Chemistry, Materials and Chemical Engineering "G. Natta", Politecnico di Milano, Via Luigi Mancinelli 7, 20131 Milan, Italy; andrea.fiorati@polimi.it (A.F.); agnese.dagostino@polimi.it (A.D.); carlomaria.pelacani@mail.polimi.it (C.M.P.); roberto.chiesa@polimi.it (R.C.); silvia.fare@polimi.it (S.F.); luigi.denardo@polimi.it (L.D.N.)
[2] National Interuniversity Consortium of Materials Science and Technology (INSTM), Via Giuseppe Giusti 9, 50121 Florence, Italy
* Correspondence: lorenzo.bonetti@polimi.it

**Abstract:** Infection is a severe complication in chronic wounds, often leading to morbidity or mortality. Current treatments rely on dressings, which frequently contain silver as a broad-spectrum antibacterial agent, although improper dosing can result in severe side effects. This work proposes a novel methylcellulose (MC)-based hydrogel designed for the topical release of silver nanoparticles (AgNPs) via an intelligent mechanism activated by the pH variations in infected wounds. A preliminary optimization of the physicochemical and rheological properties of MC hydrogels allowed defining the optimal processing conditions in terms of crosslinker (citric acid) concentration, crosslinking time, and temperature. MC/AgNPs nanocomposite hydrogels were obtained via an in situ synthesis process, exploiting MC both as a capping and reducing agent. AgNPs with a 12.2 ± 2.8 nm diameter were obtained. MC hydrogels showed a dependence of the swelling and degradation behavior on both pH and temperature and a noteworthy pH-triggered release of AgNPs (release ~10 times higher at pH 12 than pH 4). $^1$H-NMR analysis revealed the role of alkaline hydrolysis of the ester bonds (i.e., crosslinks) in governing the pH-responsive behavior. Overall, MC/AgNPs hydrogels represent an innovative platform for the pH-triggered release of AgNPs in an alkaline milieu.

**Keywords:** methylcellulose; citric acid; crosslinking; pH-responsive; silver nanoparticles (AgNPs)

## 1. Introduction

Chronic wounds represent a severe clinical problem affecting 1–2% of the population in developed countries [1,2]. Being hard to heal, chronic wounds are challenging both for patients and healthcare systems, and account for 2–3% of the healthcare budgets [3]. According to clinical literature, chronic wounds can be assigned to three main categories: vascular ulcers, diabetic ulcers, and pressure ulcers [3]. Even if no single, widely accepted definition of chronic wound exists, in most cases, a wound is defined as chronic if it has not healed within a certain amount of time, ranging from 4 to 6 weeks for most authors, or up to 3 months, as noted in standard surgical textbooks [3,4]. Regardless of the time, chronic wounds share some standard features, e.g., persistent inflammation, recurrent infections with possible formation of antibiotic-resistant biofilms, and deficiency of stem cells. Combined, these pathophysiological phenomena result in the wound's inability to heal [3].

Infection is a critical complication in chronic wounds, often leading to persistent nonhealing, significant morbidity, or even mortality. Alkalinity is associated with infection in chronic wounds, even if its cause is still unclear [5]. Bacteria metabolism could be responsible for the local increase of pH due to the release of ammonia and polyamines, which can impair the oxygenation of wound tissue and promote necrosis [6,7]. In this regard, several studies have shown how the pH of infected chronic wounds sometimes reaches values of 10 [5,6,8]. In addition, specific tissue-degrading enzymes (e.g., elastase, plasmin, matrix-metalloproteinase-2) display higher turnover rates in alkaline conditions. Consequently, an alkaline wound milieu may strongly contribute to tissue degradation and infection [6].

Topical wound therapies and dressings are primarily applied to treat chronic wounds: from standard cotton gauze dressings to highly absorbent and moisture-retaining foams, hydrocolloids, and hydrogels [9]. In many cases, due to the broad-spectrum antibacterial activity of silver, Ag-containing products provide an additional clinical option to manage bacterial growth [3,10,11]. However, silver misuse (e.g., uncontrolled release, excessive dosing) for a prolonged time can lead to potential toxicity to tissues and organs at the local and systemic levels [3,12].

Given the pH variations occurring in the wound milieu, stimuli-responsive polymers can take up the challenge of a smart (i.e., in response to pathophysiological phenomena) release of silver in high enough concentrations to ensure antibacterial activity [13]. Due to local pH variations, pH-responsive polymers undergo structural and property changes (e.g., dissolution/precipitation, degradation, swelling/collapsing, hydrophilic/hydrophobic change, change of shape, conformational change) [13,14]. The drop in pH has been investigated for the controlled release of silver nanoparticles (AgNPs) from different pH-responsive hydrogels (e.g., carboxymethyl chitosan/poly(vinyl alcohol), N-acryloyl-N'-ethyl piperazine/N-isopropylacrylamide) [15,16]. However, only a few studies reported the possibility of exploiting the alkalinity of infected wounds to trigger the release of AgNPs [12].

In this study, methylcellulose (MC) hydrogels were prepared and crosslinked with citric acid (CA) at three different crosslinking degrees: low (MC-L), medium (MC-M), and high (MC-H). Pristine hydrogel (MC) was used as a control. The thermo- and pH-responsiveness of the samples were assessed by swelling/degradation and rheological tests at different pH (4, 7, 12) and temperatures (25, 37, 50 °C). MC-H samples were further investigated due to their remarkable pH-responsive behavior. $^1$H-NMR spectrometry was exploited to disclose the mechanism of pH-responsiveness. Afterward, MC–H/AgNPs nanocomposite hydrogels were prepared via AgNPs in situ synthesis using MC both as a capping agent and reducing agent. TEM and UV–vis measurements assessed the shape, size, and distribution of AgNPs. Lastly, ICP and UV–vis measurements supported a quantitative evaluation of the pH-triggered AgNPs release mechanisms to develop systems with enhanced antibacterial activity in alkaline conditions.

## 2. Results and Discussion

### 2.1. Swelling and Degradation Tests

Swelling tests were carried out in normal saline solution (NSS) adjusted to different pH levels and at different temperatures to evaluate the pH-and thermo-responsiveness of MC hydrogels. Due to the lack of buffering ability of NSS, the solution was refreshed every 72 h to preserve the pH around the desired value. In this first part of the study, NSS (0.15 M NaCl) was selected to minimize the influence of the salt on the $T_t$. A low NaCl concentration (i.e., ~0.1 M) has been reported to affect MC gelation properties slightly. Conversely, other salts (e.g., phosphates, sulfates), strongly acting on the interactions between MC chains and water molecules, have been reported to affect the $T_t$ of MC significantly even at low concentrations (i.e., <0.2 M) [17,18].

Figure 1 reports the swelling curves for the MC specimens at 37 °C. For clarity, Figure 1 only reports the curves at pH = 4 and 12. (Refer to Supplementary Materials Figure S1 for pH = 7.) At pH 4 (Figure 1A), a decrease in the SR (%) as a function of the crosslinking degree (XLD, black arrow) can be observed. This trend agrees with previous data on the same hydrogel formulation [19,20], even if lower SR values are observed, probably due to the different swelling solutions (NSS vs. distilled water). At pH = 7 (Supplementary Materials Figure S1), the curves retrace the ones at pH = 4. At pH = 12 (Figure 1B), all the specimens reach a swelling plateau between days 1 and 7, and display no significant differences ($p > 0.05$) despite their different crosslinking degree.

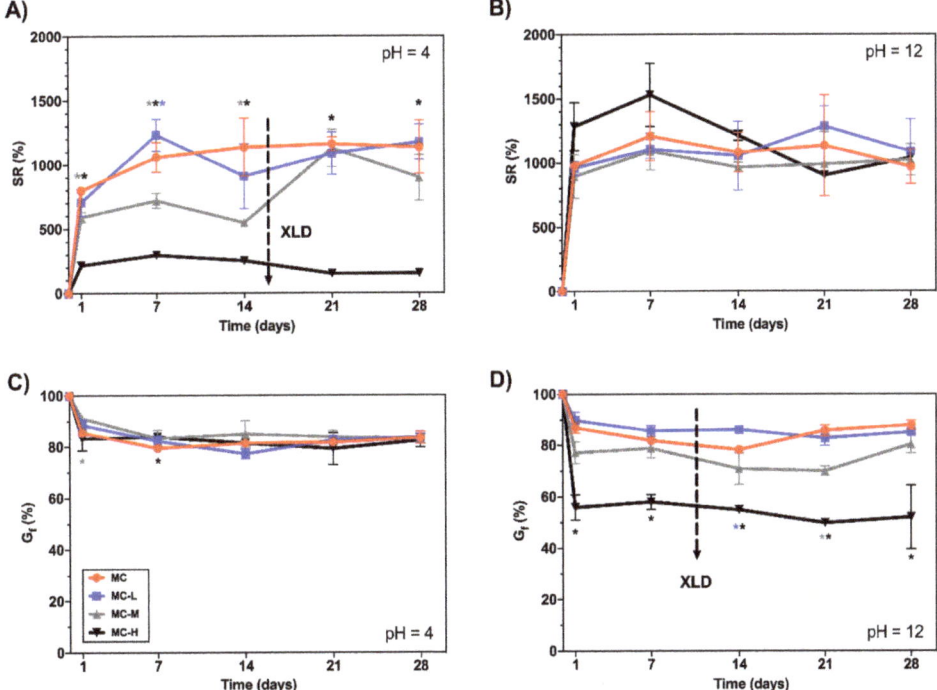

**Figure 1.** SR (%) vs. time of MC hydrogels in NSS at T = 37 °C. (**A**) pH = 4; (**B**) pH = 12. $G_f$ (%) vs. time curves of MC hydrogels samples in NSS at T = 37 °C. (**C**) pH = 4; (**D**) pH = 12. * = $p < 0.05$ compared to MC control (* = MC vs. MC-L, * = MC vs. MC-M, * = MC vs. MC-H).

Gel fraction tests were then performed to evaluate the dissolution of MC hydrogels as a function of the pH of the NSS. At pH = 4 (Figure 1C), no differences ($p > 0.05$) among MC samples can be observed. All the samples reach a plateau, with $G_f$ = 80% within the first week of the test. Such results are slightly discordant with previous outcomes on the same hydrogel formulations where higher degradation was assessed on MC samples [19]. As for swelling tests, the overall reduced degradation in this study can be attributed to the different swelling media (NSS vs. distilled water). Cl⁻ anions, destabilizing water–polymer interactions, lead to the association of MC hydrophobic groups, increasing the $G_f$ values [21]. At pH = 7 (Supplementary Materials Figure S1), the curves retrace the ones at pH = 4. At pH = 12 (Figure 1D), it is possible to distinguish a trend in the $G_f$ values as a function of the XLD of the samples. A decrease in the $G_f$ values is associated with higher crosslinking rates ($G_f$ = 80% vs. 55% at day 7 for MC and MC-H specimens, respectively).

These findings support what was obtained with the swelling tests (Figure 1B), where in the same conditions (pH = 12, T = 37 °C), the highly crosslinked sample swelled more than at lower pH. Thus, it is possible to suppose that the increased swelling (and reduced $G_f$) is due to the occurrence of two concurrent phenomena: (i) the deprotonation of the free -COOH groups in the MC-H samples, which leads to electrostatic repulsion that, in turn, increases the pore size of the hydrogel network [12]; (ii) the alkaline hydrolysis of the ester groups in the MC-H sample [19,22]. These phenomena lead to the disruption of the MC-H hydrogel network, with a resultant increase in swelling and a loss of its physical stability.

To further investigate the thermo-responsive behavior of crosslinked MC specimens, two additional swelling tests were conducted at T = 25 °C and T = 50 °C in NSS at different pHs. For clarity, Figure 2 only reports the curves at pH = 4 and 12 (refer to Supplementary Materials Figure S2 for pH = 7). As it is possible to observe in Figure 2A,B, at T = 25 °C, non-crosslinked and low-crosslinked specimens (MC, MC-L) underwent rapid swelling (up to SR = 4000–6000%), then dissolved within 72 h regardless of the pH of the NSS solution. In fact, at 25 °C (i.e., T < $T_t$), MC was in a sol state and underwent fast dissolution in the water environment [19,20]. Conversely, more crosslinked specimens (MC-M and MC-H) display different behaviors according to the pH of the NSS. At pH = 4, MC-H specimens are stable until the end of the test (168 h). In contrast, at pH = 12, they undergo fast dissolution (induced both by alkaline hydrolysis of the ester groups and low temperature, i.e., T < $T_t$) within the first 6 h of the test.

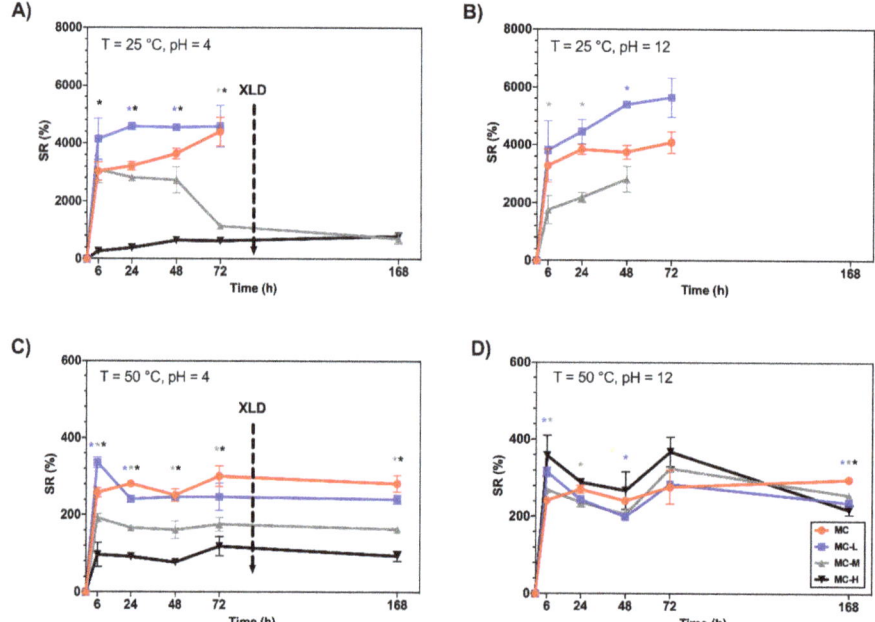

**Figure 2.** SR (%) vs. time of MC hydrogels in NSS at T = 25 °C and (**A**) pH = 4 or (**B**) pH = 12; at T = 50 °C and (**C**) pH = 4 or (**D**) pH = 12. * = $p < 0.05$ compared to MC control (* = MC vs. MC-L, * = MC vs. MC-M, * = MC vs. MC-H).

Different behavior can be observed at T = 50 °C (Figure 2C,D). First, lower SR values (SR = 100–400%) than those at T = 25 °C can be detected for both pH conditions, indicating a reduction of the swelling extent of all MC specimens induced by high temperature. This behavior occurs since, at 50 °C (i.e., T > $T_t$), MC is in a gel state governed by inter- and intramolecular hydrophobic interactions between the methoxy groups present on its backbone. In this state, hydrogen bonds between water and the hydroxyl groups of MC are least favorable, leading to reduced swelling [19,21]. Interestingly, at pH = 4 (Figure 2C), a trend in swelling can still be observed as a function of the crosslinking rate, with swelling for pristine MC higher than for the high-crosslinked specimens (SR ~300% vs. ~100% for MC and MC-H, respectively). Conversely, at pH = 12 (Figure 2D), all the specimens behaved similarly, reaching SR values of ~300%.

For the first time, the possibility of recovering the thermo-responsive behavior of crosslinked MC hydrogels after alkaline hydrolysis of the ester bonds was demonstrated. Even though this is not further investigated in this work, the recovery of the thermo-responsiveness of crosslinked MC samples would open the floodgates to the development of dual-responsive (i.e., thermo- and pH-responsive) MC hydrogels, which can offer tremendous opportunities in several fields [13].

## 2.2. Rheological Characterization

The LVR region was identified qualitatively for all the hydrogels at $\gamma < 0.1\%$ via strain sweep tests. For clarity, Figure 3 only reports the curves at pH = 4 and 12. (Supplementary Materials Figure S3 reports the graphs at pH = 7.) Within the LVR (Figure 3A,B), the storage modulus is higher than the loss modulus (G' > G", data not shown) for each pH condition and each MC hydrogel type. This behavior suggests that the hydrogels are in a gel-like state, governed by weak interactions (i.e., $R^1$-CH$_3$····H$_3$C-$R^2$) and covalent bonds (i.e., ester bonds). Beyond the LVR, G' linearity is lost, and the samples encounter a yielding phase, even if G' > G". Once the flow point (G' = G") is reached, the viscous component becomes preponderant, and the specimens start to flow [23,24]. The trend of the strain sweep curves in an acidic environment (Figure 3A) is similar to the results obtained in a previous work on the same hydrogel formulations [19]. As the crosslinking increases, the viscoelastic parameters increase. However, the different values of G' and G" here observed can be attributable to differences in the swelling medium (dH$_2$O vs. NSS), denoting an increase in the mechanical properties accountable to a salting-out anion effect [21,25,26]. At pH = 7 (Supplementary Materials Figure S3) the curves retrace the ones at pH = 4. Conversely, different behavior can be observed at pH = 12 (Figure 3B): while the pristine and low crosslinked specimens (MC and MC-L) are unaffected by an external pH variation, an increase in crosslinking (MC-M and MC-H) results in a significant decrease in G' (from 8 kPa to 1 kPa and from 30 kPa to 600 Pa for MC-M and MC-H, respectively). This change, most noticeable in the MC-H sample, is accountable for the alkaline hydrolysis of the ester bonds in the MC hydrogels network, as previously explained (Section 2.1). Despite the hydrolysis, all the hydrogel formulations maintained their gel-like state (G' > G") at pH = 12, suggesting that all the samples were above their $T_t$ at the test temperature.

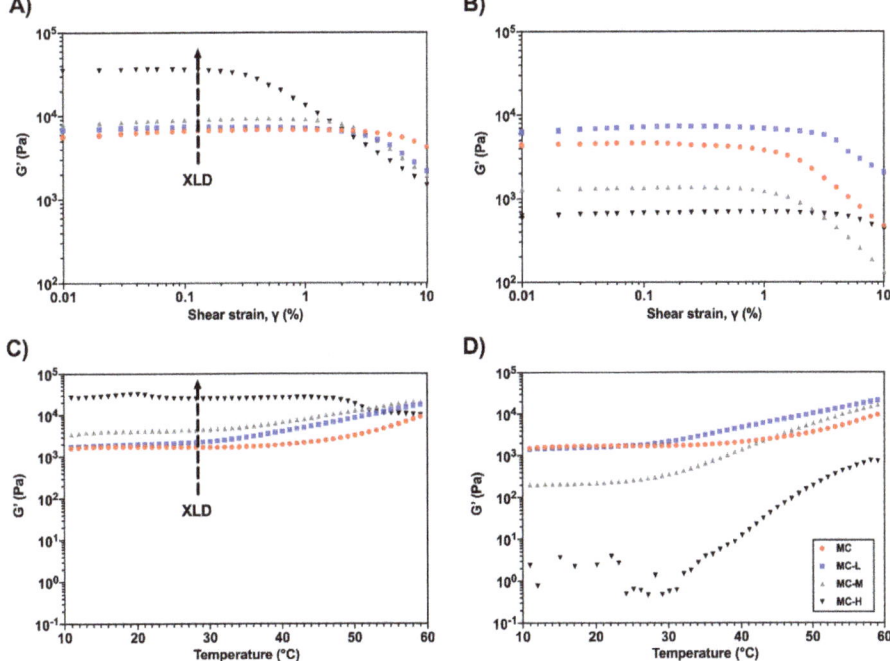

**Figure 3.** Representative G' vs. γ curves for strain sweep tests on MC samples at (**A**) pH = 4 and (**B**) pH = 12. Representative G' vs. T curves for temperature sweep tests on MC samples at (**C**) pH = 4 and (**D**) pH = 12.

Temperature sweep tests were then conducted on all the hydrogel formulations (Figure 3C,D). At pH = 4 (and pH = 7, Supplementary Materials Figure S3), it is possible to observe an increase in G' with the increase of the crosslinking degree, as previously reported [19]. A common trend was evident in all the temperature sweep curves (except for MC-H): (i) for T < $T_t$, a slight decrease of G' occurred due to the disentanglement of the MC chains; (ii) for T > $T_t$, an abrupt increase of G' occurred due to hydrophobic interaction between the MC chains (i.e., sol–gel transition) [27,28]. The MC-H specimen exhibited no sign of thermally triggering a sol–gel transition [19]. Conversely, at pH = 12, different behavior was observed. While the non-crosslinked and low crosslinked specimens (MC and MC-L) seem unaffected by the pH increase, the higher crosslinked specimens (MC-M and MC-H) displayed a significant decrease in G'. Interestingly, the MC-H samples seem to recover their temperature-induced sol–gel transition (Figure 3D), which was lost after chemical crosslinking (Figure 3C).

Table 1 reports the $T_t$ values calculated by applying a previously reported method to the G'/T curves [28]. No significant difference ($p > 0.05$) was noticed between the $T_t$ values of samples at the same pH (except for MC-L at pH = 4). This result suggests that the transition temperature was not affected by mild chemical crosslinking (MC-L and MC-M). Again, the lower $T_t$ obtained in this work compared with the literature [19] is ascribable to the presence of salting-out anions in the NSS solution [21,25,26]. Interestingly, at pH = 12, MC-H displayed a $T_t$~35 °C, which is in line with the other crosslinking conditions, confirming that the thermo-responsive behavior of this sample was recovered after hydrolysis (Section 2.1).

**Table 1.** $T_t$ calculated from temperature sweep tests for each MC hydrogel formulation. * = $p < 0.05$ compared with MC control.

|  | MC | MC-L | MC-M | MC-H |
| --- | --- | --- | --- | --- |
| pH = 4 | 36.5 ± 0.7 | 32.8 ± 0.4 * | 34.0 ± 0.7 | - |
| pH = 7 | 36.8 ± 0.4 | 33.5 ± 0.7 | 35.0 ± 1.4 | - |
| pH = 12 | 33.5 ± 4.9 | 26.5 ± 0.7 | 30.5 ± 2.1 | 34.8 ± 2.5 |

The higher crosslinking formulation (MC-H) was considered the most promising among the samples prepared for this work due to its remarkable pH responsiveness. Thus, all the subsequent characterizations were carried out only on MC-H samples.

### 2.3. $^1$H-NMR Characterization

The $^1$H-NMR spectra of MC-H samples are mainly characterized by the typical signals of MC [29]. The spectra also show -$CH_2$- peaks in the spectral region between 2.5 and 3.0 ppm (Figure 4A, yellow), correlated with chemical crosslinking. Figure 4A reports an inset of the $^1$H-NMR spectra obtained on the washed MC-H samples. The bottom spectrum was obtained for MC-H at the swelling equilibrium (i.e., 24 h in $D_2O$). A broad double doublet (dd) signal can be detected at 2.74–2.98 ppm: this set of signals can be attributed to the -$CH_2$- hydrogens of CA covalently bonded to MC (ester bond). The top spectra were then obtained after adding definite amounts of NaOD. In particular, after the addition of 35 µL of NaOD, the previously detected broad signal shifts towards a higher field (2.60–2.85 ppm) [30], and a well resolved dd signal centered on 2.59 ppm (attributable to free CA) appears. These results suggest that the ester bonds between CA and MC can be easily hydrolyzed in strongly alkaline conditions. An additional 35 µL of NaOD resulted in a sensible increase in the intensity of the dd signals at 2.59 ppm and, at the same time, an appreciable reduction of the broad shoulders attributed to the covalently bonded CA. After 24 h, such broad shoulders are no longer detectable, confirming the hypothesis of the hydrolysis of the ester bonds in increasing alkaline conditions.

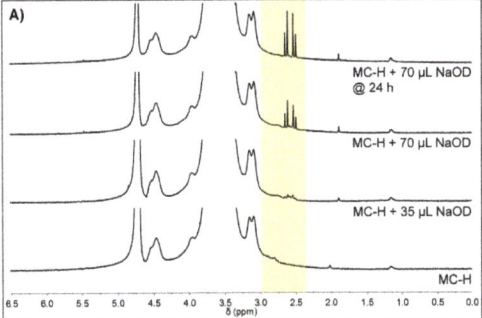
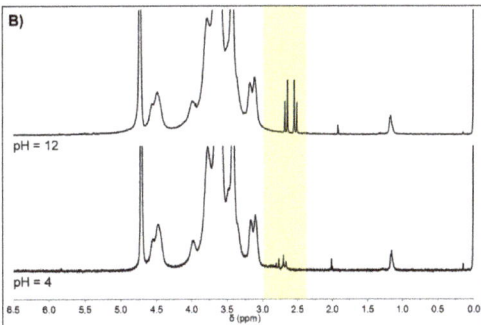

**Figure 4.** $^1$H-NMR spectrum of (**A**) washed MC-H sample swelled in $D_2O$ and then hydrolyzed and (**B**) non-washed MC-H specimens swelled at different pH (4 vs. 12). All the spectra were acquired at 30 °C. The yellow area highlights the -$CH_2$- peaks of CA.

Figure 4B reports the $^1$H-NMR spectrum of non-washed MC-H specimens in NSS at different pH (i.e., same conditions of previous physical and rheological characterization). For samples at pH = 4, the observed dd signals at 2.75 ppm were attributed to unbonded (i.e., not washed off) CA. (Similar results have been obtained for pH = 7, Supplementary Materials Figure S4). A pH increase (pH = 12) led to a remarkable increase in the signal intensity accompanied by a shift to 2.59 ppm, confirming that ester bonds were completely hydrolyzed at this pH. These observations agree with swelling tests. In particular, swelling tests at pH = 12 (Figure 1B) revealed no significant differences ($p > 0.05$) between MC-H and MC specimens for the considered time points, supporting the fact that the complete hydrolysis of the ester bonds occurred at pH = 12 within 24 h.

In addition, by integrating the CA peaks before and after hydrolysis (pH = 4 vs. pH = 12), it was also possible to estimate that the content of unbonded CA in MC-H samples was about 10 mg g$^{-1}$, and that the weight fraction of CA taking part in the ester linkage was equal to ~70%, which is in good accordance with previous findings [19].

NMR spectra were then acquired as a function of temperature, and the overall signal intensity was evaluated to assess the thermally induced transition of MC-H specimens. Figure 5 reports the $^1$H-NMR spectrum of MC-H specimens as a function of temperature, at a swelling pH = 4 and 12. At pH = 4 (Figure 5A), the overall $^1$H-NMR signal intensity increased by increasing temperature. (Similar findings were obtained with samples at pH = 7, Supplementary Materials Figure S5). On the contrary, at pH = 12, the $^1$H-NMR signals decreased the intensity by increasing the temperature due to the restricted MC chains mobility induced by the sol–gel transition [31]. The thermally induced transition of MC to a more rigid and ordered structure caused a change in the spin–spin relaxation time ($T_2$), leading to a sensible signal broadening [31]. These results confirm that the thermo-responsive behavior of the MC-H sample was retrieved after hydrolysis.

Overall, the results obtained for the MC-H samples open the floodgates to develop responsive hydrogels capable of undergoing pH-triggered, tailorable hydrolysis [32]. To the authors' best knowledge, this is the first attempt at studying the pH-responsiveness of MC hydrogels. Similar systems have already demonstrated their potential in numerous biomedical applications [13,33], particularly for drug delivery (e.g., tumor-targeted drug delivery [34], intracellular delivery of nucleic acids or proteins [35], and the treatment of inflammatory diseases [36]). Considering this, the possibility of developing a novel pH-responsive platform based on MC for the delivery of AgNPs was investigated.

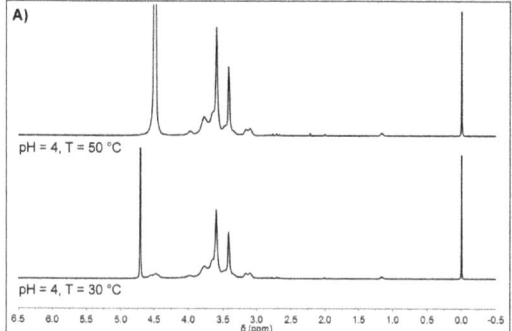
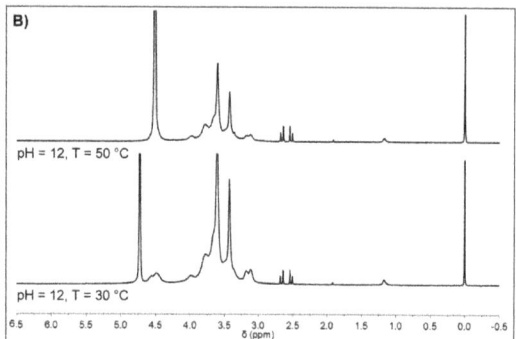

**Figure 5.** (**A**) $^1$H-NMR spectrum of MC-H swelled at (**A**) pH = 4 and (**B**) at pH = 12, as function of temperature (30 vs. 50 °C).

### 2.4. MC/AgNPs Composite Hydrogels

AgNPs were obtained by the chemical reduction of the metallic precursor (AgNO$_3$) in the polymer matrix. Several methods have been reported in the literature to obtain AgNPs (e.g., citric acid reduction, electrochemical synthesis, photochemistry, and radiation reduction) [37]. In this work, MC acted both as a capping and reducing agent, combining the advantage of a one-step reaction with the possibility of a simple control over the reaction parameters (i.e., nanoparticle size and distribution) [37,38].

TEM observations provided insight into the shape and dimensions distribution of AgNPs (Figure 6A). The in situ synthesis approach allowed for obtaining spherical nanoparticles with 12.2 ± 2.8 nm diameter and a D90 of 15 nm (dimension distribution analysis in Figure 6B). The AgNPs here showed a slightly smaller size than those previously reported by Maity et al. [38] (mean diameter: 12.2 vs. 22 nm), which is likely due to the higher concentration of MC. Indeed, the concentration of capping and reducing agents plays an essential role in the size of silver nanoparticles [39,40]. Nevertheless, the obtained dimensions aligned with other works in which MC [41] or other polymers [37,42,43] have been reported for the in situ synthesis of metal nanoparticles. Moreover, the obtainment of AgNPs with smaller sizes (in the nanometric range) can show a non-trivial effect on their antibacterial activity efficiency due to their easy binding to important functional sites on pathogens [44–46].

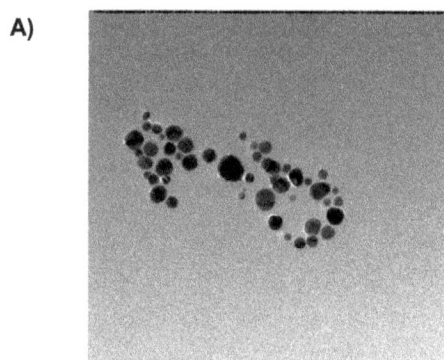
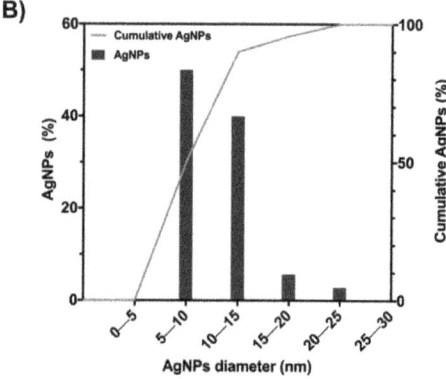

**Figure 6.** (**A**) TEM image of MC/AgNPs solution and (**B**) MC/AgNPs distribution analysis.

## 2.5. MC-H/AgNPs Characterization: UV–Vis and ICP Analyses

UV–vis analyses have been extensively reported in the literature to assess the formation and presence of AgNPs in suspension [44–47]. In this work, the as-prepared MC/AgNPs solutions (Figure 7A) displayed a well-defined absorption peak at 410 nm, which is typical of a colloidal suspension of spherical silver nanoparticles [47]. Such results are in accordance with previously reported works exploiting the dual role of MC as stabilizing and reducing agent for the in situ synthesis of AgNPs [38,41,48].

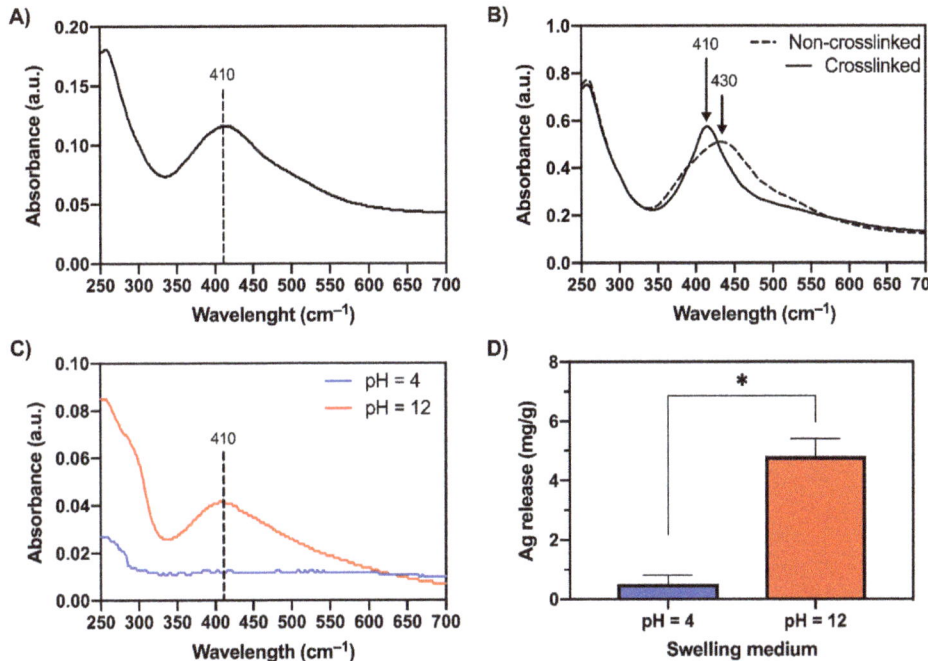

**Figure 7.** UV–vis absorption spectra of (**A**) as-synthesized MC/AgNPs, (**B**) MC-H/AgNPs dry films, (**C**) swelling media at pH = 4 and 12. (**D**) Ag release ($mg_{Ag}/g_{MC}$) from MC-H/AgNPs samples as a function of the pH obtained by ICP analyses. * = $p < 0.05$.

UV–vis measurements were then performed on MC-H/AgNPs specimens (in dry conditions, Figure 7B) to confirm the presence of AgNPs before and after crosslinking (190 °C, 15 min). For the non-crosslinked samples, a red shift (from 410 to 430 nm) occurs. This variation in the surface plasmon absorption peak position can be attributed to the different refractive index of the surrounding environment (i.e., dry film) [38,49]. After crosslinking, the characteristic resonance peak undergoes a visible sharpening, displaying a blue shift (from 430 to 410 nm) compared with the non-crosslinked conditions. Even in this case, such behavior can be ascribed to the change in the surrounding environment due to the formation of ester bonds (during the crosslinking phase) and the consequent decrease in the pH of the films [50,51].

The MC-H/AgNPs samples were immersed in two distinct buffers (pH = 4 and 12) at 37 °C to investigate the possibility of a pH-triggered release of the nanoparticles. After 1 h, UV–vis measurements (Figure 7C) reveal silver colloids' characteristic plasmon resonance peak at 410 nm at pH = 12. Conversely, no peak is detected at pH 4, suggesting no (or not detectable) release of AgNPs. A selective Ag release (Figure 7D), ~10 times higher at pH 12 (4.8 ± 0.6 $mg_{Ag}/g_{MC}$) than at pH 4 (0.5 ± 0.3 $mg_{Ag}/g_{MC}$), was quantified via ICP analyses. To better elucidate the mechanism underlying the pH-triggered AgNPs release, the theoretical physical parameters describing MC-H hydrogels microstructure (at the different pH values) were calculated by applying a simplified version of the Flory–Rehner model [19,20]. The average molecular weight between crosslinks ($\overline{M_C}$), the crosslinking density ($\rho_c$), and the mesh size ($\xi$) were calculated (Supplementary Materials Table S1). In particular, the obtained $\xi$ values were 4.09 ± 0.06, 4.33 ± 0.14, and 19.05 ± 3.07 nm for pH = 4, 7, and 12, respectively. Interestingly, the specimens' mesh size at pH = 4 and 7 is smaller than the NPs' diameter (12.2 ± 2.8 nm), while it is larger than the NPs diameter at pH = 12. These calculations shed new light on the selective (i.e., pH-triggered) NPs released from MC-H hydrogels. The AgNPs are physically entrapped in MC-H hydrogel mesh at low pHs, resulting in limited AgNPs release. At pH = 12, alkaline hydrolysis of ester bonds leads to hydrogel network expansion and release of AgNPs.

These results proved the possibility of controlling the release of AgNPs through an alkaline pH trigger. No studies evaluating the pH-triggered release of AgNPs from MC hydrogels were previously reported. Similar results have been obtained by Haidari H. et al., who showed the potential of selective release of AgNPs, at alkaline pH, from pH-responsive poly(mAA-co-AAm) hydrogels [12]. However, Haidari and co-workers achieved AgNPs loading by swelling the dry hydrogels in an AgNPs suspension after hydrogel synthesis and purification (necessary to remove toxic residual acrylamide monomers). Conversely, MC/AgNPs hydrogels were obtained via a one-step in situ synthesis process.

Overall, these outcomes are promising in light of the increasing concern about the possible impact of AgNPs misuse on human health [3,12]. Cytotoxic effects of AgNPs have been documented in various cell lines in vitro and are dependent on different factors (e.g., size, shape, dose, cell type), among which dosage is considered a significant parameter to be controlled by [52]. Additionally, it has been evidenced that there is a significant transdermal penetration of AgNPs into capillaries after dressings, textiles, and cosmetics [52]. In this regard, in vivo biodistribution studies have revealed how, following the dermal route, Ag translocation, accumulation, and toxicity can occur to various organs (e.g., spleen and liver) [52].

MC-H/AgNPs hydrogels prepared in this work can open the floodgates to the development of responsive systems capable of releasing AgNPs only at the occurrence of a pathological trigger (i.e., alkaline pH) in the wound milieu, avoiding the drawbacks associated with the misuse of AgNPs and thus representing a significant advantage compared to the current treatments. Further characterizations are undoubtedly needed to corroborate the claim of producing a pH-responsive platform for treating infections in chronic wounds. In particular, in vitro, antibacterial, and cytotoxicity tests could represent the first step in this direction. On this topic, Maity et al. [38] reported how MC/AgNPs nanocomposites exhibited strong antibacterial activity against several bacterial strains (i.e., *B. subtilis, B. cereus, P. aeruginosa, S. aureus*, and *E. coli*). Since the same AgNO$_3$ concentration was used to obtain MC-H/AgNPs hydrogels in the present study, an antibacterial efficacy comparable to the one obtained by Maity et al. [38] can be expected. Regarding the in vitro cytotoxicity assessment, the non-cytotoxic nature of MC-H specimens has already been demonstrated on L929 fibroblast cells [20], also corroborating the finding of the non-toxicity of CA for amounts up to 20% $w/w$ ($w_{CA}/w_{polymer}$) [53,54]. However, the cytotoxicity of MC-H/AgNPs specimens remains to be assessed. In this regard, tuning the AgNPs concentration could be necessary to achieve a trade-off between cytotoxicity and antibacterial activity [55].

## 3. Conclusions

In this work, citric acid crosslinked methylcellulose films were prepared and comprehensively characterized to develop pH-responsive systems helpful in treating infected chronic wounds. Highly crosslinked hydrogels (MC-H) showed a remarkable pH-responsive behavior based on selective hydrolysis in an alkaline environment. MC-H hydrogels were then disclosed as promising for the in situ synthesis of AgNPs and their subsequent pH-triggered delivery. Such a platform lends itself to the selective control of the proliferation of pathogens in infected chronic wounds characterized by an alkaline environment. Given that further characterizations are needed, such systems could be potentially exploited to reduce the adverse effects on host cells and tissues due to the uncontrolled AgNPs release.

## 4. Materials and Methods

### 4.1. Chemicals and Instruments

All chemicals were purchased from Sigma-Aldrich (Milan, Italy) and were used as received without further purification unless stated.

### 4.2. MC Hydrogels Preparation

Methylcellulose hydrogels were obtained according the method discussed in a previous work [19]. Citric acid (CA) was added to an MC solution (8% $w/v$ MC in 50 mM $Na_2SO_4$) in different amounts ([CA] = 1, 3, or 5% $w_{CA}/w_{MC}$): 15 mL of solution were cast into Petri dishes (Ø = 90 mm) and oven-dried (T = 50 °C, t = 24 h). Dry films were peeled off from the Petri dishes and crosslinked by tuning the crosslinker concentration ([CA]), the crosslinking time ($t_{XL}$) and temperature ($T_{XL}$) [19]. Three differently crosslinked MC hydrogels were hence obtained: MC-L ([CA] = 1%, $T_{XL}$ = 165 °C, $t_{XL}$ = 1 min), MC-M ([CA] = 3%, $T_{XL}$ = 178 °C, $t_{XL}$ = 8 min), and MC-H ([CA] = 5%, $T_{XL}$ = 190 °C, $t_{XL}$ = 15 min), meaning low, medium, and high crosslinking degree (XLD). Non-crosslinked MC was used as a control.

### 4.3. Synthesis of MC/AgNPs Composites

The synthesis of MC/AgNPs composites was achieved by adapting a previously reported method [38]. An MC solution (0.5% $w/v$) was prepared by dispersing MC powder in hot (55 °C) distilled water. The obtained solution was cooled to room temperature under stirring, then stored at 4 °C for 24 h. The pH of the solution was adjusted to 10 using a NaOH (1 M) solution. Cold aqueous $AgNO_3$ ($10^{-2}$ M) solution was added dropwise to the MC solution in a conical flask under mild (200 rpm) stirring, achieving a final concentration of $1.11 \times 10^{-4}$ $mol_{AgNO3}/g_{MC}$ [38]. The flask was kept in the dark, under stirring, for 24 h. The pH of the solution was then adjusted to the initial value (6.2–6.4) using an $H_2SO_4$ (0.1 M) solution. Afterward, the final $SO_4^{2-}$ concentration of 50 mM (i.e., the same as MC-H samples) was achieved by adding $Na_2SO_4$. CA (5% $w_{CA}/w_{MC}$) was added as a crosslinker. Lastly, 15 mL of the obtained solution were cast into Petri dishes (Ø = 90 mm) and oven-dried (T = 50 °C, t = 24 h). Dry films were first peeled off the Petri dish and then crosslinked (15 min, 190 °C). The obtained samples will be referred to as MC-H/AgNPs.

### 4.4. Swelling and Degradation Tests

The water absorption of CA-crosslinked MC hydrogels was evaluated by swelling tests in Normal Saline Solution (NSS, 9 g $L^{-1}$ NaCl) at different pHs (4, 7, 12) and different temperatures (25, 37, 50 °C) up to 28 days. The swelling solutions' pH was adjusted by adding a few drops of 1 M HCl or 1 M NaOH solutions. A fixed immersion ratio (MC: NSS = 20 mg: 12.5 mL) was chosen to keep the pH stable up to 72 h (the NSS solution was refreshed every 72 h). MC dry specimens were first weighted ($w_0$), then incubated at the test temperature (25, 37, or 50 °C) in NSS at different pH. At selected time points, the specimens were weighted ($w_{t,s}$: swollen weight at time t), and the swelling ratio (SW) was calculated as in equation (Equation (1)):

$$SW(\%) = \frac{w_{t,s} - w_0}{w_0} * 100 \qquad (1)$$

The stability of CA-crosslinked MC hydrogels was evaluated through degradation tests. MC dry specimens ($w_0$) were incubated at the test temperature (25, 37, or 50 °C) in NSS at different pH (4, 7, 12) containing 0.02% ($w/v$) $NaN_3$ to prevent microbial contamination. The specimens were retrieved from the solution at selected time points, dried (50 °C,

24 h), and weighted ($w_{t,d}$: dry weight at time t). The solid gel fraction ($G_f$) was then calculated [20] according to equation (Equation (2)):

$$G_f(\%) = \frac{w_{t,d}}{w_0} * 100 \qquad (2)$$

### 4.5. MC-H Gels Microstructure: Theoretical Physical Parameters

The theoretical physical parameters describing MC-H gels microstructure were calculated according to a simplified version of the Flory–Rehner model [19,20,56–58]. Average molecular weight between crosslinking points ($\overline{M_C}$), crosslinking density ($\rho_C$), and mesh size ($\xi$) were calculated at different pH values (4, 7, and 12) at 37 °C.

The average molecular weight between crosslinking points ($\overline{M_C}$) was calculated using the following equations (Equations (3)–(5)) [19,57–59]:

$$Q_v^{5/3} \cong \frac{\overline{v}\overline{M_C}}{v_l}\left(\frac{1}{2} - \chi\right) \qquad (3)$$

$$Q_v = 1 + \left(\frac{\rho_p}{\rho_s}(Q_w - 1)\right) \qquad (4)$$

$$Q_w = \frac{w_s}{w_d} \qquad (5)$$

where $w_s$ and $w_d$ are the weights of the MC-H samples in swollen (t = 24 h) and dry conditions (t = 0), respectively. $Q_w$ and $Q_v$ represent the equilibrium weight swelling ratio and the volumetric swelling ratio, respectively.

The crosslinking density ($\rho_C$) was calculated according to the following equation (Equation (6)) [19,57,58]:

$$\rho_C = \frac{1}{\overline{v}\overline{M_C}} \qquad (6)$$

The mesh size ($\xi$) of the hydrogel at the swelling equilibrium was calculated using the following equation (Equation (7)) [19,57,58,60]:

$$\xi = 0.217\sqrt{\overline{M_C}}Q_v^{1/3} \qquad (7)$$

The constant terms used in the equations (Equations (3)–(7)) are [19,20,58]:

$\rho_p$ = 0.276 g cm$^{-3}$ (density of dry polymer)
$\rho_s$ = 1 g cm$^{-3}$ (density of the solvent)
$\overline{v} = \frac{1}{\rho_p}$ = 3.623 cm$^3$ g$^{-1}$ (specific volume of dry polymer)
$v_l$ = 18 mol cm$^{-3}$ (molar volume of the solvent)
$\chi$ = 0.473 (Flory polymer-solvent interaction parameter)

### 4.6. Rheological Characterization

The rheological properties of CA-crosslinked MC hydrogels at the swelling equilibrium (24 h in NSS solutions at 37 °C) were tested using a rotational rheometer (MCR 302, Anton Paar) equipped with a parallel plate geometry (Ø = 25 mm, working gap = 1 mm). Strain sweep tests were preliminarily performed on each MC hydrogel formulation to identify the linear viscoelastic region (LVR) by applying an oscillatory strain in the 0.01–10% strain range ($\gamma$) at 37 °C and 1 Hz frequency ($\nu$). Then, to determine the transition temperature ($T_t$) of each specimen, temperature sweep tests were performed in the 10–60 °C range, applying a temperature ramp of 2 °C min$^{-1}$, $\gamma$ = 0.1% (i.e., LVR obtained by strain sweep tests), $\nu$ = 1 Hz frequency. For each hydrogel composition, the $T_t$ was identified from storage modulus ($G'$) and complex viscosity ($\eta^*$) curves. Briefly, from each curve, the $T_t$ was determined as the intersection between the interpolant of the initial T range (T < $T_t$) and the interpolant of the following T range (T > $T_t$) [27,28].

### 4.7. $^1$H-NMR Characterization

The MC-H formulation was considered the most promising for this work due to its remarkable pH-responsiveness. The pH responsivity of MC-H samples was further investigated using $^1$H-NMR spectrometry. $^1$H-NMR spectra were collected using a Bruker ARX 400 spectrometer (400 MHz $^1$H resonance frequency). All spectra were recorded in $D_2O$ containing 0.05% (w/w) of 3-(trimethylsilyl)propionic 2,2,3,3,-$d_4$ acid sodium salt as internal standard. Chemical shifts (d, ppm) are reported relative to the internal standard.

Explorative analyses were carried out on washed MC-H samples. Removal of unreacted CA from the MC-H specimens was achieved by rinsing MC-H samples several times in distilled water until reaching neutral pH [19]. The $^1$H-NMR spectra of washed MC-H samples were then collected in the following conditions: (i) at the swelling equilibrium (20 mg MC-H in 0.7 mL $D_2O$ for 24 h), (ii) after the addition of 35 µL NaOD (30 wt.% in $D_2O$), (iii) after the addition of further 35 µL NaOD, and (iv) 24 h after the addition of 70 µL NaOD. All measurements were carried out at 30 °C.

Then, the $^1$H-NMR spectra of non-washed MC-H specimens were collected to retrace the swelling and degradation tests. Briefly, MC-H were swollen in NSS at different pH (4, 7 and 12) for 24 h (i.e., swelling equilibrium), keeping fixed the immersion ratio (MC-H: NSS = 20 mg: 12.5 mL). Then, the samples were freeze-dried. After lyophilization, each specimen was swollen in 1 mL of $D_2O$ for 24 h before $^1$H-NMR spectra collection. For each pH condition, $^1$H-NMR spectra were collected at 30, 37, and 50 °C.

### 4.8. TEM Characterization

The shape, size, and distribution of silver nanoparticles in the MC/AgNPs solutions were assessed by transmission electron microscopy (TEM, Philips CM200-FEG) at an accelerating voltage of 200 kV. A drop of the sample (after 1:10 dilution in $dH_2O$) was withdrawn, deposited on a carbon-coated copper net (mesh size 200), and air-dried overnight at room temperature. To estimate the shape and size of silver nanoparticles, 50 images were analyzed by ImageJ software (ImageJ, v. 1.53, NIH). Prism 8 (GraphPad Software, La Jolla, CA, USA) was then used for the particle size distribution analysis.

### 4.9. UV–Vis Characterization

UV–vis spectroscopy (Synergy H1 spectrophotometer, BioTek) was used to monitor the intensity of the Localized Surface Plasmon Resonance (LSPR) absorption band of silver nanoparticles in the MC samples. UV–vis studies were performed (i) on MC/AgNPs solutions, (ii) on MC-H/AgNPs dry films, and (iii) on the swelling solutions of MC-H/AgNPs specimens. In the latter case, UV–vis measurements were carried out to assess the pH-triggered AgNPs release from MC-H/AgNPs specimens. To do this, MC-H/AgNPs specimens were first briefly rinsed with $dH_2O$ to remove AgNPs not embedded in the film. Then, the specimens were immersed in 10 mM acetate (pH = 4) or phosphate (pH = 12) buffers, incubated at 37 °C for 1 h, and the swelling media were analyzed. Spectra were acquired in the wavelength range 250–700 nm, with a resolution of 1 nm. The solutions were read using a Take3 Micro-Volume Plate support (BioTek), while dry films used a Slide adapter (BioTek, 1220548).

### 4.10. ICP Analysis

To quantify the amount of Ag released by MC-H/AgNPs specimens as a function of the pH, the samples were first briefly rinsed with $dH_2O$ to remove superficial AgNPs. Then, the specimens were immersed in 10 mM acetate (pH = 4) or phosphate (pH = 12) buffers and incubated at 37 °C (pH 7 was not investigated based on the outcomes of physicochemical and rheological characterization). After 1 h, the eluates were withdrawn and analyzed by inductively coupled plasma optical emission spectrometry (ICP-OES).

*4.11. Data Analysis*

Unless stated, the tests were run in triplicate ($n = 3$), and data are expressed as mean $\pm$ standard deviation (SD). Statistical analysis was performed by Prism 8 (GraphPad Software, La Jolla, CA, USA). Comparisons among the groups were performed by ANOVA (one-way or two-way), and results of $p < 0.05$ were considered statistically significant.

**Supplementary Materials:** The following supporting information can be downloaded at: https://www.mdpi.com/article/10.3390/gels8050298/s1, Supplementary Materials S1: Swelling and gel fraction; Supplementary Materials S2: Flory–Rehner model; Supplementary Materials S3: Rheology; Supplementary Materials S4: $^1$H-NMR.

**Author Contributions:** Conceptualization, methodology, formal analysis, investigation, writing—original draft, writing—review and editing, visualization, L.B.; methodology, investigation, writing—original draft, writing—review and editing, A.F.; methodology, investigation, writing—review and editing, A.D.; investigation, C.M.P.; conceptualization, writing—review and editing, R.C.; conceptualization, writing—review and editing, S.F.; conceptualization, resources, writing—review and editing, supervision, L.D.N. All authors have read and agreed to the published version of the manuscript.

**Funding:** Authors would like to acknowledge project PRIN "Multiple Advanced Materials Manufactured by Additive Technologies–PRIN 2017–prot. 20179SWLKA", project PRIN "Advanced injectable nano-composite biomaterials with dual therapeutic/regenerative behaviors for bone cancer (ACTION)–PRIN 2017–prot. 2017SZ5WZB_002", and project FESR 2014–2020 ARS01_01205 'CustOmmadeaNTibacterical/bioActive/bioCoated Prostheses' for the economic support.

**Institutional Review Board Statement:** Not applicable.

**Informed Consent Statement:** Not applicable.

**Data Availability Statement:** Not applicable.

**Conflicts of Interest:** The authors declare no conflict of interest.

# References

1. Gottrup, F. A specialized wound-healing center concept: Importance of a multidisciplinary department structure and surgical treatment facilities in the treatment of chronic wounds. *Am. J. Surg.* **2004**, *187*, S38–S43. [CrossRef]
2. Raj, V.; Kim, Y.; Kim, Y.-G.; Lee, J.-H.; Lee, J. Chitosan-gum arabic embedded alizarin nanocarriers inhibit biofilm formation of multispecies microorganisms. *Carbohydr. Polym.* **2022**, *284*, 118959. [CrossRef] [PubMed]
3. Frykberg, R.G.; Banks, J. Challenges in the Treatment of Chronic Wounds. *Adv. Wound Care* **2015**, *4*, 560–582. [CrossRef] [PubMed]
4. Siddiqui, A.R.; Bernstein, J.M. Chronic wound infection: Facts and controversies. *Clin. Dermatol.* **2010**, *28*, 519–526. [CrossRef] [PubMed]
5. Ono, S.; Imai, R.; Ida, Y.; Shibata, D.; Komiya, T.; Matsumura, H. Increased wound pH as an indicator of local wound infection in second degree burns. *Burns* **2015**, *41*, 820–824. [CrossRef] [PubMed]
6. Metcalf, D.G.; Haalboom, M.; Bowler, P.G.; Gamerith, C.; Sigl, E.; Heinzle, A.; Burnet, M.W.M. Elevated wound fluid pH correlates with increased risk of wound infection. *Wound Med.* **2019**, *26*, 100166. [CrossRef]
7. Percival, S.L.; McCarty, S.; Hunt, J.A.; Woods, E.J. The effects of pH on wound healing, biofilms, and antimicrobial efficacy. *Wound Repair Regen.* **2014**, *22*, 174–186. [CrossRef]
8. Schneider, L.A.; Korber, A.; Grabbe, S.; Dissemond, J. Influence of pH on wound-healing: A new perspective for wound-therapy? *Arch. Dermatol. Res.* **2007**, *298*, 413–420. [CrossRef]
9. Han, G.; Ceilley, R. Chronic Wound Healing: A Review of Current Management and Treatments. *Adv. Ther.* **2017**, *34*, 599–610. [CrossRef]
10. Toy, L.W.; Macera, L. Evidence-based review of silver dressing use on chronic wounds. *J. Am. Acad. Nurse Pract.* **2011**, *23*, 183–192. [CrossRef]
11. Paladini, F.; Pollini, M. Antimicrobial Silver Nanoparticles for Wound Healing Application: Progress and Future Trends. *Materials* **2019**, *12*, 2540. [CrossRef]
12. Haidari, H.; Kopecki, Z.; Sutton, A.T.; Garg, S.; Cowin, A.J.; Vasilev, K. pH-Responsive "Smart" Hydrogel for Controlled Delivery of Silver Nanoparticles to Infected Wounds. *Antibiotics* **2021**, *10*, 49. [CrossRef] [PubMed]
13. Schmaljohann, D. Thermo- and pH-responsive polymers in drug delivery. *Adv. Drug Deliv. Rev.* **2006**, *58*, 1655–1670. [CrossRef] [PubMed]
14. Kocak, G.; Tuncer, C.; Bütün, V. pH-Responsive polymers. *Polym. Chem.* **2017**, *8*, 144–176. [CrossRef]

15. Gholamali, I.; Asnaashariisfahani, M.; Alipour, E. Silver Nanoparticles Incorporated in pH-Sensitive Nanocomposite Hydrogels Based on Carboxymethyl Chitosan-Poly (Vinyl Alcohol) for Use in a Drug Delivery System. *Regen. Eng. Transl. Med.* **2020**, *6*, 138–153. [CrossRef]
16. Deen, G.; Chua, V. Synthesis and Properties of New "Stimuli" Responsive Nanocomposite Hydrogels Containing Silver Nanoparticles. *Gels* **2015**, *1*, 117–134. [CrossRef]
17. Xu, Y.; Li, L.; Zheng, P.; Lam, Y.C.; Hu, X. Controllable Gelation of Methylcellulose by a Salt Mixture. *Langmuir* **2004**, *20*, 6134–6138. [CrossRef]
18. Tang, Y.; Wang, X.; Li, Y.; Lei, M.; Du, Y.; Kennedy, J.F.; Knill, C.J. Production and characterisation of novel injectable chitosan/methylcellulose/salt blend hydrogels with potential application as tissue engineering scaffolds. *Carbohyd. Polym.* **2010**, *82*, 833–841. [CrossRef]
19. Bonetti, L.; De Nardo, L.; Variola, F.; Fare', S. Evaluation of the subtle trade-off between physical stability and thermo-responsiveness in crosslinked methylcellulose hydrogels. *Soft Matter* **2020**, *16*, 5577–5587. [CrossRef]
20. Bonetti, L.; De Nardo, L.; Farè, S. Chemically Crosslinked Methylcellulose Substrates for Cell Sheet Engineering. *Gels* **2021**, *7*, 141. [CrossRef]
21. Bonetti, L.; De Nardo, L.; Farè, S. Thermo-Responsive Methylcellulose Hydrogels: From Design to Applications as Smart Biomaterials. *Tissue Eng. Part B Rev.* **2021**, *27*, 486–513. [CrossRef] [PubMed]
22. De Cuadro, P.; Belt, T.; Kontturi, K.S.; Reza, M.; Kontturi, E.; Vuorinen, T.; Hughes, M. Cross-linking of cellulose and poly(ethylene glycol) with citric acid. *React. Funct. Polym.* **2015**, *90*, 21–24. [CrossRef]
23. Park, S.H.; Shin, H.S.; Park, S.N. A novel pH-responsive hydrogel based on carboxymethyl cellulose/2-hydroxyethyl acrylate for transdermal delivery of naringenin. *Carbohyd. Polym.* **2018**, *200*, 341–352. [CrossRef] [PubMed]
24. Chen, M.H.; Wang, L.L.; Chung, J.J.; Kim, Y.-H.; Atluri, P.; Burdick, J.A. Methods To Assess Shear-Thinning Hydrogels for Application As Injectable Biomaterials. *ACS Biomater. Sci. Eng.* **2017**, *3*, 3146–3160. [CrossRef]
25. Bain, M.K.; Bhowmick, B.; Maity, D.; Mondal, D.; Mollick, M.M.R.; Rana, D.; Chattopadhyay, D. Synergistic effect of salt mixture on the gelation temperature and morphology of methylcellulose hydrogel. *Int. J. Biol. Macromol.* **2012**, *51*, 831–836. [CrossRef]
26. Liang, H.-F.; Hong, M.-H.; Ho, R.-M.; Chung, C.-K.; Lin, Y.-H.; Chen, C.-H.; Sung, H.-W. Novel Method Using a Temperature-Sensitive Polymer (Methylcellulose) to Thermally Gel Aqueous Alginate as a pH-Sensitive Hydrogel. *Biomacromolecules* **2004**, *5*, 1917–1925. [CrossRef] [PubMed]
27. Cochis, A.; Bonetti, L.; Sorrentino, R.; Contessi Negrini, N.; Grassi, F.; Leigheb, M.; Rimondini, L.; Farè, S. 3D Printing of Thermo-Responsive Methylcellulose Hydrogels for Cell-Sheet Engineering. *Materials* **2018**, *11*, 579. [CrossRef] [PubMed]
28. Contessi, N.; Altomare, L.; Filipponi, A.; Farè, S. Thermo-responsive properties of methylcellulose hydrogels for cell sheet engineering. *Mater. Lett.* **2017**, *207*, 157–160. [CrossRef]
29. Miura, Y. Solvent isotope effect on gelation process of methylcellulose studied by NMR and DSC. *Polym. Bull.* **2018**, *75*, 4245–4255. [CrossRef]
30. Atieh, Z.; Suhre, K.; Bensmail, H. MetFlexo: An Automated Simulation of Realistic H1-NMR Spectra. *Procedia Comput. Sci.* **2013**, *18*, 1382–1391. [CrossRef]
31. Haque, A.; Morris, E.R. Thermogelation of methylcellulose. Part I: Molecular structures and processes. *Carbohyd. Polym.* **1993**, *22*, 161–173. [CrossRef]
32. Song, J.; Hwang, E.; Lee, Y.; Palanikumar, L.; Choi, S.-H.; Ryu, J.-H.; Kim, B.-S. Tailorable degradation of pH-responsive all polyether micelles via copolymerisation with varying acetal groups. *Polym. Chem.* **2019**, *10*, 582–592. [CrossRef]
33. Chassenieux, C.; Tsitsilianis, C. Recent trends in pH/thermo-responsive self-assembling hydrogels: From polyions to peptide-based polymeric gelators. *Soft Matter* **2016**, *12*, 1344–1359. [CrossRef] [PubMed]
34. Sun, C.-Y.; Shen, S.; Xu, C.-F.; Li, H.-J.; Liu, Y.; Cao, Z.-T.; Yang, X.-Z.; Xia, J.-X.; Wang, J. Tumor Acidity-Sensitive Polymeric Vector for Active Targeted siRNA Delivery. *J. Am. Chem. Soc.* **2015**, *137*, 15217–15224. [CrossRef] [PubMed]
35. Ren, J.; Zhang, Y.; Zhang, J.; Gao, H.; Liu, G.; Ma, R.; An, Y.; Kong, D.; Shi, L. pH/Sugar Dual Responsive Core-Cross-Linked PIC Micelles for Enhanced Intracellular Protein Delivery. *Biomacromolecules* **2013**, *14*, 3434–3443. [CrossRef] [PubMed]
36. Ninan, N.; Forget, A.; Shastri, V.P.; Voelcker, N.H.; Blencowe, A. Antibacterial and Anti-Inflammatory pH-Responsive Tannic Acid-Carboxylated Agarose Composite Hydrogels for Wound Healing. *ACS Appl. Mater. Interfaces* **2016**, *8*, 28511–28521. [CrossRef]
37. Maity, D.; Kanti Bain, M.; Bhowmick, B.; Sarkar, J.; Saha, S.; Acharya, K.; Chakraborty, M.; Chattopadhyay, D. In situ synthesis, characterization, and antimicrobial activity of silver nanoparticles using water soluble polymer. *J. Appl. Polym. Sci.* **2011**, *122*, 2189–2196. [CrossRef]
38. Maity, D.; Mollick, M.M.R.; Mondal, D.; Bhowmick, B.; Bain, M.K.; Bankura, K.; Sarkar, J.; Acharya, K.; Chattopadhyay, D. Synthesis of methylcellulose–silver nanocomposite and investigation of mechanical and antimicrobial properties. *Carbohyd. Polym.* **2012**, *90*, 1818–1825. [CrossRef]
39. Restrepo, C.V.; Villa, C.C. Synthesis of silver nanoparticles, influence of capping agents, and dependence on size and shape: A review. *Environ. Nanotechnol. Monit. Manag.* **2021**, *15*, 100428. [CrossRef]
40. Yaqoob, A.A.; Umar, K.; Ibrahim, M.N.M. Silver nanoparticles: Various methods of synthesis, size affecting factors and their potential applications–a review. *Appl. Nanosci.* **2020**, *10*, 1369–1378. [CrossRef]

41. Kolarova, K.; Samec, D.; Kvitek, O.; Reznickova, A.; Rimpelova, S.; Svorcik, V. Preparation and characterization of silver nanoparticles in methyl cellulose matrix and their antibacterial activity. *Jpn. J. Appl. Phys.* **2017**, *56*, 06GG09. [CrossRef]
42. Shen, J.; Cui, C.; Li, J.; Wang, L. In Situ Synthesis of a Silver-Containing Superabsorbent Polymer via a Greener Method Based on Carboxymethyl Celluloses. *Molecules* **2018**, *23*, 2483. [CrossRef] [PubMed]
43. Martínez-Higuera, A.; Rodríguez-Beas, C.; Villalobos-Noriega, J.M.A.; Arizmendi-Grijalva, A.; Ochoa-Sánchez, C.; Larios-Rodríguez, E.; Martínez-Soto, J.M.; Rodríguez-León, E.; Ibarra-Zazueta, C.; Mora-Monroy, R.; et al. Hydrogel with silver nanoparticles synthesized by Mimosa tenuiflora for second-degree burns treatment. *Sci. Rep.* **2021**, *11*, 11312. [CrossRef] [PubMed]
44. Haidari, H.; Goswami, N.; Bright, R.; Kopecki, Z.; Cowin, A.J.; Garg, S.; Vasilev, K. The interplay between size and valence state on the antibacterial activity of sub-10 nm silver nanoparticles. *Nanoscale Adv.* **2019**, *1*, 2365–2371. [CrossRef]
45. Quintero-Quiroz, C.; Acevedo, N.; Zapata-Giraldo, J.; Botero, L.E.; Quintero, J.; Zárate-Triviño, D.; Saldarriaga, J.; Pérez, V.Z. Optimization of silver nanoparticle synthesis by chemical reduction and evaluation of its antimicrobial and toxic activity. *Biomater. Res.* **2019**, *23*, 27. [CrossRef]
46. Rai, M.; Yadav, A.; Gade, A. Silver nanoparticles as a new generation of antimicrobials. *Biotechnol. Adv.* **2009**, *27*, 76–83. [CrossRef]
47. Pallavicini, P.; Arciola, C.R.; Bertoglio, F.; Curtosi, S.; Dacarro, G.; D'Agostino, A.; Ferrari, F.; Merli, D.; Milanese, C.; Rossi, S.; et al. Silver nanoparticles synthesized and coated with pectin: An ideal compromise for anti-bacterial and anti-biofilm action combined with wound-healing properties. *J. Colloid Interface Sci.* **2017**, *498*, 271–281. [CrossRef]
48. Kim, M.H.; Park, H.; Nam, H.C.; Park, S.R.; Jung, J.-Y.; Park, W.H. Injectable methylcellulose hydrogel containing silver oxide nanoparticles for burn wound healing. *Carbohydr. Polym.* **2018**, *181*, 579–586. [CrossRef]
49. Mulvaney, P. Surface Plasmon Spectroscopy of Nanosized Metal Particles. *Langmuir* **1996**, *12*, 788–800. [CrossRef]
50. Sharma, R.; Dhillon, A.; Kumar, D. Mentha-Stabilized Silver Nanoparticles for High-Performance Colorimetric Detection of Al(III) in Aqueous Systems. *Sci. Rep.* **2018**, *8*, 5189. [CrossRef]
51. Riaz, M.; Mutreja, V.; Sareen, S.; Ahmad, B.; Faheem, M.; Zahid, N.; Jabbour, G.; Park, J. Exceptional antibacterial and cytotoxic potency of monodisperse greener AgNPs prepared under optimized pH and temperature. *Sci. Rep.* **2021**, *11*, 2866. [CrossRef] [PubMed]
52. Ferdous, Z.; Nemmar, A. Health Impact of Silver Nanoparticles: A Review of the Biodistribution and Toxicity Following Various Routes of Exposure. *Int. J. Mol. Sci.* **2020**, *21*, 2375. [CrossRef] [PubMed]
53. Shi, R.; Bi, J.; Zhang, Z.; Zhu, A.; Chen, D.; Zhou, X.; Zhang, L.; Tian, W. The effect of citric acid on the structural properties and cytotoxicity of the polyvinyl alcohol/starch films when molding at high temperature. *Carbohydr. Polym.* **2008**, *74*, 763–770. [CrossRef]
54. Jiang, Q.; Reddy, N.; Zhang, S.; Roscioli, N.; Yang, Y. Water-stable electrospun collagen fibers from a non-toxic solvent and crosslinking system. *J. Biomed. Mater. Res. Part A* **2013**, *101*, 1237–1247. [CrossRef] [PubMed]
55. Bruna, T.; Maldonado-Bravo, F.; Jara, P.; Caro, N. Silver Nanoparticles and Their Antibacterial Applications. *Int. J. Mol. Sci.* **2021**, *22*, 7202. [CrossRef]
56. Flory, P.J.; Rehner, J. Statistical Mechanics of Cross-Linked Polymer Networks II. Swelling. *J. Chem. Phys.* **1943**, *11*, 521–526. [CrossRef]
57. Baier Leach, J.; Bivens, K.A.; Patrick, C.W., Jr.; Schmidt, C.E. Photocrosslinked hyaluronic acid hydrogels: Natural, biodegradable tissue engineering scaffolds. *Biotechnol. Bioeng.* **2003**, *82*, 578–589. [CrossRef]
58. Gold, G.T.; Varma, D.M.; Taub, P.J.; Nicoll, S.B. Development of crosslinked methylcellulose hydrogels for soft tissue augmentation using an ammonium persulfate-ascorbic acid redox system. *Carbohydr. Polym.* **2015**, *134*, 497–507. [CrossRef]
59. Stalling, S.S.; Akintoye, S.O.; Nicoll, S.B. Development of photocrosslinked methylcellulose hydrogels for soft tissue reconstruction. *Acta Biomater.* **2009**, *5*, 1911–1918. [CrossRef]
60. Rowe, R.C.; Sheskey, P.J.; Quinn, M.E. *Handbook of Pharmaceutical Excipients*, 6th ed.; Pharmaceutical Press: London, UK, 2009; ISBN 9780853697923.

Article

# In Situ Growth of Zeolitic Imidazolate Framework-L in Macroporous PVA/CMC/PEG Composite Hydrogels with Synergistic Antibacterial and Rapid Hemostatic Functions for Wound Dressing

Hang Yang, Xianyu Lan and Yuzhu Xiong *

College of Materials and Metallurgy, Guizhou University, Guiyang 550025, China;
gs.mjsong19@gzu.edu.cn (H.Y.); gs.lanxy21@gzu.edu.cn (X.L.)
* Correspondence: yzxiong@gzu.edu.cn

**Abstract:** Although many advances have been made in medicine, traumatic bleeding and wound infection are two of the most serious threats to human health. To achieve rapid hemostasis and prevent infection by pathogenic microbes, the development of new hemostatic and antibacterial materials has recently gained significant attention. In this paper, safe, non-toxic, and biocompatible polyvinyl alcohol (PVA); carboxymethyl cellulose (CMC), which contains several carboxyl and hydroxyl groups; and polyethylene glycol (PEG), which functions as a pore-forming agent, were used to prepare a novel PVA/CMC/PEG-based composite hydrogel with a macroporous structure by the freeze-thaw method and the phase separation technique. In addition, a PVA/CMC/PEG@ZIF-L composite hydrogel was prepared by the in situ growth of zeolitic imidazolate framework-L (ZIF-L). ZIF-L grown in situ on hydrogels released $Zn^{2+}$ and imidazolyl groups. They elicited a synergistic antibacterial effect in hemostasis with PVA and CMC, rendering the PVA/CMC/PEG@ZIF-L hydrogel with a good antibacterial effect against *Staphylococcus aureus*. At the same time, the macroporous structure enabled the rapid release of $Zn^{2+}$ and imidazolyl groups in ZIF-L and promoted cell proliferation at an early stage, enhancing the coagulation efficiency. A rat liver injury model was used to confirm its rapid hemostasis capacity.

**Keywords:** in situ growth; macroporous structure; synergistic antibacterial; rapid hemostatic; carboxymethyl cellulose; wound dressing; hydrogel; ZIF-L; polyvinyl alcohol; phase separation technique

Citation: Yang, H.; Lan, X.; Xiong, Y. In Situ Growth of Zeolitic Imidazolate Framework-L in Macroporous PVA/CMC/PEG Composite Hydrogels with Synergistic Antibacterial and Rapid Hemostatic Functions for Wound Dressing. *Gels* 2022, 8, 279. https://doi.org/10.3390/gels8050279

Academic Editors: Christian Demitri, Lorenzo Bonetti and Laura Riva

Received: 4 March 2022
Accepted: 27 April 2022
Published: 1 May 2022

**Publisher's Note:** MDPI stays neutral with regard to jurisdictional claims in published maps and institutional affiliations.

**Copyright:** © 2022 by the authors. Licensee MDPI, Basel, Switzerland. This article is an open access article distributed under the terms and conditions of the Creative Commons Attribution (CC BY) license (https://creativecommons.org/licenses/by/4.0/).

## 1. Introduction

Traumatic bleeding is one of the main causes of civilian and military deaths [1]. According to a previous report, more than 5.8 million deaths each year are caused by excessive bloodshed worldwide [2]. As such, the rapid control of bleeding is the key to reducing mortality. However, secondary tissue damage caused by bacterial infections is a major problem associated with the rapid control of bleeding [3–5]. Although cotton gauzes, zeolite-based Quik Clot, and other hemostatic products are commercially available, they show hemostatic effects and possess different shortcomings in that they do not meet the requirements for antibacterial control, which limits their wide clinical application [6]. Therefore, it is of great significance to develop safe and rapid hemostatic agents or wound dressings with an excellent antibacterial ability to treat severe bleeding and subsequent trauma [7].

Many investigators continue to conduct important research on hemostatic and antibacterial hydrogel wound dressings. According to the hydrogel mechanism of action and the antibacterial mechanism of action, the currently available antibacterial hydrogels can be divided into three categories; namely, hydrogels with antibacterial activity, hydrogels loaded with antibacterial agents, and stimulus-responsive hydrogels [8–11]. Chitosan, gelatin, cellulose, polyols, and alginates are the main representatives of their own antibacterial active substances. There are single hydrogels that use the antibacterial substance as the matrix as well as composite hydrogels such as alginate-based hydrogels [12,13], chitosan-based hydrogels [14], and PVA/chitosan-based hydrogels [15]. Amongst them, polyvinyl alcohol (PVA) hydrogels can be formed by physical cross-linking through safe, non-toxic, and simple freeze-thaw means; it is biodegradable, non-toxic, biocompatible, and hydrophilic [16]. These characteristics are favored by researchers. Hydrogels loaded with antibacterial agents have recently received much attention. Antibiotics (e.g., vancomycin hydrochloride [17] and gentamicin [18]) have successfully been loaded onto hydrogels [12,13,19]; the antibacterial agents generally are metal ions (e.g., $Ag^+$, $Zn^{2+}$, and $Cu^{2+}$ [20]), metal–organic framework (MOF) materials [21], antibacterial polypeptides [22–24], and biological extracts [25,26]. A variety of antibacterial hydrogels with a controlled release of metal ions can be prepared through metal coordination bonds between metal ions and polymer matrices such as $Zn^{2+}$ and $Sr^{2+}$ double-ion cross-linked hydrogel membranes [27] and $Ag^+$-loaded hydrogels [28]. At the same time, the introduction of metal coordination bonds can enhance the mechanical strength of hydrogels [29]. Other studies have reported that zeolite-like imidazole ester framework materials (ZIFs) have excellent antibacterial properties. For example, air filters that are combined with ZIF-8 particles exhibit an excellent inactivation efficiency against airborne bacteria [30]. ZIFs can not only slowly release $Zn^{2+}$ during wound healing and effectively kill bacteria by destroying microbial cell membranes, but also accelerate wound healing by promoting cell migration, angiogenesis, and collagen deposition [31]. There are many studies on final stimulus-responsive antibacterial hydrogels in the field of wound healing [32–36]. Although several hemostatic and antibacterial hydrogels have been developed, many hydrogels are not applicable to the rapid control of bleeding and further research is needed.

According to the ideal state of wound dressing [37] and the theory of wet wound healing [38], this paper selected safe, non-toxic, and biocompatible PVA as well as carboxymethyl cellulose (CMC) with several carboxyl and hydroxyl groups as the hydrogel matrices. Both PVA and CMC not only have certain antibacterial properties, but the carboxyl and hydroxyl groups in CMC can also chelate with zinc ions. A zeolitic imidazolate framework-L(ZIF-L) was then formed in situ in the hydrogel by combining $Zn^{2+}$ with the imidazolyl groups. ZIF-L has antibacterial properties and provides rapid hemostasis in hydrogels [39,40]. Polyethylene glycol (PEG) was used as the pore-former of the gel. The hydrogel was prepared by phase separation using PEG to create its macroporous structure. In the prepared PVA/CMC/PEG@ZIF-L composite hydrogel, the $Zn^{2+}$ and imidazolyl groups released by ZIF-L, PVA, and CMC exhibited synergistic antibacterial effects in hemostasis, indicating that the PVA/CMC/PEG@ZIF-L composite hydrogel had a good antibacterial effect. At the same time, the macroporous structure of the hydrogel not only provided the hydrogel with a good gas exchange capacity and the rapid absorption of blood to allow platelets and blood cells to aggregate to cause blood coagulation, but also allowed the $Zn^{2+}$ and imidazolyl groups in ZIF-L to be rapidly released, allowing the hydrogel to quickly control bleeding.

## 2. Results and Discussion

### 2.1. Structure Analysis

The cross-sectional SEM images of the PVA/CMC/PEG@ZIF-L composite hydrogel are shown in Figure 1a–d. The hydrogel had a three-dimensional, network-like, and porous structure, and the average pore size of the PVA/CMC/PEG@ZIF-L composite hydrogel was 1.76 µm, which was close to 2 µm. The standard deviation was 0.6436 µm. It indicated that PEG fully exerted its role as a pore-forming agent. A macroporous structure was prepared by phase separation. The pore wall of the pure PVA/CMC/PEG composite hydrogel without ZIF-L was relatively smooth whereas the surface of the PVA/CMC/PEG@ZIF-L composite hydrogel was rough, with leaf-like crystals observed on the gel network. The leaf-like crystals were ZIF-L [41,42]. There were many ZIF-L nanosheets in the pore wall of the hydrogel, indicating that ZIF-L was successfully formed in situ in the PVA/CMC/PEG hydrogel. The in situ growth process of ZIF-L on the hydrogel network is illustrated in Figure 2. When the PVA/CMC/PEG hydrogel was immersed in the Zn $(NO_3)_2 \cdot 6H_2O$ aqueous solution, a large amount of $Zn^{2+}$ was adsorbed by the carboxymethyl and hydroxyl groups in the hydrogel and attached to the surface and pore walls of the hydrogel. It was soaked in a 2-methylimidazole aqueous solution; the imidazole group and $Zn^{2+}$ were combined and ZIF-L grew in situ on the hydrogel network.

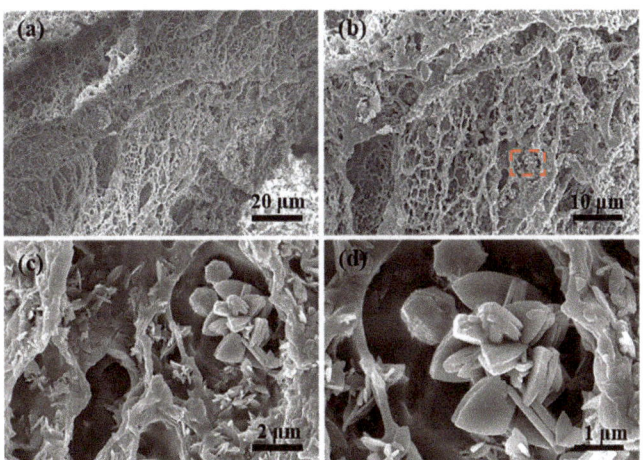

**Figure 1.** SEM images of the PVA/CMC/PEG@ZIF-L hydrogel: (**a**) magnified 1000 times, scale bar is 20 µm; (**b**) magnified 2000 times, scale bar is 20 µm; (**c**) magnified 10,000 times, scale bar is 2 µm; (**d**) magnified 30,000 times, scale bar is 1 µm.

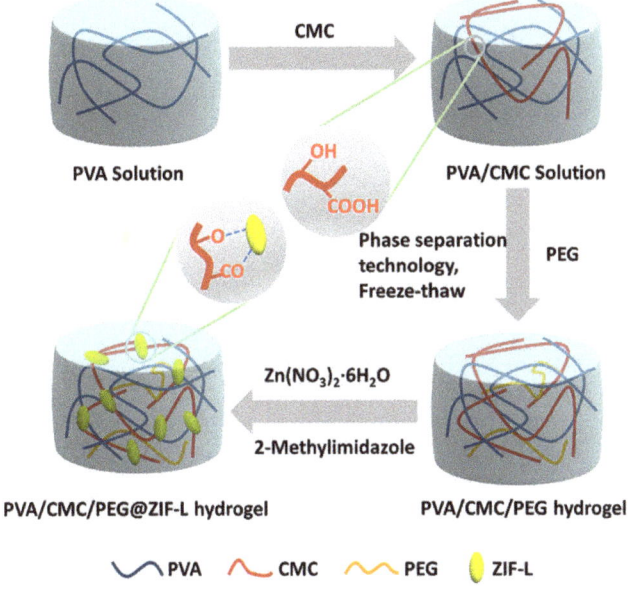

**Figure 2.** The formation of the PVA/CM/PEG@ZIF-L hydrogel and the in situ growth process of ZIF-L on the hydrogel network.

Figure 3a is the infrared spectrum of the PVA/CMC/PEG hydrogel. The absorption peaks at 848 cm$^{-1}$ and 1090 cm$^{-1}$ corresponded with the characteristic peaks of the ether bond (-O-) in PEG [43], and the corresponding characteristic peaks were also clearly observed in the PVA/CMC/PEG composite hydrogel. This indicated that PEG was entangled with the PVA molecular chain during the freezing-thawing process and not completely precipitated, which was subsequently condensed into water. The mechanical properties of the glue played a certain enhancement role. In addition, 1631 cm$^{-1}$ in the CMC spectrum corresponded with the stretching vibration peak of COO- in the carboxymethyl group of CMC [44] and 3460 cm$^{-1}$ corresponded with the stretching vibration peak of -OH of CMC. In the PVA/CMC/PEG hydrogel spectrum, the stretching vibration peaks of COO- and -OH corresponded with 1709 cm$^{-1}$ and 3278 cm$^{-1}$, respectively. In the PVA/CMC/PEG hydrogel spectrum, the stretching vibration peak of -OH at 3278 cm$^{-1}$ broadened, indicating hydrogen bonding between the hydroxyl group of PVA and the carboxymethyl group of CMC [45]. Therefore, physical cross-links were formed between PVA and CMC during the freezing-thawing process, confirming the successful preparation of the PVA/CMC/PEG hydrogels. COO- of CMC was a broad peak; it was possible that COO- also formed intermolecular hydrogen bonds. Furthermore, COO- shifted in the PVA/CMC/PEG hydrogel spectrum, which might be the result of the interaction between PVA and CMC [45]. Although the stretching vibration peaks of COO- and -OH shifted in the PVA/CMC/PEG hydrogel spectrum, these findings indicated that COO- and -OH existed in the PVA/CMC/PEG hydrogel, providing a basis for the chelation of carboxyl, hydroxyl, and $Zn^{2+}$.

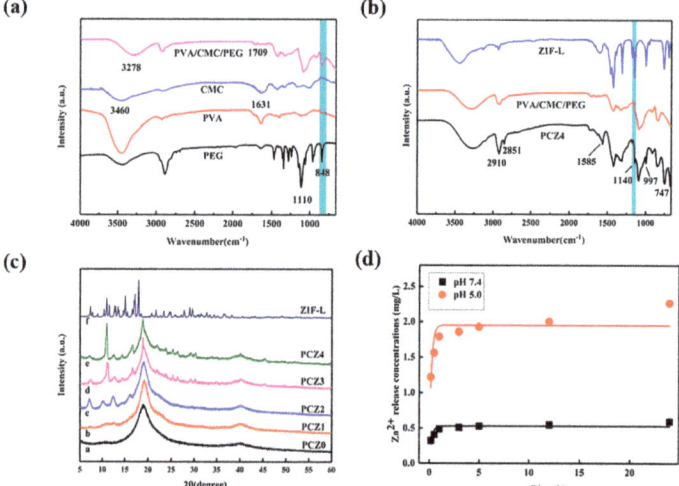

Figure 3. (a) Fourier infrared spectra of PVA, CMC, PEG, and PVA/CMC/PEG hydrogels; (b) Fourier infrared spectra of PVA/CMC/PEG, ZIF-L, and PCZ4 hydrogels; (c) XRD spectra of ZIF-L and PCZ with different amounts of ZIF-L; (d) fitted curve of the first-order kinetic model of the $Zn^{2+}$ release of the PCZ4 hydrogel at different pH values.

The infrared spectrum of the PCZ4 hydrogel is shown in Figure 3b. The peak at 2910 cm$^{-1}$ corresponded with the stretching vibration peak of methylene (-CH$_2$-) of PCZ4 and the new peak at 2851 cm$^{-1}$ was the water peak caused by the residual water. The peaks at 1585, 1140, 997, 747, and 423 cm$^{-1}$ in the PCZ4 spectrum corresponded with the characteristic peaks of the ZIF-L crystal [46].

The XRD spectrum of ZIF-L and PCZ with different amounts of ZIF-L is shown in Figure 3c. The PVA/CMC/PEG hydrogel (PCZ0, which was without ZIF-L) had a strong diffraction peak at 2θ = 19.5°, which was similar to the (101) crystal in PVA. The diffraction peaks of ZIF-L at the (110), (200), (211), (220), (310), and (222) planes were observed at $2\theta$ values of 7.30°, 10.95°, 12.71°, 15.11°, 17.2°, and 17.99°, respectively. This was similar to previous research [47–50]. Therefore, this confirmed that ZIF-L had been successfully synthesized. The sample containing ZIF-L (PCZ1, PCZ2, PCZ3, PCZ4) had a few new peaks, similar to the characteristic peaks of the ZIF-L crystal plane. These findings indicated that ZIF-L successfully grew in situ in the PVA/CMC/PEG hydrogel. In addition, the peak intensity in the PCZ sample increased with the increase in the ZIF-L crystal content. However, it might have been due to the influence of the PVA crystal planes on the ZIF-L crystal planes, resulting in a significant increase in the peak intensity only at $2\theta \cong 11°$. Additionally, the lower the content of ZIF-L, the greater the influence of other substances (in this research, it was mainly reflected in the effect of PVA on ZIF-L) on its peak [51,52]. Hence, the characteristic peaks of ZIF-L in PCZ1 and PCZ2 were not significantly obvious and there was an apparent shift from $2\theta \cong 10°$ to $2\theta \cong 11°$ between PCZ2 and PCZ3.

### 2.2. Zinc Ion Release

The fitted curve of the first-order kinetic model of the $Zn^{2+}$ release of the PCZ4 hydrogel at different pH values is shown in Figure 3d. During the first hour, the release rate of $Zn^{2+}$ was fast, which was beneficial to the rapid hemostasis of the wound. Afterwards, the release rate of $Zn^{2+}$ was relatively slow and stable, allowing it to act on the wound. The PVA, CMC, and imidazolyl groups produced by the hydrolysis of ZIF-L in the hydrogel also had certain antibacterial effects [11–13]. $Zn^{2+}$ cooperated with other substances in the hydrogel for a synergistic antibacterial effect. Furthermore, the $Zn^{2+}$ release of the PCZ4 hydrogel increased with the decrease in pH. ZIF-L in the PCZ4 hydrogel rapidly degraded under acidic conditions (pH 5.0) whereas the degradation was relatively slow under physiological conditions (pH 7.4) [34,53,54]. These results showed that the $Zn^{2+}$ release from the PVA/CMC/PEG@ZIF-L composite hydrogel was pH-responsive and $Zn^{2+}$ was easily released under acidic conditions. Human skin is finely acidic [55]; the pH responsiveness of the hydrogel was beneficial to hemostasis and antibacterial control. In addition, a first-order kinetic equation was fitted to the zinc ion release. Although the release of $Zn^{2+}$ could be up to four-fold higher depending on the pH, the mechanism of the $Zn^{2+}$ release was not affected as suggested by the model. The kinetic behavior of zinc ion release at different pH levels requires further research.

### 2.3. Biocompatibility Analysis

An ideal wound dressing requires a good compatibility with the vasculature and should not lyse red blood cells when it encounters blood. The blood compatibility of the dressing was evaluated by the hemolysis rate. The lower the hemolysis rate, the better the blood compatibility. The hemolysis rate of the PCZ hydrogel with different amounts of ZIF-L is shown in Figure 4a. The hemolysis rate of the PCZ hydrogel was less than 5%; as the amount of ZIF-L continued to increase, the hemolysis rate continued to decrease because ZIF-L generated in situ in the PVA/CMC/PEG composite hydrogel was superhydrophobic [56–59]. ZIF-L in the pores reduced the surface energy of the material, thereby reducing damage to the red blood cells [60]. Thus, the hemolysis rate was low. These results showed that the PVA/CMC/PEG@ZIF-L composite hydrogel had excellent blood compatibility and it met the criteria of wound dressings.

Cytotoxicity is another indicator of biocompatibility. After adding PCZ hydrogels with different amounts of ZIF-L to an NIH-3T3 cell suspension, the relative cell viability of the NIH-3T3 cells incubated for 24 h, 72 h, and 120 h was calculated and is shown in Figure 4b–d. In the hydrogels with different amounts of ZIF-L incubated for 24 h, we found that the relative cell viability of the NIH-3T3 cells was close to 100%. This showed that the PCZ hydrogels had excellent blood compatibility. In addition, PCZ1 and PCZ2 both promoted cell proliferation to varying degrees, which is conducive to wound healing. After 72 h of incubation, the relative cell viability in the hydrogel decreased and the relative cell viability of PCZ3 and PCZ4 was less than 80% because ZIF-L released $Zn^{2+}$ and free

2-methylimidazole during hydrolysis. When $Zn^{2+}$ in the leaching solution is too high, there is excessive generation of reactive oxygen species, which damages DNA. Furthermore, 2-methylimidazole is not conducive to cell survival. There was no difference in relative activity after 72 h and 120 h of incubation. The hydrolysis of ZIF-L slowed after 72 h, which was suggestive of complete hydrolysis. This had no effect on the cell survival. According to the biological evaluation standard of medical devices, the PVA/CMC/PEG@ZIF-L composite hydrogel met this standard and could be used as a wound dressing.

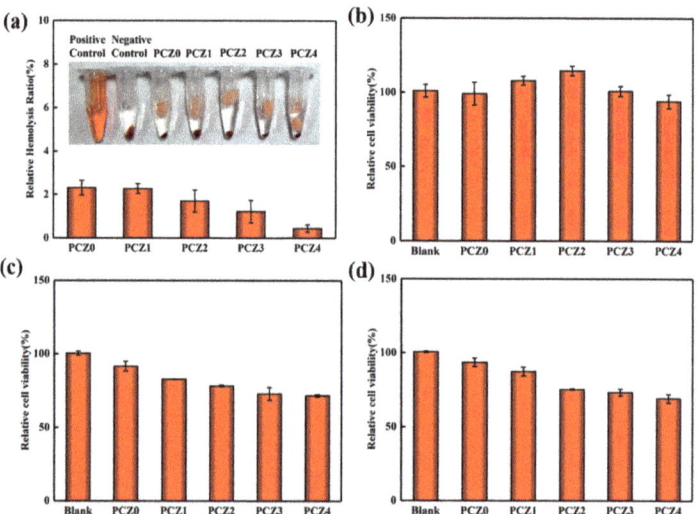

**Figure 4.** (a) Hemolysis rate of PCZ hydrogels with different amounts of ZIF-L. Relative cell viability of NIH-3T3 cells incubated for (b) 24 h; (c) 72 h; and (d) 120 h.

### 2.4. Hemostatic Analysis

The hemostatic efficiency of a wound dressing is related to its blood-clotting ability. A PVA/CMC/PEG@ZIF-L composite hydrogel can promote blood clotting based on two factors. On the one hand, because of its interconnected porous structure that can quickly absorb a large amount of blood, platelets and blood cells can aggregate to cause blood clotting. On the other hand, when a PVA/CMC/PEG@ZIF-L composite hydrogel contacts blood, ZIF-L gradually decomposes into $Zn^{2+}$ and imidazolyl. $Zn^{2+}$ can activate coagulation factors XII and VII to trigger the coagulation cascade and accelerate the production of thrombin and fibrin, thereby promoting coagulation [31]. ZIF-L may also slowly release $Zn^{2+}$ through an exchange with $Na^+$, activating the coagulation cascade. In addition, ZIF-L has a high porosity and large surface area, which can accelerate blood absorption and red blood cell/platelet coagulation. These two processes have a synergistic effect in the promotion of blood clotting.

Theoretically, the blood coagulation ability is inversely proportional to the coagulation index (BCI value). The coagulation index of the PVA/CMC/PEG@ZIF-L composite hydrogel with different amounts of ZIF-L is shown in Figure 5. The PVA/CMC/PEG hydrogel had a certain coagulation effect whereas the BCI value of the PVA/CMC/PEG@ZIF-L composite hydrogel was lower than that of the PVA/CMC/PEG hydrogel. PCZ1, PCZ2, PCZ3, and PCZ4 displayed coagulation effects with BCI values of 39.7 ± 2.5%, 42 ± 2.1%, 24.2 ± 2.9%, and 40.1 ± 3.1%, respectively. These findings indicated that ZIF-L played a role in promoting blood coagulation. In addition, as shown in Figure 5, it could be seen that the BCI value of PCZ3 was lower than PCZ4, indicating that the hemostasis efficiency of PCZ3 was higher than PCZ4. An excessive zinc ion concentration can produce too many reactive oxygen species (ROS) [30,31]. ROS not only cause oxidative damage to nucleic acid, but also damage proteins and biological cell membranes [61]. In the process of hemostasis, excess ROS might destroy blood cells, thrombin, and blood fibrin, which affect coagulation. Therefore, it was possible that the ZIF-L content of PCZ4 was too high, which produced too many reactive oxygen species and affected the hemostasis efficiency. In general, the PVA/CMC/PEG@ZIF-L composite hydrogel could promote blood coagulation.

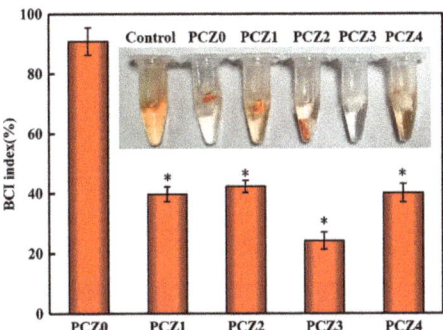

**Figure 5.** The coagulation index of PCZ hydrogels with different amounts of ZIF-L (* $p < 0.05$; $n = 3$).

The coagulation properties of the samples were further assessed using a rat liver injury model by recording the time to hemostasis and the amount of blood loss leading to hemostasis (Figure 6). As shown in Figure 6b, the amount of blood loss (67 mg) caused by PCZ3 was compared with a medical gauze and PCZ0 without ZIF-L. The hemostatic time followed the same trend in the overall loss of blood. In Figure 6c, it can be seen that the hemostasis time of the medical gauze, PCZ0, and PCZ3 was 204 s, 202 s, and 120 s, respectively. PCZ3 had the shortest hemostatic time. The coagulation ratio was consistent with the image after a hemorrhag (Figure 6c). The results showed that the PVA/CMC/PEG @ZIF-L hydrogel had a fast hemostatic ability.

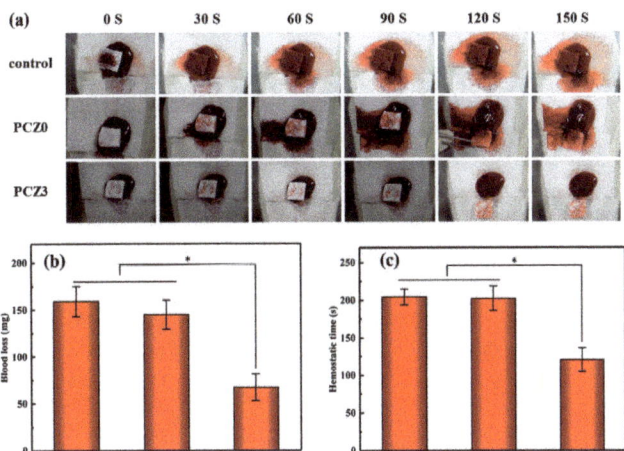

**Figure 6.** In vivo evaluation of the hemostatic capacity of different hydrogels in a rat liver trauma model. (**a**) Macroscopic images; (**b**) blood loss; and (**c**) time to hemostasis in a hemostatic application (* $p < 0.05$; $n = 3$).

## 2.5. In Vitro Antibacterial Assays

The in vitro antibacterial effects of the PCZ hydrogel with different amounts of ZIF-L are shown in Figure 7. As the content of ZIF-L increased, the content of released zinc ions also increased and the antibacterial effect of the hydrogel gradually improved, especially for *Staphylococcus aureus*. ZIF-L has inherent antibacterial properties [39,40] due to the release of $Zn^{2+}$ and imidazole groups. The imidazole groups can destroy the liposome structure of bacteria. In addition, $Zn^{2+}$ can interact with the cell membrane of bacteria and penetrate the cell wall to destroy its protein and DNA, leading to cell death [31,39,40]. Compared with *Escherichia coli*, the hydrogel had a better antibacterial effect against *Staphylococcus aureus*. This might be because the main component of the cell wall of Gram-positive bacteria is peptidoglycan; peptidoglycan is located in the surface layer [62], so the antibacterial components in the hydrogel could act on it more quickly. As Gram-negative bacteria contain less peptidoglycan and are located in the inner layer [62], the antibacterial ingredients killed less *E. coli* than *Staphylococcus aureus* in the same time duration.

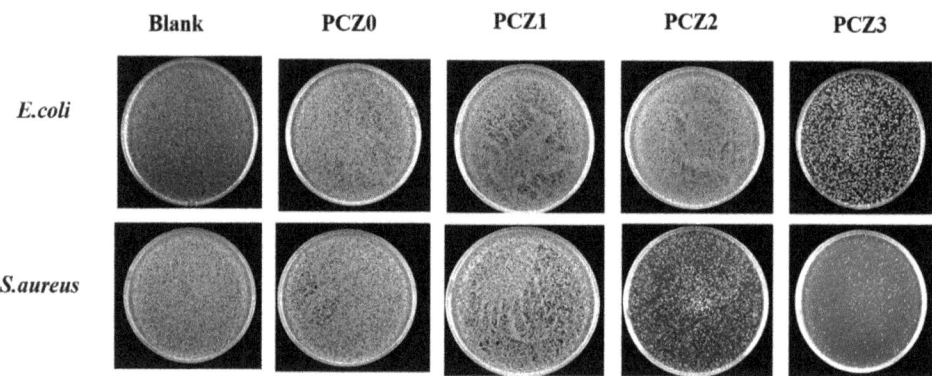

**Figure 7.** In vitro antibacterial colonies of PCZ hydrogels with different amounts of ZIF-L.

PVA and CMC also have certain antibacterial properties and they synergistically kill bacteria through the release of $Zn^{2+}$ and imidazole groups by the hydrolysis of ZIF-L, thus giving PCZ excellent antibacterial effects against *Staphylococcus aureus*. Although the PVA/CMC/PEG@ZIF-L composite hydrogel had poor antibacterial effects against *E. coli*, requiring further research and improvement, its excellent antibacterial effects against *Staphylococcus aureus* could make the PVA/CMC/PEG@ZIF-L composite hydrogel a good wound dressing, exceeding many hemostatic products on the market.

## 3. Conclusions

Although several advancements have been made in medicine, traumatic bleeding and bacterial wound infections still pose a serious threat to human health. To solve this problem, many investigators have performed important research and developed new hemostatic and antibacterial materials, but most of these hydrogels still have shortcomings with respect to the rapid control of bleeding. In this paper, a PVA/CMC/PEG composite hydrogel was prepared using PVA and CMC as the raw materials and PEG as the pore-forming agent. The PVA/CMC/PEG@ZIF-L composite hydrogel was prepared by the in situ growth of ZIF-L on the hydrogel network. The morphologies and microstructures of the prepared hydrogels were characterized by SEM, FTIR, and XRD. The results showed that the PVA/CMC/PEG composite hydrogel and the PVA/CMC/PEG@ZIF-L composite hydrogel with a macroporous structure were successfully prepared by the freeze-thaw method and phase separation technology. In addition, to explore the biocompatibility of the PVA/CMC/PEG@ZIF-L composite hydrogel and the hemostatic and antibacterial mechanisms, a $Zn^{2+}$ release test, blood compatibility test, cytotoxicity test, in vitro coagulation test, in vivo hemostasis test, and in vitro antibacterial test were carried out. The experimental results showed that the PVA/CMC/PEG@ZIF-L composite hydrogel was in line with the index of wound dressing and the various components in the PVA/CMC/PEG@ZIF-L composite hydrogel had synergistic antibacterial effects. As such, the gel had good antibacterial effects against *Staphylococcus aureus*. At the same time, the macroporous structure enabled the rapid release of $Zn^{2+}$ and imidazolyl groups in ZIF-L and promoted cell proliferation at an early stage, thereby enhancing the coagulation efficiency.

## 4. Experimental Section

### 4.1. Materials

Polyvinyl alcohol (PVA1799, 99% degree of hydrolysis, 44.05 Mw, AR), carboxymethyl cellulose (CMC, USP), 2-methylimidazole (reagent purity: 98%), and methanol (AR) were purchased from the Aladdin Reagent Co., Ltd. (Shanghai, China). Polyethylene glycol (PEG2000, AR) and tert–butanol (AR) were purchased from the McLean Biochemical Technology Co., Ltd. (Shanghai, China). $Zn(NO_3)_2 \cdot 6H_2O$ (AR) was purchased from the Comeo Chemical Reagent Co., Ltd. (Tianjin, China).

### 4.2. Preparation of PVA/CMC/PEG Composite Hydrogel

In brief, 2 g of PVA and 20 mL of deionized water were added to a blue cap reagent bottle. The PVA was dissolved in the deionized water and heated at 90 °C for 1 h. Thereafter, 0.2 g of CMC was added and stirred, followed by 1.5 g of PEG2000 with stirring and heating at 75 °C. After cooling to room temperature, the turbid liquid was poured into a mold (cylindrical PTFE, 3 cm diameter, about 2 cm height poured) and stored at −20 °C for 12 h. The frozen sample was carefully demolded and melted at room temperature to form the hydrogel. This was frozen and thawed once. The hydrogel was soaked in deionized water to remove PEG precipitated by the phase change to generate the PVA/CMC/PEG composite hydrogel.

### 4.3. In Situ Growth of ZIF-L on PVA/CMC/PEG Composite Hydrogel

The PVA/CMC/PEG hydrogel was soaked in 0.119 g of $Zn(NO_3)_2 \cdot 6H_2O$ in 30 mL of an aqueous solution for 24 h and then soaked in 0.263 g of 2-methylimidazole in 30 mL of an aqueous solution for 24 h to realize the in situ generation of ZIF-L in the hydrogel, which was denoted as PCZ1. The above experimental steps were repeated to achieve the soaking of the PVA/CMC/PEG composite hydrogel according to a two-fold concentration gradient of PCZ1, which was denoted as PCZ2. Likewise, PCZ3 followed a two-fold concentration gradient of PCZ2. The above steps were repeated for PCZ4. The resulting hydrogel was solvent-exchanged in a tert–butanol solution. To avoid the excessive shrinkage of the hydrogel, the content of tert–butanol in the solvent exchange process was gradually increased, followed by soaking in 50%, 75%, and 100% tert–butanol aqueous solutions for 1 h. The resulting hydrogel was then freeze-dried (vacuum degree 15 Pa, condensing temperature −50 °C, drying for 48 h) to obtain the product.

### 4.4. Testing and Characterization

#### 4.4.1. Scanning Electron Microscopy (SEM)

Scanning electron microscopy (SEM; SU8010, HITACHI Corporation, Tokyo, Japan) was used to observe the morphology of the PVA/CMC/PEG@ZIF-L composite hydrogel. The gel samples were freeze-dried and made brittle with liquid nitrogen; their microstructures were then observed by SEM. All samples had to be sprayed with gold before being tested. In the SEM images, 200 holes were selected for the pore size measurement. After the measurement and calculation, the average pore size was obtained.

#### 4.4.2. Fourier Transform Infrared Spectroscopy (FTIR)

Fourier transform infrared spectroscopy (FTIR; NEXUS6700, Thermo Company, Waltham, MA, USA) was used for the group characterization of the sample powder. KBr was added to

1–2 mg of sample solids and the samples were pressed and tested. The scanning beam range was 500–4000 cm$^{-1}$. The gel samples were lyophilized and directly tested by ATR-FTIR.

### 4.4.3. X-ray Diffraction (XRD)

X-ray diffraction (XRD; Empyrean PANalytical, Netherlands) was used to analyze the state of in situ growth of ZIF-L on the hydrogels with different ZIF-L contents. The characterization of hydrogels was undertaken after freeze-drying. The scanning speed was 0.2 s/step and the scanning angle range was 5~80°.

### 4.4.4. Zinc Ion Release

Inductively coupled plasma spectrometry was used to determine the release of $Zn^{2+}$ from the PVA/CMC/PEG@ZIF-L composite hydrogel. A total of 0.2 g of PCZ4 hydrogel was dipped into 30 mL of PBS at a pH of 5.0 and a pH of 7.4 and the stirring speed was adjusted to 100 rpm. According to the crystallization output method, the solubility of ZIF-L in PBS with a pH of 5 and with a pH of 7.4 was 30 mg/100 g and 6 mg/100 g, respectively. After 10 and 30 min as well as 1, 3, 5, 12, and 24 h, a 2 mL aliquot was used to determine the $Zn^{2+}$ release. After removing each 2 mL aliquot, 2 mL of PBS was added to ensure that the hydrogel soaking solution was always 30 mL.

### 4.4.5. Biocompatibility Test

The biocompatibility test was divided into the blood compatibility test and the cytotoxicity test. For the blood compatibility test, 4.0 mL of peripheral blood was drawn from a healthy volunteer using a vacuum blood collection tube for a total of two tubes with 3.2% sodium citrate anticoagulation. After a tube of anticoagulated whole blood was centrifuged for 10 min, 0.4 mL of a red blood cell pellet was drawn into a 1.5 mL centrifuge tube. A total of 1.0 mL of normal saline was then added to the centrifuge tube and mixed with the red blood cells, centrifuged for 10 min, and the supernatant was aspirated. Subsequently, 0.5 mL of physiological saline was added to the centrifuge tube to dilute the red blood cells to obtain a red blood cell suspension. After obtaining the red blood cell suspension, 10 mg of the sample was placed in a 1.5 mL centrifuge tube, 1 mL of physiological saline was added, and then 20 µL of the red blood cell suspension was added, followed by shaking at 37 °C for 4 h, centrifugation for 5 min, and the aspiration of the supernatant. A microplate reader was used to measure the absorbance at 545 nm. Thereafter, 20 mL of the erythrocyte suspension was added to 1 mL of physiological saline as a negative control and 20 mL of the erythrocyte suspension was added to 1 mL of tri-distilled water as a positive control. The centrifuge tubes were imaged. The relative hemolysis rate of the sample was calculated according to the following formula:

$$\text{Hemolysis Rate (\%)} = \frac{D_t - D_{nc}}{D_{pc} - D_{nc}} \times 100\% \quad (1)$$

where $D_t$, $D_{nc}$, and $D_{pc}$ are the adsorptions, negative control, and positive control of samples, respectively. A relative hemolysis rate > 5% was regarded as a hemolysis phenomenon.

The in vitro cytotoxicity test used the MTT method. The ratio of calf serum:sodium pyruvate:glutamine:non-essential amino acid was 87:10:1:1 and a DMCM high-glucose medium was prepared. The NIH-3T3 cells were cultured at $4 \times 10^3$/well in 96-well plates for 24 h at 37 °C. The samples were cut into small pieces after sterilization by UV light for 30 min; 100 mg of the samples was added to 1 mL of a DMEM high-glucose medium and placed in a 37 °C, 5% $CO_2$ incubator for 24 h. After centrifugation, the supernatant was taken and filtered with a 0.22 µm membrane to obtain a sample extract with a concentration of 100 mg/mL. The leaching solution was diluted to 10%, 20%, 50%, 80%, and 100% and added to a 96-well plate at 100 µL/well, respectively. The control group was the DMEM medium of the NIH-3T3 cells without an extracting solution. Three replicate wells were made for each treatment and cultured for 24 h, 72 h, and 120 h, respectively. The medium containing the samples was then removed, each well was washed 3 times with PBS, and

100 µL of an MTT medium containing 0.5 mg/mL was added to each well. This was incubated in a 5% $CO_2$, 37 °C constant temperature incubator for 4 h and 100 µL of DMSO was added to each well of the supernatant. Finally, it was shaken for 10 min to completely dissolve the crystal and the absorbance at 570 nm was detected. The absorbances of the experimental group and the control group at 570 nm were recorded as $OD_E$ and $OD_C$, respectively. The following is the formula used to calculate the relative viability of cells:

$$\text{Relative Viability (\%)} = \frac{OD_E - \text{Background values}}{\sum OD_c/n - \text{Background values}} \times 100\% \tag{2}$$

The experimental group used a hydrogel extract (100 mg/mL) with different amounts of ZIF-L and the control group did not contain a hydrogel extract.

### 4.4.6. In Vitro Coagulation Test

The coagulation index BCI (blood-clotting index) was used for the in vitro coagulation test. PCZ0, PCZ1, PCZ2, PCZ3, and PCZ4 were cut and 10 mg was placed in a 1.5 mL centrifuge tube and kept at 37 °C for 5 min. A total of 5 µL of recalcified anticoagulated whole blood was pipetted and dropped onto the surface of the sample and incubated at 37 °C for 20 min. A total of 1 mL of deionized water was then added to the centrifuge tube along the edge of the tube and the non-coagulated red blood cells were coagulated by shaking and were incubated for 1 min. Subsequently, 0.8 mL of the supernatant was placed into the centrifuge tube and the absorbance of the solution was measured at 545 nm with a microplate reader. The measurement of each sample was repeated three times. Thereafter, 4.55 mL of anticoagulated whole blood was directly added to 1 mL of deionized water and incubated at 37 °C with shaking for 1 min, which served as a positive control. Subsequently, the BCI was calculated according to the following formula:

$$BCI(\%) = \frac{I_a}{I_w} \times 100\% \tag{3}$$

where $I_a$ represents the absorbance of the solution after the recalcified anticoagulated whole blood contacted the sample for a set time and $I_w$ represents the absorbance of the solution after the anticoagulated whole blood was mixed with deionized water.

### 4.4.7. In Vivo Hemostasis Test

A rat liver injury model was used to evaluate the hemostatic ability of the composite hydrogel samples. The rats were randomly divided into three groups (medical gauze, PCZ0, and PCZ3) of three rats in each group. The rats were anesthetized and fixed. After an abdominal incision, the liver was exposed and the fluid around the liver was cleaned. A wound (5 × 5 mm, depth: 3 mm) was created on the anterior lobe of the rat liver using a scalpel. A pre-weighed gauze and samples of the required size (length: 1.5 cm) were immediately administered to the site of the incision on the anterior lobe of the liver with a scalpel. Observations were recorded every 30 s until the bleeding completely stopped to measure the time to hemostasis. Photographs were taken of the wound. Subsequently, the hemostatic materials were weighed to calculate the amount of bleeding after they completely absorbed the blood.

### 4.4.8. In Vitro Antibacterial Test

A medium preparation of an LB liquid medium was composed of 100 mL of distilled water in a graduated cylinder and poured into a 250 mL reagent bottle. Subsequently, 1 g of tryptone, 0.5 g of yeast powder, and 1 g of sodium chloride were weighed by an analytical electronic balance. The above weighed reagents were added to the reagent bottle and mixed and then sterilized at a high temperature in a high-pressure steam sterilizer at 121 °C for 15 min. An LB solid medium was prepared by measuring 100 mL of distilled water in a graduated cylinder, which was then poured into a 250 mL reagent bottle. A total of 1 g of tryptone, 0.5 g of yeast powder, 1 g of sodium chloride, and 1.5 g of agar powder were weighed by an analytical electronic balance. The above weighed reagents were added to the reagent bottle and mixed and then sterilized at a high temperature in a high-pressure steam sterilizer at 121 °C for 15 min.

For the experiment procedure, PCZ0, PCZ1, PCZ2, and PCZ3 were weighed to 0.1 g, respectively, and sterilized under UV irradiation for 30 min. Before dilution, the concentration of *Escherichia coli* was $1.31 \times 10^9$ CFU/mL and the concentration of *Staphylococcus aureus* was $1.01 \times 10^9$ CFU/m. The bacterial solution was diluted to $2 \times 10^6$ CFU/mL with the LB liquid medium and 10 mL of the diluted bacterial solution was added to a disposable bacterial culture tube. The samples were incubated with *Staphylococcus aureus* (ID: ATCC29213) and *Escherichia coli* (ID: ATCC25922) for 6 h at 37 °C. Thereafter, 10 μL of the suspension was diluted with 1 mL of sterile PBS and 100 mL of the diluted solution was evenly spread onto the solid LB medium. The plates were incubated in an incubator set at 37 °C for 20 h. The plates were imaged. The control was the liquid medium without a sample. The experiment was repeated three times.

### 4.4.9. Statistical Analysis

All experiments were carried out three times and the data were expressed as a mean ± standard deviation. The results were analyzed with a one-way ANOVA by Origin 2017. A value of $p < 0.05$ was considered to be statistically significant.

**Author Contributions:** Conceptualization, H.Y.; data curation, X.L.; formal analysis, H.Y.; investigation, H.Y.; methodology, H.Y.; project administration, Y.X.; resources, Y.X.; visualization, H.Y.; writing—original draft, X.L.; writing—review and editing, Y.X. All authors have read and agreed to the published version of the manuscript.

**Funding:** This research was funded by the National Nature Science Foundation of China, grant number 52063006.

**Institutional Review Board Statement:** The study was conducted according to the guidelines of the Declaration of Helsinki, and approved by the Medical Ethical Committee of Yongchuan Hospital Affiliated to Chongqing Medical University (Approval number: 20210315-1).

**Informed Consent Statement:** Written informed consent was obtained from the patient(s) to publish this paper.

**Data Availability Statement:** Not applicable.

**Acknowledgments:** The authors greatly appreciate the financial support from the National Nature Science Foundation of China, grant number 52063006.

**Conflicts of Interest:** The authors declare no conflict of interest.

## References

1. Wang, Y.; Fu, Y.; Li, J.; Mu, Y.; Zhang, X.; Zhang, K.; Liang, M.; Feng, C.; Chen, X. Multifunctional chitosan/dopamine/diatom-biosilica composite beads for rapid blood coagulation. *Carbohydr. Polym.* **2018**, *200*, 6–14. [CrossRef]
2. Zhao, X.; Guo, B.; Hao, W.; Liang, Y.; Ma, P.X. Injectable antibacterial conductive nanocomposite cryogels with rapid shape recovery for noncompressible hemorrhage and wound healing. *Nat. Commun.* **2018**, *9*, 1–17. [CrossRef] [PubMed]
3. Feng, Y.; Wang, Q.; He, M.; Zhao, W.; Zhao, C. Nonadherent Zwitterionic Composite Nanofibrous Membrane with Halloysite Nanocarrier for Sustained Wound Anti-Infection and Cutaneous Regeneration. *ACS Biomater. Sci. Eng.* **2019**, *6*, 621–633. [CrossRef] [PubMed]
4. Liu, M.; Wang, X.; Li, H.; Xia, C.; Wu, J. Magnesium oxide incorporated electrospun membranes inhibit bacterial infections and promote healing process of infected wounds. *J. Mater. Chem. B* **2021**, *9*, 3727–3744. [CrossRef]
5. Zhang, L.; Liu, M.; Zhang, Y.; Pei, R. Recent Progress of Highly Adhesive Hydrogels as Wound Dressings. *Biomacromolecules* **2020**, *21*, 3966–3983. [CrossRef]
6. Kozen, B.G.; Kircher, S.J.; Henao, J.; Godinez, F.S.; Johnson, A.S. An Alternative Hemostatic Dressing: Comparison of CELOX, HemCon, and QuikClot. *Acad. Emerg. Med.* **2014**, *15*, 74–81. [CrossRef] [PubMed]
7. Pourshahrestani, S.; Zeimaran, E.; Kadri, N.A.; Gargiulo, N.; Samuel, S.; Naveen, S.V.; Kamarul, T.; Towler, M.R. Gallium-Containing mesoporous bioactive glass with potent hemostatic activity and antibacterial efficacy. *J. Mater. Chem. B. Mater. Biol. Med.* **2016**, *4*, 71–86. [CrossRef] [PubMed]

8. Quan, C.; Zhang, L.; Gao, Y.; Tang, P. Research progress of antibacterial hydrogel dressings. *Acad. J. Chin. PLA Med. Sch.* **2017**, *38*, 4.
9. Deng, Z.; Guo, Y.; Zhao, X.; Ma, P.X.; Guo, B. Multifunctional Stimuli-Responsive Hydrogels with Self-Healing, High Conductivity, and Rapid Recovery through Host-Guest Interactions. *Chem. Mater. A Publ. Am. Chem. Soc.* **2018**, *30*, 1729–1742. [CrossRef]
10. Gan, D.; Xing, W.; Jiang, L.; Fang, J.; Zhao, C.; Ren, F.; Fang, L.; Wang, K.; Lu, X. Plant-inspired adhesive and tough hydrogel based on Ag-Lignin nanoparticles-triggered dynamic redox catechol chemistry. *Nat. Commun.* **2019**, *10*, 1–10. [CrossRef]
11. Zhong, Y.; Xiao, H.; Seidi, F.; Jin, Y. Natural Polymer-Based Antimicrobial Hydrogels without Synthetic Antibiotics as Wound Dressings. *Biomacromolecules* **2020**, *21*, 2983–3006. [CrossRef]
12. Mirean, V. Current Trends in Advanced Alginate-Based Wound Dressings for Chronic Wounds. *J. Pers. Med.* **2021**, *11*, 890.
13. Barbu, A.; Neamu, M.B.; Zhan, M.; Mirean, V. Trends in alginate-based films and membranes for wound healing. *Rom. Biotechnol. Lett.* **2020**, *25*, 1683–1689.
14. Sharma, S.; Kumar, R.; Kumari, P.; Kharwar, R.N.; Yadav, A.K.; Saripella, S. Mechanically magnified chitosan-based hydrogel as tissue adhesive and antimicrobial candidate. *Int. J. Biol. Macromol.* **2018**, *125*, 109–115. [CrossRef] [PubMed]
15. Lotfipour, F.; Alami-Milani, M.; Salatin, S.; Hadavi, A.; Jelvehgari, M. Freeze-Thaw-Induced cross-linked PVA/chitosan for oxytetracycline-loaded wound dressing: The experimental design and optimization. *Res. Pharm. Sci.* **2019**, *14*, 175.
16. Chaudhuri, B.; Mondal, B.; Ray, S.K.; Sarkar, S.C. A novel biocompatible conducting polyvinyl alcohol (PVA)-polyvinylpyrrolidone (PVP)-hydroxyapatite (HAP) composite scaffolds for probable biological application. *Colloids Surf. B Biointerfaces* **2016**, *143*, 71–80. [CrossRef]
17. Chuenbarn, T.; Sirirak, J.; Tuntarawongsa, S.; Okonogi, S.; Phaechamud, T. Design and Comparative Evaluation of Vancomycin HCl-Loaded Rosin-Based In Situ Forming Gel and Microparticles. *Gels* **2022**, *8*, 231. [CrossRef]
18. Posadowska, U.; Brzychczy-Włoch, M.; Drożdż, A.; Krok-Borkowicz, M.; Włodarczyk-Biegun, M.; Dobrzyński, P.; Chrzanowski, W.; Pamuła, E. Injectable hybrid delivery system composed of gellan gum, nanoparticles and gentamicin for the localized treatment of bone infections. *Expert Opin. Drug Deliv.* **2016**, *13*, 613–620. [CrossRef]
19. Carpa, R.; Remizovschi, A.; Culda, C.A.; Butiuc-Keul, A.L. Inherent and Composite Hydrogels as Promising Materials to Limit Antimicrobial Resistance. *Gels* **2022**, *8*, 70. [CrossRef]
20. Li, S.; Dong, S.; Xu, W.; Tu, S.; Yan, L.; Zhao, C.; Ding, J.; Chen, X. Antibacterial hydrogels. *Adv. Sci.* **2018**, *5*, 1700527. [CrossRef]
21. Han, D.; Li, Y.; Liu, X.; Li, B.; Han, Y.; Zheng, Y.; Yeung, K.W.K.; Li, C.; Cui, Z.; Liang, Y. Rapid bacteria trapping and killing of metal-organic frameworks strengthened photo-responsive hydrogel for rapid tissue repair of bacterial infected wounds. *Chem. Eng. J.* **2020**, *396*, 125194. [CrossRef]
22. Hou, S.; Liu, Y.; Feng, F.; Zhou, J.; Feng, X.; Fan, Y. Polysaccharide-Peptide Cryogels for Multidrug-Resistant-Bacteria Infected Wound Healing and Hemostasis. *Adv. Healthc. Mater.* **2020**, *9*, 1901041. [CrossRef]
23. Shi, Y.; Li, D.; Ding, J.; He, C.; Chen, X. Physiologically relevant pH-and temperature-responsive polypeptide hydrogels with adhesive properties. *Polym. Chem.* **2021**, *12*, 2832–2839. [CrossRef]
24. Wan, Y.; Liu, L.; Yuan, S.; Sun, J.; Li, Z. pH-Responsive peptide supramolecular hydrogels with antibacterial activity. *Langmuir* **2017**, *33*, 3234–3240. [CrossRef] [PubMed]
25. Jing, J.; Liang, S.; Yan, Y.; Tian, X.; Li, X. Fabrication of hybrid hydrogels from silk fibroin and tannic acid with enhanced gelation and antibacterial activities. *ACS Biomater. Sci. Eng.* **2019**, *5*, 4601–4611. [CrossRef] [PubMed]
26. Yang, W.; Fortunati, E.; Bertoglio, F.; Owczarek, J.; Bruni, G.; Kozanecki, M.; Kenny, J.; Torre, L.; Visai, L.; Puglia, D. Polyvinyl alcohol/chitosan hydrogels with enhanced antioxidant and antibacterial properties induced by lignin nanoparticles. *Carbohydr. Polym.* **2018**, *181*, 275–284. [CrossRef] [PubMed]
27. Li, Z.; Chen, S.; Wu, B.; Liu, Z.; Cheng, L.; Bao, Y.; Ma, Y.; Chen, L.; Tong, X.; Dai, F. Multifunctional dual ionic-covalent membranes for wound healing. *ACS Biomater. Sci. Eng.* **2020**, *6*, 6949–6960. [CrossRef] [PubMed]
28. Dacrory, S.; Abou-Yousef, H.; Abouzeid, R.E.; Kamel, S.; Abdel-Aziz, M.S.; El-Badry, M. Antimicrobial cellulosic hydrogel from olive oil industrial residue. *Int. J. Biol. Macromol.* **2018**, *117*, 179–188. [CrossRef]
29. Dai, C.; Zhou, Z.; Guan, Z.; Wu, Y.; Liu, Y.; He, J.; Yu, P.; Tu, L.; Zhang, F.; Chen, D. A Multifunctional Metallohydrogel with Injectability, Self-Healing, and Multistimulus-Responsiveness for Bioadhesives. *Macromol. Mater. Eng.* **2018**, *303*, 1800305. [CrossRef]
30. Li, P.; Li, J.; Feng, X.; Li, J.; Hao, Y.; Zhang, J.; Wang, H.; Yin, A.; Zhou, J.; Ma, X. Metal-Organic frameworks with photocatalytic bactericidal activity for integrated air cleaning. *Nat. Commun.* **2019**, *10*, 2177. [CrossRef]
31. Tubek, S.; Grzanka, P.; Tubek, I. Role of zinc in hemostasis: A review. *Biol. Trace Elem. Res.* **2008**, *121*, 1–8. [CrossRef] [PubMed]
32. Davidson-Rozenfeld, G.; Stricker, L.; Simke, J.; Fadeev, M.; Vázquez-González, M.; Ravoo, B.J.; Willner, I. Light-responsive arylazopyrazole-based hydrogels: Their applications as shape-memory materials, self-healing matrices and controlled drug release systems. *Polym. Chem.* **2019**, *10*, 4106–4115. [CrossRef]
33. Fan, X.; Yang, F.; Huang, J.; Yang, Y.; Nie, C.; Zhao, W.; Ma, L.; Cheng, C.; Zhao, C.; Haag, R. Metal–organic-framework-derived 2D carbon nanosheets for localized multiple bacterial eradication and augmented anti-infective therapy. *Nano Lett.* **2019**, *19*, 5885–5896. [CrossRef] [PubMed]
34. Liang, Y.; Wang, M.; Zhang, Z.; Ren, G.; Liu, Y.; Wu, S.; Shen, J. Facile synthesis of ZnO QDs@ GO-CS hydrogel for synergetic antibacterial applications and enhanced wound healing. *Chem. Eng. J.* **2019**, *378*, 122043. [CrossRef]

35. Liang, Y.; Zhao, X.; Hu, T.; Chen, B.; Yin, Z.; Ma, P.X.; Guo, B. Adhesive hemostatic conducting injectable composite hydrogels with sustained drug release and photothermal antibacterial activity to promote full-thickness skin regeneration during wound healing. *Small* **2019**, *15*, 1900046. [CrossRef]
36. Patra, P.; Rameshbabu, A.P.; Das, D.; Dhara, S.; Panda, A.B.; Pal, S. Stimuli-responsive, biocompatible hydrogel derived from glycogen and poly (N-isopropylacrylamide) for colon targeted delivery of ornidazole and 5-amino salicylic acid. *Polym. Chem.* **2016**, *7*, 5426–5435. [CrossRef]
37. Joorabloo, A.; Khorasani, M.T.; Adeli, H.; Mansoori-Moghadam, Z.; Moghaddam, A. Fabrication of heparinized nano ZnO/poly (vinylalcohol)/carboxymethyl cellulose bionanocomposite hydrogels using artificial neural network for wound dressing application. *J. Ind. Eng. Chem.* **2019**, *70*, 253–263. [CrossRef]
38. Winter, G.D. Formation of the scab and the rate of epithelization of superficial wounds in the skin of the young domestic pig. *Nature* **1962**, *193*, 293–294. [CrossRef]
39. Gao, L.; Zhang, H.; Yu, B.; Li, W.L.; Cong, H. Chitosan composite hydrogels cross-linked by multifunctional diazo resin as antibacterial dressings for improved wound healing. *J. Biomed. Mater. Res. Part A* **2020**, *108*, 1890–1898. [CrossRef]
40. Jiao, C.; Gao, L.; Zhang, H.; Yu, B.; Shen, Y. Dynamic covalent C=C bond crosslinked injectable and self-healable hydrogels via Knoevenagel condensation. *Biomacromolecules* **2020**, *21*, 1234–1242. [CrossRef]
41. Chen, R.; Yao, J.; Gu, Q.; Smeets, S.; Baerlocher, C.; Gu, H.; Zhu, D.; Morris, W.; Yaghi, O.M.; Wang, H. A two-dimensional zeolitic imidazolate framework with a cushion-shaped cavity for $CO_2$ adsorption. *Chem. Commun.* **2013**, *49*, 9500–9502. [CrossRef] [PubMed]
42. Zhang, J.; Zhang, T.; Yu, D.; Xiao, K.; Hong, Y. Transition from ZIF-L-Co to ZIF-67: A new insight into the structural evolution of zeolitic imidazolate frameworks (ZIFs) in aqueous systems. *CrystEngComm* **2015**, *17*, 8212–8215. [CrossRef]
43. Zhao, H.; Feng, Y.; Guo, J. Grafting of poly (ethylene glycol) monoacrylate onto polycarbonateurethane surfaces by ultraviolet radiation grafting polymerization to control hydrophilicity. *J. Appl. Polym. Sci.* **2011**, *119*, 3717–3727. [CrossRef]
44. Li, Y.; Zhu, C.; Fan, D.; Fu, R.; Ma, P.; Duan, Z.; Li, X.; Lei, H.; Chi, L. Construction of porous sponge-like PVA-CMC-PEG hydrogels with pH-sensitivity via phase separation for wound dressing. *Int. J. Polym. Mater. Polym. Biomater.* **2020**, *69*, 505–515. [CrossRef]
45. Wang, L.-Y.; Wang, M.-J. Removal of heavy metal ions by poly (vinyl alcohol) and carboxymethyl cellulose composite hydrogels prepared by a freeze–thaw method. *ACS Sustain. Chem. Eng.* **2016**, *4*, 2830–2837. [CrossRef]
46. Yang, X.-H.; Yao, Y.-Q.; Huang, M.-H.; Chai, C.-P. Preparation and characterization of poly (vinyl alcohol)/ZIF-8 porous composites by ice-templating method with high ZIF-8 loading amount. *Chin. J. Polym. Sci.* **2020**, *38*, 638–643. [CrossRef]
47. Wang, J.; Li, Y.; Lv, Z.; Xie, Y.; Shu, J.; Alsaedi, A.; Hayat, T.; Chen, C. Exploration of the adsorption performance and mechanism of zeolitic imidazolate framework-8@ graphene oxide for Pb (II) and 1-naphthylamine from aqueous solution. *J. Colloid Interface Sci.* **2019**, *542*, 410–420. [CrossRef]
48. Shi, Z.; Yu, A.Y.; Fu, A.C.; Wanga, B.L.; Lia, X. Water-based synthesis of zeolitic imidazolate framework-8 for $CO_2$ capture. *RSC Adv.* **2017**, *7*, 29227–29232. [CrossRef]
49. Wijaya, C.J.; Ismadji, S.; Aparamarta, H.W.; Gunawan, S. Facile and Green Synthesis of Starfruit-Like ZIF-L, and Its Optimization Study. *Molecules* **2021**, *26*, 4416. [CrossRef]
50. Fu, H.; Wang, Z.; Wang, X.; Wang, P.; Wang, C.C. Formation mechanism of rod-like ZIF-L and fast phase transformation from ZIF-L to ZIF-8 with morphology changes controlled by polyvinylpyrrolidone and ethanol. *CrystEngComm* **2018**, *20*, 1473–1477. [CrossRef]
51. He, Y.; Zeng, L.; Feng, Z.; Zhang, Q.; Zhao, X.; Ge, S.; Hu, X.; Lin, H. Preparation, characterization, and photocatalytic activity of novel AgBr/ZIF-8 composites for water purification—ScienceDirect. *Adv. Powder Technol.* **2020**, *31*, 439–447. [CrossRef]
52. Khan, I.U.; Othman, M.; Ismail, A.F.; Ismail, N.; Jaafar, J.; Hashim, H.; Rahman, M.A.; Jilani, A. Structural transition from two-dimensional ZIF-L to three-dimensional ZIF-8 nanoparticles in aqueous room temperature synthesis with improved $CO_2$ adsorption. *Mater. Charact.* **2018**, *136*, 407–416. [CrossRef]
53. Yao, X.; Zhu, G.; Zhu, P.; Ma, J.; Chen, W.; Liu, Z.; Kong, T. Omniphobic ZIF-8@Hydrogel Membrane by Microfluidic-Emulsion-Templating Method for Wound Healing. *Adv. Funct. Mater.* **2020**, *30*, 1909389. [CrossRef]
54. Taheri, M.; Ashok, D.; Sen, T.; Enge, T.G.; Tsuzuki, T. Stability of ZIF-8 nanopowders in bacterial culture media and its implication for antibacterial properties. *Chem. Eng. J.* **2020**, *413*, 127511. [CrossRef]
55. du Plessis, J.L.; Stefaniak, A.B.; Wilhelm, K.P. Measurement of Skin Surface pH. *Curr. Probl. Dermatol.* **2018**, *54*, 19–25. [PubMed]
56. Yuan, S.; Zhu, J.; Li, Y.; Zhao, Y.; Li, J.; Van Puyvelde, P.; Bart, V.D.B. Structure architecture of micro/nanoscale ZIF-L on a 3D printed membrane for a superhydrophobic and underwater superoleophobic surface. *J. Mater. Chem. A* **2019**, *7*, 2723–2729. [CrossRef]
57. Yin, X.; Mu, P.; Wang, Q.; Li, J. Superhydrophobic ZIF-8-Based Dual-Layer Coating for Enhanced Corrosion Protection of Mg Alloy. *ACS Appl. Mater. Interfaces* **2020**, *12*, 35453–35463. [CrossRef]
58. Miao, W.; Wang, J.; Liu, J.; Zhang, Y. Zeolitic Imidazolate Framework: Self-Cleaning and Antibacterial Zeolitic Imidazolate Framework Coatings (Adv. Mater. Interfaces 14/2018). *Adv. Mater. Interfaces* **2018**, *5*, 1870068. [CrossRef]
59. Wang, J.; Yu, S.; Yin, X.; Wang, L.; Zhu, G.; Wang, K.; Li, Q.; Li, J.; Yang, X. Fabrication of cross-like ZIF-L structures with water repellency and self-cleaning property via a simple in-situ growth strategy. *Colloids Surf. A Physicochem. Eng. Asp.* **2021**, *623*, 126731. [CrossRef]

60. Bartlet, K.; Movafaghi, S.; Dasi, L.P.; Kota, A.K.; Popat, K.C. Antibacterial activity on superhydrophobic titania nanotube arrays. *Colloids Surf. B-Biointerfaces* **2018**, *166*, 179–186. [CrossRef]
61. Xie, Y. Talking about active oxygen and human diseases. *Mod. Agric. Sci. Technol.* **2009**, 285–286+288. [CrossRef]
62. Yin, S. *Environmental Microbiology*; Academic Press: Cambridge, MA, USA, 2006.

*Review*

# Recent Progress of Cellulose-Based Hydrogel Photocatalysts and Their Applications

Jinyu Yang [1,2,†], Dongliang Liu [1,†], Xiaofang Song [1,2], Yuan Zhao [1,2], Yayang Wang [1,2], Lu Rao [1], Lili Fu [1], Zhijun Wang [1], Xiaojie Yang [1], Yuesheng Li [1,*] and Yi Liu [2,3,*]

1 Hubei Key Laboratory of Radiation Chemistry and Functional Materials, Non-Power Nuclear Technology Collaborative Innovation Center, Hubei University of Science and Technology, Xianning 437100, China; yjyxjj@wust.edu.cn (J.Y.); ldl142325@163.com (D.L.); yuebanwan68@163.com (X.S.); zhyf308@hbust.edu.cn (Y.Z.); wyy1750934313@163.com (Y.W.); rl13908440364@163.com (L.R.); fll151852538462022@163.com (L.F.); qwertasdkl@163.com (Z.W.); mailyangxiaojie@126.com (X.Y.)
2 Key Laboratory of Coal Conversion and New Carbon Materials of Hubei Province, School of Chemistry and Chemical Engineering, Wuhan University of Science and Technology, Wuhan 430081, China
3 College of Chemistry and Chemical Engineering, Tiangong University, Tianjin 300387, China
\* Correspondence: frank78929@163.com (Y.L.); yiliuchem@whu.edu.cn (Y.L.)
† These authors contributed equally to this work.

**Abstract:** With the development of science and technology, photocatalytic technology is of great interest. Nanosized photocatalysts are easy to agglomerate in an aqueous solution, which is unfavorable for recycling. Therefore, hydrogel-based photocatalytic composites were born. Compared with other photocatalytic carriers, hydrogels have a three-dimensional network structure, high water absorption, and a controllable shape. Meanwhile, the high permeability of these composites is an effective way to promote photocatalysis technology by inhibiting nanoparticle photo corrosion, while significantly ensuring the catalytic activity of the photocatalysts. With the growing energy crisis and limited reserves of traditional energy sources such as oil, the attention of researchers was drawn to natural polymers. Like almost all abundant natural polymer compounds in the world, cellulose has the advantages of non-toxicity, degradability, and biocompatibility. It is used as a class of reproducible crude material for the preparation of hydrogel photocatalytic composites. The network structure and high hydroxyl active sites of cellulose-based hydrogels improve the adsorption performance of catalysts and avoid nanoparticle collisions, indirectly enhancing their photocatalytic performance. In this paper, we sum up the current research progress of cellulose-based hydrogels. After briefly discussing the properties and preparation methods of cellulose and its descendant hydrogels, we explore the effects of hydrogels on photocatalytic properties. Next, the cellulose-based hydrogel photocatalytic composites are classified according to the type of catalyst, and the research progress in different fields is reviewed. Finally, the challenges they will face are summarized, and the development trends are prospected.

**Keywords:** cellulose; cellulose derivatives; hydrogels; photocatalytic composites

## 1. Introduction

With the economy's progress and the improvement in living standards, environmental and energy problems are becoming increasingly prominent. The emission of dyes, heavy metals, pesticides, and the emergence of a vast number of microbial pathogens is extremely hazardous to human health and can even cause system disorders, cancer, and other diseases [1,2]. Researchers have carried out the disposal of organic pollutants and heavy metal ions by employing adsorption [3], reverse osmosis [4], and ion exchange [5]. Still, these techniques have limitations, such as high cost and low efficiency. For the threat of pathogenic bacteria, antibiotics [6], ultraviolet light, and high-temperature sterilization [7] are commonly used. However, antibiotic sterilization tends to cause the development of drug-resistant strains, and UV and high-temperature sterilization also have limitations. Meanwhile, energy shortages affect socioeconomic development and human living standards. Therefore, it is urgent to seek a green, safe, and sustainable energy technology. Photocatalytic technology uses sunlight to irradiate photocatalytic materials in order to degrade organic pollutants, reduce heavy metal ions, inactivate bacteria, and produce hydrogen, which is highly efficient, green, safe, and cheap, and is one of the ideal ways to solve environmental and energy problems.

Nanosized photocatalysts are easily agglomerated in water and are not suitable for recovery. Activated carbon [8], molecular sieve [9], hydrogel [10], and other materials are usually used as carriers to immobilize photocatalysts and improve the utilization rate. Hydrogels, with high permeability, adsorption, and insolubility, are a type of bionic photocatalytic reactor that has received wide attention from researchers [11]. Hydrogels are three-dimensional network structures formed by electrostatic interaction, an entanglement of molecular chains, and cross-linking of chemical bonds [12]. Due to their absorption of large amounts of water and certain flexibility, they can form various hydrogels, such as gel columns, gel spheres, gel films, etc. [13]. Meanwhile, supercritical drying, freeze-drying, and evaporative drying can all be used to obtain porous solid aerogels [14].

With the increasing emphasis on the development and utilization of renewable resources, bio-based polymers have received a lot of attention from researchers. The biocompatibility and degradability of bio-based hydrogels have led to their application in agriculture [15], medical [16], and environmental fields [17]. Cellulose is a linear polymer composed of many D-glucopyranose units interconnected by β-glycosidic bonds. It is one of the most widely distributed polysaccharides in nature, and is present in several plants, bacteria, and algae such as cotton, rice straw, trees, and Chlorophyta. [18–20] (Figure 1). The exposed hydroxyl groups of cellulose can undergo more chemical reactions, providing the possibility of preparing cellulose derivatives, optimizing their disadvantages such as water solubility and poor mechanical properties [21]. With their non-toxicity, and easy degradability, hydrogels of cellulose and its descendants can be used in agriculture (storage and continuous release of water and fertilizers [22]), water treatment (adsorption of heavy metal ions [23], desalination of seawater [24], photocatalytic degradation of organic pollutants [25], or photocatalytic sterilization [26]), and biomedicine (drug delivery [27], wound dressings [28], tissue engineering [29], health care hygiene [30], and smart materials [31]), among other areas.

Cellulose hydrogels are cited for use in both the water treatment [32] and antimicrobial fields [33], emphasizing their adsorption properties as well as their photocatalytic properties. This paper mainly summarizes the properties, preparation methods, and classification of cellulose-based hydrogel materials and their different applications in the direction of photocatalysis over the past 15 years, providing a good basis for future development.

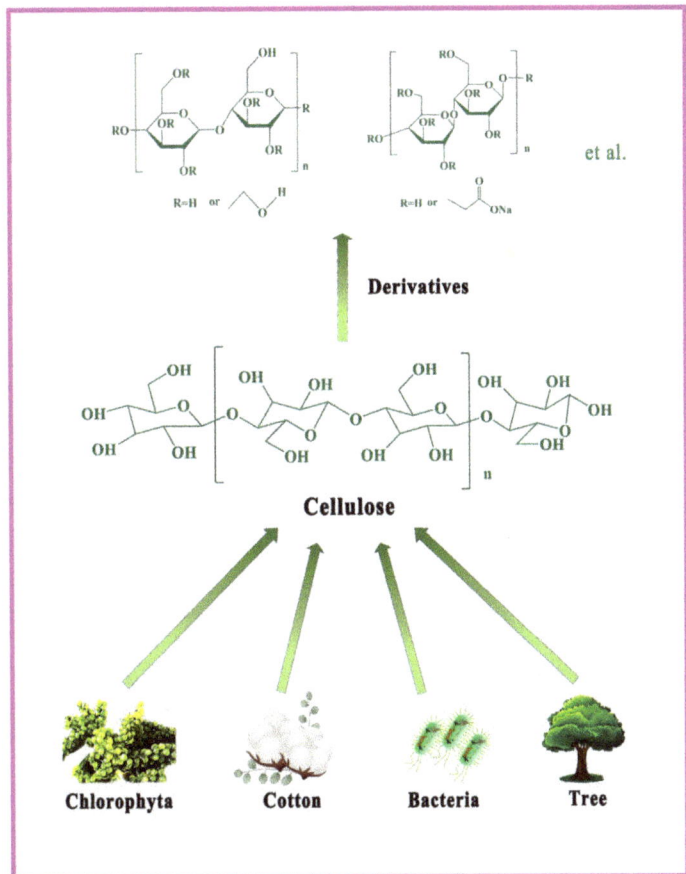

Figure 1. Molecular structure of cellulose and its origin in nature.

## 2. Characteristics of Cellulose-Based Hydrogel Photocatalytic Composites

Cellulose hydrogels as semiconductor carriers have three main characteristics: high adsorption, dispersibility, and morphological auxiliary. The adjustment of these characteristics can change the structural features of the semiconductor, which in turn, can effectively improve photocatalytic efficiency.

### 2.1. High Adsorption

Cellulose molecules form intramolecular and intermolecular hydrogen bonds, and the molecular chains are coiled to form highly crystalline fibers that are more difficult to dissolve. At the same time, hydroxyl groups are blocked within, which affects their adsorption properties for water, oil, and heavy metal ions [34]. In practical applications, cellulose is chemically modified (oxidation, esterification, grafting, etc.) by introducing specific groups to avoid its insolubility and enhance its adsorption properties [35,36]. Therefore, cellulose derivatives are common materials for preparing cellulose-based hydrogels, specifically: hydroxy cellulose (HEC), carboxy cellulose (CMC), and amino cellulose. Han et al. prepared titanium dioxide hydrogel cages using HEC and CMC. The hydrogel cages showed good adsorption performance: within 5 min, the hydrogel cages adsorbed 43% more dye than titanium dioxide nanoparticles, which greatly enhanced the photocatalytic performance of the composites [37]. Adsorption is an essential part of the photocatalytic link, and a high stirring speed is usually used to reduce the mass transfer resistance. Hydroxy cellulose and carboxy cellulose hydrogel photocatalytic composites enhance the adsorption of dyes in wastewater through electrostatic interaction, which in turn promotes photocatalytic reaction activity. In short, a synergistic adsorption–photocatalytic system was constructed to enhance the photocatalytic effect [38,39]. Amino cellulose has an amino group at the end, which is similar to the structure of chitosan. The introduction of the amino group makes it very soluble, film-forming, and adsorbent of heavy metal ions, yielding good prospects for biological applications such as wound dressing, immunofluorescence, and drug release [40–42]. However, the complex synthesis process of amino cellulose and the poor selectivity and economy of the synthesis process hinders the production and limits the application of amino cellulose. No research on hydrogel-type photocatalytic composites has been done in photocatalysis.

## 2.2. Dispersibility

Nanophotocatalysts have high specific surface energy and are thermodynamically unstable systems. The nanoparticles agglomerate to form soft and hard agglomerates due to van der Waals and Coulomb forces between the particles during preparation or post-processing, affecting the adequate performance of their photocatalytic properties [43]. Therefore, cellulose can be a suitable carrier for improving the dispersion of nanophotocatalysts to expose more active sites to capture light or change the semiconductor bandgap to participate in the reaction. The atoms and ions of semiconductors are anchored by functional groups in cellulose through chemical or hydrogen bonding. For example, based on the chemical interaction between $Zn^{2+}$ and $COO^-$ of ZnO, $Zn^{2+}$ was adsorbed in floatable carboxy methyl cellulose/polyphenyl amide hydrogel (PAM/CMC/DDM), which enabled the practical separation of heavy metal ions in sewage. Subsequently, PAM/CMC/DDM-ZnO photocatalytic composite was obtained by processing $Zn^{2+}$ into ZnO nanoparticles using an in situ precipitation method. The nanoparticles in this composite avoided agglomeration during the preparation processes, creating a highly efficiently degradation of the dye under visible light [25]. The hydroxyl groups of cellulose can form strong hydrogen-bonding interactions with titanium dioxide nanoparticles, which can be made to adhere to the cellulose surface using a hydrothermal method to obtain cellulose nanofiber/titanium dioxide (P25) aerogel (CNFT) (Figure 2a). The P25 in CNFT2 is uniformly dispersed, and the transmission electron micrographs show that its average diameter is around 6.8 nm (Figure 2b). Moreover, the spectral red-shift of the composites was obtained by UV-vis diffuse reflection (Figure 2c), narrowing the bandgap of P25 and favoring the photocatalytic reaction [44].

## 2.3. Morphological Adjuvants

Usually, controlling the photocatalyst morphology is also a meaningful means of enhancing the catalyst activity. The structure, specific surface area, crystal shape, and crystal defect of photocatalyst are the factors that affect the separation of photogenerated electrons from holes [45]. In contrast to the existing morphology-modulating auxiliaries, cellulose-based hydrogels have the advantages of being green, simple, and efficient.

Cellulose hydrogels can be used as reactors. Qin et al. successfully prepared flower-like ZnO nanoparticles with a crystalline form of hexagonal fibrous zincite from sodium hydroxide and zinc acetate. In this study, there is a chemical bonding between the hydrogel reactor and sodium hydroxide and water, which makes $Zn^{2+}$ and $OH^-$ slowly combine into $[Zn(OH)_4]^{2-}$ ions in the three-dimensional pores. Finally, the nanoflowers are generated by dehydration, induction into nanosheets, and self-assembly, yielding homogeneous size of the flower-like ZnO nanosheets in cellulose hydrogel pores and a high surface area (39.18 $m^2$/g) after calcination. It accelerated a decrease in rhodamine B concentration under UV light [46].

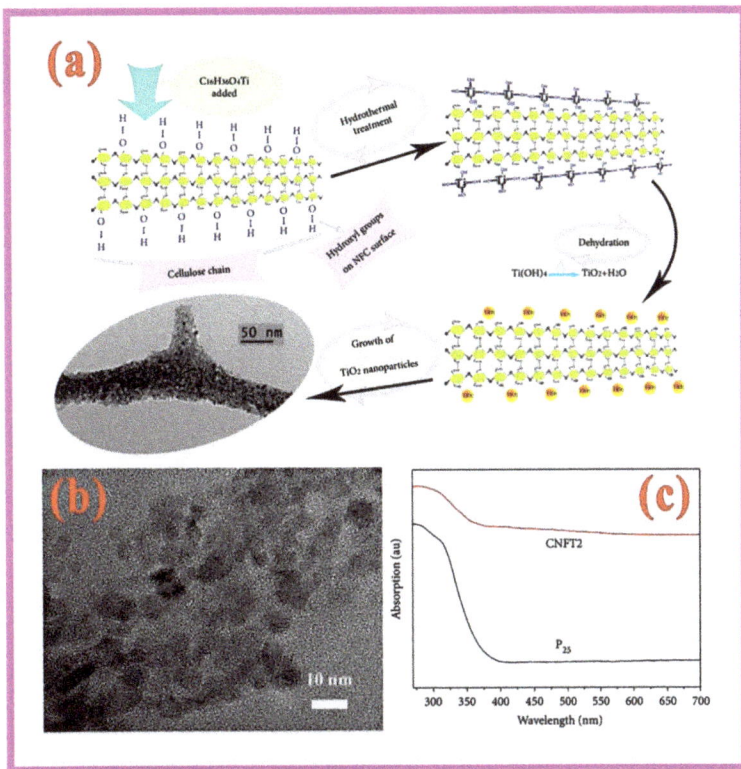

**Figure 2.** (a) Synthesis mechanism of cellulose nanotitanium dioxide aerogel (CNFT); (b) TEM micrograph of CNFT2; (c) UV-Vis spectra of P25 and CNFT2 [44].

Cellulose hydrogels act as green capping agents to guide the semiconductor shape change. As a typical example, Sabbaghan et al. selectively prepared cellulose oxide gel membranes of different shapes (NFC/Cu$_2$O) (spherical, cubic, and truncated cubic) using the reduced end groups of cellulose for Cu$^{2+}$ ion binding. The bandgap of the NFC/Cu$_2$O films with different shapes was shown to be in the range of 2.02–2.25 eV by inspection, and this optical property opens up new applications for cellulose gel films [47].

### 3. Preparation of Cellulose-Based Hydrogel Photocatalytic Composites

There is a cross-linking of the hydroxyl, acetyl, and carboxyl functional groups of cellulose and a loading of photocatalyst to obtain hydrogel photocatalytic composites. Standard cross-linking methods can be categorized as physical, chemical, and radiation cross-linking methods (Table 1).

Table 1. Classification of cellulose-based hydrogel photocatalytic composites.

| Categories | Photocatalysts | Hydrogel Materials | Characteristics | Preparation Methods | Specific Surface Area/m$^2$g$^{-1}$ | References |
|---|---|---|---|---|---|---|
| Metal Oxide Semiconductor Composites | ZnO | PAM/CMC/DDM | Suspended hydrogels, adsorb heavy metal ions, and degrade dyes efficiently | Mechanical foaming and in situ polymerization | - | Zhao et al. 2021 |
| | | Cellulose | Dispersion framework for nanomaterials | Physical crosslinking | - | Jiao et al. 2018 |
| | | Bamboo fiber | High specific surface area | Chemical crosslinking | 39.18 | Qin et al. 2017 |
| | Cu$_2$O | Cellulose/AA/AM | High adsorption | Chemical crosslinking | 89.56 | Su et al. 2017 |
| | TiO$_2$ | CMCNa/HEC | Superabsorbent, biodegradable, and photocatalytic degradation crosslinker | Chemical crosslinking | - | Marcı et al. 2006 |
| | | Cotton cellulose | High temperature resistant | Physical crosslinking | 6.10 | Melone et al. 2013 |
| | | α-Cellulose | TiO$_2$ in situ generators, excellent strength and good toughness | Chemical crosslinking | 550 | Wang et al. 2017 |
| | | TOCNs/PAM | Super-tough | Chemical crosslinking | - | Yue et al. 2020 |
| | | BC | Self-cleaning, antibacterial, and UV shielding | Chemical crosslinking | - | Rahman et al. 2021 |
| | | CNFs | Good adsorption, photocatalytic degradation ability, low density, and easy recovery | Chemical crosslinking | 330 | Li et al. 2021 |
| | Na$_2$Ti$_3$O$_7$ | Sisal cellulose | High specific surface area | Physical crosslinking | 248.93 | Liu et al. 2021 |
| Metal sulfide (chloride) semiconductor composites | MoS$_2$ | BC | Bifunctional adsorbent/photocatalyst membranes | Chemical crosslinking | 137 | Ferreira-Neto et al. 2022 |
| | CdS | Straw cellulose | Green recyclable | Chemical crosslinking | - | Qian et al. 2020 |
| | Cd$_x$Zn$_{1-x}$S | CMC | High yield of hydrogen, good stability, easy recovery | Chemical crosslinking | - | Wu et al. 2018 |
| | Ag/AgCl | CMC | Hydrogel beads, photocatalytic degradation of RhB | Chemical crosslinking | - | Heidarpour et al. 2020 |
| Organic semiconductor composites | g-C$_3$N$_4$ | Polyester fiber/cotton wool | High specific surface area, impact resistant | Chemical crosslinking | - | Chen et al. 2019 |
| | | Cotton linter | Enhanced carrier separation | - | - | Bai et al. 2019; Yao et al. 2019 |
| | | CMC/β-Cyclodextrin | | | | |
| | GO | MCC | Adsorption–photocatalytic synergy | Chemical crosslinking | 48.6 | Liu et al. 2021 |
| | MIL-100(Fe) | CMC/β-Cyclodextrin | Good water retainability | Chemical crosslinking | - | Zhang et al. 2021 |

## 3.1. Physical Cross-Linking Method

The physical cross-linking method is mainly based on hydrogen bonding, crystallization, and van der Waals forces to obtain hydrogels with a three-dimensional network structure [48]. Its operation is simple, but the reversible connection between the chains means that heat will return the gel the solution state [49]. Su et al. heated cellulose, carrageenan, and titanium dioxide in 1-ethyl-3-methylimidazole acetate solution, then cooled and washed it to obtain hydrogel photocatalytic membranes by hydrogen bonding [50].

*3.2. Chemical Cross-Linking Method*

The chemical cross-linking method uses covalent bonds between molecules to form hydrogels with desirable stable structures and mechanical strength. This method is accessible using a broad range of monomers and mild conditions, usually with cross-linking agents such as acrylic acid, polyethylene glycol, ammonium persulfate, polyethyleneimine, etc. Su et al. used ammonium persulfate to generate free radicals in cellulose, which cross-linked with acrylic acid (AA) and acrylamide (AM). Meanwhile, the Cu source was added and freeze-dried to gain $Cu_2O$/cellulose-based aerogels. The adsorption of large amounts of molecular oxygen at the surface of aerogels promotes the separation of $Cu_2O$ photoelectrons and holes, thus enhancing the catalytic activity [51].

*3.3. Radiation Cross-Linking Method*

Radiation technology uses the interaction between rays, accelerated electrons, ions, and substances to ionize and excite in order to produce free radicals, and initiate cross-linking reactions [52]. The radiation cross linking method has the advantages of simple operation, room temperature reaction, high efficiency, and green and non-polluting properties when compared to the above methods [53]. Liu and his colleagues used the electron beam radiation method to develop poly-N-isopropyl acrylamide/highly substituted hydroxypropyl cellulose/carbon nitride (NIPAAm/HHPC/g-$C_3N_4$) intelligent hydrogels, which is a thermally driven property photocatalyst. The high specific surface area, porosity, and large specific surface area of this hydrogel enhance the contact of rhodamine (RhB) dyes. The networked three-dimensional structure of the hydrogel effectively adsorbed RhB dye ions, achieving the combination of adsorption and photocatalysis [54].

## 4. Classification of Cellulose-Based Hydrogel Photocatalytic Materials

Based on the current research results, the cellulose-based hydrogel photocatalytic materials can be classified into metal oxide semiconductor composites, metal sulfide (chloride) semiconductor composites, and organic semiconductor composites according to the types of photocatalysts (Table 1).

*4.1. Metal Oxide Semiconductor Composites*

Currently, most of the metal oxide semiconductors applied in cellulose hydrogels are titanium dioxide, due to their stable nature, non-toxicity, and cheapness [55]. Earlier, researchers considered that toxic cross linking agents (dialkyl sulfone) would remain in the synthesis of cellulose hydrogels, causing water contamination [56]. Therefore, a new highly absorbent and biodegradable hydrogel was synthesized using sodium carboxymethyl cellulose (CMCNa), hydroxyethyl cellulose (HEC), and titanium dioxide nanoparticles ($TiO_2$). This hydrogel was entirely degraded by dialkyl sulfone under 5 h of light [57]. Subsequently, several hydrogels with unique properties have been developed. For example, high-temperature resistant cotton fiber aerogel [58]; α-cellulose hydrogel as $TiO_2$ in situ reactors with excellent strength and good toughness [59]; high stiffness titanium dioxide/polyacrylamide/chitin oxide nanofiber hydrogel ($TiO_2$-TOCNs-PAM), its compressive strength at 70% strain is 1.46 MPa, tensile stress is 316 kPa, tensile strain is 310%, and toughness is 47.25 $kJ/m^3$ [60]; multifunctional flexible bacterial cellulose gel film with self-cleaning, photocatalytic, and UV protection properties [61]; and cellulose nanofiber aerogels loaded with $TiO_2$, with good adsorption properties, high photocatalytic degradation, low density, and easy recycling [62].

ZnO nanoparticles and cuprous oxide are also materials of interest to researchers. Hasanpour et al. prepared six different shapes of cellulose/ZnO (CA/ZnO) heterogeneous aerogels using microcrystalline cellulose (MCC) and zinc nitrate hexahydrate as the primary raw materials by means of hydrothermal, sol-gel, and impregnation methods. Among these, the highest degradation rate of MO was 94.78% for the CA/ZnO heterogeneous aerogel in plate shape [63]. However, iron trioxide (α-$Fe_2O_3$), sodium trititanate ($Na_2Ti_3O_7$), silver phosphate ($Ag_3PO_4$), and bismuth vanadate ($BiVO_4$) semiconductors are relatively rare composites combined with cellulose hydrogels due to their high price or complexity of preparation [64–68].

### 4.2. Metal Sulfide (Chloride) Semiconductor Composites

Metal sulfides have a narrower bandgap compared to metal oxides. Currently, cadmium sulfide (CdS) and molybdenum sulfide ($MoS_2$) are mostly studied in cellulose-based hydrogel photocatalytic materials [69]. CdS crystals are one of the best visible light-reactive photocatalysts. Its forbidden bandwidth is 2.4 eV [70]. CdS nanoparticles are combined with cellulose to form hydrogel composites, and their strong adsorption ability on MB molecules indirectly improves photocatalytic activity [23,71]. Cadmium sulfide solid solution ($Cd_xZn_{1-x}S$) can be used to improve the optical properties of the catalyst by modulating the elemental composition. Wu et al. used in situ chemistry to embed $Cd_xZn_{1-x}S$ particles with a dimension of about 3 nm into carboxymethyl cellulose hydrogels. Among them, the maximum hydrogen yield of $Cd_{0.2}Zn_{0.8}S$ gel composites was 1762.5 $\mu mol\ g^{-1}\ h^{-1}$. This is 104 times the hydrogen production rate of pure cadmium sulfide. This hydrogel photocatalytic complex is stable and easily recyclable, meeting the criteria for green hydrogen production [72]. The surface of noble metal nanoparticles can absorb visible light and has a surface plasmon effect [73]. On the path of surface plasmon photocatalyst exploration, Ag/AgCl has been the most studied by scientists. Heidarpour et al. wrapped Ag/AgCl in Al(III) and Fe(III) crosslinked cellulose hydrogel beads (Ag/AgCl@Al-CMC and Ag/AgCl@Fe-CMC, respectively). Experimental tests showed that the gel beads have good photocatalytic properties. The diverse cases on the photocatalytic performance was also explored, and the photocatalytic reaction rate constants are shown in Table 2 [74].

Table 2. Photocatalytic rate constants of Ag/AgCl@Al-CMC and Ag/AgCl@Fe-CMC under different conditions ((copied from Reference [74]).

|  | Ag/AgCl@Ag-CMC | | | | AgCl@Fe-CMC | | | |
|---|---|---|---|---|---|---|---|---|
| Catalyst Dosage | 1 (g/L) | 2 (g/L) | 4 (g/L) | 6 (g/L) | 1 (g/L) | 2 (g/L) | 4 (g/L) | 6 (g/L) |
| $K_{app}$ | 0.0101 | 0.0223 | 0.0517 | 0.0711 | 0.0073 | 0.0152 | 0.0304 | 0.0395 |
| $R^2$ | 0.98 | 0.99 | 0.95 | 0.98 | 0.98 | 0.99 | 0.99 | 0.99 |
| RhB concentration | 10 (ppm) | 15 (ppm) | 20 (ppm) | 25 (ppm) | 10 (ppm) | 15 (ppm) | 20 (ppm) | 25 (ppm) |
| $K_{app}$ | 0.0517 | 0.0318 | 0.0233 | 0.0141 | 0.0304 | 0.0224 | 0.0170 | 0.0103 |
| $R^2$ | 0.95 | 0.99 | 0.97 | 0.99 | 0.99 | 0.99 | 0.98 | 0.99 |
| pH | 4 | 7 | 9 |  | 4 | 7 | 9 |  |
| $K_{app}$ | 0.0295 | 0.0517 | 0.0673 |  | 0.0198 | 0.0304 | 0.0352 |  |
| $R^2$ | 0.99 | 0.95 | 0.99 |  | 0.99 | 0.99 | 0.95 |  |

### 4.3. Organic Semiconductor Composites

Graphitic phase carbon nitride (g-$C_3N_4$), graphene oxide (GO), and organic metal frameworks are representative materials for organic conjugated semiconductors. Graphitic-phase carbon nitride is a layered material consisting of triazine and tri-s-triazine rings as basic units [75]. It is of interest because of its advantages, such as being non-toxic, cheap, and responsive in visible light. However, the disadvantages of carbon nitride, such as a small specific surface area, easy polymerization, and few active sites, affect its photocatalytic performance [76]. Combining it with cellulose to form aerogel photocatalytic materials can expand the specific surface area and upgrade carrier separation, thus improving the photocatalytic ability [77–79]. GO has a large specific surface area compared to g-$C_3N_4$. Its large number of hydroxyl and carboxyl groups can be used for adsorption. However, GO is soluble in water and difficult to use as an adsorbent [80]. Combining it with MCC and polyaniline (PANI) perpetuates the adsorption performance and achieves sound synergistic adsorption-photocatalytic degradation [81]. Metal-organic frameworks (MOFs) are composed of metal units and organic ligands combined in a framework by coordination to form an open network with high porosity, a stable network, and a massive surface area. MIL-100(Fe) is merged with CMC and cyclodextrin to form a hydrogel with catalytic and water fixation capabilities. It has a good hydrophilicity, with a swelling rate of 363%, which allows it to be used in environmental applications [82].

## 5. Application of Cellulose-Based Hydrogel Photocatalytic Materials

The advantages of cellulose-based hydrogel photocatalytic composites and photocatalysts and hydrogels are combined to promote cellulose with unique properties for different applications. In this section, the latest applications of cellulose-based hydrogel photocatalysts in wastewater treatment and energy will be briefly outlined.

### 5.1. Wastewater Treatment

The insolubility and hydrophilicity of most types of cellulose make cellulose-based hydrogel photocatalytic composites widely used in wastewater treatment.

#### 5.1.1. Removal of Dyes and Heavy Metal Ions

The degradation of dyes and heavy metals are the two most frequent methods used to evaluate photocatalytic performance. Every year, printing and dyeing processes produce hundreds of millions of tons of highly concentrated wastewater containing different types of dyes, in addition to heavy metals, acids, and bases, causing severe environmental problems [83,84]. The dyes and heavy metals commonly used for photocatalytic degradation are rhodamine B [85,86], methyl orange [87,88], methylene blue [89], carmine [90], and hexavalent chromium ions [91]. Cellulose hydrogels carry functional groups that enhance the adsorption and induce the photocatalytic degradation of dyes [92,93]. Table 3 lists the degradation efficiency of methyl orange (MO) by different photocatalytic materials.

Table 3. Degradation efficiency of MO by different photocatalytic materials.

| Dye | Catalysts | Dye Concentration (mg/L) | Time (min) | Degradation (%) | References |
|---|---|---|---|---|---|
| MO | $TiO_2$-TOCNs-PAM | 10 | 90 | 97.3 | Yue et al. 2020 |
| | CA/ZnO | 20 | 120 | 94.78 | Hasanpour et al. 2021 |
| | g-$C_3N_4$ Cellulose aerogel | 20 | 180 | 99 | Ma et al. 2021 |
| | Ag@AgCl-contained cellulose hydrogel | 10 | 70 | 93 | Tang et al. 2018 |
| | $Cu_2O$/$TiO_2$/CNF/rGH | 20 | 120 | 85.62 | Zheng et al. 2022 |
| | Cu@$Cu_2O$/RGO/cellulose hybrid aerogel | 10 | 120 | 92.8 | Du et al. 2019 |

Two representative cases are presented in particular. Du et al. synthesized layered stomatal Cu/doped $Cu_2O$/reduced graphene oxide/cellulose (Cu@$Cu_2O$/RGO/CE) catalytic materials using the in situ deposition method (Figure 3a). It had a high photocatalytic performance for the degradation of MO in visible light. A reasonable photocatalytic degradation diagram was obtained by EPR tests showing that hydroxyl radicals and superoxide radicals take effect in the photocatalytic process (Figure 3b) [94]. $NH_2$-MIL-88B(Fe) (NM88) and g-$C_3N_4$ loaded onto aerogels, combined with natural cellulose and polyacrylonitrile fibers (BMFAs), achieved a 99% reduction of Cr(VI) within 20 min. At the same time, this composite has excellent memory properties and shape deformability (Figure 3c) [95].

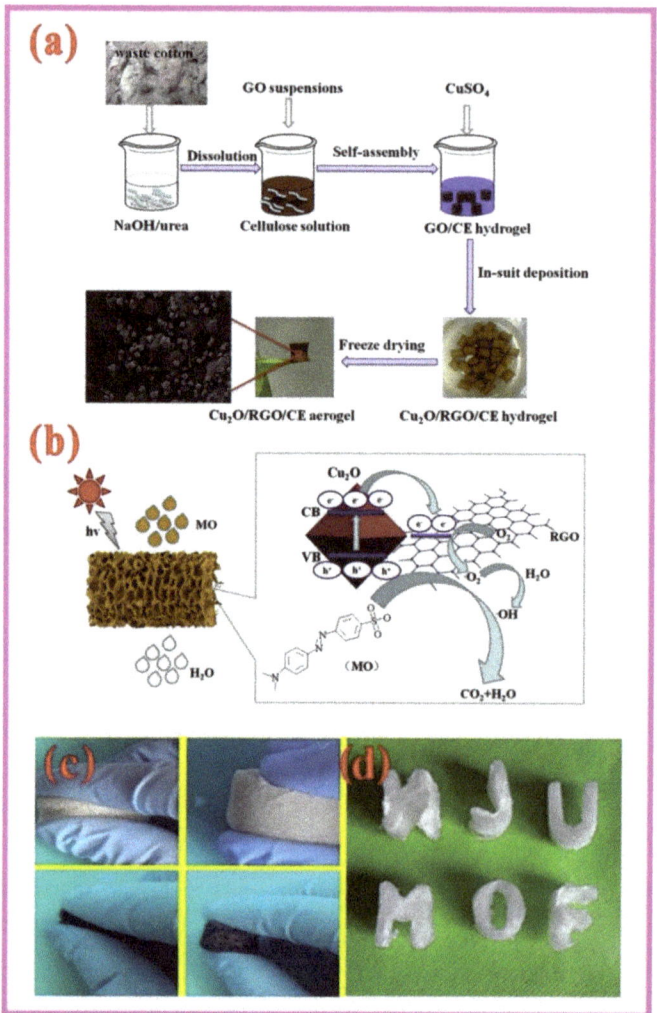

**Figure 3.** (a) Synthesis process of Cu@Cu$_2$O/RGO/CE hybrid catalysts; (b) Mechanism diagram of Cu@Cu$_2$O/RGO/CE photocatalytic degradation of MO; (c) Elasticity test map; and (d) plasticity of BMFAs [94,95].

5.1.2. Degradation of Antibiotics

Antibiotics are remarkably effective in treating infectious diseases and are in high demand in the livestock and aquaculture industries. However, residual antibiotics can also have severe ecological and public health impacts [96]. Therefore, the problem of antibiotic reprocessing is among the urgent issues to be addressed. Currently, efficient, mild, and non-polluting photocatalytic technology shows good prospects for degrading antibiotic wastewater. Tetracycline, as a spectral antibiotic, has received significant attention from researchers. Recent literature reported that the combination of photocatalyst and cellulose to form an open porous three-dimensional structure and high specific surface area enhanced its adsorption, thus improving the degradation efficiency [97]. The high mechanical strength of hydrogels allows the reusability of the composites [98].

### 5.1.3. Antibacterial Properties

After an assessment by the World Health Organization (WHO), the following information was obtained. In developing countries, 80% of diseases originate from water sources contaminated with pathogenic microorganisms, including fungi, bacteria, and viruses [99]. These microorganisms can cause diseases such as diarrhea, typhoid, and pneumonia. Photocatalysis has received attention from researchers as an effective and inexpensive method for sterilization. Zhang et al. synthesized multifunctional cellulose/$TiO_2$/β-CD hydrogels with extreme photocatalytic antibacterial properties and drug release capacity. Their excellent photocatalytic antibacterial activity was verified by the inhibition circle method under dark and light conditions [100].

## 5.2. Energy
### 5.2.1. Hydrogen Energy

In today's society, the development of green energy plays a vital role in economic development and human living standards. Hydrogen energy with high calorific value and no secondary pollution is becoming a hot spot for research. In contrast to electrochemical hydrogen generation [101] and anaerobic microbial fermentation [102], photocatalytic hydrogen generation is essential for the development of hydrogen energy by converting sunlight into hydrogen energy using water as a raw material. Kang et al. used an in situ method to combine nanoscale CdS and CMC, while doping with trace amounts of Pt, to produce highly efficient hydrogen-producing photocatalytic hydrogels, with a hydrogen generation efficiency of 1365 $\mu mol\ h^{-1} g^{-1}$. The cellulose hydrogel enabled a better dispersion of CdS nanoparticles and avoided the secondary contamination of nanoparticles, which has practical implications for the development of hydrogen energy [103].

### 5.2.2. Food Packaging

Food packaging bags mainly include two types: plastic and paper bags. Plastic bags make up the bulk of packaging materials, prepared by the polymerization of ethylene in petroleum cracking; these are not easily degradable, and doing so will cause secondary pollution. At the same time, due to energy constraints, the transformation of plastic packaging bags is imminent. Cellulose is easily accessible and biodegradable, laying the foundation for the development of green packaging materials. Xie et al. prepared a cellulose gel film containing zinc oxide nanoparticles on the surface by chemical cross-linking and hydrothermal methods. This food packaging film has specific mechanical properties to block oxygen and water vapor and ensure the freshness of food. At the same time, it has some antibacterial effects under both dark and UV light irradiation conditions. Under UV light, the bacteria were inactivated more efficiently by synergistic photocatalytic oxidation and mechanical rupture [104].

## 6. Conclusions and Prospect

Cellulose is widely distributed in nature, has the largest reserves, and is easily degradable. Therefore, the development of functional cellulose materials is of significant meaning to the progress of green chemistry and the reduction of dependence on fossil resources. The functional properties of cellulose, including hydrophilicity and good biocompatibility, make it a suitable carrier for photocatalysts. In this paper, we review the research progress of cellulose-based hydrogel photocatalytic materials, detailing the properties, preparation methods, classification, and applications of the composites in the environmental and energy fields. Combined with the above literature, it is concluded that cellulose hydrogels as carriers exhibit the following main advantages. First, the three-dimensional network of cellulose hydrogels increases the specific surface area, which dramatically improves the adsorption performance of composites. Second, the negative ions on cellulose can bind metal cations, which can immobilize the photocatalyst on the surface and improve the dispersibility of nanoparticles. Finally, cellulose can improve the carrier separation efficiency.

Although the research on cellulose-based hydrogel photocatalytic composites continues to progress, there are still some issues of concern: (1) From the preparation method of composites, radiation preparation has the advantages of green qualities and high efficiency, but relatively little research has been reported on this method. (2) Whether or not the mechanical strength of a hydrogel carrier will be affected during the photocatalytic cycle. Therefore, further research is needed. (3) Most of the applications of the composites are concentrated in the environmental field and very few are developed in the energy field. We have not seen any research in the medical fields, such as photodynamic therapy. Therefore, future research can focus on this area and fully expand its biocompatibility to enable its application in the medical field.

**Author Contributions:** Writing—original draft, J.Y.; Resources, D.L. and X.S.; Methodology, Y.Z.; Formal analysis, Y.W.; Visualization, L.R.; Investigation, L.F. and Z.W.; Supervision, X.Y.; Writing—review & editing, Y.L. (Yuesheng Li) and Y.L. (Yi Liu). All authors have read and agreed to the published version of the manuscript.

**Funding:** The authors extend their appreciation to the National Natural Science Foundation of China (No. 11405050), the Hubei Provincial Colleges and Universities Outstanding Young and Middle-aged Technological Innovation Team Project (No. T2020022), Xianning City Key Program of Science and Technology (No. 2021GXYF021), and the Science Development Foundation of the Hubei University of Science and Technology (No. 2020TD01, 2021ZX01).

**Institutional Review Board Statement:** Not applicable.

**Informed Consent Statement:** Not applicable.

**Acknowledgments:** This work was supported by the National Natural Science Foundation of China (No. 11405050), the Hubei Provincial Colleges and Universities Outstanding Young and Middle-aged Technological Innovation Team Project (No. T2020022), Xianning City Key Program of Science and Technology (No. 2021GXYF021), and the Science Development Foundation of the Hubei University of Science and Technology (No. 2020TD01, 2021ZX01).

**Conflicts of Interest:** The author declares no conflict of interest.

## References

1. Yakamercan, E.; Ari, A.; Aygün, A. Land application of municipal sewage sludge: Human health risk assessment of heavy metals. *J. Clean. Prod.* **2021**, *319*, 128568. [CrossRef]
2. Su, H.C.; Liu, Y.S.; Pan, C.G.; Chen, J.; He, L.Y.; Ying, G.G. Persistence of antibiotic resistance genes and bacterial community changes in drinking water treatment system: From drinking water source to tap water. *Sci. Total Environ.* **2018**, *616*, 453–461. [CrossRef] [PubMed]
3. Chen, X.; Song, Z.; Yuan, B.; Li, X.; Li, S.; Nguyen, T.T.; Guo, M.; Guo, Z. Fluorescent carbon dots crosslinked cellulose nanofibril/chitosan interpenetrating hydrogel system for sensitive detection and efficient adsorption of Cu (II) and Cr (VI). *Chem. Eng. J.* **2022**, *430*, 133154. [CrossRef]
4. Alhalili, Z.; Romdhani, C.; Chemingui, H.; Smiri, M. Removal of dithioterethiol (DTT) from water by membranes of cellulose acetate (AC) and AC doped ZnO and $TiO_2$ nanoparticles. *J. Saudi Chem. Soc.* **2021**, *25*, 101282. [CrossRef]
5. Cui, J.; Li, J.F.; Cui, J.; Wang, W.; Wu, Y.; Xu, B.; Chang, Y.J.; Liu, X.J.; Li, H.; Yao, D.R. Removal effects of a biomass bottom ash composite on tailwater phosphate and its application in a rural sewage treatment plant. *Sci. Total Environ.* **2021**, *812*, 152549. [CrossRef]
6. Calero-Cáceres, M.; Muniesa, M. Persistence of naturally occurring antibiotic resistance genes in the bacteria and bacteriophage fractions of wastewater. *Water Res.* **2016**, *95*, 11–18. [CrossRef]
7. Matsunaga, T.; Tomoda, R.; Nakajima, T.; Wake, H. Photoelectrochemical sterilization of microbial cells by semiconductor powders. *FEMS Microbiol. Lett.* **1985**, *29*, 211–214. [CrossRef]
8. Khan, M.E. State-of-the-art developments in carbon-based metal nanocomposites as a catalyst: Photocatalysis. *Nanoscale Adv.* **2021**, *3*, 1887–1900. [CrossRef]
9. He, X.; Antonelli, D. Recent Advances in Synthesis and Applications of Transition Metal Containing Mesoporous Molecular Sieves. *Angew. Chem. Int. Ed.* **2002**, *41*, 214–229. [CrossRef]
10. Chaubey, S.; Singh, P.; Singh, C.; Shambhavi; Sharma, K.; Yadav, R.K.; Kumar, A.; Baeg, J.; Dwivedi, D.K.; Singh, A.P. Self-assembled protein/carbon nitride/sulfur hydrogel photocatalyst for highly selective solar chemical production. *Mater. Lett.* **2020**, *259*, 126752. [CrossRef]

11. Koo, H.J.; Velev, O.D. Biomimetic photocatalytic reactor with a hydrogel-embedded microfluidic network. *J. Mater. Chem. A* **2013**, *1*, 11106–11110. [CrossRef]
12. Bao, Z.T.; Xian, C.H.; Yuan, Q.J.; Liu, G.T.; Wu, J. Natural Polymer-Based Hydrogels with Enhanced Mechanical Performances: Preparation, Structure, and Property. *Adv. Healthc. Mater.* **2019**, *8*, 1900670. [CrossRef] [PubMed]
13. Yao, B.W.; Wang, H.Y.; Zhou, Q.Q.; Wu, M.M.; Zhang, M.; Li, C.; Shi, G.Q. Ultrahigh-Conductivity Polymer Hydrogels with Arbitrary Structures. *Adv. Mater.* **2017**, *29*, 1700974. [CrossRef]
14. Job, N.; Théry, A.; Pirard, R.; Marien, J.; Kocon, L.; Rouzaud, J.-N.; Béguin, F.; Pirard, J.-P. Carbon aerogels, cryogels and xerogels: Influence of the drying method on the textural properties of porous carbon materials. *Carbon* **2005**, *43*, 2481–2494. [CrossRef]
15. Michalik, R.; Wandzik, I. A Mini-Review on Chitosan-Based Hydrogels with Potential for Sustainable Agricultural Applications. *Polymers* **2020**, *12*, 2425. [CrossRef]
16. Zhu, T.; Mao, J.; Cheng, Y.; Liu, H.; Lv, L.; Ge, M.; Li, S.; Huang, J.; Chen, Z.; Li, H.; et al. Recent Progress of Polysaccharide-Based Hydrogel Interfaces for Wound Healing and Tissue Engineering. *Adv. Mater. Interfaces* **2019**, *6*, 1900761. [CrossRef]
17. Dragan, E.S.; Dinu, M.V. Progress in Polysaccharide/Zeolites and Polysaccharide Hydrogel Composite Sorbents and Their Applications in Removal of Heavy Metal Ions and Dyes. *Curr. Green Chem.* **2015**, *2*, 342–353. [CrossRef]
18. Hemmati, F.; Jafari, S.M.; Taheri, R.A. Optimization of homogenization-sonication technique for the production of cellulose nanocrystals from cotton linter. *Int. J. Biol. Macromol.* **2019**, *137*, 374–381. [CrossRef]
19. Marin, D.C.; Vecchio, A.; Ludueña, L.N.; Fasce, D.; Alvarez, V.A.; Stefani, P.M. Revalorization of Rice Husk Waste as a Source of Cellulose and Silica. *Fibers Polym.* **2015**, *16*, 285–293. [CrossRef]
20. Gullo, M.; La China, S.; Falcone, P.M.; Giudici, P. Biotechnological production of cellulose by acetic acid bacteria: Current state and perspectives. *Appl. Microbiol. Biotechnol.* **2018**, *102*, 6885–6898. [CrossRef]
21. Rincon-Iglesias, M.; Lizundia, E.; Lanceros-Méndez, S. Water-Soluble Cellulose Derivatives as Suitable Matrices for Multifunctional Materials. *Biomacromolecules* **2019**, *20*, 2786–2795. [CrossRef] [PubMed]
22. Kareem, S.A.; Dere, I.; Gungula, D.T.; Andrew, F.P.; Saddiq, A.M.; Adebayo, E.F.; Tame, V.T.; Kefas, H.M.; Joseph, J.; Patrick, D.O. Synthesis and Characterization of Slow-Release Fertilizer Hydrogel Based on Hydroxy Propyl Methyl Cellulose, Polyvinyl Alcohol, Glycerol and Blended Paper. *Gels* **2021**, *7*, 262. [CrossRef] [PubMed]
23. Wang, F.D.; Li, J.; Su, Y.; Li, Q.; Gao, B.Y.; Yue, Q.Y.; Zhou, W.Z. Adsorption and recycling of Cd(II) from wastewater using straw cellulose hydrogel beads. *J. Ind. Eng. Chem.* **2019**, *80*, 361–369. [CrossRef]
24. Sun, Z.Z.; Li, Z.X.; Li, W.Z.; Bian, F.G. Mesoporous cellulose/TiO$_2$/SiO$_2$/TiN-based nanocomposite hydrogels for efficient solar steam evaporation: Low thermal conductivity and high light-heat conversion. *Cellulose* **2020**, *27*, 481–491. [CrossRef]
25. Zhao, H.; Li, Y. Removal of heavy metal ion by floatable hydrogel and reusability of its waste material in photocatalytic degradation of organic dyes. *J. Environ. Chem. Eng.* **2021**, *9*, 105316. [CrossRef]
26. Luo, Y.; Huang, J.G. Hierarchical-Structured Anatase-Titania/Cellulose Composite Sheet with High Photocatalytic Performance and Antibacterial Activity. *Chem. A Eur. J.* **2015**, *21*, 2568–2575. [CrossRef]
27. Dhanya, G.; Palanisamy Uma, M.; Khadar Mohamed Meera Sherifffa, B.; Gangasalam, A. Biomass-Derived Dialdehyde Cellulose Cross-linked Chitosan-Based Nanocomposite Hydrogel with Phytosynthesized Zinc Oxide Nanoparticles for Enhanced Curcumin Delivery and Bioactivity. *J. Agric. Food Chem.* **2019**, *67*, 10880–10890.
28. Azarniya, A.; Tamjid, E.; Eslahi, N.; Simchi, A. Modification of bacterial cellulose/keratin nanofibrous mats by a tragacanth gum-conjugated hydrogel for wound healing. *Int. J. Biol. Macromol.* **2019**, *134*, 280–289. [CrossRef]
29. Dorishetty, P.; Balu, R.; Athukoralalage, S.S.; Greaves, T.L.; Mata, J.; De Campo, L.; Saha, N.; Zannettino, A.C.W.; Dutta, N.K.; Choudhury, N.R. Tunable biomimetic hydrogels from silk fibroin and nanocellulose. *ACS Sustain. Chem. Eng.* **2020**, *8*, 2375–2389. [CrossRef]
30. Enawgaw, H.; Tesfaye, T.; Yilma, K.T.; Limeneh, D.Y. Synthesis of a Cellulose-Co-AMPS Hydrogel for Personal Hygiene Applications Using Cellulose Extracted from Corncobs. *Gels* **2021**, *7*, 236. [CrossRef]
31. Arias, S.L.; Cheng, M.K.; Civantos, A.; Devorkin, J.; Jaramillo, C.; Allain, J.P. Ion-Induced Nanopatterning of Bacterial Cellulose Hydrogels for Biosensing and Anti-Biofouling Interfaces. *ACS Appl. Nano Mater.* **2020**, *3*, 6719–6728. [CrossRef]
32. Yu, J.; Wang, A.C.; Zhang, M.; Lin, Z. Water treatment via non-membrane inorganic nanoparticles/cellulose composites. *Mater. Today* **2021**, *50*, 329–357. [CrossRef]
33. Bao, Y.; He, J.; Song, K.; Guo, J.; Zhou, X.; Liu, S. Functionalization and Antibacterial Applications of Cellulose-Based Composite Hydrogels. *Polymers* **2022**, *14*, 769. [CrossRef]
34. Gnanasambandam, R.; Proctor, A. Soy hull as an adsorbent source in processing soy oil. *J. Am. Oil Chem. Soc.* **1997**, *74*, 685–692. [CrossRef]
35. Bendahou, A.; Hajlane, A.; Dufresne, A.; Boufi, S.; Kaddami, H. Esterification and amidation for grafting long aliphatic chains on to cellulose nanocrystals: A comparative study. *Res. Chem. Intermed.* **2015**, *41*, 4293–4310. [CrossRef]
36. Sim, K.; Youn, H.J.; Jo, Y. Surface modification of cellulose nanofibrils by carboxymethylation and tempo-mediated oxidation. *Palpu Chongi Gisul/J. Korea Tech. Assoc. Pulp Pap. Ind.* **2015**, *47*, 42–52. [CrossRef]
37. Han, S.; Wang, T.; Li, B. Preparation of a hydroxyethyl-titanium dioxide-carboxymethyl cellulose hydrogel cage and its effect on the removal of methylene blue. *J. Appl. Polym. Sci.* **2017**, *134*, 44925. [CrossRef]

38. Thomas, M.; Naikoo, G.A.; Sheikh, M.U.D.; Bano, M.; Khan, F. Effective photocatalytic degradation of Congo red dye using alginate/carboxymethyl cellulose/TiO$_2$ nanocomposite hydrogel under direct sunlight irradiation. *J. Photochem. Photobiol. A Chem.* **2016**, *327*, 33–43. [CrossRef]
39. Jiao, Y.; Wan, C.; Li, J. Hydrothermal synthesis of SnO$_2$-ZnO aggregates in cellulose aerogels for photocatalytic degradation of rhodamine b. *Cellul. Chem. Technol.* **2018**, *52*, 481–491.
40. Hu, Y.; Li, N.; Yue, P.; Chen, G.; Hao, X.; Bian, J.; Peng, F. Highly antibacterial hydrogels prepared from amino cellulose, dialdehyde xylan, and Ag nanoparticles by a green reduction method. *Cellulose* **2022**, *29*, 1055–1067. [CrossRef]
41. Obst, M.; Heinze, T. Simple synthesis of reactive and nanostructure forming hydrophobic amino cellulose derivatives. *Macromol. Mater. Eng.* **2016**, *301*, 65–70. [CrossRef]
42. Ansari, M.M.; Ahmad, A.; Kumar, A.; Alam, P.; Khan, T.H.; Jayamurugan, G.; Raza, S.S.; Khan, R. Aminocellulose-grafted-polycaprolactone coated gelatin nanoparticles alleviate inflammation in rheumatoid arthritis: A combinational therapeutic approach. *Carbohydr. Polym.* **2021**, *258*, 117600. [CrossRef]
43. Stöhr, M.; Sadhukhan, M.; Al-Hamdani, Y.S.; Hermann, J.; Tkatchenko, A. Coulomb interactions between dipolar quantum fluctuations in van der waals bound molecules and materials. *Nat. Commun.* **2021**, *12*, 1–9. [CrossRef]
44. Li, S.; Hao, X.; Dai, X.; Tao, T. Rapid photocatalytic degradation of pollutant from water under UV and sunlight via cellulose nanofiber aerogel wrapped by TiO$_2$. *J. Nanomater.* **2018**, *2018*, 1–12. [CrossRef]
45. Habibi, S.; Jamshidi, M. Synthesis of TiO$_2$ nanoparticles coated on cellulose nanofibers with different morphologies: Effect of the template and sol-gel parameters. *Mater. Sci. Semicond. Process.* **2020**, *109*, 104927. [CrossRef]
46. Qin, C.C.; Li, S.J.; Jiang, G.Q.; Cao, J.; Guo, Y.L.; Li, J.W.; Zhang, B.; Han, S.S. Preparation of Flower-like ZnO Nanoparticles in a Cellulose Hydrogel Microreactor. *BioResources* **2017**, *12*, 3182–3191. [CrossRef]
47. Sabbaghan, M.; Argyropoulos, D.S. Synthesis and characterization of nano fibrillated cellulose/Cu$_2$O films; micro and nano particle nucleation effects. *Carbohydr. Polym.* **2018**, *197*, 614–622. [CrossRef]
48. Qi, X.; Guan, Y.; Chen, G.; Zhang, B.; Ren, J.; Peng, F.; Sun, R. A non-covalent strategy for montmorillonite/xylose self-healing hydrogels. *RSC Adv.* **2015**, *5*, 41006–41012. [CrossRef]
49. Gu, Y.; Zhang, L.; Du, X.; Fan, Z.; Wang, L.; Sun, W.; Cheng, Y.; Zhu, Y.; Chen, C. Reversible physical crosslinking strategy with optimal temperature for 3D bioprinting of human chondrocyte-laden gelatin methacryloyl bioink. *J. Biomater. Appl.* **2018**, *33*, 609–618. [CrossRef]
50. Jo, S.; Oh, Y.; Park, S.; Kan, E.; Lee, S.H. Cellulose/carrageenan/TiO$_2$ nanocomposite for adsorption and photodegradation of cationic dye. *Biotechnol. Bioprocess Eng.* **2017**, *22*, 734–738. [CrossRef]
51. Su, X.; Liao, Q.; Liu, L.; Meng, R.; Qian, Z.; Gao, H.; Yao, J. Cu2O nanoparticle -functionalized cellulose-based aerogel as high-performance visible-light photocatalyst. *Cellulose* **2017**, *24*, 1017–1029. [CrossRef]
52. Volokhova, A.A.; Kudryavtseva, V.L.; Spiridonova, T.I.; Kolesnik, I.; Goreninskii, S.I.; Sazonov, R.V.; Remnev, G.E.; Tverdokhlebov, S.I. Controlled drug release from electrospun PCL non-woven scaffolds via multi-layering and e-beam treatment. *Mater. Today Commun.* **2021**, *26*, 102134. [CrossRef]
53. Ishak, W.H.W.; Jia, O.Y.; Ahmad, I. pH-responsive gamma-irradiated poly(acrylic acid)-cellulose-nanocrystal-reinforced hydrogels. *Polymers* **2020**, *12*, 1932. [CrossRef]
54. Liu, G.; Li, T.T.; Song, X.F.; Yang, J.Y.; Qin, J.T.; Zhang, F.F.; Wang, Z.X.; Chen, H.G.; Wu, M.H.; Li, Y.S. Thermally driven characteristic and highly photocatalytic activity based on N-isopropyl acrylamide/high-substituted hydroxypropyl cellulose/g-C$_3$N$_4$ hydrogel by electron beam pre-radiation method. *J. Thermoplast. Compos. Mater.* **2020**, *33*, 089270572094421. [CrossRef]
55. Mhsin, I.; Maria, Z. Titanium dioxide nanostructures as efficient photocatalyst: Progress, challenges and perspective. *Int. J. Energy Res.* **2021**, *45*, 3569–3589.
56. Dearfield, K.L.; Harrington-Brock, K.; Doerr, C.L.; Rabinowitz, J.R.; Moore, M.M. Genotoxicity in mouse lymphoma cells of chemicals capable of Michael addition. *Mutagenesis* **1991**, *6*, 519–525. [CrossRef]
57. Marcı, G.; Mele, G.; Palmisano, L.; Pulito, P.; Sannino, A. Environmentally sustainable production of cellulose-based superabsorbent hydrogels. *Green Chem.* **2006**, *8*, 439–444.
58. Melone, L.; Altomare, L.; Alfieri, I.; Lorenzi, A.; De Nardo, L.; Punta, C. Ceramic aerogels from TEMPO-oxidized cellulose nanofibre templates: Synthesis, characterization, and photocatalytic properties. *J. Photochem. Photobiol. A Chem.* **2013**, *261*, 53–60. [CrossRef]
59. Wang, Q.; Wang, Y.; Chen, L.; Cai, J.; Zhang, L. Facile construction of cellulose nanocomposite aerogel containing TiO$_2$ nanoparticles with high content and small size and their applications. *Cellulose* **2017**, *24*, 2229–2240. [CrossRef]
60. Yue, Y.; Wang, X.; Wu, Q.; Han, J.; Jiang, J. Highly recyclable and super-tough hydrogel mediated by dual-functional TiO$_2$ nanoparticles toward efficient photodegradation of organic water pollutants. *J. Colloid Interface Sci.* **2020**, *564*, 99–112. [CrossRef]
61. Rahman, K.U.; Ferreira-Neto, E.P.; Rahman, G.U.; Parveen, R.; Monteiro, A.S.; Rahman, G.; Van Le, Q.; Domeneguetti, R.R.; Ribeiro, S.J.L.; Ullah, S. Flexible bacterial cellulose-based BC-SiO2-TiO2-Ag membranes with self-cleaning, photocatalytic, antibacterial and UV-shielding properties as a potential multifunctional material for combating infections and environmental applications. *J. Environ. Chem. Eng.* **2021**, *9*, 104708. [CrossRef]
62. Li, K.; Zhang, X.; Qin, Y.; Li, Y. Construction of the Cellulose Nanofibers (CNFs) Aerogel Loading TiO$_2$ NPs and Its Application in Disposal of Organic Pollutants. *Polymers* **2021**, *13*, 1841. [CrossRef]

63. Hasanpour, M.; Motahari, S.; Jing, D.; Hatami, M. Investigation of the Different Morphologies of Zinc Oxide (ZnO) in Cellulose/ZnO Hybrid Aerogel on the Photocatalytic Degradation Efficiency of Methyl Orange. *Top. Catal.* **2021**, *64*, 1–14. [CrossRef]
64. Wang, J.; Li, X.; Cheng, Q.; Lv, F.; Chang, C.; Zhang, L. Construction of β-FeOOH@tunicate cellulose nanocomposite hydrogels and their highly efficient photocatalytic properties. *Carbohydr. Polym.* **2019**, *229*, 115470. [CrossRef]
65. Zhu, Z.; Qu, J.; Hao, S.; Han, S.; Jia, K.; Yu, Z. Alpha-Fe2O3 Nanodisk/Bacterial Cellulose Hybrid Membranes as High-Performance Sulfate-Radical-Based Visible Light Photocatalysts under Stirring/Flowing States. *ACS Appl. Mater. Interfaces* **2018**, *10*, 30670–30679. [CrossRef]
66. Liu, Y.; Chen, Y.; Chen, Z.; Qi, H. A novel cellulose-derived carbon aerogel@Na$_2$Ti$_3$O$_7$ composite for efficient photocatalytic degradation of methylene blue. *J. Appl. Polym. Sci.* **2021**, *138*, 5134. [CrossRef]
67. Hoa, N.T.; Tien, N.T.; Phong, L.H.; Tai, V.V.; Hung, N.V.; Nui, P.X. Photocatalytic activity of Ag-Ag$_3$PO$_4$/Cellulose aerogel composite for degradation of dye pollutants under visible light irradiation. *Vietnam. J. Catal. Adsorpt.* **2021**, *10*, 6–17.
68. Wang, L.; Chen, S.; Wu, P.; Wu, K.; Wu, J.; Meng, G.; Hou, J.; Liu, Z.; Guo, X. Enhanced optical absorption and pollutant adsorption for photocatalytic performance of three-dimensional porous cellulose aerogel with BiVO$_4$ and PANI. *J. Mater. Res.* **2020**, *35*, 1316–1328. [CrossRef]
69. Ferreira-Neto, E.P.; Ullah, S.; Silva, T.C.A.; Domeneguetti, R.R.; Perrissinotto, A.P.; Vicente, F.S.; Rodrigues-Filho, U.P.; Ribeiro, S.J.L. Bacterial Nanocellulose/MoS$_2$ Hybrid Aerogels as Bifunctional Adsorbent/Photocatalyst Membranes for in-Flow Water Decontamination. *ACS Appl. Mater. Interfaces* **2020**, *12*, 41627–41643. [CrossRef]
70. Wang, S.; Zhu, B.; Liu, M.; Zhang, L.; Yu, J.; Zhou, M. Direct Z-scheme ZnO/CdS hierarchical photocatalyst for enhanced photocatalytic H2-production activity. *Appl. Catal. B: Environ.* **2019**, *243*, 19–26. [CrossRef]
71. Qian, X.; Xu, Y.; Yue, X.; Wang, C.; Liu, M.; Duan, C.; Xu, Y.; Zhu, C.; Dai, L. Microwave-assisted solvothermal in-situ synthesis of CdS nanoparticles on bacterial cellulose matrix for photocatalytic application. *Cellulose* **2020**, *27*, 5939–5954. [CrossRef]
72. Wu, Y.; Gao, J.; Hao, C.; Mei, S.; Yang, J.; Wang, X.; Zhao, R.; Zhai, X.; Liu, Y. Easily recoverable CdxZn1-xS-Gel photocatalyst with a tunable band structure for efficient and stable H2 production mediated by visible light. *Cellulose* **2018**, *25*, 167–177. [CrossRef]
73. Ghosh, S.K.; Pal, T. Interparticle Coupling Effect on the Surface Plasmon Resonance of Gold Nanoparticles: From Theory to Applications. *Chem. Rev.* **2007**, *107*, 4797–4862. [CrossRef] [PubMed]
74. Heidarpour, H.; Golizadeh, M.; Padervand, M.; Karimi, A.; Vossoughi, M.; Tavakoli, M.H. In-situ formation and entrapment of Ag/AgCl photocatalyst inside cross-linked carboxymethyl cellulose beads: A novel photoactive hydrogel for visible-light-induced photocatalysis. *J. Photochem. Photobiol. A Chem.* **2020**, *398*, 112559. [CrossRef]
75. Ong, W.J.; Tan, L.L.; Ng, Y.H.; Yong, S.T.; Chai, S.P. Graphitic Carbon Nitride (g-C3N4)-Based Photocatalysts for Artificial Photosynthesis and Environmental Remediation: Are We a Step Closer To Achieving Sustainability? *Chem. Rev.* **2016**, *116*, 7159–7329. [CrossRef]
76. Jiang, T.; Liu, S.; Gao, Y.; Rony, A.H.; Fan, M.; Tan, G. Surface modification of porous g-C3N4 materials using a waste product for enhanced photocatalytic performance under visible light. *Green Chem.* **2019**, *21*, 5934–5944. [CrossRef]
77. Chen, S.; Lu, W.; Han, J.; Zhong, H.; Xu, T.; Wang, G.; Chen, W. Robust three-dimensional g-C3N4@cellulose aerogel enhanced by cross-linked polyester fibers for simultaneous removal of hexavalent chromium and antibiotics. *Chem. Eng. J.* **2019**, *359*, 119–129. [CrossRef]
78. Bai, W.; Yang, X.; Du, X.; Qian, Z.; Zhang, Y.; Liu, L.; Yao, J. Robust and recyclable macroscopic g-C3N4/cellulose hybrid photocatalysts with enhanced visible light photocatalytic activity. *Appl. Surf. Sci.* **2019**, *504*, 144179. [CrossRef]
79. Qi, H.; Ji, X.; Shi, C.; Ma, R.; Huang, Z.; Guo, M.; Li, J.; Guo, Z. Bio-templated 3D porous graphitic carbon nitride hybrid aerogel with enhanced charge carrier separation for efficient removal of hazardous organic pollutants. *J. Colloid Interface Sci.* **2019**, *556*, 366–375. [CrossRef]
80. Yao, M.; Wang, Z.; Liu, Y.; Yang, G.; Chen, J. Preparation of dialdehyde cellulose graftead graphene oxide composite and its adsorption behavior for heavy metals from aqueous solution. *Carbohydr. Polym.* **2019**, *212*, 345–351. [CrossRef]
81. Liu, T.; Wang, Z.; Wang, X.; Yang, G.; Liu, Y. Adsorption-photocatalysis performance of polyaniline/dicarboxyl acid cellulose@graphene oxide for dye removal. *Int. J. Biol. Macromol.* **2021**, *182*, 492–501. [CrossRef] [PubMed]
82. Zhang, H.; Zhou, L.; Li, J.; Rong, S.; Jiang, J.; Liu, S. Photocatalytic Degradation of Tetracycline by a Novel (CMC)/MIL-101(Fe)/β-CDP Composite Hydrogel. *Front. Chem.* **2021**, *8*, 593730. [CrossRef] [PubMed]
83. Levec, J.; Pintar, A. Catalytic wet-air oxidation processes: A review. *Catal. Today* **2007**, *124*, 172–184. [CrossRef]
84. Li, Y.; Zhao, K.; Yang, W.; Chen, G.; Zhang, X.; Zhao, Y.; Liu, L.; Chen, M. Efficient removal of Cd$^{2+}$ ion from water by calcium alginate hydrogel filtration membrane. *Water Sci. Technol.* **2017**, *75*, 2322–2330. [CrossRef]
85. Ma, Z.; Zhou, P.; Zhang, L.; Zhong, Y.; Sui, X.; Wang, B.; Ma, Y.; Feng, X.; Xu, H.; Mao, Z. A recyclable 3D g-C3N4 based nanocellulose aerogel composite for photodegradation of organic pollutants. *Cellulose* **2021**, *28*, 3531–3547. [CrossRef]
86. Lu, Y.; Sun, Q.F.; Li, J.; Liu, Y. Fabrication, Characterization and Photocatalytic Activity of TiO$_2$/Cellulose Composite Aerogel. *Key Eng. Mater.* **2014**, *609–610*, 542–546. [CrossRef]
87. Tang, L.; Fu, T.; Li, M.; Li, L. Facile synthesis of Ag@AgCl-contained cellulose hydrogels and their application. *Colloids Surf. A Physicochem. Eng. Asp.* **2018**, *553*, 618–623. [CrossRef]
88. Hasanpour, M.; Motahari, S.; Jing, D.; Hatami, M. Statistical analysis and optimization of photodegradation efficiency of methyl orange from aqueous solution using cellulose/zinc oxide hybrid aerogel by response surface methodology (RSM). *Arab. J. Chem.* **2021**, *14*, 103401. [CrossRef]

89. Brandes, R.; Trindade, E.C.A.; Vanin, D.F.; Vargas, V.M.M.; Carminatti, C.A.; AL-Qureshi, H.A.; Recouvreux, D.O. Spherical Bacterial Cellulose/TiO$_2$ Nanocomposite with Potential Application in Contaminants Removal from Wastewater by Photocatalysis. *Fibers Polym.* **2018**, *19*, 1861–1868. [CrossRef]
90. Jiao, Y.; Wan, C.; Li, J. Room-temperature embedment of anatase titania nanoparticles into porous cellulose aerogels. *Appl. Phys.* **2015**, *120*, 341–347. [CrossRef]
91. Li, M.; Qiu, J.; Xu, J.; Yao, J. Cellulose/TiO$_2$-based carbonaceous composite film and aerogel for high-efficient photocatalysis under visible light. *Ind. Eng. Chem. Res.* **2020**, *59*, 13997–14003. [CrossRef]
92. Zhang, Q.; Cheng, Y.; Fang, C.; Shi, J.; Chen, J.; Han, H. Novel and multifunctional adsorbent fabricated by Zeolitic imidazolate framworks-8 and waste cigarette filters for wastewater treatment: Effective adsorption and photocatalysis. *J. Solid State Chem.* **2021**, *299*, 122190. [CrossRef]
93. Zheng, A.L.T.; Sabidi, S.; Ohno, T.; Maeda, T.; Andou, Y. Cu$_2$O/TiO$_2$ decorated on cellulose nanofiber/reduced graphene hydrogel for enhanced photocatalytic activity and its antibacterial applications. *Chemosphere* **2022**, *286*, 131731. [CrossRef] [PubMed]
94. Du, X.; Wang, Z.; Pan, J.; Gong, W.; Liao, Q.; Liu, L.; Yao, J. High photocatalytic activity of Cu@Cu$_2$O/RGO/cellulose hybrid aerogels as reusable catalysts with enhanced mass and electron transfer. *React. Funct. Polym.* **2019**, *138*, 79–87.
95. Qiu, J.; Fan, P.; Yue, C.; Liu, F.; Li, A. Multi-networked nanofibrous aerogel supported by heterojunction photocatalysts with excellent dispersion and stability for photocatalysis. *J. Mater. Chem. A* **2019**, *7*, 7053–7064. [CrossRef]
96. Xue, C.; Zheng, C.; Zhao, Q.; Sun, S. Occurrence of antibiotics and antibiotic resistance genes in cultured prawns from rice-prawn co-culture and prawn monoculture systems in China. *Sci. Total Environ.* **2022**, *806*, 150307. [CrossRef]
97. Amaly, N.; EL-Moghazy, A.Y.; Nitin, N.; Sun, G.; Pandey, P.K. Synergistic adsorption-photocatalytic degradation of tetracycline by microcrystalline cellulose composite aerogel dopped with montmorillonite hosted methylene blue. *Chem. Eng. J.* **2022**, *430*, 133077. [CrossRef]
98. Yue, Y.; Shen, S.; Cheng, W.; Han, G.; Wu, Q.; Jiang, J. Construction of mechanically robust and recyclable photocatalytic hydrogel based on nanocellulose-supported CdS/MoS$_2$/Montmorillonite hybrid for antibiotic degradation. *Colloids Surf. A Physicochem. Eng. Asp.* **2022**, *636*, 128035. [CrossRef]
99. Montgomery, M.A.; Elimelech, M. Water and Sanitation in Developing Countries: Including Health in the Equation. *Environ. Sci. Technol.* **2007**, *41*, 17–24. [CrossRef]
100. Zhang, H.; Zhu, J.; Hu, Y.; Chen, A.; Zhou, L.; Gao, H.; Liu, Y.; Liu, S. Study on Photocatalytic Antibacterial and Sustained-Release Properties of Cellulose/TiO$_2$/β-CD Composite Hydrogel. *J. Nanomater.* **2019**, *2019*, 1–12. [CrossRef]
101. Li, W.; Zhang, H.; Hong, M.; Zhang, L.; Feng, X.; Shi, M.; Hu, W.; Mu, S. Defective RuO$_2$/TiO$_2$ nano-heterostructure advances hydrogen production by Electrochemical Water Splitting. *Chem. Eng. J.* **2021**, *431*, 134072. [CrossRef]
102. Zhong, D.; Li, J.; Ma, W.; Xin, H. Magnetite nanoparticles enhanced glucose anaerobic fermentation for bio-hydrogen production using an expanded granular sludge bed (EGSB) reactor. *Int. J. Hydrog. Energy* **2020**, *45*, 10664–10672. [CrossRef]
103. Kang, H.; Gao, J.; Xie, M.; Sun, Y.; Wu, F.; Gao, C.; Liu, Y.; Qiu, H. Carboxymethyl cellulose gel membrane loaded with nanoparticle photocatalysts for hydrogen production. *Int. J. Hydrog. Energy* **2019**, *44*, 13011–13021. [CrossRef]
104. Xie, Y.; Pan, Y.; Cai, P. Cellulose-based antimicrobial films incorporated with ZnO nanopillars on surface as biodegradable and antimicrobial packaging. *Food Chem.* **2021**, *368*, 130784. [CrossRef] [PubMed]

Article

# Properties of Cellulose Pulp and Polyurethane Composite Films Fabricated with Curcumin by Using NMMO Ionic Liquid

Chaehyun Jo [1,†], Sam Soo Kim [1,†], Balasubramanian Rukmanikrishnan [1,*], Srinivasan Ramalingam [2], Prabakaran D. S. [3,4] and Jaewoong Lee [1,*]

1. Department of Fiber System Engineering, Yeungnam University, 280 Daehak-Ro, Gyeongsan 38541, Korea; jo1114@ynu.ac.kr (C.J.); sskim@yu.ac.kr (S.S.K.)
2. Department of Food Science and Technology, Yeungnam University, 280 Daehak-Ro, Gyeongsan 38541, Korea; sribt27@gmail.com
3. Department of Radiation Oncology, College of Medicine, Chungbuk National University, Chungdae-ro 1, Seowon-gu, Cheongju 28644, Korea; prababio@gmail.com
4. Department of Biotechnology, Ayya Nadar Janaki Ammal College (Autonomous), Srivilliputhur Main Road, Sivakasi 626124, Tamil Nadu, India
* Correspondence: rukmibala@gmail.com (B.R.); jaewlee@yu.ac.kr (J.L.); Tel.: +82-53-810-2786 (J.L.); Fax: +82-53-810-4685 (J.L.)
† These authors contributed equally to this work.

**Abstract:** Cellulose pulp (CP), polyurethane (PU), and curcumin-based biocompatible composite films were prepared using a simple cost-effective method. Significant structural and microstructural changes were studied using FT-IR spectroscopy, XRD, and SEM. The 5% and 10% gravimetric losses of the CP/PU/curcumin composite were found to be in the range 87.2–182.3 °C and 166.7–249.8 °C, respectively. All the composites exhibited single $T_g$ values in the range 147.4–154.2 °C. The tensile strength of CP was measured to be 93.2 MPa, which dropped to 14.1 MPa for the 1:0.5 CP/PU composite and then steadily increased to 30.5 MPa with further addition of PU. The elongation at the break of the composites decreased from 8.1 to 3.7% with the addition of PU. The addition of PU also improved the water vapor permeability ($3.96 \times 10^{-9}$ to $1.75 \times 10^{-9}$ g m$^{-1}$ s$^{-1}$ Pa$^{-1}$) and swelling ratio (285 to 202%) of the CP composite films. The CP/PU/curcumin composite exhibited good antioxidant activity and no cytotoxicity when tested on the HaCat cell line. The visual appearance and UV transmittance (86.2–32.9% at 600 nm) of the CP composite films were significantly altered by the incorporation of PU and curcumin. This study demonstrates that CP/PU/curcumin composites can be used for various packaging and biomedical applications.

**Keywords:** cellulose pulp; polyurethane; curcumin; antioxidant properties; non-cytotoxicity; barrier properties; mechanical properties

## 1. Introduction

In recent years, sustainable and environmentally friendly materials have garnered significant attention in both scientific and industrial research fields. This is because they are abundant; low-cost; present low risk to animals, humans, and other living organisms; carbon neutral, biodegradable and biocompatible [1–5]. These concerns and challenges have led to the development of novel environment-friendly and biodegradable packaging materials. Among natural biopolymer resources, biomass has evolved as a value-added product for various industrial applications. Biomass-based nanocomposites are novel composites consisting of biodegradable biopolymers and bio-nanofillers. Recently, several researchers fabricated metal nanocomposites using different plant-based and natural biopolymers and revealed the least cytotoxicity and higher antimicrobial properties. Cellulose pulp (CP) is the most abundant, renewable, and inexpensive lignocellulosic material consisting of D-glucopyranose units, which is widely used in the paper and agricultural industries [6–9].

However, the brittleness and solubility of cellulose pulp restrict its application. Owing to its partly crystalline and partly amorphous structure, cellulose pulp is poorly soluble in many organic solvents [10–17]. A combination of 10% NaOH, NaOH/urea, and DMA/LiCl can dissolve cellulose pulp, but it requires a relatively lengthy pre-treatment process. Ionic liquids are promising green solvents for dissolving cellulose pulp and are rapidly replacing organic solvents. Moreover, they have been used to prepare various composites by blending cellulose, wood pulp, and chitosan [18,19].

The addition of organic polymers, inorganic nanoparticles, and natural fillers to polymer backbones has gained popularity because they help improve the thermal, mechanical, and water barrier properties of composites. For instance, polyurethane (PU) is a block copolymer consisting of both soft and hard segments. PU-based composites have been extensively studied and widely used in polymer and materials sciences owing to their excellent mechanical, gas barrier, and water barrier properties [20–25]. They have several biomedical applications, such as in the field of tissue engineering; they also have important medical and industrial applications because of their exceptional biocompatibility and biodegradability. A synergistic combination of cellulose pulp and PU has been shown to reduce the dependence on synthetic resources to a certain degree; in addition, such plant-based natural fiber-reinforced polymer matrix composites promote the use of renewable resources [26–30].

Turmeric (Curcuma longa) is a rhizomatous perennial plant. It is a rich source of phenolic compounds, such as curcuminoids. Curcumin (1,7-bis(4-hydroxy-3-methoxyphenyl)-1,6-heptadiene-3,5-dione) is a natural bioactive compound extracted from turmeric powder. It has a low molecular weight and a phenolic structure consisting of $\alpha,\beta$-unsaturated carbonyl groups. Curcumin has various therapeutic applications because of its antimicrobial, antioxidant, and anti-inflammatory properties. Moreover, it is compatible with different types of polymers [31–36]. We hypothesize that the incorporation of PU and curcumin into CP can improve the antioxidant, mechanical, and thermal properties of the resulting composite films.

To test our hypothesis, in this study, we prepared composites based on a synergistic novel combination of CP, PU, and curcumin by simple dissolution in N-methylmorpholine N-oxide (NMMO) solvent. We thoroughly investigated the thermal, mechanical, and water barrier properties of these sustainable composites by performing Fourier transform infrared (FT-IR) spectroscopy, X-ray diffraction (XRD), scanning electron microscopy (SEM), and UV spectroscopy. In addition, we examined the antioxidant activity and cytotoxicity of these CP/PU/curcumin composites. In this study, we highlight the properties that make these composites suitable for packaging applications.

## 2. Experimental Section

### 2.1. Materials

Cellulose pulp was purchased from Moorim P & P Co., Ltd., (Ulsan, South Korea) and curcumin was purchased from Sigma-Aldrich (Yongin, South Korea). NMMO and curcumin powder were purchased from Dajang Chemicals (Siheung, South Korea). Polyurethane (Elastollan® 1283 D11 U) was purchased from BASF (Seoul, South Korea). Water-soluble tetrazolium salt (WST-1) cat. #EZ-1000 cytox was purchased from Daeil Lab Service Co. (Seoul, South Korea), to measure cytotoxicity. Note that all the reagents were used as received without further purification.

### 2.2. Preparation of CP/PU/Curcumin Composite Films

First, CP (4 g), PU (4 g), and curcumin (10 wt%) were mixed thoroughly using a hand blender. Next, the CP/PU/curcumin mixture was dispersed in a round-bottom (RB) flask containing NMMO (92 g) and heated to 100 °C. The RB flask was tightly closed and dipped in an oil bath to maintain uniform temperature and avoid recrystallization of the NMMO solvent at the top of the flask. After 2.3 h, the RB flask was removed and kept in a vacuum oven at 100 °C for another 6 h to completely remove air bubbles. Subsequently, a viscous

CP/PU/curcumin solution at 80 °C was spread uniformly on a glass surface. The glass plate was immediately transferred to a coagulation bath containing water, where it was kept for 6 h and fresh water was transferred every 1 h without disturbing the film. Following the complete removal of the solvent from the CP/PU/curcumin composite film, it was transferred to another glass plate and dried at room temperature for 48 h. The thickness of the resulting transparent films was in the range of 0.07–0.10 mm. The prepared composite films were stored at 30 °C and 60 ± RH. All CP/PU/Curcumin composites were prepared using the above-mentioned procedure. A series of composite films were prepared using a similar method, such that they consisted of only CP; 1:0.5 CP/PU; 1:1 CP/PU and 1:1 CP/PU with 10 wt% curcumin (Table S1). The characterization methods of these composite films are provided in the Supporting Information

## 3. Results and Discussion

### 3.1. FT-IR Spectroscopy

The FT-IR analysis of the CP/PU/curcumin blend films was carried out to study the interaction between the polymeric materials, and the resulting FT-IR spectra were shown in (Figure 1). The broadband observed at 3100–3450 $cm^{-1}$ (for both CP and CP/PU composites) corresponds to the stretching vibration of the O–H bond. The peaks at 2893 $cm^{-1}$ and 1367 $cm^{-1}$ correspond to the stretching vibrations of the C–H bond and bending vibrations of the O–H bond in cellulose, respectively. The peak at 858 $cm^{-1}$ results from the vibrations of the C–H bond, which is commonly observed in neat cellulose pulp [37,38]. The band at 895 $cm^{-1}$ can be attributed to the stretching vibrations of C–O–C in the β-(1-4)-glycosidic linkages of cellulose, which is considered to be in the amorphous phase. PU added CP composites showed characteristic bands at 1697 $cm^{-1}$, 1313 $cm^{-1}$, and 1231 $cm^{-1}$ owing to the stretching vibrations of C=O, C–N, and C–O in the urethane group, respectively [39]. It also exhibited bands corresponding to the carbonyl (1650–1850 $cm^{-1}$), amide (1420–1650 $cm^{-1}$) and ether (910–1250 $cm^{-1}$) groups. The characteristic peaks of curcumin at 1652 $cm^{-1}$, 1532 $cm^{-1}$, and 1446 $cm^{-1}$ correspond to the vibrations of the carbonyl group C=O, stretching vibrations of the C–C bond in the benzene ring, and olefinic bending vibrations of the C–H bond in the benzene ring exhibited in 1:1 CP/PU with 10 wt% curcumin composites, respectively. The intensity and positions of the peaks in the 1:1 CP/PU composite spectrum were observed to change when 10 wt% curcumin was added, indicating that the CP, PU, and curcumin bonded with each other in the CP/PU/curcumin composite. Characteristic peaks at 1357 $cm^{-1}$ (C–O–H bending), 1017 $cm^{-1}$ (C–O stretching), and 1704 $cm^{-1}$ (PU-carbonyl group) shifted to lower wavenumbers, namely, 1372, 1056, and 1697 $cm^{-1}$, respectively. These results suggest that hydrogen bond interactions were present in the CP/PU/curcumin composite films.

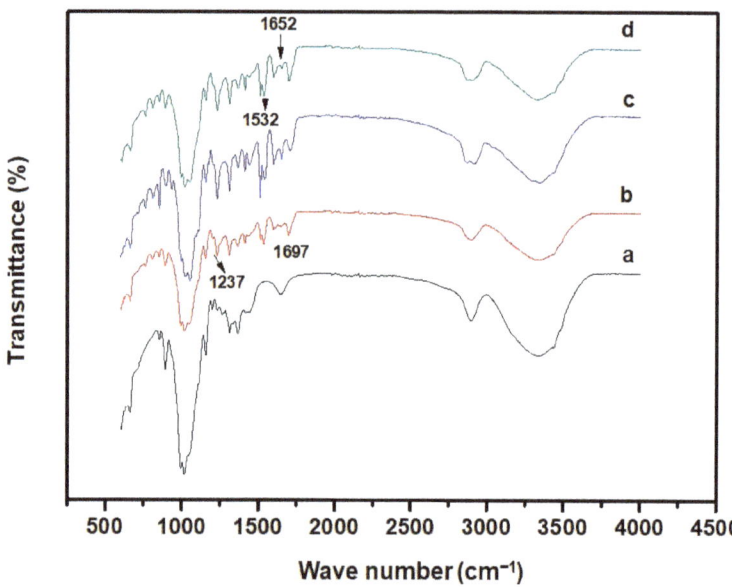

**Figure 1.** FT–IR spectra of composite films having: (**a**) only CP, (**b**) 1:0.5 CP/PU, (**c**) 1:1 CP/PU, and (**d**) 1:1 CP/PU with 10 wt% curcumin.

*3.2. XRD Analysis*

The miscibility and compatibility of the materials of each film were studied by XRD pattern. The reinforcement effect of PU on CP was characterized using XRD analysis, and the corresponding XRD patterns are shown in Figure 2. It is well known that the width and intensity of the peaks in an XRD pattern are related to the crystalline structure of the underling material. The XRD pattern of neat CP exhibited a peak of about $2\theta = 16–23°$, which indicates a crystalline phase. However, after the dissolution of PU, the structure of CP changed, and the diffraction peaks of the 1:0.5 CP/PU and 1:1 CP/PU composites were observed to shift. The amorphous nature of PU significantly altered the intensity of the XRD peaks and a broad peak was exhibited by the 1:1 CP/PU composite with 10 wt% curcumin. The crystallinity of the CP/PU composites gradually decreased owing to the dissolution of PU in the CP matrix. Broader peaks indicate an increase in the amorphous phase of the CP composites, which in turn increases the solubility of the composites in the NMMO solvent and helps form intermolecular hydrogen bonds between the components. The addition of PU to CP increased its amorphous nature such that the XRD peaks of the CP/PU composites were relatively broad with high intensity. Furthermore, peak shifts were observed in the XRD patterns of the CP/PU and CP/PU/curcumin composite films. The XRD pattern of curcumin showed intense diffraction peaks between 10° and 30° owing to its highly crystalline structure. The substantial decrease in the peak intensity of the CP/PU/curcumin composite clearly indicates the interaction between all the components in the polymer backbone. The absence of peaks corresponding to curcumin in the XRD pattern of 1:1 CP/PU composite with 10 wt% curcumin indicates that curcumin is properly combined with the CP/PU biopolymer matrix and is present in a low-order or amorphous state. These results confirm the proper dispersion of materials in the solvent and are in good agreement with the results of previous studies [6,35,40–43].

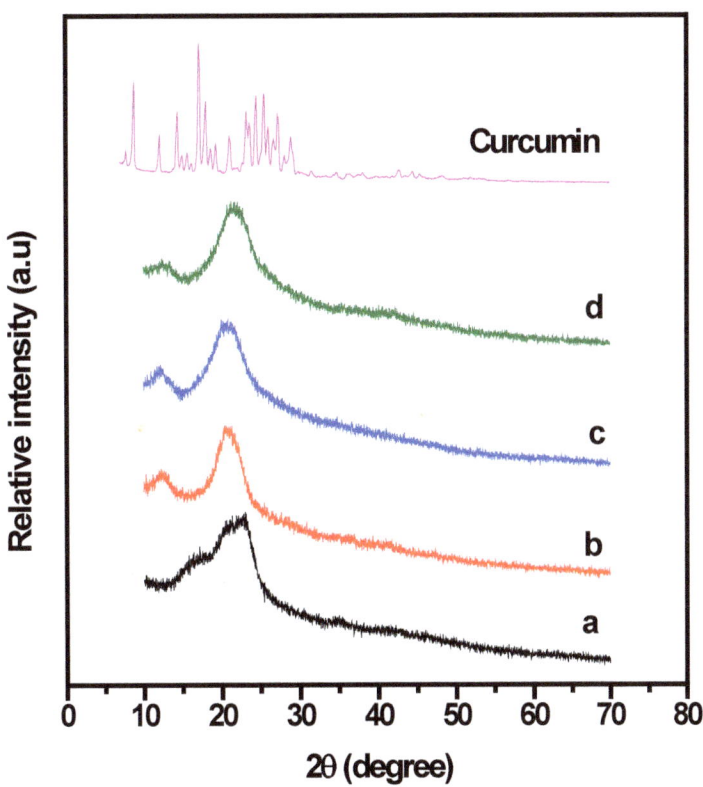

**Figure 2.** XRD patterns of composite films having: (**a**) only CP, (**b**) 1:0.5 CP/PU, (**c**) 1:1 CP/PU, and (**d**) 1:1 CP/PU with 10 wt% curcumin.

### 3.3. UV Spectroscopy

UV light protection is one of the properties in great demand in multifunctional materials. The thickness of the CP/PU/curcumin composite films was in the range of 0.07–0.1. The surface color and visual appearance of the CP films were affected owing to the addition of PU. The neat CP films appeared translucent, whereas the 1:0.5 CP/PU, 1:1 CP/PU, and 1:1 CP/PU with 10 wt% curcumin films appeared slightly opaque. The Hunter L* (lightness) value of the composites decreased, whereas the Hunter a* (redness) and b* (yellowness) values increased with increasing PU content. The neat CP composite film was colorless and transparent; however, the composite films containing PU and curcumin exhibited a yellow color. The UV transmittance of the CP/PU/curcumin composites as a function of wavelength is shown in Figure 3 (also, see Table 1). The UV transmittance of neat CP was 86.2% at 600 mm; however, it decreased significantly after the incorporation of PU. The transmittance of 1:0.5 CP/PU, 1:1 CP/PU, and 1:1 CP/PU with 10 wt% curcumin was 65.1, 41.2, and 32.9%, respectively. Note that the 1:1 CP/PU composite and 1:1 CP/PU composite with 10 wt% curcumin almost completely blocked the UV light below 400 mm. This indicates that the materials were well dispersed in the solvent and the resulting composites are suitable for packaging applications.

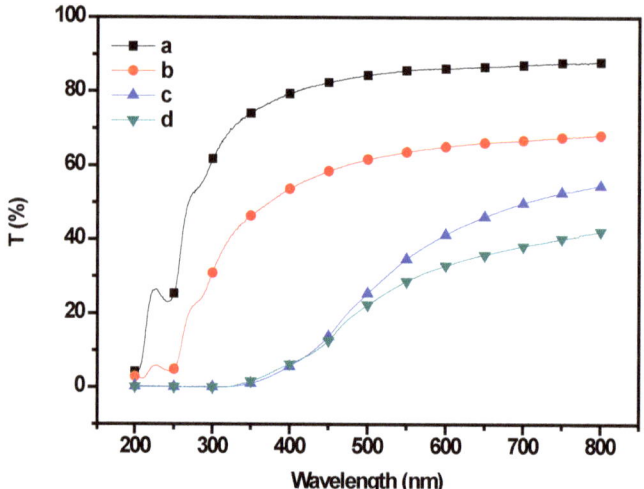

**Figure 3.** UV transmittance spectra of composite films having: (**a**) only CP, (**b**) 1:0.5 CP/PU, (**c**) 1:1 CP/PU, and (**d**) 1:1 CP/PU with 10 wt% curcumin.

**Table 1.** UV-transmittance, thickness, and color values of the CP/PU/curcumin composite films.

| Sample | Thickness (mm) | T (%) at 600 nm | Water Vapor Permeability ($\times 10^{-9}$ gm/m² Pas) | Water Contact Angle (°) | Swelling Ratio (%) |
|---|---|---|---|---|---|
| Only CP | 0.07 | 86.2 | 3.96 | 51.4 | 285 |
| 1:0.5 CP/PU | 0.09 | 65.1 | 2.34 | 55.6 | 262 |
| 1:1 CP/PU | 0.10 | 41.2 | 1.81 | 58.2 | 215 |
| 1:1 CP/PU with 10 wt% of curcumin | 0.09 | 32.9 | 1.75 | 60.1 | 202 |

*3.4. Water Vapor Permeability*

The water vapor permeability (WVP) is the most important ertyin food packaging application to control the moisture transfer between the food and the surrounding atmosphere. The WVP of the CP/PU/curcumin composites was measured to determine their water barrier properties, and the corresponding data are summarized in Table 1. The WVP of neat CP was measured to be $3.96 \times 10^{-9}$ g m$^{-1}$ s$^{-1}$ Pa$^{-1}$, which decreased to $2.34 \times 10^{-9}$ g m$^{-1}$ s$^{-1}$ Pa$^{-1}$ on the addition of PU. Note that both PU and curcumin are hydrophobic in nature; moreover, when PU and curcumin are dispersed throughout the CP matrix, a tortuous path for the diffusion of water molecules is created that increases the diffusion path length. The WVP of the 1:1 CP/PU composite was $1.81 \times 10^{-9}$ g m$^{-1}$ s$^{-1}$ Pa$^{-1}$, which decreased to $1.75 \times 10^{-9}$ g m$^{-1}$ s$^{-1}$ Pa$^{-1}$ with the addition of curcumin powder. Note, that the reduced hydrophilicity of the CP/PU/curcumin composites makes them suitable for packaging applications.

*3.5. Swelling Ratio*

The swelling behavior of the CP/PU/curcumin composites was studied in water at 25 °C. A low swelling ability is attributed to a high cross-link density of the composite films. This is because, at higher cross-link densities, water permeation into the matrix becomes more difficult, which in turn influences the swelling capacity of the gel. The SR of neat CP decreased from 285% to 262% with the addition of PU. Thus, the SR of neat CP was 32% higher than that of the 1:1 CP/PU composite. The SR further decreased to 202% for the CP/PU-1:1 composite with 10 wt% curcumin. Therefore, the hydrophobic nature of PU and curcumin fillers hinders the diffusion of the water molecules through the composite films and significantly influences the swelling properties of CP.

### 3.6. Contact Angle

The surface hydrophilicity/hydrophobicity of the composite films is an important determining factor for packaging applications. It is determined by measuring the water CA, whose values are listed in Table 1. The results indicate that the addition of PU increased the hydrophobicity of the composite films. The CA of neat CP was 51.4°, which increased to 55.6° and 58.2° for the 1:0.5 CP/PU and 1:1 CP/PU composites, respectively. Furthermore, the addition of curcumin slightly increased the CA of the 1:1 CP/PU composite film. This is owing to the hydrophobic nature of the curcumin fillers, which hinder the diffusion of water molecules through the composite films. This result is consistent with the SR values of the CP/PU/curcumin composites.

### 3.7. Thermogravimetric Analysis

The characteristic thermal stability of the CP/PU/curcumin composites was studied using TGA in an $O_2$ atmosphere and the results are presented in Figure 4 and Table 2. All composites exhibited three stages of weight loss. PU significantly improved the thermal stability of CP. The 5% and 10% gravimetric losses of the composites were in the range 87.2–182.3 °C and 166.7–249.8 °C, respectively. The initial degradation temperature of neat CP was 87.2 °C, which increased to 136.8, 172.1, and 182.3 °C for the 1:0.5 CP/PU, 1:1 CP/PU, and 1:1 CP/PU with 10 wt% curcumin composites, respectively. We found that adding PU to the CP matrix caused a shift in the first and third degradation temperatures towards higher temperatures (approximately 130–230 °C and 250–450 °C, respectively). However, the weight loss was nearly constant during the second degradation stage. The char yield of neat CP was 7.8%, which increased to 15.7% for the 1:1 CP/PU with 10 wt% curcumin composites. The improvement in the thermal stability of the CP/PU/curcumin composite can be ascribed to a confined network structure and uniform dispersion of the components in the polymer matrix. Furthermore, the incorporation of PU leads to synergistic interfacial interactions, which slows down the thermal decomposition of these composite films [28,39].

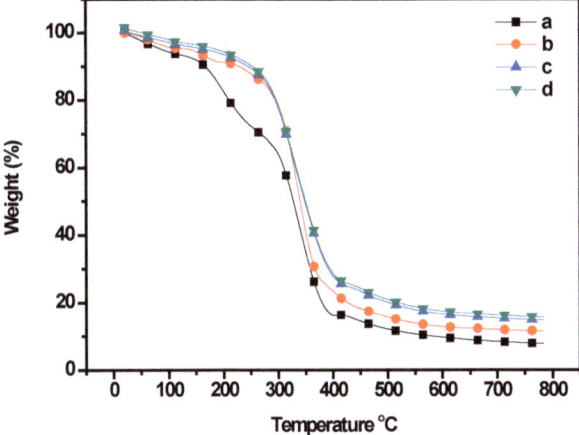

**Figure 4.** TGA curves of composites having: (**a**) only CP, (**b**) 1:0.5 CP/PU, (**c**) 1:1 CP/PU, and (**d**) 1:1 CP/PU with 10 wt% curcumin.

**Table 2.** Thermal properties, mechanical properties, WVP, and CA of the CP/PU/curcumin composite films.

| Sample | TGA | | | DSC | | Tensile Strength (MPa) | Elongation at Break (%) |
|---|---|---|---|---|---|---|---|
| | $T_{5\%}$ | $T_{10\%}$ | CY (%) | $T_g$ (°C) | $T_m$ (°C) | | |
| Only CP | 87.2 | 166.7 | 7.8 | 147.4 | 187.3 | 93.2 | 8.1 |
| 1:0.5 CP/PU | 136.8 | 227.7 | 11.5 | 151.2 | 191.1 | 14.1 | 3.7 |
| 1:1 CP/PU | 172.1 | 247.5 | 14.8 | 153.4 | 202.4 | 29.5 | 3.9 |
| 1:1 CP/PU with 10 wt% of curcumin | 182.3 | 249.8 | 15.7 | 154.2 | 204.2 | 30.5 | 3.7 |

*3.8. Differential Scanning Calorimetry*

The DSC thermogram revealed $T_g$ and $T_m$ of the CP/PU/curcumin composites and the results are summarized in Table 2 and Figure 5. The addition of PU seemed to improve $T_g$ and $T_m$ of neat CP. The 1:1 CP/PU with 10 wt% curcumin composite exhibited the highest $T_g$ (154.2 °C) and $T_m$ (204.2 °C) values, which is owing to the higher order interactions and relaxation mechanisms in this composite. All composite films exhibited single $T_g$ and $T_m$ peaks. However, the intensity of the $T_m$ peaks decreased with the addition of PU to the composites. The significant improvement in the $T_g$ and $T_m$ values indicates the restricted molecular motion in the composite network, which is formed by the covalent cross-linking and strong hydrogen bonding between the CP and PU networks. These results are consistent with that of the XRD analysis and the mechanical properties of the composites [26,39,44,45].

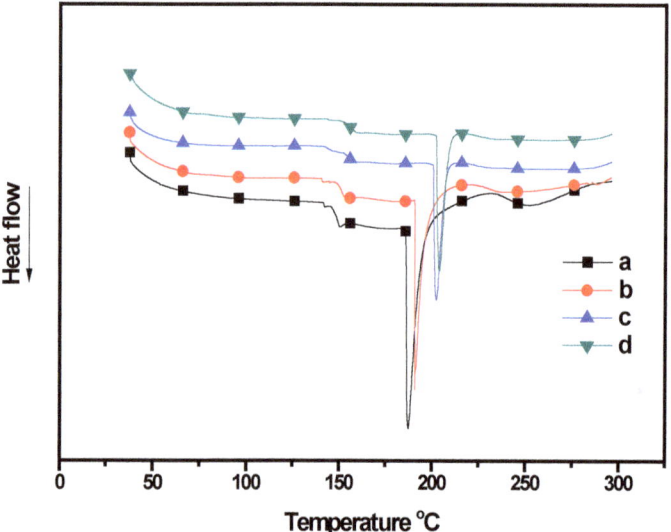

**Figure 5.** DSC thermogram of composites having: (**a**) only CP, (**b**) 1:0.5 CP/PU, (**c**) 1:1 CP/PU, and (**d**) 1:1 CP/PU with 10 wt% curcumin.

*3.9. Mechanical Properties*

The mechanical properties of the hydrogel films are important for the food packaging application during shipping, handling and storage. The mechanical properties of the CP/PU/curcumin composites are listed in Table 2. The tensile strength (TS) of neat CP was 93.2 MPa; however, it decreased to 14.1 MPa in the 1:0.5 CP/PU composite. The TS increased again to 29.5 MPa in the 1:1 CP/PU composite. The sudden decrease in TS was due to the dilution effect, poor adhesion between the polymer matrices leads to the formation of numerous voids at the matrix interface. The elongation at break (EB) of the CP was 8.1% which decreased to 3.7 and 3.9 with the addition of polyurethane. This decrease in EB on the addition of PU can be attributed to a decrease in polymer extensibility, which is caused by the formation of microphase separations in the composite films. These results are consistent with those of the SEM analysis, which revealed the CP/PU/curcumin composites to have an uneven and rough surface that may be responsible for the low EB [46]. Note that a low aspect ratio and particulate nature can also lead to a low EB [47]. Overall, the CP/PU/curcumin composites were found to be hard and brittle in nature.

### 3.10. SEM Analysis

Scanning electron microscopy (SEM) analysis was carried out to characterize the microstructure morphology and homogeneity of the CP/PU/Curcumin composite films with different ratios. Figures 6 and 7 show the surface and cross-sectional SEM images of the composite films, respectively. The neat CP film had a homogenous surface without any agglomerates. However, the 1:0.5 CP/PU and 1:1 CP/PU composites had less homogenous surfaces; in addition, they exhibited small, uneven, and discontinuous voids. The cross-sectional image of the neat CP film shows that it has a slightly rough and uneven surface. The addition of PU increased the surface roughness of the composites. Note, that these results are consistent with the mechanical properties of the composite films. The voids and cavities revealed by the cross-sectional SEM images are the main cause of the low TS and EB of the CP/PU/curcumin composite films. These results suggest that cellulose pulp, polyurethane and curcumin based composite films possess reasonably good mechanical properties. The SEM results were consistent with the mechanical and water absorption properties of the composite films.

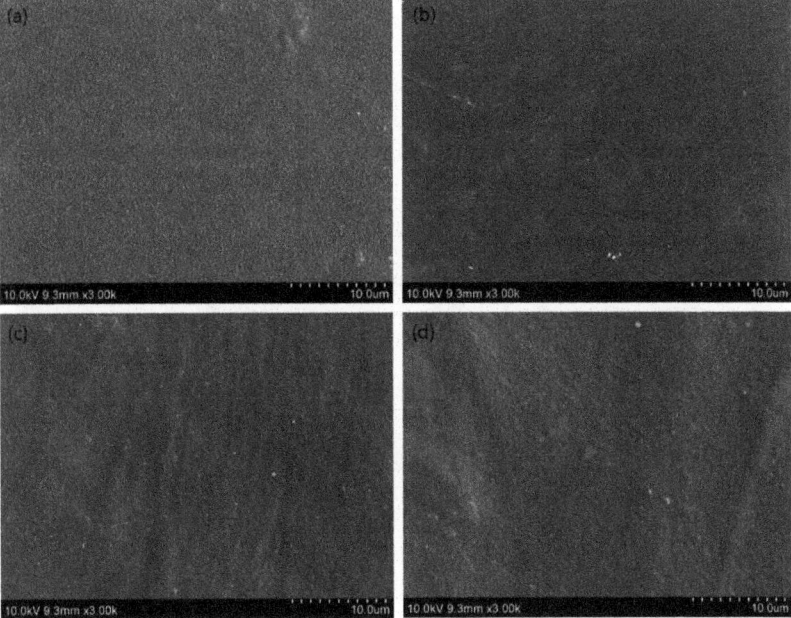

**Figure 6.** SEM images of composites having: (**a**) only CP, (**b**) 1:0.5 CP/PU, (**c**) 1:1 CP/PU, and (**d**) 1:1 CP/PU with 10 wt% curcumin.

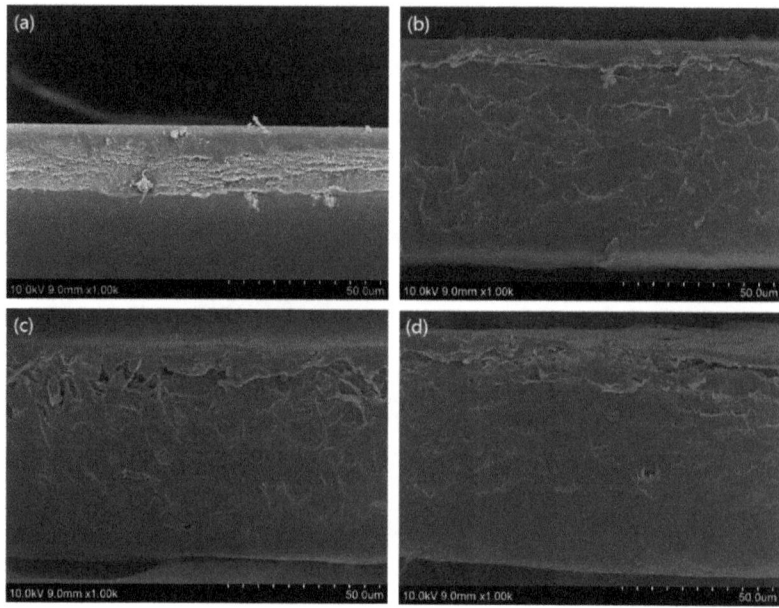

**Figure 7.** Cross-sectional SEM images of composites having: (**a**) only CP, (**b**) 1:0.5 CP/PU (**c**) 1:1 CP/PU, and (**d**) 1:1 CP/PU with 10 wt% curcumin.

*3.11. Antioxidant Activity*

Packaging films that have antioxidant properties are extremely effective in protecting food materials. The antioxidant activity of the CP/PU/curcumin composite films was determined using the DPPH radical scavenging activity (see Figure 8). As expected, the neat CP, 1:0.5 CP/PU, and 1:1 CP/PU composites showed no antioxidant activity. Curcumin is well known for its strong antioxidant activity, as it acts as a superoxide radical scavenger and singlet oxygen quencher. The 1:1 CP/PU with 10 wt% curcumin composite showed good antioxidant activity (24.8%) after an incubation period of 1 h. Curcumin is reported to have good thermal stability, as heating does not affect its active compounds, such as curcumin and phenolic hydroxyl groups. Some studies have pointed out that the β-diketone moiety plays an important role in the antioxidant mechanism of curcumin; for instance, the donation of an H-atom from the β-diketone moiety to a lipid alkyl or a lipid peroxyl radical has been shown to contribute to the antioxidant activity of curcumin. The high antioxidant activity of the CP/PU/curcumin composites makes them potential candidates for antioxidant food packaging material.

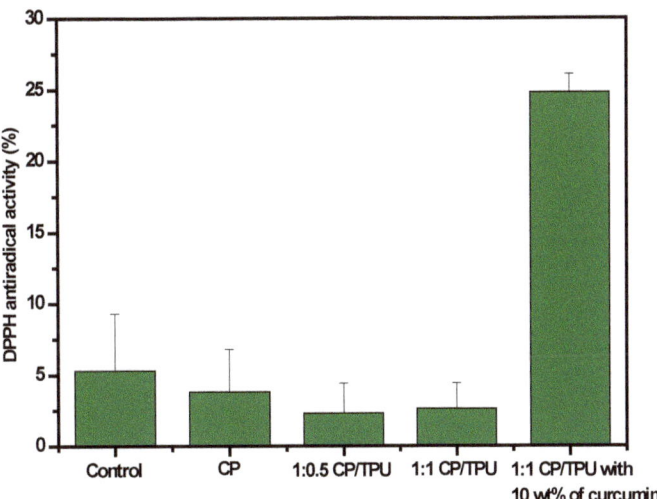

**Figure 8.** DPPH radical scavenging activity of composites.

*3.12. Cytotoxicity*

Cell viability is an essential parameter to test the biosafety of a material. The cytotoxicity of the synthesized CP, CP/PU, and CP/PU/curcumin composites was evaluated using the WST-1 assay method on human keratinocyte cells (i.e., the HaCaT cell line), and the results are summarized in Figures 9 and 10. All the different composites were found to be non-cytotoxic. There was no difference in cell viability between the control and neat CP films. However, cell viability decreased in the 1:0.5 CP/PU and 1:1 CP/PU composites. Note that the cell viability of the 1:1 CP/PU with 10 wt% curcumin composite was better than that of the other CP/PU composites. The decrease in cell viability was caused by cellular stress owing to the presence of a foreign element in the cell medium. The cell viability of the 1:1 CP/PU with 10 wt% curcumin composite was 91%, which can be attributed to the disappearance of cellular growth that is favorable for cell proliferation. This result is in good agreement with those of previous studies [35,48]. The above results show that the CP/PU/curcumin composites are biocompatible and safe to be used for packaging applications.

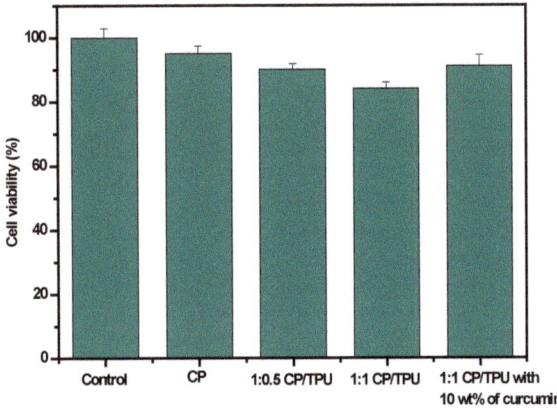

**Figure 9.** Cell viability of HaCaT cell line for composites.

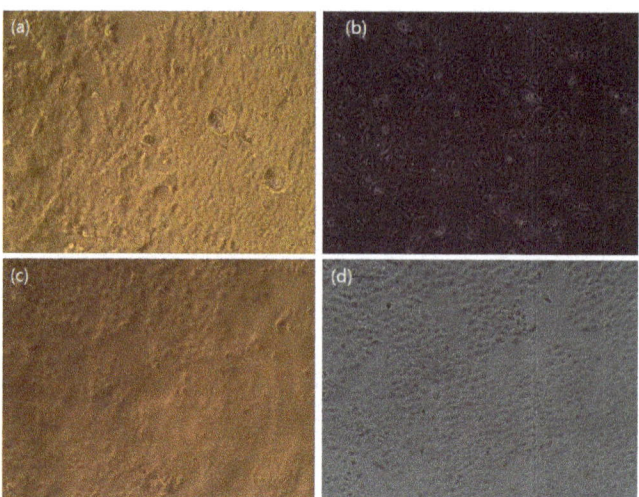

**Figure 10.** Light microscopy images of HaCaT cell line for composites having: (**a**) only CP, (**b**) 1:0.5 CP/PU, (**c**) 1:1 CP/PU, and (**d**) 1:1 CP/PU with 10 wt% curcumin.

## 4. Conclusions

In this study, we successfully blended two sustainable materials, namely, CP and curcumin, with a commercially available synthetic material, namely, PU, using an ionic solvent, namely, NMMO. The apparent color and UV transmittance of the composite films were significantly changed owing to the addition of PU. XRD analysis showed that PU decreased the overall crystallinity of CP. SEM analysis revealed that although all the components were uniformly dispersed in the polymer matrix, the composites had a rough and uneven surface that affected their mechanical properties. The TS of neat CP was 93.2 MPa, which decreased to 14.1 MPa for the 1:0.5 CP/PU composite and increased to 29.5 MPa for the 1:1 CP/PU composite. However, the EB of CP declined monotonically with the addition of PU. The CP/PU/curcumin composite also exhibited excellent thermal stability. The $T_g$ and $T_{5\%}$ values of the composites were in the range 147.4–154.2 °C and 87.2–182.3 °C, respectively. Moreover, the addition of PU and curcumin to the CP matrix decreased the hydrophilicity of the CP/PU/curcumin composite, which in turn improved the water barrier properties and SR of the CP/PU/curcumin composite. The CP/PU/curcumin composite exhibited excellent antioxidant properties and all the composites were found to be non-cytotoxic when tested on the HaCaT cell line. Nevertheless, further research is required to better understand the mechanical properties of these composites and to establish their utility in packaging applications.

**Supplementary Materials:** The following supporting information can be downloaded at: https://www.mdpi.com/article/10.3390/gels8040248/s1, Table S1: Cellulose pulpn (CP)/polyurethane (PU)/Curcumin composite composition.

**Author Contributions:** Conceptualization, C.J. and J.L.; Data curation, B.R., S.R., P.D.S. and J.L.; Formal analysis, P.D.S. and J.L.; Funding acquisition, S.S.K. and J.L.; Investigation, C.J., S.S.K., B.R., S.R., P.D.S. and J.L.; Methodology, B.R., S.R., P.D.S. and J.L.; Project administration, J.L.; Resources, B.R.; Software, C.J. and B.R.; Supervision, S.S.K., B.R. and J.L.; Validation, S.S.K., S.R. and P.D.S.; Writing—original draft, C.J., S.S.K., B.R. and J.L.; Writing—review & editing, C.J., S.S.K., B.R. and J.L. All authors have read and agreed to the published version of the manuscript.

**Funding:** This research received no external funding.

**Institutional Review Board Statement:** Not applicable.

**Informed Consent Statement:** Not applicable.

**Data Availability Statement:** Not applicable.

**Acknowledgments:** This work was supported by Korea Institute for Advancement of Technology (KIAT) grant (P0012770) funded by the Korea Government (Ministry of Trade, Industry and Energy-MOTIE).

**Conflicts of Interest:** The authors declare no conflict of interest.

## References

1. Sha, D.; Jinrui, M.; Yilan, G.; Feng, C.; Qiang, F. One-step modification and nanofibrillation of microfibrillated cellulose for simultaneously reinforcing and toughening of poly(ε-caprolactone). *Compos. Sci. Technol.* **2018**, *157*, 168–177.
2. Antonio, N.N.; Hiroyuki, Y. The effect of fiber content on the mechanical and thermal expansion properties of biocomposites based on microfibrillated cellulose. *Cellulose* **2008**, *15*, 555–559.
3. Natalia, H.; Aji, P.M.; Kristiina, O. Plasticized polylactic acid/cellulose nanocomposites prepared using melt-extrusion and liquid feeding: Mechanical, thermal and optical properties. *Compos. Sci. Technol.* **2015**, *106*, 149–155.
4. Juho, A.S.; Miikka, V. Highly Transparent Nanocomposites Based on Poly(vinyl alcohol) and Sulfated UV-Absorbing Wood Nanofibers. *Biomacromolecules* **2019**, *20*, 2413–2420.
5. Rong-Lan, W.; Xiu-Li, W.; Fang, L.; Hui-Zhang, L.; Yu-Zhong, W. Green composite films prepared from cellulose, starch and lignin in room-temperature ionic liquid. *Bioresour. Technol.* **2009**, *100*, 2569–2574.
6. Quanqing, H.; Xin, G.; Heng, Z.; Keli, C.; Lincai, P.; Qingmin, J. Preparation and comparative assessment of regenerated cellulose films from corn (Zea mays) stalk pulp fines in DMAc/LiCl solution. *Carbohydr. Polym.* **2019**, *218*, 315–323.
7. Cortés-Triviño, E.; Valencia, C.; Delgado, M.A.; Franco, J.M. Rheology of epoxidized cellulose pulp gel-like dispersions in castor oil: Influence of epoxidation degree and the epoxide chemical structure. *Carbohydr. Polym.* **2018**, *199*, 563–571. [CrossRef]
8. Sara, R.L.; Jari, S.K.; Katja, S.; Janne, A.; Ilkka, K. Reactive dissolution of cellulose and pulp through acylation in pyridine. *Cellulose* **2012**, *19*, 1295–1304.
9. Anwar, J.S.; Lalaso, V.M.; Niteen, A.D.; Dipak, V.P. Structural characterization of cellulose pulp in aqueous NMMO solution under the process conditions of lyocell slurry. *Carbohydr. Polym.* **2019**, *206*, 220–228.
10. Wil, V.S.I.; Srikanth, P.; Zachary, C.W.; Cecily, A.R.; Joseph, P.G.; Curtis, W.F.; Sarah, L.B. Mechanisms and impact of fiber–matrix compatibilization techniques on the material characterization of PHBV/oak wood flour engineered biobased composites. *Compos. Sci. Technol.* **2012**, *72*, 708–715.
11. Xinping, L.; Xin, Z.; Shuangquan, Y.; Hui, C.; Yaoyu, W.; Zhao, Z. UV-blocking, transparent and hazy cellulose nanopaper with superior strength based on varied components of poplar mechanical pulp. *Cellulose* **2020**, *27*, 6563–6576.
12. Anamaria, S.; Raluca, N.D.; Marian, T.; Georgeta, C.; Cornelia, V. Low density polyethylene composites containing cellulose pulp fibers. *Compos. Part B Eng.* **2012**, *43*, 1873–1880.
13. Helena, N.; Sylvain, G.; Larsson, P.T.; Kristofer, G.E.; Tommy, I. Compression molded wood pulp biocomposites: A study of hemicellulose influence on cellulose supramolecular structure and material properties. *Cellulose* **2012**, *19*, 751–760.
14. Qiyang, W.; Jie, C.; Lina, Z.; Min, X.; He, C.; Charles, C.H.; Shigenori, K.; Jun, X.; Rui, X. A bioplastic with high strength constructed from a cellulose hydrogel by changing the aggregated structure. *J. Mater. Chem. A* **2013**, *1*, 6678–6686.
15. Alcala, M.; Gonza'lez, I.; Boufi, S.; Vilaseca, F.; Mutje, P. All-cellulose composites from unbleached hardwood kraft pulp reinforced with nanofibrillated cellulose. *Cellulose* **2013**, *20*, 2909–2921. [CrossRef]
16. Hasan, M.; Tze, K.L.; Deepu, A.G.; Jawaid, M.; Owolabi, F.A.T.; Mistar, E.M.; Tata, A.; Noriman, N.Z.; Haafiz, M.K.M.; Abdul Khalil, H.P. Micro Crystalline Bamboo Cellulose Based Seaweed Biodegradable Composite Films for Sustainable Packaging Material. *J. Polym. Environ.* **2019**, *27*, 1602–1612. [CrossRef]
17. Prodyut, D.; Debashis, T.; Amit, K.; Vimal, K. Effect of cellulose nanocrystal polymorphs on mechanical, barrier and thermal properties of poly(lactic acid) based bionanocomposites. *RSC Adv.* **2015**, *5*, 60426–60440.
18. Ronny, W.; Alistair, K.; Arno, P.; Kristiina, K.; Anna, S. Cellulose hydrolysis with thermo- and alkali-tolerant cellulases in cellulose-dissolving superbase ionic liquids. *RSC Adv.* **2013**, *3*, 20001–20009.
19. Xiaofei, L.; Zongbao, L.; Li, W.; Shengsheng, Z.; Hai, Z. Preparation and performance of composite films based on 2-(2-aminoethoxy) ethyl chitosan and cellulose. *RSC Adv.* **2017**, *7*, 13707–13713.
20. Sanches, A.O.; Ricco, L.H.S.; Malmonge, L.F.; da Silva, M.J.; Sakamoto, W.K.; Malmonge, J.A. Influence of cellulose nanofibrils on soft and hard segments of polyurethane/cellulose nanocomposites and effect of humidity on their mechanical properties. *Polym. Testing.* **2014**, *40*, 99–105.
21. He, L.; Shuqin, C.; Shibin, S.; Dan, W.; Jie, S. Properties of rosin-based waterborne polyurethanes/cellulose nanocrystals composites. *Carbohydr. Polym.* **2013**, *96*, 510–515.
22. Zhenzhong, L.; Jun, P.; Tuhua, Z.; Jin, S.; Xiaobo, W.; Chao, Y. Biocompatible elastomer of waterborne polyurethane based on castor oil and polyethylene glycol with cellulose nanocrystals. *Carbohydr. Polym.* **2012**, *87*, 2068–2075.
23. Dan, C.; Pingdong, W.; Lina, Z.; Jie, C. New Approach for the Fabrication of Carboxymethyl Cellulose Nanofibrils and the Reinforcement Effect in Water-Borne Polyurethane. *ACS Sustain. Chem. Eng.* **2019**, *7*, 11850–11860.

24. Andres, I.C.; Javier, I.A.; Elena, F.; Jose, M.K.; Leonel, M.C. The role of nanocrystalline cellulose on the microstructure of foamed castor-oil polyurethane nanocomposites. *Carbohydr. Polym.* **2015**, *134*, 110–118.
25. Seydibeyoglu, M.O.; Oksman, K. Novel nanocomposites based on polyurethane and micro fibrillated cellulose. *Compos. Sci. Technol.* **2008**, *68*, 908–914. [CrossRef]
26. Kai, L.; Pingdong, W.; Junchao, H.; Duoduo, X.; Yi, Z.; Lei, H.; Lina, Z.; Jie, C. Mechanically Strong Shape-Memory and Solvent-Resistant Double-Network Polyurethane/Nanoporous Cellulose Gel Nanocomposites. *ACS Sustain. Chem. Eng.* **2019**, *7*, 15974–15982.
27. Sang, H.; Park, K.; Wha, O.; Seong, H.K. Reinforcement effect of cellulose nanowhisker on bio-based polyurethane. *Compos. Sci. Technol.* **2013**, *86*, 82–88.
28. Yao, X.; Qi, X.; He, Y.; Tan, D.; Cheng, F.; Fu, Q. Simultaneous Reinforcing and Toughening of Polyurethane via Grafting on the Surface of Microfibrillated Cellulose. *ACS Appl. Mater. Interfaces* **2014**, *6*, 2497–2507.
29. Saralegi, A.; Rueda, L.; Martin, L.; Arbelaiz, A.; Eceiza, A.; Corcuera, M.A. From elastomeric to rigid polyurethane/cellulose nanocrystal bionanocomposites. *Compos. Sci. Technol.* **2013**, *88*, 39–47. [CrossRef]
30. Izaskun, L.; Julen, V.; Arantzazu, S.E.; Alvaro, T.; Maider, A.; Eneritz, V.; Ander, O.; Ainara, S.; Aitor, A.; Arantxa, E. The effect of the carboxylation degree on cellulose nanofibers and waterborne polyurethane/cellulose nanofiber nanocomposites properties. *Polym. Degrad. Stab.* **2020**, *173*, 109084.
31. Kangkang, Z.; Huilan, Y.; Zhongguo, S.; Kerang, H.; Min, Z.; Xiaodong, X. Dual-modified starch nanospheres encapsulated with curcumin by self-assembly: Structure, physicochemical properties and anti-inflammatory activity. *Int. J. Biol. Macromol.* **2021**, *191*, 305–314.
32. Gao, Y.Z.; Chen, J.C.; Wu, Y.X. Amphiphilic Graft Copolymers of Quaternized Alginate-g-Polytetrahydrofuran for Anti-protein Surfaces, Curcumin Carriers, and Antibacterial Materials. *ACS Appl. Polym. Mater.* **2021**, *3*, 3465–3477. [CrossRef]
33. Balasubramanian, R.; Srinivasan, R.; Sam, S.K.; Jaewoong, L. Rheological and anti-microbial study of silica and silver nanoparticles-reinforced k-carrageenan/hydroxyethyl cellulose composites for food packaging applications. *Cellulose* **2021**, *28*, 5577–5590.
34. Balasubramanian, R.; Srinivasan, R.; Jaewoong, L. Quaternary ammonium silane-reinforced agar/polyacrylamide composites for packaging applications. *Int. J. Biol. Macromol.* **2021**, *182*, 1301–1309.
35. Chaehyun, J.; Balasubramanian, R.; Prabakaran, D.S.; Srinivasan, R.; Jaewoong, L. Cellulose pulp-based stretchable composite film with hydroxyethyl cellulose and turmeric powder for packaging applications. *ACS Sustain. Chem. Eng.* **2021**, *9*, 13653–13662.
36. Balasubramanian, R.; Srinivasan, R.; Jaewoong, L. Barrier, rheological and antimicrobial properties of sustainable nanocomposites based on gellan gum/polyacrylamide/zinc oxide. *Polym. Eng. Sci.* **2021**, *61*, 2477–2486. [CrossRef]
37. Ahmed, A.O.; Jong-Whan, R. Preparation and characterization of sodium carboxymethyl cellulose/cotton linter cellulose nanofibril composite films. *Carbohydr. Polym.* **2015**, *127*, 101–109.
38. Shiv, S.; Jong-Whan, R. Preparation of nanocellulose from micro-crystalline cellulose: The effect on the performance and properties of agar-based composite films. *Carbohydr. Polym.* **2016**, *135*, 18–26.
39. Mehran, G.; Fugen, D.; Elena, P.I.; Billy, J.M.; Benu, A. Use of Synergistic Interactions to Fabricate Transparent and Mechanically Robust Nanohybrids Based on Starch, Non-Isocyanate Polyurethanes, and Cellulose Nanocrystals. *ACS Appl. Mater. Interfaces* **2020**, *12*, 47865–47878.
40. Qiuju, W.; Marielle, H.; Xiaohui, L.; Lars, A.B. A High Strength Nanocomposite Based on Microcrystalline Cellulose and Polyurethane. *Biomacromolecules* **2007**, *8*, 3687–3692.
41. Ni, Z.; Peng, T.; Yanxv, L.; Shuangxi, N. Effect of lignin on the thermal stability of cellulose nanofibrils produced from bagasse pulp. *Cellulose* **2019**, *26*, 7823–7835.
42. Yanxu, L.; Peng, T.; Ni, Z.; Shuangxi, N. Preparation and thermal stability evaluation of cellulose nanofibrils from bagasse pulp with differing hemicelluloses contents. *Carbohydr. Polym.* **2020**, *245*, 116463.
43. Balasubramanian, R.; Chaehyun, J.; Seungjin, C.; Srinivasan, R.; Jaewoong, L. Flexible ternary combination of gellan gum sodium carboxymethyl cellulose and silicon dioxide nanocompositesfabricated by quaternary ammonium silane: Rheological, thermal and antimicrobial properties. *ACS Omega* **2020**, *5*, 28767–28775.
44. Francisco, L.; Dejin, J.; Jiaqi, G.; Daniel, H.; Alexander, E.; Andreas, W. Outstanding Synergies in Mechanical Properties of Bioinspired Cellulose Nanofibril Nanocomposites using Self-Cross-Linking Polyurethanes. *ACS Appl. Polym. Mater.* **2019**, *1*, 3334–3342.
45. Ru, L.; Yao, P.; Jinzhen, C.; Yu, C. Comparison on properties of lignocellulosic flour/polymer composites by using wood, cellulose, and lignin flours as fillers. *Compos. Sci. Technol.* **2014**, *103*, 1–7.
46. Wu, G.M.; Liu, G.F.; Chena, J.; Kong, Z.W. Preparation and properties of thermoset composite films from two-component waterborne polyurethane with low loading level nanofibrillated cellulose. *Prog. Org. Coat.* **2017**, *106*, 170–176. [CrossRef]
47. Aji, P.M.; Kristiina, O.; Mohini, S. Mechanical Properties of Biodegradable Composites from Poly Lactic Acid (PLA) and Microcrystalline Cellulose (MCC). *J. Appl. Polym. Sci.* **2005**, *97*, 2014–2025.
48. Prince, C.; Agnieszka, N.; Aarti, B.; Renata, N.W.; Ravinder, K.; Mansuri, M.T. Potential of Gum Arabic Functionalized Iron Hydroxide Nanoparticles Embedded Cellulose Paper for Packaging of Paneer. *Nanomaterials* **2021**, *11*, 1308.

*Review*

# Cellulosic-Based Conductive Hydrogels for Electro-Active Tissues: A Review Summary

Esubalew Kasaw Gebeyehu [1,2], Xiaofeng Sui [1,3,*], Biruk Fentahun Adamu [2,4], Kura Alemayehu Beyene [2] and Melkie Getnet Tadesse [5,*]

1. Key Lab of Science and Technology of Eco-Textile, Ministry of Education, College of Chemistry, Chemical Engineering and Biotechnology, Donghua University, Shanghai 201620, China; esubalewtsega@gmail.com
2. Textile Engineering Department, Ethiopian Institute of Textile and Fashion Technology, Bahir Dar University, Bahir Dar 1037, Ethiopia; birukfentahun2009@gmail.com (B.F.A.); kuraalemayehu@gmail.com (K.A.B.)
3. Innovation Center for Textile Science and Technology, Donghua University, Shanghai 201620, China
4. College of Textiles, Donghua University, Shanghai 201620, China
5. Textile Chemical Process Engineering Department, Ethiopian Institute of Textile and Fashion Technology, Bahir Dar University, Bahir Dar 1037, Ethiopia
* Correspondence: suixf@dhu.edu.cn (X.S.); melkiegetnet23@gmail.com (M.G.T.)

**Abstract:** The use of hydrogel in tissue engineering is not entirely new. In the last six decades, researchers have used hydrogel to develop artificial organs and tissue for the diagnosis of real-life problems and research purposes. Trial and error dominated the first forty years of tissue generation. Nowadays, biomaterials research is constantly progressing in the direction of new materials with expanded capabilities to better meet the current needs. Knowing the biological phenomenon at the interaction among materials and the human body has promoted the development of smart bio-inert and bio-active polymeric materials or devices as a result of vigorous and consistent research. Hydrogels can be tailored to contain properties such as softness, porosity, adequate strength, biodegradability, and a suitable surface for adhesion; they are ideal for use as a scaffold to provide support for cellular attachment and control tissue shapes. Perhaps electrical conductivity in hydrogel polymers promotes the interaction of electrical signals among artificial neurons and simulates the physiological microenvironment of electro-active tissues. This paper presents a review of the current state-of-the-art related to the complete process of conductive hydrogel manufacturing for tissue engineering from cellulosic materials. The essential properties required by hydrogel for electro-active-tissue regeneration are explored after a short overview of hydrogel classification and manufacturing methods. To prepare hydrogel from cellulose, the base material, cellulose, is first synthesized from plant fibers or generated from bacteria, fungi, or animals. The natural chemistry of cellulose and its derivatives in the fabrication of hydrogels is briefly discussed. Thereafter, the current scenario and latest developments of cellulose-based conductive hydrogels for tissue engineering are reviewed with an illustration from the literature. Finally, the pro and cons of conductive hydrogels for tissue engineering are indicated.

**Keywords:** conductive hydrogel; cellulose; tissue engineering; hydrogel design and characterization; electro-active tissues

## 1. Introduction

Every year, millions of people lose tissue or organs as a result of accidents or illnesses [1]. Tissue or organ transplantation is used to treat these patients. This approach, however, is constrained by the lack of donors. To solve the problem of the severe shortage of organ transplants, intensive research work, a review [2], has been performed in the last four decades to develop artificial organs and tissue for diagnosis and research purposes. Cells and their extracellular subassemblies are used to develop biological tissues for body repair, primarily with bio-based material scaffolds. The scaffolds support cellular attachment and regulate tissue shape. Some of the strategies used to develop scaffolds were tri-dimensional textiles [3,4], aerogel [5,6], hydrogels [1], nanofibers [7,8], and composites [9]. In all cases, scaffolds should ideally be sufficiently porous to enable the growth of cells, nutritional diffusion, and physiologic waste extraction [10]; have adequate tensile strength and elasticity [11]; have controlled degradation [12]; and possess suitable chemistry for cell adhesion [13]. Other desirable properties, such as electrical conductivity, in polymers have also been reported to accelerate the nerve regeneration in artificial nerve grafts [14]. However, the importance of electrically conductive hydrogels in tissue engineering has received insufficient attention as yet [15].

A hydrogel is a tri-dimensional polymeric material that can take the form of a matrix, film, liquid, or microsphere [16] which is water insoluble and has the ability to swell and preserve a significant amount of water, typically greater than the mass ratio of the polymer materials in their interstitial structures [17]. Due to the presence of hydrophilic groups, such as $-NH_2$, $-OH$, $-COOH$, and $-SO_3H$, in their polymer networks and osmotic pressure, hydrogels continue to absorb and swell upon contact with water to form 3D structures. Physical or chemical crosslinking causes the ability of hydrogels to maintain an unaltered 3D structure during swelling, and this also helps to prevent hydrogels from dissolving in the solvent [18]. Upon hydration in an aqueous environment, hydrophilic groups or domains of polymeric networks form the hydrogel structure presented in Figure 1 [19].

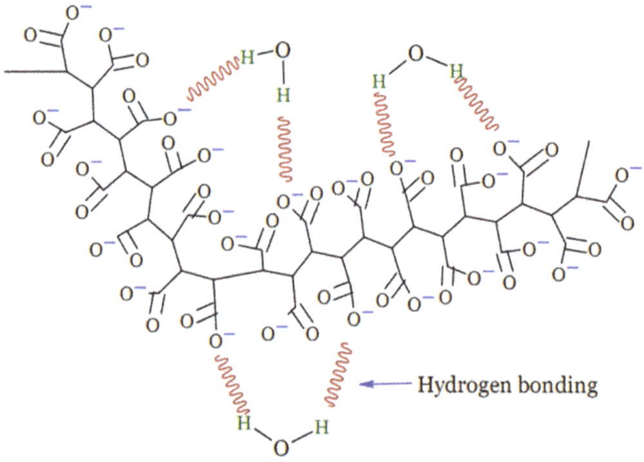

**Figure 1.** Structure of hydrogel.

Hydrogel products are classified based on a variety of criteria, as shown in Figure 2. The origins of the polymeric constituent of hydrogel can be classified as synthetic [20], hybrid [21], and natural [22]. Hydrogels from natural sources [22] can be derived from polysaccharide-based materials, such as cellulose [23]; chitosan [24]; glycosaminoglycans [25]; alginate [26,27]; protein-based materials, such as, silk fibroin [28,29], collagen [30], elastin [31], gelatin [32], and fibrin [33]; and decellularized hydrogels [34,35]. The integrity of the hydrogel is maintained through chemical crosslinking, physical crosslinking, or both [17,19,36–38]. Physical crosslinking, such as heating/cooling, hydrophobic interactions, freezing–thawing, complex coacervation, ionic interactions, and hydrogen bonding, have resulted in temporary networks of hydrogels, whereas permanent junctions exist in chemically crosslinked networks [16] obtained by employing grafting, chemical crosslinkers, radiation crosslinking, enzymatic reactions, click chemistry, radical polymerization, and thermo-gelation. A crosslinked network is a set of one, two, or three, and maybe more, types of monomers referred to as homopolymer, copolymer, and multipolymer, respectively [17]. Perhaps the hydrogel network is arranged in a network to swell up in a monomer; afterward, it reacts by forming a second intermeshing network structure to form interpenetrating polymeric composition [20].

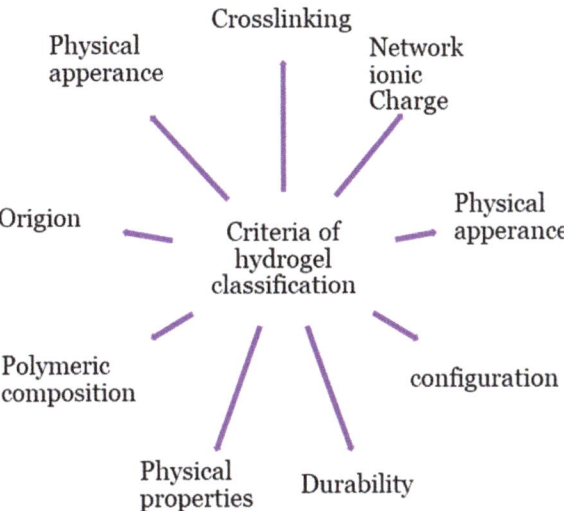

Figure 2. Classification bases of hydrogel.

The spatiotemporal control of hydrogel physicochemical characteristics is essential for monitoring their dynamics, such as durability and orientation [39]. The orientational dynamic appearance of biomolecules may be either crystalline or amorphous.

Hydrogels can be biodegradable or non-biodegradable in terms of durability [20]. Hydrogel physical properties have been advanced from conventional to smart [16]. Smart hydrogels are stimuli-sensitive and change the volume of the system structure in response to various stimuli, such as electric, light, temperature, and pH [17,40–42].

The work presented in this review aims to demonstrate the possibilities for producing conductive hydrogel of cellulosic-based materials. The organic compound cellulose was preferred for the hydrogel base material in this review, due to its excellent inherent properties, such as renewability [43,44], absorption ability [45], hygroscopicity [46], air permeability [47], biocompatibility [48–50], stability [51], bioactivity [50], and biodegradability [52]. In contrast to some of the other studies discussed in this review, however, design parameters' consideration and characterization of hydrogel scaffolds for electroactive tissues, in general, and the preparation techniques of cellulosic-based conductive hydrogel, in particular, are given a place of due attention on this list.

## 2. Cellulosic-Based Hydrogel as Biomaterials

Biomaterials should be biocompatible and still have specialized biochemical, mechanical, and physical attributes in order to mimic fundamental aspects of the in vivo conditions. The need to assess the biocompatibility of biomaterial is to examine the biological responses that may cause harm or unwanted side effects to the host. In tissue engineering, biomaterials need to be biocompatible to have the ability to function as a support for tissue regeneration without eliciting any adverse local or systemic response in the eventual host. Biocompatibility is the potential of a material to coexist and interact in the presence of specific tissues or biological functions without causing excessive harm. The purpose of making a hydrogel biocompatible is to assess its toxic impact on the body [53]. Another attribute of a biomaterial is its bio-activity. A bioactive material is one that has been designed to induce specific biological activity, for example, tissue uptake, metabolism, or physiological response [54,55]. Cell arrangement, viability, and function are all influenced by biomechanical interactions between cells and biomaterial scaffolds. Biomechanics is the analysis of the interaction of biomedical scaffold and biological structures of cells, as well as the effects of such forces during stem-cell differentiation and molecular transport [56].

Tissue function in tissue engineering is affected by cell adhesion, proliferation, differentiation, and maturation. Biocompatibility, bioactivity, and biomechanics are critical requirements for any biomaterial, in general, and tissue engineering, in particular; cellulose-based biomaterials meet all three of these requirements. Cell adhesion facility, in conjunction with cellulose's hydrophilic hydroxyl substituent, as well as specialized β-d environments, helps cells to attach to cellulose [57]. Due to the nature variety of chemical structures and functional properties, cellulose-based hydrogels are nontoxic. The patterned porosity of hydrogel scaffolds was found to promote the cellular adhesion, growth, proliferation, and infiltration of cells [58].

Cellulose is a polysaccharide polymer that is made of a linear chain of glucose molecules [59]. Five thousand to fifteen thousand glucose molecules with the molecular formula $(C_6H_{10}O_5)_n$ are covalently bonded together to form cellulose by acetal oxygen via covalent bonding of $C_1$ of glucose ring and $C_4$ of the adjacent ring, as shown in Figure 3 [59,60].

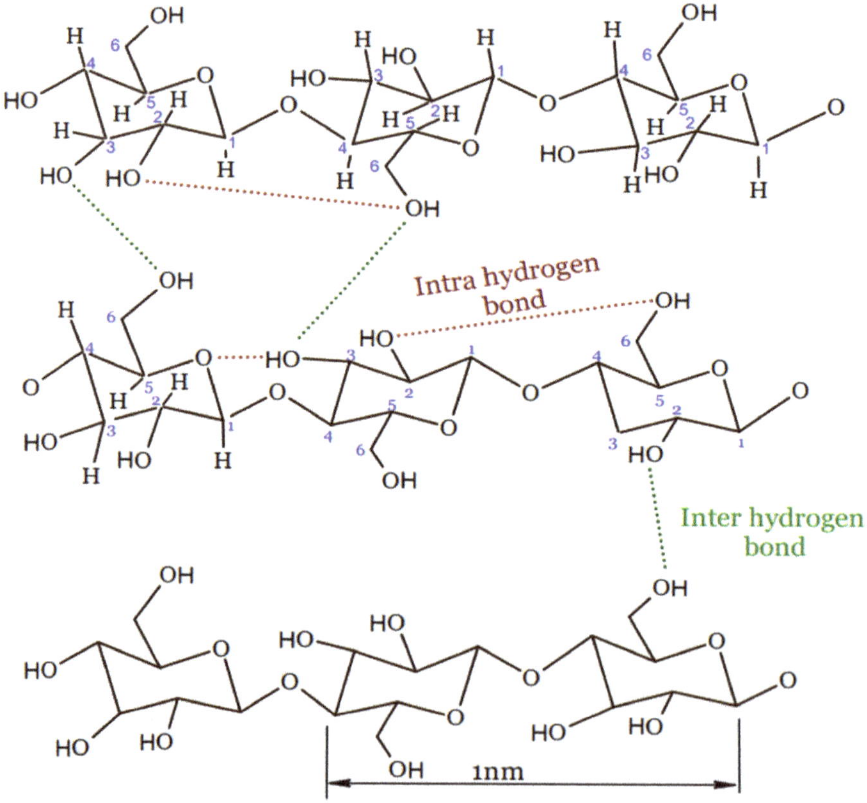

**Figure 3.** Cellulose structure.

The electrical characteristic of the body is essential in the physiological aspects of life [61]. Electric-potential stimulation occurs in a number of stem-cell functions, including cell interaction and the stimulation of signal transduction involved in cell-cycle progression. For instance, neurological rehabilitation and cell birth occur when nerve cells access the voltage-gated route liable for receptor activation in the presence of electrical activity, and cardiomyocytes are electrically active cells that generate contractile force when heart tissue works in tandem with bone tissue in a living heart. Even bone cells can be electrically stimulated as a result of the stress exerted on them by muscle contractions. Hence, the cells make use of electrical characteristics for a number of different purposes, and electrical stimulation of tissue in a controlled and targeted manner can enhance vascularization and differentiation of stem cells into different types of cells [61].

## 3. Classification of Cellulose Hydrogels

### 3.1. Source

Cellulose is abundantly found in plant-cell walls or synthesis from bacteria (Figure 4) [57]. Cellulose is extracted from reinforcing the polymer of the cell walls of plants via chemical, mechanical, and biological extraction [62]. It is also synthesized from extracellular polysaccharides that are produced as protective envelopes around the cells of bacteria [38,63,64]. Diverse bacteria produce celluloses with varying morphological features, structures, characteristics, and functionalities [64]. Plant cellulose and bacterial cellulose differ in terms of macromolecular properties [65]. Plant cellulose contains impurities, such as hemicellulose and lignin; has a moderate water-holding capacity of 60%; and possesses a moderate level of tensile strength and crystallinity. Meanwhile, bacterial cellulose is chemically pure, which is hydrophilic and has a high water-holding capacity (100%), as well as high crystallinity and tensile strength [66]. A wide range of studies have been conducted on the difference of bacterial and plant cellulose as potential biomaterials [65,66].

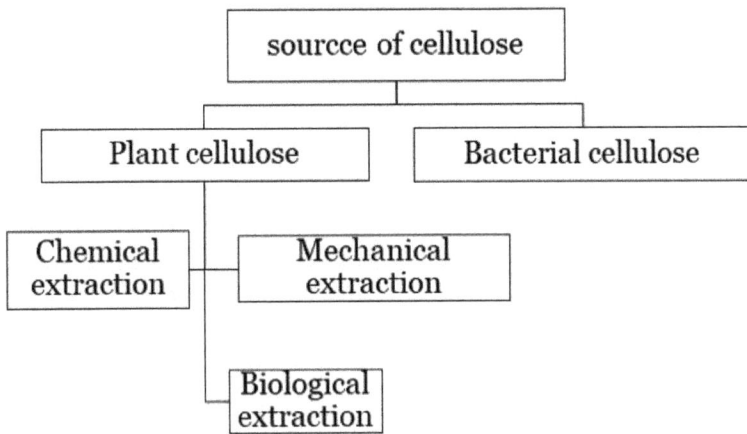

**Figure 4.** Cellulose source.

To broaden its applicability, the esterification, etherification, and electrolytic dissociation reactions are used to modify the parent cellulose structure by substituting the hydroxyl group via organic species, such as methyl and ethyl units [23].

### 3.2. Crosslinking

Cellulosic hydrogel scaffolds can be naturally derived or synthesized. The stabilization process of polymer to create the multidimensional extension of a polymeric chain to produce a network structure is achieved by performing either chemical or physical crosslinking. Linking of macromolecular chains together changes a liquid polymer into solid or gel. The rheological measurements of a liquid polymer is monitored by using the cone and plate geometry [67]. This transition from a structure with finite branched polymer to an infinite molecule system is referred to as gelation, and it results in an insoluble network [19,37,68] (Figure 5).

Permanent and temporary junctions of cellulose-based hydrogels can be manufactured by crosslinking aqueous solutions of cellulose derivatives, such as ethyl cellulose (EC), hydroxyethyl methylcellulose (HEMC), methylcellulose (MC), hydroxypropyl methylcellulose, Hydroxypropylcellulose (HPC), and sodium carboxymethylcellulose (CMC), which are the most common forms of etherified modified cellulose (Figure 6) [18,69]. A majority of water-soluble cellulose derivatives are produced through cellulose etherification, which occurs when the active hydroxyl groups of cellulose react with organic species such as methyl and ethyl units. The average number of etherified hydroxyl groups in a glucose unit determines the degree of substitution, which is controlled so that cellulose derivatives have the desired solubility and viscosity in water solutions [18]. The arrangement of functional chemical groups and their subsequent physicochemical characteristics are affected by the original material, as well as the fabrication method [57].

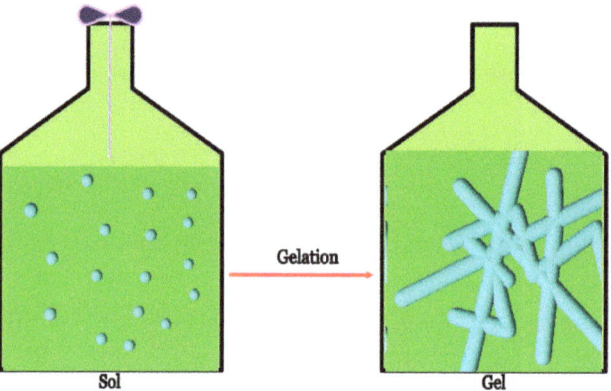

Figure 5. Sol–gel transition.

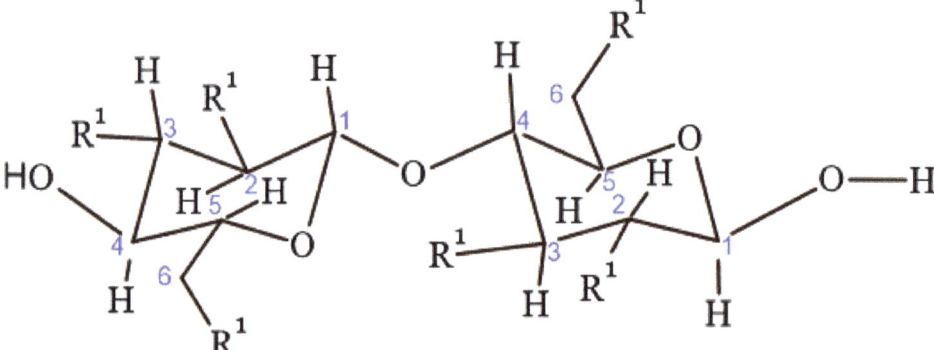

Figure 6. Chemical structure of ether derivatives, where $R^1$ is $OCH_3$, $OCH_2CH_3$, $OCH_3$, $[CH_2CH_2O]_nH$, and $O[CH_2CH(CH_3)O]H$ $OCH_2COONa$ for MC, EC, HEMC, HPC, and CMC, respectively.

Due to the presence of numerous hydroxyl groups, which can link the polymer network via hydrogen bonding, hydrogels based on natural cellulose can be prepared from a pure cellulose solution via physical crosslinking [18,58]. The highly extended hydrogen-bonded structure of cellulose results in a compactness that is difficult to dissolve in common solvents. Hence, different solvents have been used to dissolve cellulose [58]. Table 1 shows a list of cellulose derivatives, as well as their solvents and processing methods.

Table 1. Summary of some cellulose derivatives and their corresponding hydrogel processing methods (copied from Reference [58]).

| Cellulose and Cellulose Derivatives | Nature of Solvent | Solvent Systems | Corresponding Hydrogel Preparation Methods |
|---|---|---|---|
| Cellulose form wood | Polar solvents | NMMO | Solution polymerization at 85 °C |
| Cellulose from cotton pulp | Polar solvents | LiCl/DMAc | Solution polymerization at 75–90 °C |
| Filter paper | Ionic solvents | [Amim]Cl | Solution polymerization at 70 °C, 2 h |
| Tunicate cellulose | Alkali aqueous | Alkali/urea | Polymerization at −12 to −10 °C, 5–10 min |
| Cotton linter | systems | Alkali/thiourea | Polymerization at −5 °C, 2–10 min |
| Carboxymethylcellulose | Alkali aqueous | $H_2O$ | Solution polymerization, in situ polymerization |
| (CMC) | systems | DCM/DMSO | Solution polymerization, in situ polymerization |
| Methyl cellulose (MC) | Polar solvents | $H_2O$ | Solution polymerization, cryogenic treatment |
| Hydroxyethyl cellulose (HEC) | Polar solvents | $H_2O$/ethanol | Solution polymerization, inverse-phase suspension polymerization |
| Hydroxypropyl methyl cellulose | Polar solvents | Acetone/ $H_2O$ | Chemical crosslinking |

NMMO, N-methylmorpholine-N-oxide; LiCl/DMAc, lithium chloride/dimethylacetamide; [Amim]Cl, 1-allyl-3-methylimidazolium chloride; $H_2O$, water; DCM/DMSO, dichloromethane/dimethyl sulfoxide.

## 4. Design and Characterization of Hydrogel Scaffolds

Tissue engineering integrates engineering and cell science principles and consists of three elements, namely scaffolds, cells, and growth factors, as shown in Figure 7 [36]. Scaffolds play an integral role in the development of tissue regeneration by providing structure support in tri-dimensional space to accommodate and guide their growth into a particular tissue [70]. Hydrogels can be used as scaffolds that imitate extracellular matrices to encapsulate and deliver cells, to provide structural integrity and bulk for cellular organization and morphogenic guidance, to act as tissue barriers and bio adhesives, to serve as drug depots, and to deliver bioactive moieties that promote natural reparative mechanisms [71]. Forming hydrogels for cellular experiments typically entails either encapsulating viable cells within the material or fabricating substrates, using molds that are later seeded with cells [72].

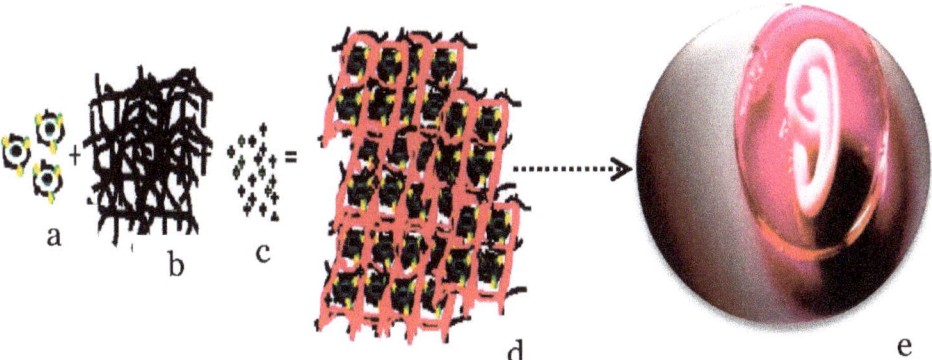

Figure 7. Tissue generation from cell to organ; cell (**a**), scaffold (**b**), bioactive factors (**c**), tissue engineering construct (**d**), and synthetic organ (**e**).

A variety of hydrogel properties, such as mechanics, swelling, mesh size, and degradation, may be of interest to characterize. When purchasing commercial kits or following specific hydrogel recipes, these may already be known and do not need to be described by every user. However, it is critical to understand how these characteristics are defined and how they may affect the utility of hydrogels in cell culture applications [72]. Hence, a proper designing of engineering hydrogel scaffolds considering all the possible factors is essential and a prerequisite for controlling cell orientation and tissue growth, and a few are listed in Table 2.

Table 2. Design parameters consideration and characterization of hydrogel scaffolds for electroactive tissues.

| Parameters | The Study Significance for Tissue Engineering | Instrument/Test Method | References |
|---|---|---|---|
| Molecular structures | To know the formation of hydrogel through investigating functional group reaction and intermolecular bonding. | Fourier-transform infrared (FTIR) spectroscopy | [73,74] |
| Morphologies | To justify suitability of hydrogel for cell adhesion by examining the mechanical toughness of hydrogel. | Scanning electron microscopy (SEM) | [73] |
| Polymer Morphology | To understand suitability of hydrogel for adhesivity to cells by studying the crystalline nature of polymeric hydrogels. | X-ray diffraction (XRD) | [73] |
| Cross polarization | Enables to know local magnetic fields around atomic nuclei/magnetic angle spinning by examining the molecular identity and structure. | Nuclear magnetic resonance (NMR) | [75] |
| Thermal stability | To Establish a connection between temperatures decomposition properties of substances through weight loss by studying the thermal property of the material. | Thermogravimetric (TGA) analysis | [73] |
| Thermal properties | To investigate the correlation between temperature and particular physical properties of the materials to use an aqueous phase diagram and the study of material physicochemical parameters in a composite formation. | Differential scanning calorimetry (DSC) | [74] |
| Swelling kinetics | Control of the most efficient way to transfer nutrients to cells and absorb wound exudates for rapid wound healing. The swelling properties can be used to detect batch-to-batch variations and consistency in hydrogel fabrication properties, as well as to determine whether the hydrogel mechanics are changing over time. | Soaking and swelling ratio calculation | [72–74] |
| Hydrophility | Enables to know the number of active hydrophilic groups. | Contact angle measurement by drop-shape analyzer | [75] |
| Electrical conductivities | Capable of delivering the electrical stimulation of nerve cells by measuring the electrical conductivity of scaffold. | Four-probe method, two-point probe, insulation resistance tester | [73,76,77] |
| Electrical and dielectric investigations | To investigate the correlation between temperature and electrical conductivity properties of the materials. | Broadband dielectric spectroscopy (BDS-40) | [78] |
| Electro mechanical properties | To simulate electrical properties of nerve cell by studying the dielectric behavior of gel through actuation test. | Laser displacement sensor | [74,79] |

Table 2. Cont.

| Parameters | The Study Significance for Tissue Engineering | Instrument/Test Method | References |
|---|---|---|---|
| Mechanical properties | The durability and stability of the material in culture influence cellular mechanotransduction, which has consequences for cellular behaviors such as spreading, migration, and stem cell differentiation. It is investigated by using stress–strain measurements, elastic modulus, break elongation, and tensile strength. | Tensile strength tester Atomic force microscopy (AFM) | [72–74,76,79] |
| Degradation kinetics | Understanding culture stability and biophysical properties such as hydrogel elastic modulus are made possible with the degradation kinetics analysis. Changes in mechanics and swelling that may affect cell behaviors such as motility, spreading, and traction force generation is correlated with degradation kinetics. Stability to a certain timescales is useful even for degradable hydrogels mechanical and or enzymatic disruption may require in isolating cells from hydrogels that require kinetic degradation analysis. | Buffer degradation profile, changes in mechanical properties | [72] |
| Antimicrobial activity | Enables us to understand tissue infections through bacterial surface adhesion and subsequent colonization. | The agar plate method Disc agar diffusion method | [80,81] |
| Purity | Rather than extracting cells for analysis, some hydrogel studies will require in-situ cell imaging to visualize cells and biomolecules in hydrogels, necessitating knowledge of hydrogel transparency. Neat hydrogel has a high degree of transparency. | UV–Vis Spectroscopy | [72,74] |
| Porosity | Influence nutrient flux throughout the matrix is studied by the measurement of the diffusion of fluorescently tagged polymers entrapped within the hydrogel. The ability of hydrogel to allow nutrients, oxygen, and metabolic products to diffuse easily into their matrices need to be studied. | SEM, Fluorescence recovery after photo-bleaching (FRAP), DNA electrophoresis | [72,79] |
| Self-healing activity | Considering the strong penetrability to biological systems, examining the reversible melting process and recrystallization under heating and cooling cycle of hydrogel is essential. | Healing efficiency calculation by tracking optical microscopy. | [82] |
| Electro stimulated Cell Culture | To examine cell viability through electro stimulating potentials | Fluorescence staining and a MTT assay. | [77] |
| Animal Experiments | The electro-active hydrogels combined with electrical fields, mimicking the electro-physiological environment in native tissues for proof of concept in skin tissue regeneration. | in vitro biological evaluation | [77] |

## 5. Hydrogel Conductivity Inclusion

### 5.1. Electro-Active Tissues

Electrical conductivity is an integral component of the human body [83]. Neurons function as a result of interacting networks woven by nerve cells. The nervous system is thought to contain approximately a trillion neurons. These highly irregularly shaped cells have the basic properties of the nervous system, such as intrinsic electrical conductivity. The ability of neurons to transmit signals from one neuron to another, as well as from a neuron to muscles and glands, is referred to as conductivity. The cell membrane allows a relatively large amounts of potassium ions to diffuse out of the cell, while allowing only a small amount of sodium ions to enter. These diffusive movements are simply the result of these ions moving down concentration gradients, following active transport by the sodium–potassium pump. When a voltage-gated ion channel opens, positively charged sodium ions diffuse into the axon, changing the membrane potential from −70 mv to zero and even higher, frequently reaching +35 mv. The membrane is said to have depolarized at that point. It happens in about a half-millisecond. The sodium gate then closes, and the usual outward diffusion of potassium occurs, causing the membrane potential to return to −70 and possibly lower to −73, due to a temporary overshoot in outward diffusion of potassium. This return to resisting is referred to as repolarization. Repolarization takes approximately half a millisecond. Thus, an action potential is a depolarization that is followed by a repolarization that takes about a millisecond to complete for a set of cells and tissues to function [55,84].

Electrical stimulation is a concern that is specific to a subset of cell types, including neurons and myocytes, in nerve-tissue engineering [85] (Table 3). As a result, electro-active biomaterials are required [61]. To meet these performance requirements, cellulose scaffolds coated with conductive materials can be used. Such materials have defined pore sizes, physicochemical characteristics, and electrical conductivities; they are also biocompatible and promote neurological differentiation [57].

**Table 3.** Conductivity of human tissue (Siemens per meter ($Sm^{-1}$); copied from Reference [85].

| Tissues | $Sm^{-1}$ | Tissues | $Sm^{-1}$ |
| --- | --- | --- | --- |
| Cerebellum | 0.10 | Pancreas | 0.35 |
| C.S.F. | 2.00 | Prostate | 0.40 |
| Cornea | 0.40 | Small intestine | 0.50 |
| Eye humor | 1.50 | Spleen | 0.10 |
| Grey matter | 0.10 | Stomach | 0.50 |
| Hypothalamus | 0.08 | Stomach contents | 0.35 |
| Eye lens | 0.25 | Tendon | 0.30 |
| Pineal body | 0.08 | Testis | 0.35 |
| Pituitary | 0.08 | Thyroid gland | 0.50 |
| Salivary gland | 0.35 | Trachea | 0.35 |
| Thalamus | 0.08 | Urine | 0.70 |
| Tongue | 0.30 | Blood | 0.70 |
| White matter | 0.06 | Cortical bone | 0.02 |
| Adrenals | 0.35 | Bone marrow | 0.06 |
| Bladder | 0.20 | Cartilage | 0.18 |
| Large intestine | 0.10 | Fat | 0.04 |
| Duodenum | 0.50 | Muscle | 0.35 |
| Esophagus | 0.50 | Nerve (Spinal cord) | 0.03 |
| Bile | 1.40 | Skin | 0.10 |
| Gall bladder | 0.20 | Tooth | 0.02 |
| Heart | 0.10 | Ligament | 0.30 |

*5.2. Electro-Active Hydrogel*

Cellulose scaffolds are an excellent material for nerve neurogenesis, due to their customizable surface chemistry and mechanical characteristics. Perhaps, to improve integrin-based attachment and cell–scaffold interactions, cellulose materials can be chemically modified and protein-coated [57]. Electro-active biomaterial-mediated stem-cell differentiation into specific cell lineages is of great significance for tissue regeneration. Although the underlying molecular events and mechanism of electro activation are not fully understood, there are some general guidelines for designing conductive hydrogels. Aside from matching the morphology and mechanical properties of hydrogels to the tissue microenvironment, it is critical to mimic the tissue electrophysiological environment. Neurons form synapses to transmit electrical signals and integrate into neuronal circuits in the mature nervous system. Neurons switch from an active electrical transmission state to an electrically silent and growth-competent state after axonal injury. When a cell shifts a single cell to multicellular collections and tissues, a striking parallel is found. Cells are regulated not only by their own potential, but also the potential of their neighboring cells via gap junctions [84,86].

Electrically conductive materials and crosslinked hydrogel networks are used to create conductive hydrogels through co-networks, blends, and self-assembly. This can be achieved through post-polymerization of a conducting monomer in a prefabricated hydrogel; composite strategies involving the mixing of conductive materials/monomers and hydrophilic polymers/monomers, followed by simultaneous or stepwise crosslinking to produce conductive hydrogels; and in situ polymerization involving the self-assembly of the modified electrically conductive materials [87]. The pros and cons of the strategies are given in Table 4.

**Table 4.** Advantages and disadvantages of different design strategies for preparing conductive hydrogels (copied from Reference [87]).

| Design Strategies | Advantages | Disadvantages |
|---|---|---|
| In situ polymerization | • Barrier-free preparation<br>• Uniform polymerization | • Potentially cytotoxic unreactive oxidants and monomers<br>• Need for chemical synthesis process design |
| Post-polymerization | • Adding conductive materials to synthesized hydrogels<br>• Possibility of the conductive coating method | • Cytotoxic unreactive oxidants and monomers<br>• Additional polymerization step |
| Composite strategies | • Adjustable conductivity<br>• No cytotoxic unreactive oxidants or monomers | • Non-uniform additive distribution<br>• Conductive additive toxicity |

Different types of conductive materials exhibit varying properties. Research (a review by Rong et al., 2018) [88] shows that three classes of materials are used on hydrogels for conductive purposes: metals, carbon allotropes, and intrinsically conductive polymers (ICPs).

For semi-conductor hydrogels, ionomers and silicones may be used as conducting materials as well [88]. As just an instance, when cellulose dissolved in an aqueous solution of benzyltrimethyl ammonium hydroxide (BzMe3NOH), ionic conductive cellulose hydrogels (CCHs) with anti-freezing properties were directly fabricated by chemical crosslinking without further treatment [89].

## 6. Incorporation of Conductive Materials on to Cellulose Hydrogels

### 6.1. Intrinsically Conductive Polymers (ICPs)

ICPs are conjugated polymers that have an extended delocalized system of $\pi$ electrons that generally runs along the polymer backbone and is made conductive through doping [90]. The free motion of the loosely held $\pi$ electrons within the unsaturated segments can open an electrical path for itinerant charge carriers. However, the changes in surface zeta potential and polymer surface properties, such as wettability and spatial conformation, can affect the cell behavior of electrical stimulation behavior [91]. It was subsequently understood that several polymers, such as polyacetylene, polypyrrole (PPy), polyaniline, polythiophene, poly (p-phenylene), and poly (3,4-ethylenedioxythiophene) polystyrene sulfonate (PEDOT/PSS), are conjugated polymers whose electrical conductivity is dramatically increased by doping. Doping involves the addition of a small amount of a chemical agent, which alters the electronic structure. The doping process, on the other hand, is reversible and involves a redox process.

Two major fabrication routes have already been investigated for the development of conductive polymer hydrogels by using ICP: gelation of CPs and hydrophilic polymers/monomers by self-assembly or the introduction of cross-linkable elements, as well as chemical oxidation; and electrochemical polymerization in a prefabricated hydrogel [88]. In a specific instance, X. Liang et al. developed a conductive hydrogel by polymerizing PPy through a prefabricated of chemically crosslinked microcrystalline cellulose (MMC) [73].

Gelation and chemical physical polymerization were employed after mixing the bacterial cellulose (BC) and PEDOT/PSS [92] to enhance the conductivity also.

## 6.2. Carbon Allotropes

Carbon-based materials are regarded as promising conductive materials for the fabrication of conductive hydrogels, due to their unique properties of high electrical conductivities, excellent environmental stability, and low production costs [88]. Materials that consist of only carbon atoms can have a wide range of conductivities, from the insulator diamond to conductors such as charcoal [93], carbon black (CB), graphene, and carbon nanotubes (CNTs) [94,95]. The level of conductivity will depend on the degree of delocalized electrons, thus making the graphitization and purity of the carbon compounds important factors [96]. Carbon-based biomaterials are commonly used as reinforcing agents in tissue-engineering applications to improve the mechanical performance and conductivity of the polymer matrix. Along with their unique mechanical properties, chemical stability, large specific surface area, and high electrical conductivity, graphene and carbon nanotube-based materials are the most widely used in tissue engineering. Furthermore, their large surface area and abundance of functional groups aid in the loading and release of bioactive species, such as chemical drugs, growth factors, genes, and proteins [83].

Blending with various polymers and self-assembly after modification are the two most common ways of preparing carbon-based conductive hydrogels [88]. Cellulose nanocrystals were grafted in to phenylboronic acid (CNCs-ABA) and multi-walled carbon nanotubes (MWCNTs) to develop electrical conductivity [97]. Another illustration is the post-polymerization of MWCNTs, with graphene powder (r-GOx) to adhere to pure regenerated cellulose-based electrolyte membranes [79].

Sometimes more than one conductive material may be used to enhance the conductivity of hydrogel. In a specific instance, to develop a conductive hydrogel, bacterial cellulose (rBC) slurry was mixed with PPy and single-walled carbon nanotubes (SWCNTs) and crosslinked in a stepwise manner. After preparing the rBC/PPy hydrogel, CNTs were added to the prepared rBC/PPy solution and dispersed before physical crosslinking had occurred [75]. The different preparation techniques of cellulosic-based conductive hydrogels is illustrated in Table 5.

## 6.3. Metals

Metals' exceptional features, such as high conductivity, optical, magnetic, and catalytic properties, as well as metallic nanoparticles/nanowires, such as Al, Au, Ag, Cu, etc., have been widely used in the fabrication of conductive hydrogels [98]. Due to their high mechanical properties, fatigue resistance, and conductivity, bulk metals, such as titanium, magnesium, and stainless steel, have been used as bone-repair implants [83]. Although metals have some drawbacks, such as lack of flexibility, toxicity, cost, and negative environmental effects, they remain the only viable alternative for applications requiring high conductivity [91,99].

The common methods to develop metal-based conductive hydrogels are UV crosslinking and the in situ polymerization of hydrogel monomer and reduction of metal ions, using reducing agent [88]. An illustration of in situ polymerization through simultaneous crosslinking was performed by blending a precursor cellulose microcrystalline (CMC) solution; a monomer acrylic acid, initiator ammonium persulfate, catalyst N,N,N',N' tetramethylethylenediamine and crosslinker aluminum hexahydrate ($AlCl_3 \cdot 6H_2O$), and the conductive materials of metallic ions of $Al^{3+}$ produced conductive hydrogels [82]. Another example is grafting of acrylonitrile (AN) and acrylamide copolymers onto the hydroxypropyl methylcellulose (HPMC) chains in the presence of zinc chloride ($ZnCl_2$), using ceric ammonium nitrate (AM) as the initiator [76]. In situ polymerization to form nanocomposite hydrogels of tannic acid–coated cellulose nanocrystal (TA@CNCs) ionic gel and then immersion in $Al^{3+}$ solution to produce ionic coordination [100] have also been reported to develop cellulosic-based conductive hydrogel.

Table 5. Preparation techniques of cellulosic-based conductive hydrogel.

| Hydrogel Features | Method of Crosslinking | Hydrogel Material | Conductivity (S/m) | (Potential) Application | Reference |
|---|---|---|---|---|---|
| Electro-active | Composite strategies | rBC/PPy and rBC/PPy/CNT | $6.2 \times 10^{-2}$ | Cell proliferation | [75] |
| Conductive | Post-Polymerization | MCC/PPy | 0.783 | Electrochemical biosensors, electro-stimulated controlled drug release, and neural prosthetics | [73] |
| Conductive, self-healing, and strain- and thermal-sensitive performance | In situ polymerization | PAA-CMC-Al$^{3+}$ | 162 | Flexible and wearable temperature-sensing devices | [82] |
| Self-healing, shape memory, and biocompatible | Composite strategies | CNCs-ABA | $3.8 \times 10^{-2}$ | Strain sensors | [97] |
| Ultra-stretchable, tough, anti-freezing, and conductive | Composite strategies via graft polymerization | HPMC-g-P (AN-co-AM) | 1.54 | Strain Sensor | [76] |
| Transparent, anti-freezing, and ionic conductive | Chemical crosslinking | CCHs | 2.37 | Sensor | [89] |
| Thermally stable, crystalline, and electroactive | Composite strategies | Polyvinyl alcohol cellulose (PC) | | Actuator | [74] |
| Anisotropic and conductive, with high water content | Composite strategies | BC-PEDOT/PSS | Conductivity is proved by light emitting diode | Scaffolds, implantable biosensors, and smart soft electronic devices | [92] |
| Tough, stretchable, self-adhesive, self-healing, and strain-sensitive | In situ polymerization | TA@CNCs | | Wearable electronic sensors and healthcare monitoring | [100] |
| Electroactive and ultrafast for electro-mechanical response | Post-polymerization | Cellulose-based all-hydrogel artificial muscles membrane. | 0.83–2.49 | Transportation of nerve impulses from human muscle | [79] |

## 7. Conclusions and Future Outlook

Tissue engineering and regeneration are growing fields that have the potential to revolutionize biomedical engineering. On the other hand, the translation of laboratory findings to the clinic has, indeed, been weak. The electrical stimulation of tissue can improve vascularization and differentiation of stem cells into various cells, but it is difficult to achieve targeted and controlled electrical stimulation. Electrically, conduction hydrogels hold great potential for conquering such barriers. In order to reconstruct completely operational tissue, the physicochemical and biological characteristics of hydrogel must enable cell generation and differentiation. The papers discussed in this article have consolidated research in the area of cellulose-based biomaterials in the broad sense of characteristics that regulate cellular functions and scaffold practicability for tissue engineering. Due to their diverse and adaptable physicochemical and biological properties, cellulose-based materials clearly have a high potential to become the next generation of standard biomaterials. Hydrogel properties are constantly evolving in an attempt to match the sophistication of native tissues.

In tissue construct, cell and tissue microenvironments vary at different periods throughout human life, notably during organ development and tissue repair, and designing an electro-active scaffold hydrogel that accommodate the changes over the period is a big challenge. Despite considerable advances in tissue engineering, neither material fully conveys the intricacies of native tissue or restores function to an ideal level. Conductive hydrogels have attracted a lot of attention for their widespread use in biomedical engineering, due to their structural similarity to soft tissue. However, designing conductive materials that combine biocompatibility with good mechanical and electrical properties remains difficult. AS a practical matter, the vestiges difficulty is just to develop new materials-design approaches in order to achieve actual biomimetic tissues. As the complexity of the application increases, such as in highly dynamic tissues, active remodeling of the scaffolding will be required. As a result, the complex interaction between cells and the artificial matrix will be critical. Perhaps cellulose-based hydrogels are difficult to prepare because cellulose is insoluble in water and common organic solvents, and the use of ionic liquid is evolving.

**Author Contributions:** Conceptualization and original draft writing, E.K.G.; supervision, conceptualization, writing, and reviewing and editing, X.S.; resources, validation, reviewing, and editing, B.F.A.; resources, reviewing, and editing, K.A.B.; proofreading, editing, and budget administration, M.G.T. All authors contributed equally to the work reported. All authors have read and agreed to the published version of the manuscript.

**Funding:** This research received no external funding.

**Institutional Review Board Statement:** Not applicable.

**Informed Consent Statement:** Not applicable.

**Data Availability Statement:** The data used to support the review summary of this paper are included within the article.

**Conflicts of Interest:** The authors declare that they have no conflict of interest.

## References

1. Lee, K.Y.; Mooney, D.J. Hydrogels for tissue engineering. *Chem. Rev.* **2001**, *101*, 1869–1880. [CrossRef]
2. Berthiaume, F.; Maguire, T.J.; Yarmush, M.L. Tissue engineering and regenerative medicine: History, progress, and challenges. *Annu. Rev. Chem. Biomol. Eng.* **2011**, *2*, 403–430. [CrossRef]
3. Liao, I.C.; Moutos, F.T.; Estes, B.T.; Zhao, X.; Guilak, F. Composite three-dimensional woven scaffolds with interpenetrating network hydrogels to create functional synthetic articular cartilage. *Adv. Funct. Mater.* **2013**, *23*, 5833–5839. [CrossRef]
4. Guilak, F.; Moutos, F. Three-Dimensional Fiber Scaffolds for Tissue Engineering. U.S. Patent 11/406,519, 25 February 2007.
5. Wei, Z.; Wu, C.; Li, R.; Yu, D.; Ding, Q. Nanocellulose based hydrogel or aerogel scaffolds for tissue engineering. *Cellulose* **2021**, *28*, 7497–7520. [CrossRef]

6. Yahya, E.B.; Amirul, A.A.; Abdul Khalil, P.S.A.; Olaiya, N.G.; Iqbal, M.O.; Jummaat, F.; Atty Sofea, K.A.; Adnan, A.S. Insights into the Role of Biopolymer Aerogel Scaffolds in Tissue Engineering and Regenerative Medicine. *Polymers* **2021**, *13*, 1612. [CrossRef] [PubMed]
7. Jayaraman, K.; Kotaki, M.; Zhang, Y.; Mo, X.; Ramakrishna, S. Recent advances in polymer nanofibers. *J. Nanosci. Nanotechnol.* **2004**, *4*, 52–65. [PubMed]
8. Mo, X.; Xu, C.; Kotaki, M.; Ramakrishna, S. Electrospun P (LLA-CL) nanofiber: A biomimetic extracellular matrix for smooth muscle cell and endothelial cell proliferation. *Biomaterials* **2004**, *25*, 1883–1890. [CrossRef] [PubMed]
9. Burg, K.J. Tissue Engineering Composite. US Patent No. 6,991,652, 31 January 2006.
10. Hutmacher, D.W. Scaffolds in tissue engineering bone and cartilage. In *The Biomaterials: Silver Jubilee Compendium*; Elsevier: Amsterdam, The Netherlands, 2000; pp. 175–189.
11. Al-Munajjed, A.A.; Hien, M.; Kujat, R.; Gleeson, J.P.; Hammer, J. Influence of pore size on tensile strength, permeability and porosity of hyaluronan-collagen scaffolds. *J. Mater. Sci. Mater. Med.* **2008**, *19*, 2859–2864. [CrossRef]
12. Pina, S.; Oliveira, J.M.; Reis, R.L. Natural-based nanocomposites for bone tissue engineering and regenerative medicine: A review. *Adv. Mater.* **2015**, *27*, 1143–1169. [CrossRef]
13. Liu, X.; Ma, P.X. Polymeric scaffolds for bone tissue engineering. *Ann. Biomed. Eng.* **2004**, *32*, 477–486. [CrossRef]
14. Huang, J.; Hu, X.; Lu, L.; Ye, Z.; Zhang, Q.; Luo, Z. Electrical regulation of Schwann cells using conductive polypyrrole/chitosan polymers. *J. Biomed. Mater. Res. Part A Off. J. Soc. Biomater. Jpn. Soc. Biomater. Aust. Soc. Biomater. Korean Soc. Biomater.* **2010**, *93*, 164–174. [CrossRef] [PubMed]
15. Jiang, L.; Wang, Y.; Liu, Z.; Ma, C.; Yan, H.; Xu, N.; Gang, F.; Wang, X.; Zhao, L.; Sun, X. Three-Dimensional Printing and Injectable Conductive Hydrogels for Tissue Engineering Application. *Tissue Eng. Part B Rev.* **2019**, *25*, 398–411. [CrossRef] [PubMed]
16. Ebara, M.; Kotsuchibashi, Y.; Uto, K.; Aoyagi, T.; Kim, Y.-J.; Narain, R.; Idota, N.; Hoffman, J.M. Introductory Guide to Smart Biomaterials. In *Smart Biomaterials*; Springer: Tokyo, Japan, 2014; pp. 1–7.
17. Ahmed, E.M. Hydrogel: Preparation, characterization, and applications: A review. *J. Adv. Res.* **2015**, *6*, 105–121. [CrossRef]
18. Kabir, S.M.F.; Sikdar, P.P.; Haque, B.; Bhuiyan, M.A.R.; Ali, A.; Islam, M.N. Cellulose-based hydrogel materials: Chemistry, properties and their prospective applications. *Prog. Biomater.* **2018**, *7*, 153–174. [CrossRef]
19. Gulrez, S.K.; Al-Assaf, S.; Phillips, G.O. Hydrogels: Methods of preparation, characterisation and applications. In *Progress in Molecular and Environmental Bioengineering—From Analysis and Modeling to Technology Application*; Books on Demand: Norderstedt, Germany, 2011; pp. 117–150.
20. Gibas, I.; Janik, H. Synthetic polymer hydrogels for biomedical applications. *Chem. Chem. Technol.* **2010**, *4*, 297–304. [CrossRef]
21. Wang, C.; Stewart, R.J.; Kopeček, J. Hybrid hydrogels assembled from synthetic polymers and coiled-coil protein domains. *Nature* **1999**, *397*, 417–420. [CrossRef] [PubMed]
22. Catoira, M.C.; Fusaro, L.; Di Francesco, D.; Ramella, M.; Boccafoschi, F. Overview of natural hydrogels for regenerative medicine applications. *J. Mater. Sci. Mater. Med.* **2019**, *30*, 115. [CrossRef] [PubMed]
23. Pal, D.; Nayak, A.K.; Saha, S. *Cellulose-Based Hydrogels: Present and Future*; Springer: Singapore, 2019; pp. 285–332. [CrossRef]
24. Bhattarai, N.; Gunn, J.; Zhang, M. Chitosan-based hydrogels for controlled, localized drug delivery. *Adv. Drug Deliv. Rev.* **2010**, *62*, 83–99. [CrossRef] [PubMed]
25. Atallah, P.; Schirmer, L.; Tsurkan, M.; Limasale, Y.D.P.; Zimmermann, R.; Werner, C.; Freudenberg, U. In situ-forming, cell-instructive hydrogels based on glycosaminoglycans with varied sulfation patterns. *Biomaterials* **2018**, *181*, 227–239. [CrossRef]
26. Pereira, R.; Carvalho, A.; Vaz, D.C.; Gil, M.; Mendes, A.; Bártolo, P. Development of novel alginate based hydrogel films for wound healing applications. *Int. J. Biol. Macromol.* **2013**, *52*, 221–230. [CrossRef]
27. Augst, A.D.; Kong, H.J.; Mooney, D.J. Alginate hydrogels as biomaterials. *Macromol. Biosci.* **2006**, *6*, 623–633. [CrossRef] [PubMed]
28. Kim, U.-J.; Park, J.; Li, C.; Jin, H.-J.; Valluzzi, R.; Kaplan, D.L. Structure and properties of silk hydrogels. *Biomacromolecules* **2004**, *5*, 786–792. [CrossRef] [PubMed]
29. Wang, H.Y.; Zhang, Y.Q. Processing silk hydrogel and its applications in biomedical materials. *Biotechnol. Prog.* **2015**, *31*, 630–640. [CrossRef] [PubMed]
30. Griffanti, G.; Jiang, W.; Nazhat, S.N. Bioinspired mineralization of a functionalized injectable dense collagen hydrogel through silk sericin incorporation. *Biomater. Sci.* **2019**, *7*, 1064–1077. [CrossRef]
31. Wright, E.R.; McMillan, R.A.; Cooper, A.; Apkarian, R.P.; Conticello, V.P. Self-Assembly of Hydrogels from Elastin-Mimetic Block Copolymers. *MRS Online Proc. Libr.* **2002**, *724*. [CrossRef]
32. Yamamoto, M.; Ikada, Y.; Tabata, Y. Controlled release of growth factors based on biodegradation of gelatin hydrogel. *J. Biomater. Sci. Polym. Ed.* **2001**, *12*, 77–88. [CrossRef]
33. Ye, K.Y.; Sullivan, K.E.; Black, L.D. Encapsulation of cardiomyocytes in a fibrin hydrogel for cardiac tissue engineering. *J. Vis. Exp. JoVE* **2011**, *55*, e3251. [CrossRef]
34. Wolf, M.T.; Daly, K.A.; Brennan-Pierce, E.P.; Johnson, S.A.; Carruthers, C.A.; D'Amore, A.; Nagarkar, S.P.; Velankar, S.S.; Badylak, S.F. A hydrogel derived from decellularized dermal extracellular matrix. *Biomaterials* **2012**, *33*, 7028–7038. [CrossRef]
35. Giobbe, G.G.; Crowley, C.; Luni, C.; Campinoti, S.; Khedr, M.; Kretzschmar, K.; De Santis, M.M.; Zambaiti, E.; Michielin, F.; Meran, L. Extracellular matrix hydrogel derived from decellularized tissues enables endodermal organoid culture. *Nat. Commun.* **2019**, *10*, 5658. [CrossRef]

36. Yue, S.; He, H.; Li, B.; Hou, T. Hydrogel as a Biomaterial for Bone Tissue Engineering: A Review. *Nanomaterials* **2020**, *10*, 1511. [CrossRef]
37. Maitra, J.; Shukla, V.K. Cross-linking in hydrogels-a review. *Am. J. Polym. Sci.* **2014**, *4*, 25–31.
38. Esa, F.; Tasirin, S.M.; Abd Rahman, N. Overview of bacterial cellulose production and application. *Agric. Agric. Sci. Procedia* **2014**, *2*, 113–119. [CrossRef]
39. Kharkar, P.M.; Kiick, K.L.; Kloxin, A.M. Designing degradable hydrogels for orthogonal control of cell microenvironments. *Chem. Soc. Rev.* **2013**, *42*, 7335–7372. [CrossRef] [PubMed]
40. Guarino, V.; Alvarez-Perez, M.A.; Borriello, A.; Napolitano, T.; Ambrosio, L. Conductive PANi/PEGDA macroporous hydrogels for nerve regeneration. *Adv. Healthc. Mater.* **2013**, *2*, 218–227. [CrossRef]
41. Fu, F.; Wang, J.; Zeng, H.; Yu, J. Functional Conductive Hydrogels for Bioelectronics. *ACS Mater. Lett.* **2020**, *2*, 1287–1301. [CrossRef]
42. Zhang, Y.S.; Khademhosseini, A. Advances in engineering hydrogels. *Science* **2017**, *356*, eaaf3627. [CrossRef] [PubMed]
43. De Oliveira Barud, H.G.; da Silva, R.R.; da Silva Barud, H.; Tercjak, A.; Gutierrez, J.; Lustri, W.R.; de Oliveira Junior, O.B.; Ribeiro, S.J. A multipurpose natural and renewable polymer in medical applications: Bacterial cellulose. *Carbohydr. Polym.* **2016**, *153*, 406–420. [CrossRef]
44. Bhat, A.; Khan, I.; Usmani, M.A.; Umapathi, R.; Al-Kindy, S.M. Cellulose an ageless renewable green nanomaterial for medical applications: An overview of ionic liquids in extraction, separation and dissolution of cellulose. *Int. J. Biol. Macromol.* **2019**, *129*, 750–777. [CrossRef]
45. Bashari, A.; Rouhani Shirvan, A.; Shakeri, M. Cellulose-based hydrogels for personal care products. *Polym. Adv. Technol.* **2018**, *29*, 2853–2867. [CrossRef]
46. Nandakumar, D.K.; Ravi, S.K.; Zhang, Y.; Guo, N.; Zhang, C.; Tan, S.C. A super hygroscopic hydrogel for harnessing ambient humidity for energy conservation and harvesting. *Energy Environ. Sci.* **2018**, *11*, 2179–2187. [CrossRef]
47. Filipova, I.; Irbe, I.; Spade, M.; Skute, M.; Dāboliņa, I.; Baltiņa, I.; Vecbiskena, L. Mechanical and air permeability performance of novel biobased materials from fungal hyphae and cellulose fibers. *Materials* **2021**, *14*, 136. [CrossRef]
48. Fricain, J.; Granja, P.; Barbosa, M.; De Jéso, B.; Barthe, N.; Baquey, C. Cellulose phosphates as biomaterials. In vivo biocompatibility studies. *Biomaterials* **2002**, *23*, 971–980. [CrossRef]
49. Torres, F.G.; Commeaux, S.; Troncoso, O.P. Biocompatibility of bacterial cellulose based biomaterials. *J. Funct. Biomater.* **2012**, *3*, 864–878. [CrossRef] [PubMed]
50. Credou, J.; Berthelot, T. Cellulose: From biocompatible to bioactive material. *J. Mater. Chem. B* **2014**, *2*, 4767–4788. [CrossRef]
51. Teeri, T.T.; Brumer, H., III; Daniel, G.; Gatenholm, P. Biomimetic engineering of cellulose-based materials. *Trends Biotechnol.* **2007**, *25*, 299–306. [CrossRef] [PubMed]
52. Torgbo, S.; Sukyai, P. Biodegradation and thermal stability of bacterial cellulose as biomaterial: The relevance in biomedical applications. *Polym. Degrad. Stab.* **2020**, *179*, 109232. [CrossRef]
53. Naahidi, S.; Jafari, M.; Logan, M.; Wang, Y.; Yuan, Y.; Bae, H.; Dixon, B.; Chen, P. Biocompatibility of hydrogel-based scaffolds for tissue engineering applications. *Biotechnol. Adv.* **2017**, *35*, 530–544. [CrossRef]
54. Goodnow, R.A. *Platform Technologies in Drug Discovery and Validation*; Academic Press: Cambridge, MA, USA, 2017.
55. Bohner, M.; Lemaitre, J. Can bioactivity be tested in vitro with SBF solution? *Biomaterials* **2009**, *30*, 2175–2179. [CrossRef]
56. Butler, D.L.; Goldstein, S.A.; Guldberg, R.E.; Guo, X.E.; Kamm, R.; Laurencin, C.T.; McIntire, L.V.; Mow, V.C.; Nerem, R.M.; Sah, R.L. The impact of biomechanics in tissue engineering and regenerative medicine. *Tissue Eng. Part B Rev.* **2009**, *15*, 477–484. [CrossRef]
57. Hickey, R.J.; Pelling, A.E. Cellulose Biomaterials for Tissue Engineering. *Front. Bioeng. Biotechnol.* **2019**, *7*, 45. [CrossRef]
58. Dutta, S.D.; Patel, D.K.; Lim, K.T. Functional cellulose-based hydrogels as extracellular matrices for tissue engineering. *J. Biol. Eng.* **2019**, *13*, 55. [CrossRef]
59. Hebeish, A.; Guthrie, T. *The Chemistry and Technology of Cellulosic Copolymers*; Springer Science & Business Media: Berlin/Heidelberg, Germany, 2012; Volume 4.
60. Klemm, D.; Philipp, B.; Heinze, T.; Heinze, U.; Wagenknecht, W. *Comprehensive Cellulose Chemistry*; Wiley-VCH Verlag GmbH: Weinheim, Germany, 1998; Volume 1.
61. Hu, X.; Ricci, S.; Naranjo, S.; Hill, Z.; Gawason, P. Protein and Polysaccharide-Based Electroactive and Conductive Materials for Biomedical Applications. *Molecules* **2021**, *26*, 4499. [CrossRef] [PubMed]
62. Ng, H.-M.; Sin, L.T.; Tee, T.-T.; Bee, S.-T.; Hui, D.; Low, C.-Y.; Rahmat, A. Extraction of cellulose nanocrystals from plant sources for application as reinforcing agent in polymers. *Compos. Part B Eng.* **2015**, *75*, 176–200. [CrossRef]
63. Picheth, G.F.; Pirich, C.L.; Sierakowski, M.R.; Woehl, M.A.; Sakakibara, C.N.; de Souza, C.F.; Martin, A.A.; da Silva, R.; de Freitas, R.A. Bacterial cellulose in biomedical applications: A review. *Int. J. Biol. Macromol.* **2017**, *104*, 97–106. [CrossRef] [PubMed]
64. Wang, J.; Tavakoli, J.; Tang, Y. Bacterial cellulose production, properties and applications with different culture methods—A review. *Carbohydr. Polym.* **2019**, *219*, 63–76. [CrossRef] [PubMed]
65. Naomi, R.; Bt Hj Idrus, R.; Fauzi, M.B. Plant- vs. Bacterial-Derived Cellulose for Wound Healing: A Review. *Int. J. Environ. Res. Public Health* **2020**, *17*, 6803. [CrossRef]

66. Khalil, H.; Jummaat, F.; Yahya, E.B.; Olaiya, N.G.; Adnan, A.S.; Abdat, M.; Nam, N.; Halim, A.S.; Kumar, U.S.U.; Bairwan, R.; et al. A Review on Micro- to Nanocellulose Biopolymer Scaffold Forming for Tissue Engineering Applications. *Polymers* **2020**, *12*, 2043. [CrossRef]
67. Agrawal, S.K.; Sanabria-DeLong, N.; Tew, G.N.; Bhatia, S.R. Rheological characterization of biocompatible associative polymer hydrogels with crystalline and amorphous endblocks. *J. Mater. Res.* **2006**, *21*, 2118–2125. [CrossRef]
68. Hennink, W.E.; van Nostrum, C.F. Novel crosslinking methods to design hydrogels. *Adv. Drug Deliv. Rev.* **2012**, *64*, 223–236. [CrossRef]
69. Sannino, A.; Demitri, C.; Madaghiele, M. Biodegradable Cellulose-based Hydrogels: Design and Applications. *Materials* **2009**, *2*, 353–373. [CrossRef]
70. Murugan, R.; Ramakrishna, S. Design strategies of tissue engineering scaffolds with controlled fiber orientation. *Tissue Eng.* **2007**, *13*, 1845–1866. [CrossRef]
71. El-Sherbiny, I.M.; Yacoub, M.H. Hydrogel scaffolds for tissue engineering: Progress and challenges. *Glob. Cardiol. Sci. Pract.* **2013**, *2013*, 316–342. [CrossRef] [PubMed]
72. Caliari, S.R.; Burdick, J.A. A practical guide to hydrogels for cell culture. *Nat. Methods* **2016**, *13*, 405–414. [CrossRef] [PubMed]
73. Liang, X.; Qu, B.; Li, J.; Xiao, H.; He, B.; Qian, L. Preparation of cellulose-based conductive hydrogels with ionic liquid. *React. Funct. Polym.* **2015**, *86*, 1–6. [CrossRef]
74. Jayaramudu, T.; Ko, H.-U.; Zhai, L.; Li, Y.; Kim, J. Preparation and characterization of hydrogels from polyvinyl alcohol and cellulose and their electroactive behavior. *Soft Mater.* **2016**, *15*, 64–72. [CrossRef]
75. Wang, L.; Hu, S.; Ullah, M.W.; Li, X.; Shi, Z.; Yang, G. Enhanced cell proliferation by electrical stimulation based on electroactive regenerated bacterial cellulose hydrogels. *Carbohydr. Polym.* **2020**, *249*, 116829. [CrossRef]
76. Chen, D.; Zhao, X.; Wei, X.; Zhang, J.; Wang, D.; Lu, H.; Jia, P. Ultrastretchable, Tough, Antifreezing, and Conductive Cellulose Hydrogel for Wearable Strain Sensor. *ACS Appl. Mater. Interfaces* **2020**, *12*, 53247–53256. [CrossRef]
77. Gan, D.; Han, L.; Wang, M.; Xing, W.; Xu, T.; Zhang, H.; Wang, K.; Fang, L.; Lu, X. Conductive and Tough Hydrogels Based on Biopolymer Molecular Templates for Controlling in Situ Formation of Polypyrrole Nanorods. *ACS Appl. Mater. Interfaces* **2018**, *10*, 36218–36228. [CrossRef]
78. Abd El-Aziz, M.E.; Youssef, A.M.; Kamel, S.; Turky, G. Conducting hydrogel based on chitosan, polypyrrole and magnetite nanoparticles: A broadband dielectric spectroscopy study. *Polym. Bull.* **2018**, *76*, 3175–3194. [CrossRef]
79. Sun, Z.; Yang, L.; Zhao, J.; Song, W. Natural Cellulose-Full-Hydrogels Bioinspired Electroactive Artificial Muscles: Highly Conductive Ionic Transportation Channels and Ultrafast Electromechanical Response. *J. Electrochem. Soc.* **2020**, *167*, 047515. [CrossRef]
80. Youssef, A.M.; Abdel-Aziz, M.E.; El-Sayed, E.S.A.; Abdel-Aziz, M.S.; Abd El-Hakim, A.A.; Kamel, S.; Turky, G. Morphological, electrical & antibacterial properties of trilayered Cs/PAA/PPy bionanocomposites hydrogel based on Fe3O4-NPs. *Carbohydr. Polym.* **2018**, *196*, 483–493. [CrossRef] [PubMed]
81. Abd El-Aziz, M.E.; Morsi, S.M.M.; Salama, D.M.; Abdel-Aziz, M.S.; Abd Elwahed, M.S.; Shaaban, E.A.; Youssef, A.M. Preparation and characterization of chitosan/polyacrylic acid/copper nanocomposites and their impact on onion production. *Int. J. Biol. Macromol.* **2019**, *123*, 856–865. [CrossRef] [PubMed]
82. Pang, J.; Wang, L.; Xu, Y.; Wu, M.; Wang, M.; Liu, Y.; Yu, S.; Li, L. Skin-inspired cellulose conductive hydrogels with integrated self-healing, strain, and thermal sensitive performance. *Carbohydr. Polym.* **2020**, *240*, 116360. [CrossRef] [PubMed]
83. Liu, Z.; Wan, X.; Wang, Z.L.; Li, L. Electroactive Biomaterials and Systems for Cell Fate Determination and Tissue Regeneration: Design and Applications. *Adv. Mater.* **2021**, *33*, e2007429. [CrossRef] [PubMed]
84. Llinas, R.R. Intrinsic electrical properties of mammalian neurons and CNS function: A historical perspective. *Front. Cell. Neurosci.* **2014**, *8*, 320. [CrossRef]
85. Hirata, A.; Takano, Y.; Kamimura, Y.; Fujiwara, O. Effect of the averaging volume and algorithm on the in situ electric field for uniform electric- and magnetic-field exposures. *Phys. Med. Biol.* **2010**, *55*, N243–N252. [CrossRef]
86. Hardy, J.G.; Cornelison, R.C.; Sukhavasi, R.C.; Saballos, R.J.; Vu, P.; Kaplan, D.L.; Schmidt, C.E. Electroactive Tissue Scaffolds with Aligned Pores as Instructive Platforms for Biomimetic Tissue Engineering. *Bioengineering* **2015**, *2*, 15–34. [CrossRef]
87. Xu, J.; Tsai, Y.L.; Hsu, S.H. Design Strategies of Conductive Hydrogel for Biomedical Applications. *Molecules* **2020**, *25*, 5296. [CrossRef]
88. Rong, Q.; Lei, W.; Liu, M. Conductive Hydrogels as Smart Materials for Flexible Electronic Devices. *Chemistry* **2018**, *24*, 16930–16943. [CrossRef]
89. Wang, Y.; Zhang, L.; Lu, A. Transparent, Antifreezing, Ionic Conductive Cellulose Hydrogel with Stable Sensitivity at Subzero Temperature. *ACS Appl. Mater. Interfaces* **2019**, *11*, 41710–41716. [CrossRef]
90. Tomczykowa, M.; Plonska-Brzezinska, M.E. Conducting Polymers, Hydrogels and Their Composites: Preparation, Properties and Bioapplications. *Polymers* **2019**, *11*, 350. [CrossRef] [PubMed]
91. Saberi, A.; Jabbari, F.; Zarrintaj, P.; Saeb, M.R.; Mozafari, M. Electrically Conductive Materials: Opportunities and Challenges in Tissue Engineering. *Biomolecules* **2019**, *9*, 448. [CrossRef]
92. Qian, C.; Higashigaki, T.; Asoh, T.A.; Uyama, H. Anisotropic Conductive Hydrogels with High Water Content. *ACS Appl. Mater. Interfaces* **2020**, *12*, 27518–27525. [CrossRef] [PubMed]

93. Kasaw, E.; Haile, A.; Getnet, M. Conductive Coatings of Cotton Fabric Consisting of Carbonized Charcoal for E-Textile. *Coatings* **2020**, *10*, 579. [CrossRef]
94. Hu, X.; Tian, M.; Qu, L.; Zhu, S.; Han, G. Multifunctional cotton fabrics with graphene/polyurethane coatings with far-infrared emission, electrical conductivity, and ultraviolet-blocking properties. *Carbon* **2015**, *95*, 625–633. [CrossRef]
95. Rahman, M.J.; Mieno, T. Conductive Cotton Textile from Safely Functionalized Carbon Nanotubes. *J. Nanomater.* **2015**, *2015*, 1–10. [CrossRef]
96. Coeuret, F. Electrical Conductivity of Carbon or Graphite Felts. Available online: https://www.electrochem.org/dl/ma/203/pdfs/2277 (accessed on 29 January 2022).
97. Xiao, G.; Wang, Y.; Zhang, H.; Zhu, Z.; Fu, S. Cellulose nanocrystal mediated fast self-healing and shape memory conductive hydrogel for wearable strain sensors. *Int. J. Biol. Macromol.* **2021**, *170*, 272–283. [CrossRef]
98. Peng, Q.; Chen, J.; Wang, T.; Peng, X.; Liu, J.; Wang, X.; Wang, J.; Zeng, H. Recent advances in designing conductive hydrogels for flexible electronics. *InfoMat* **2020**, *2*, 843–865. [CrossRef]
99. Holmström, E.; Lizárraga, R.; Linder, D.; Salmasi, A.; Wang, W.; Kaplan, B.; Mao, H.; Larsson, H.; Vitos, L. High entropy alloys: Substituting for cobalt in cutting edge technology. *Appl. Mater. Today* **2018**, *12*, 322–329. [CrossRef]
100. Shao, C.; Wang, M.; Meng, L.; Chang, H.; Wang, B.; Xu, F.; Yang, J.; Wan, P. Mussel-Inspired Cellulose Nanocomposite Tough Hydrogels with Synergistic Self-Healing, Adhesive, and Strain-Sensitive Properties. *Chem. Mater.* **2018**, *30*, 3110–3121. [CrossRef]

*Article*

# Soft Stretchable Conductive Carboxymethylcellulose Hydrogels for Wearable Sensors

Kyuha Park [1,†], Heewon Choi [1,†], Kyumin Kang [1], Mikyung Shin [2,3,*] and Donghee Son [1,4,*]

1. Department of Electrical and Computer Engineering, Sungkyunkwan University (SKKU), Suwon 16419, Korea; telos6063@gmail.com (K.P.); chwchw97@gmail.com (H.C.); zseqqq@gmail.com (K.K.)
2. Department of Biomedical Engineering, Sungkyunkwan University (SKKU), Suwon 16419, Korea
3. Department of Intelligent Precision Healthcare Convergence, Sungkyunkwan University (SKKU), Suwon 16419, Korea
4. Department of Superintelligence Engineering, Sungkyunkwan University (SKKU), Suwon 16419, Korea
* Correspondence: mikyungshin@g.skku.edu (M.S.); daniel3600@g.skku.edu (D.S.)
† These authors contributed equally to this work.

**Abstract:** Hydrogels that have a capability to provide mechanical modulus matching between time-dynamic curvilinear tissues and bioelectronic devices have been considered tissue-interfacing ionic materials for stably sensing physiological signals and delivering feedback actuation in skin-inspired healthcare systems. These functionalities are totally different from those of elastomers with low ionic conductivity and higher stiffness. Despite such remarkable progress, their low conductivity remains limited in transporting electrical charges to internal or external terminals without undesired information loss, potentially leading to an unstable biotic–abiotic interfaces in the wearable electronics. Here, we report a soft stretchable conductive hydrogel composite consisting of alginate, carboxymethyl cellulose, polyacrylamide, and silver flakes. This composite was fabricated via sol–gel transition. In particular, the phase stability and low dynamic modulus rates of the conductive hydrogel were confirmed through an oscillatory rheological characterization. In addition, our conductive hydrogel showed maximal tensile strain (≈400%), a low deformations of cyclic loading (over 100 times), low resistance (≈8.4 Ω), and a high gauge factor (≈241). These stable electrical and mechanical properties allowed our composite hydrogel to fully support the operation of a light-emitting diode demonstration under mechanical deformation. Based on such durable performance, we successfully measured the electromyogram signals without electrical malfunction even in various motions.

**Keywords:** carboxymethylcellulose; alginate; polyacrylamide; silver flake composite; conductive hydrogel; soft hydrogel; stretchable hydrogel; electromyogram

## 1. Introduction

Soft and stretchable devices that can intimately interface with various biological tissues, such as skin, brain, heart as well as peripheral nerves have attracted huge attention in realizing the ultimate closed-loop bioelectronics capable of personalized healthcare monitoring and feedback precise treatment [1–6]. In the beginning of such devices, their stretchability with low stiffness was achieved by adopting some deformable designs such as a buckled structure in a neutral mechanical plane or rigid active cell island with wavy interconnects [7–9]. However, this approach remains critical due to the challenges associated with high-cost microfabrication processes and low areal density. To overcome such limitations, a materials-driven approach, "intrinsically stretchable electronics", was newly introduced [10–12]. The point is that conductive stretchable composite materials can replace the wavy interconnects/electrodes fabricated through the previous structural deformable designs. The conductive stretchable composites are generally formed by mixing elastomers (e.g., polydimethylsiloxane (PDMS), styrene ethylene/butylene styrene (SEBS)) or hydrogels with conductive fillers (metal particles, carbon materials, or conjugated polymers) [13–16]. Despite the remarkable progresses of the elastomer-based conductive composites, their stiffness values that are higher than those of the living tissues may lead to chronic tissue irritation or electrical performance degradation.

To solve these issues, hydrogels with high water contents are good candidates due to their tissue-like stiffness, ionic conductivity, and even long-term biocompatibility. Most hydrogels are fabricated through chemical crosslinking or the physical entanglement of natural polymers [17–24]. From the aspect of fabrication cost, the use of natural polymers can also provide advantages for developing cheap and large-scale bioelectronics [25]. Nevertheless, the current hydrogels cannot be applied to tissue-interfacing electrodes or interconnects due to their low ionic conductivity, compared to those of metal or carbon materials [26–29]. Thus, a significant improvement of the electrical conductivity of those hydrogels is required for utilizing them as electrode and electrical circuit materials. Although much effort to enhance their conductivity has been toward embedding a versatile conductive filler in the polymeric network, such an approach result in heterogenous mechanical/electrical properties of the final products. When the conductive fillers are incorporated in pre-gel solution, their sol–gel transition via generally light-induced crosslinking can be inhibited. For homogenous gelation, the polymeric concentration as well as the amounts of initiators and monomers should be carefully optimized.

In this study, we optimized the methodology for a homogenous fabrication of conductive hydrogel composite in a way that mixes three different polymers (alginate, polyacrylamide (PAM), and carboxymethyl cellulose (CMC)) with the conductive filler (e.g., silver flake; Gaff) (Figure 1). Specifically, the CMC that is synthesized via the alkali-catalyzed reaction of cellulose partially contains hydroxy or carboxyl groups. The CMC is highly tough, owing to its triple networks (Figure 1a,b). The mechanical modulus ($\approx$50 kPa) thanks to the conductive composite hydrogel (Alg/PAM/CMC-CHs) was effectively reduced via two issues: (i) decrease in the number of binding sites between the carboxyl and amine groups and (ii) the plasticizing effect of AgF (Figure 1c). Such low stiffness allowed our conductive hydrogel composite to be further suitable for interfacing with skin tissue. To better optimize its mechanical and electrical characteristics, the AgF concentration in the hydrogel was optimized while analyzing the rheological behaviors, mechanical strength, and stretching durability. Based on the reliable properties, we successfully demonstrated the stable operations of a light-emitting diode and electromyogram signal monitoring without electrical malfunction even in various deformations.

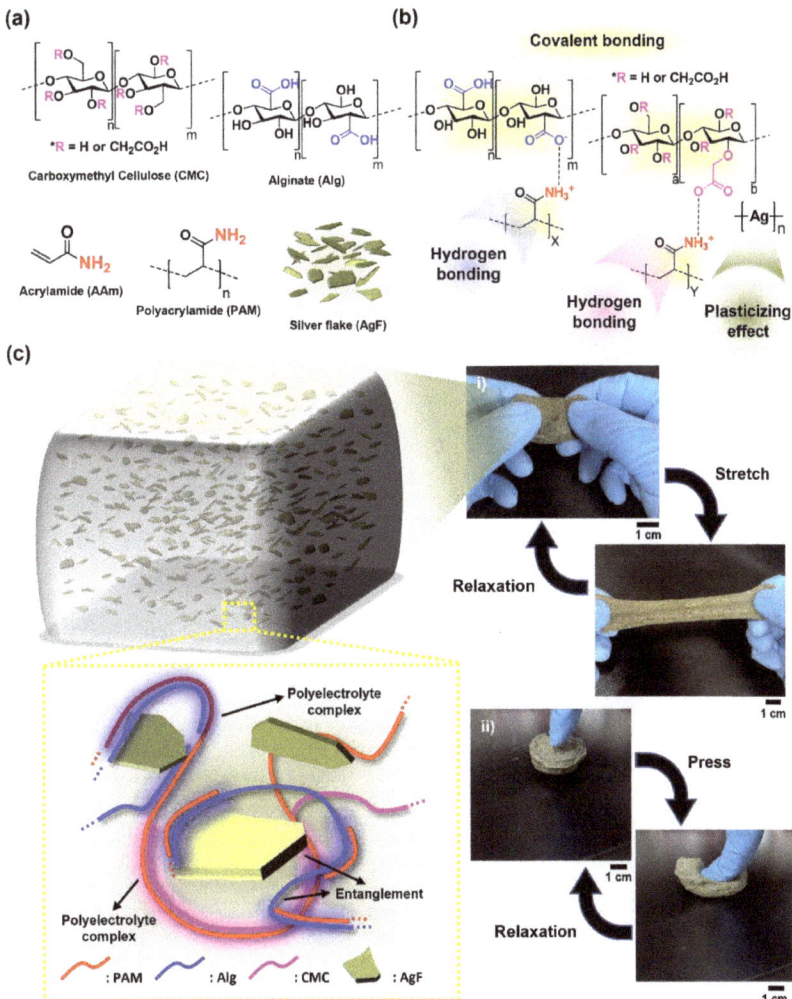

**Figure 1.** Schematic description of stretchable Alg-PAM-CMC conductive hydrogel. (**a**,**b**) Chemical structures of carboxyl cellulose (CMC), alginate (Alg), acrylamide (AAm), polyacrylamide (PAM), and silver flake (AgF) (**a**) and Alg-CMC-PAM-AgF composite-based conductive hydrogel (**b**). (**c**) Illustration describes the network structure of Alg-CMC-PAM conductive hydrogel. Macroscopic images show the flexibility (**c**, (**i**)) and stretchability (**c**, (**ii**)).

## 2. Materials and Methods

### 2.1. Preparation of Alg/PAM/CMC-CHs

Sodium alginate, acrylamide (AAm) (suitable for electrophoresis, ≥99%), carboxymethylcellulose sodium salt (CMC) (low viscosity), N,N-methylenebisacrylamide (MBAA) (powder, for molecular biology, suitable for electrophoresis, ≥99.5%), and ammonium persulfate (APS) (ACS reagent, ≥98.0%) were purchased from Sigma-Aldrich. Silver flakes (AgF) (DSF-500 MWZ-S) were purchased from Daejoo Electronics. Gallium indium tin eutectic, 99.99% (metal basis) (liquid metal) was purchased from Alfa Aesar.

The Alg/PAM/CMC-CH was synthesized using the sol–gel transition of Alg/PAM tough hydrogel synthesis protocols from previous studies [15,30]. First, the CMC, alginate, and AAm were dissolved in deionized water (ratio between the CMC, alginate, and AAm was 1:1:8 wt %). Then, the MBAA (0.25 × 10$^{-2}$% of the total weight of AAm) and AgF (0%, 10%, 20%, 40%, and 80% of the total DI water wt%) were added to the mixture. The entire synthesis was performed by stirring at 400 rpm. Finally, the APS (bleaching agent) (0.03 wt % of the total weight of AAm) was added to the mixture solution, which was subsequently cast into cylindrical molds and placed in an oven at 70 °C for 5 h at approximately 40% humidity. Each completed sample was labeled: 'Alg/PAM/CMC' (w/o AgF = pristine), 'AgF-1' (10%), 'AgF-2' (20%), 'AgF-4' (40%), and 'AgF-8' (80%) (Table 1).

Table 1. The precursor compositions of different Alg-PAM-CMC conductive hydrogels.

|  | Di Water (g) | Alg (g) | CMC (g) | AAm (g) | AgF (g) | MBAA (g) | APS (g) |
|---|---|---|---|---|---|---|---|
| Alg/PAM | 10 | 1 | 0 | 4 | 0 | 0.01 | 0.12 |
| Alg/PAM/CMC | 10 | 0.5 | 0.5 | 4 | 0 | 0.01 | 0.12 |
| AgF-1 | 10 | 0.5 | 0.5 | 4 | 1 | 0.01 | 0.12 |
| AgF-2 | 10 | 0.5 | 0.5 | 4 | 2 | 0.01 | 0.12 |
| AgF-4 | 10 | 0.5 | 0.5 | 4 | 4 | 0.01 | 0.12 |
| AgF-8 | 10 | 0.5 | 0.5 | 4 | 8 | 0.01 | 0.12 |

Alg = alginate; CMC = carboxymethyl cellulose; AAm = acrylamide; AgF = silver flake; MBAA = N' methylenebisacrylamide; APS = ammonium persulfate.

### 2.2. Rheological Characterization

The oscillation frequency sweep and continuous step strain measurements of the rheological characteristics of Alg/PAM/CMC-CHs were conducted using a TA Instruments Discovery Hybrid Rheometer 2 (TA Instruments, USA). The Alg/PAM/CMC-CHs were fabricated using bulky cylindrical molds (1.25 × 1.25 × 2 mm$^3$). All rheology measurements were conducted using a parallel-plate geometry (diameter: 20 mm). Continuous step strain measurements were used to evaluate the disruption and recovery of storage (G′) and loss (G″) modulus of conductive hydrogels. These measurements were conducted on alternate strains with 0.5% and 1000% storage and loss modulus, respectively. The oscillation frequency sweep measurement was used to evaluate the change in G′ and G″ rates per frequency rate from 0.01 to 10 Hz. The gap size and axial force were maintained at 2 mm and 1 N, respectively, during all rheological characterization. The two moduli were calculated using the following equations:

$$\varepsilon = \varepsilon_0 \sin(\omega t), \sigma = \sigma_0 \sin(\omega t + \delta) \tag{1}$$

$$\text{Storage modulus } (G') = \sigma_0/\varepsilon_0 (\cos \delta) \tag{2}$$

$$\text{Loss modulus } (G'') = \sigma_0/\varepsilon_0 (\sin \delta) \tag{3}$$

($\sigma$ is stress (Pa), $\varepsilon$ is strain (mm/mm), $\omega$ is frequency of strain oscillation (rad(°)/s), $t$ is time (s), and $\delta$ is phase lag between stress and strain (rad(°)/s)).

### 2.3. Universal Tensile Machine Measurement

To measure the tensile stress per stretching of the Alg/PAM/CMC-CHs controlled via different volumes of AgF, linear-shaped Alg/PAM/CMC-CHs were fabricated using narrow cylindrical molds (2.5 × 2.5 × 20 mm$^3$). The experiments were performed using a universal tensile machine (UTM, INSTRON 6800), which performed measurements of two types: continuous stretching at a speed rate of 10 mm/min and cyclic stretching test (about 100 times) from 0% to 50% tensile strain at a speed rate of 1 mm/s.

### 2.4. Resistance per Stretching Measurement

To measure the resistance per stretching of the Alg/PAM/CMC-CHs, which can be used to compare the conductive performance between the different volumes of AgF and organic precursors, linear-shaped Alg/PAM/CMC-CHs were fabricated using narrow cylindrical molds (similar to the tensile strength measurement samples). The experiments were performed using a probe station (MST 5500 B, MSTECH Inc., Hwaseong, Korea); an LCR meter basic support program (4284A, Agilent Technology Inc., Santa Clara, CA, USA) with 1 kHz frequency, 0.1 V DC voltage bias, and a step motor controller (SMC-100, Ecopia Corp., Anyang, Korea); and an automatic stretch-testing machine (Stretching Tester, Jaeil Optical System Corp., Incheon, Korea). Two types of experiments were performed: continuous stretching at a speed of 0.015 mm/s and cyclic stretching test (100 times) from 0% to 50% strain at a speed of 0.3 mm/s. During the measurements, a small amount of liquid metal was dripped onto the contact area of the samples for stable electrical contact between the samples and the probe station. Conductivity was calculated using the following equation [19]:

$$\sigma = L/(\rho \times A) \tag{4}$$

where $\sigma$ is the conductivity (S/mm), $\rho$ is the resistance ($\Omega$), $A$ is the cross-section of the hydrogels (mm$^2$)), and $L$ is the length of the hydrogel (mm).

### 2.5. Light-Emitting Diode and Electromyogram Demonstration

To assess the performance of the Alg/PAM/CMC-CH as an electrode for a wearable sensor platform, electromyogram (EMG) signal monitoring and a demonstration using an LED were performed. First, the LED demonstration proceeded via a light-emitting diodes Bulb LED Lamp, 5 mm (white color) with a basic breadboard and a Keithley 2450 source meter (Tektronix, Inc., Clackamas, OR, USA) as a power supply. Cyclic stretching was performed from 0% to 200% tensile strain at a speed of 0.3 mm/s. Second, two samples were connected to fine wire and hook-type electrodes, and the EMG signals were monitored in real time via data acquisition equipment (DAQ, PowerLab 8/35). Transparent skin patches (3M Tegaderm, Minnesota, MN, USA) were used to establish contact between the skin and electrodes with a sampling frequency of 200 kHz.

## 3. Results and Discussion
### 3.1. Rheological Behaviors of Alg/PAM/CMC-CHs

Rheological behaviors are typically evaluated through the rate distinctions and the absolute values between G' and G''. These moduli indicate the relative energy loss with varying frequency rate, which is observable in the oscillatory behavior of parallel-plate geometry with viscoelastic material. Thus, the dynamic modulus rates from the viscoelastic material can be easily calculated (Equations (2) and (3)). The oscillation frequency sweep results show each optimum ratio between the fixed organic precursors and AgF via the gaps between G' and G'' (Figure 2a,b). The results showed that the Alg/PAM/CMC-CHs have low dynamic modulus rates and clear rate distinctions between G' and G''. Usually, most hydrogels made through sol–gel transitions barely satisfy both low absolute values and clear rate distinctions of G' and G''. However, all the Alg/PAM/CMC-CHs overcome these issues. In addition, continuous step strain results showed the disruption and recovery of G' and G'' of all samples. The results showed that Alg/PAM/CMC-CHs have a similar dynamic modulus compared to that of the pristine hydrogel (Alg/PAM/CMC) (Figure 2c,d).

### 3.2. Evaluating Tensile Stress of Alg/PAM/CMC-CHs

All the candidate samples of the hydrogels for UTM measurement were protected via sandpaper for reducing deviations from mechanical mismatching with jigs (Figure 3a,b). The continuous tensile strain test was conducted to measure the maximum tensile strain rates of the hydrogel with varying toughness. The mechanical behavior of Alg/PAM could not endure more than 200% elongation, but the Alg/PAM/CMC hydrogel endured more than 400% elongation. On the other hand, compare with w/o AgF samples, in the case of Alg/PAM/CMC-CHs containing different volumes of AgF, the elongation value was significantly decreased, while the overall toughness value was decreased because of the plasticizing effect of AgF. In addition, in the case of the AgF-4 condition in which an appropriate amount of AgF was added, it looks as though the mechanical strengths were rapidly increased due to the linear energy dispersion of AgF. Furthermore, evaluating the mechanical performance of samples as a conductor of stretchable electronics, the results of AgF-4 and AgF-8 are excellent, but other Alg/PAM/CMC-CHs quickly reached the mechanical fatigue point and were thus unreasonable to use as a conductor of stretchable electronics (Figure 3c). The mechanical behaviors of cyclic stretching showed that all of the Alg/PAM/CMC-CHs could endure cyclic strains more than 100 times (Figure 3d).

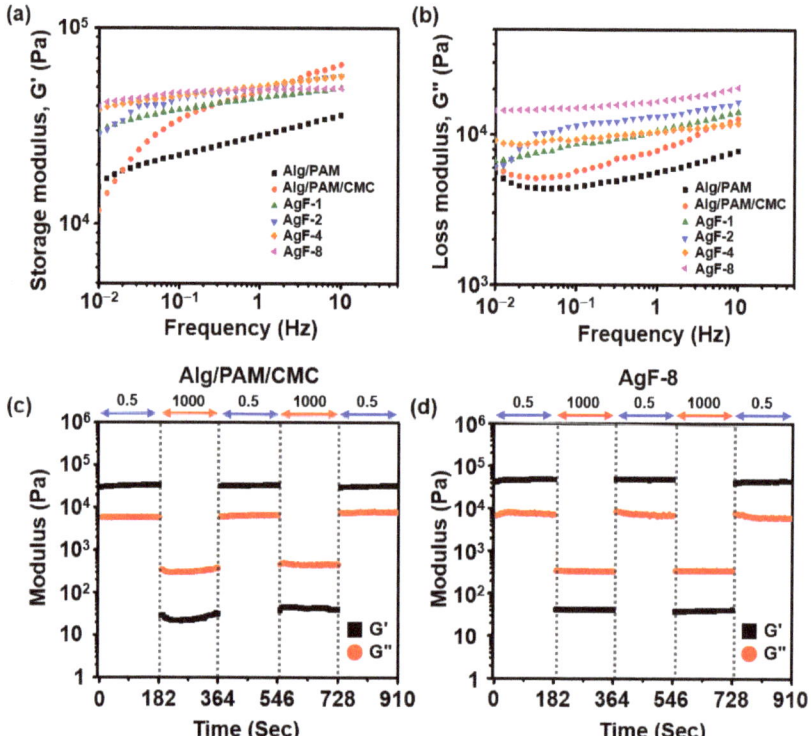

**Figure 2.** Rheological characterization of stretchable Alg- PAM-CMC conductive hydrogels containing different amounts of AgF. (a,b) Oscillation frequency sweep measurements of Alg/PAM dual-network hydrogel (Alg/PAM), Alg/PAM/CMC triple-network hydrogel (Pristine), and various Alg/PAM/CMC-CHs with different AgF ratio. (c,d) Continuous step strain measurement of Alg/PAM/CMC and Alg/PAM/CMC-CHs.

### 3.3. Electrical Performance per Tensile Strain of Alg/PAM/CMC-CHs

To analyze the varying resistance per tensile strain of the Alg/PAM/CMC-CHs, the method employed in a previous report was followed (Figure 4a,b) [19]. The results show that initially, the resistance of the pristine samples decreased exponentially when the amount of AgF in the samples increased. Similarly, varying resistance rates per tensile strain were also affected by the ratio of AgF (Figure 4c). In addition, the results of the calculated gauge factor (GF) showed that AgF-4 and AgF-8 were excellent candidates (maximum GF = AgF-4: 20, AgF-8: 241) for electrodes of the strain sensor. In contrast, the AgF-8 (8.4 $\Omega$ to 8.1 k$\Omega$ per 0% to 391% strain) will be the best for use as an integrated circuit for fabricating wearable sensor devices via 3D printing. The cyclic stretching test (0% to 100% strain repeated 100 times) showed the denaturation of resistance (Figure 4d). The results showed that AgF-4 and AgF-8 quickly denatured the strain resistance, during which significant resistance noise was detected. The results in this section suggest appropriate electrical demonstrations for each of the different Alg/PAM/CMC-CHs.

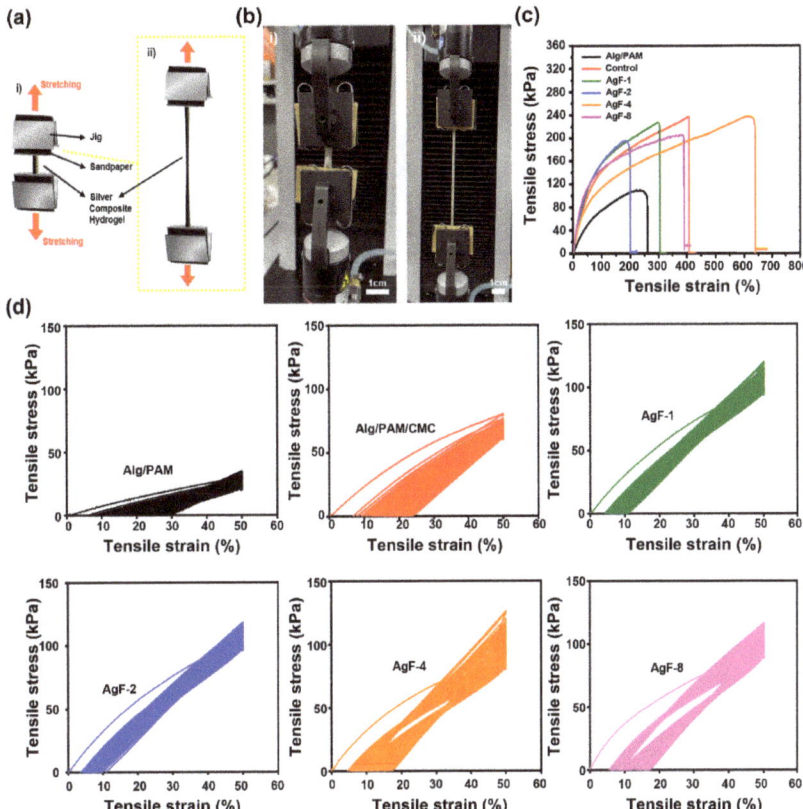

**Figure 3.** Tensile strength measurement via tensile stress per stretching. (**a**) Schematic illustration of measuring tensile strain–stress curve and (**b**) instrument (UTM) photograph. (**c**,**d**) Strain–stress curves of Alg/PAM/CMC-CH. Each measurement was proceeded via continuously strain (**c**), and cyclic strain was about 0% to 50% at a speed of 1 mm/s at 100 times (**d**).

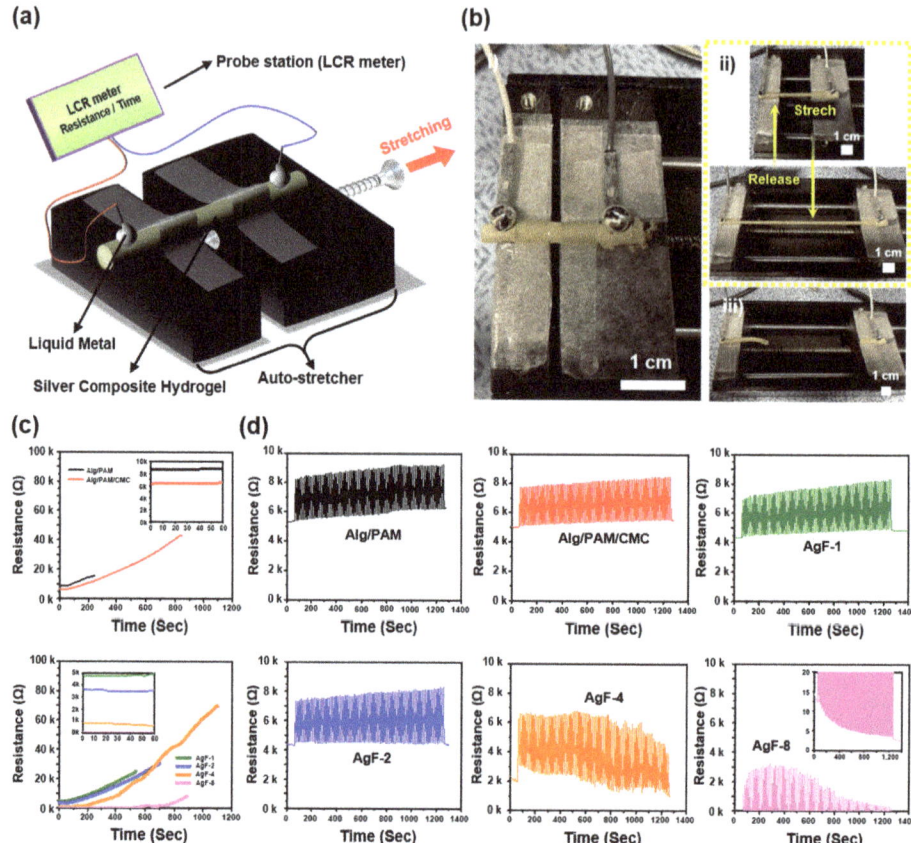

**Figure 4.** Measurements of resistance per tensile strain via different Alg/PAM/CMC-CHs. (**a**) Schematic illustration of resistance per strain test with automatic stretch-testing machine and LCR meter and (**b**) photographs showing pristine (i), stretched/released (ii), and mechanically broken Alg/PAM/CMC-CHs. (**c**) Continuous tensile strain of the samples at a speed of 0.015 mm/s. (**d**) Cyclic stretching test from 0% to 100% repetitively for strains at a speed of 0.3 mm/s at 100 times.

### 3.4. Electrical Performance Demonstration of Alg/PAM/CMC-CHs

Two types of electrical performance demonstrations were performed. First, LED demonstrations were performed using the method described in our previous report (Figure 5a,b) [31]. The results show a luminous change in the LEDs between the stretch state and release state. In addition, the tensile strained state of AgF-4 and AgF-8 was optically confirmed (Figure 5c,d). Similarly, EMG demonstrations were performed using the method described in our previous report (Figure 6a,b) [31]. The patch type of the AgF-8 based electrode exhibited excellent signal transmission efficiency of EMG signals generated by the repeated clenching and opening of the human fist (Figure 6c).

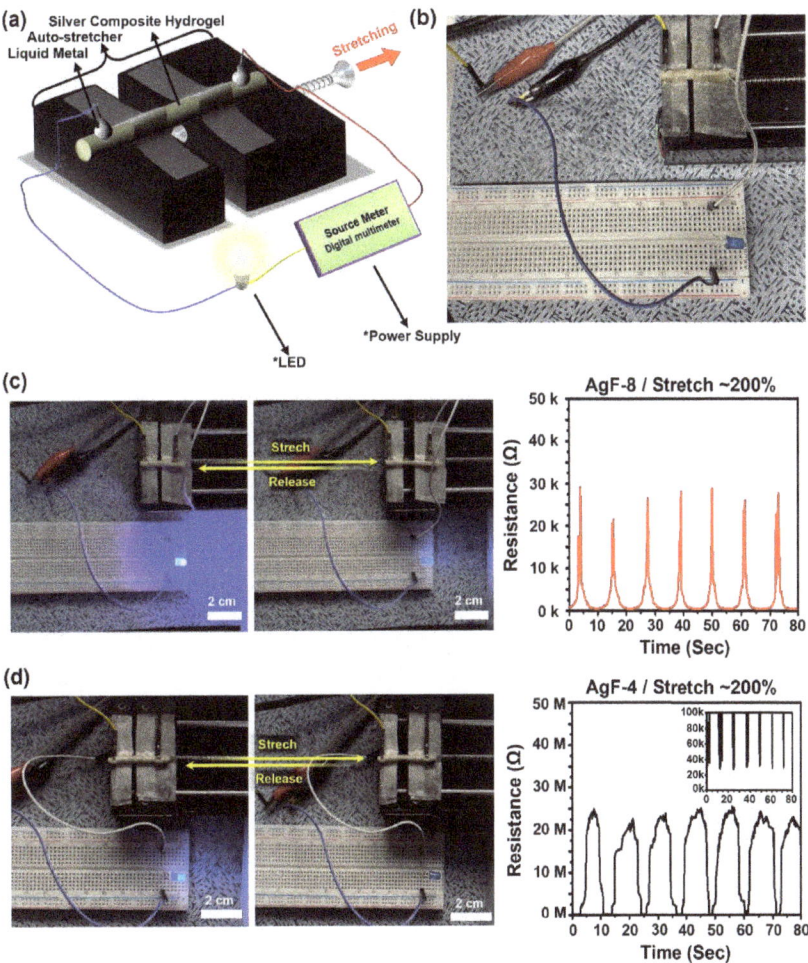

**Figure 5.** LED demonstration of Alg/PAM/CMC-CHs (AgF-8 and AgF-4) as a strain sensor electrode. Schematic illustration of LED demonstration (**a**,**b**) photograph. (**c**,**d**) Resistance per repeatedly stretch (≈200%) curve and macroscopic images of stretched and released Alg/PAM/CMC-CHs.

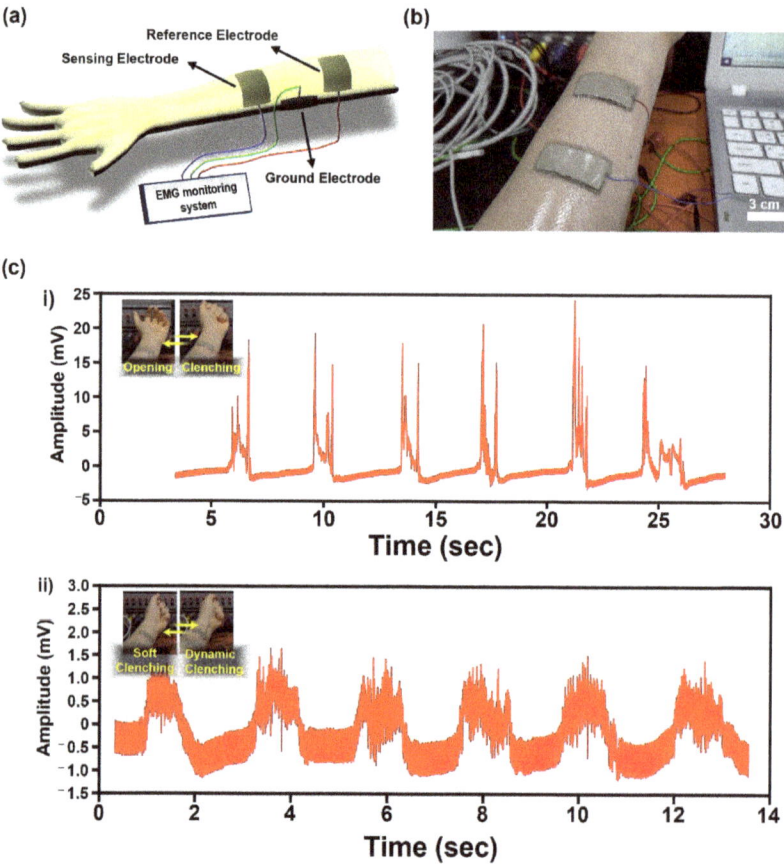

**Figure 6.** Schematic description of real-time EMG signal monitoring with Alg/PAM/CMC-CH (AgF-8). (**a**) Illustration describes the three-electrode (sensing, reference, and ground) EMG recording system and (**b**) photograph. (**c**) Amplitude changes in response to clenching-unfolding fist (**c**, (**i**)) and soft-dynamic repeated clenching fist (**c**, (**ii**)).

## 4. Conclusions

In this study, we evaluated the Alg/PAM/CMC-CHs, which can be utilized as a conductor in soft electronics. The Alg, CMC, and PAM are capable of forming mild and stretchable hydrogels via sol–gel transition, which was classified as a triple-network-based composite with both strong and weak bonding. The UTM measurements and rheological characterization showed that the Alg/PAM/CMC-based triple-network hydrogel had lower toughness and higher stretchability than the Alg/PAM-based dual-network hydrogel. In addition, the UTM measurements of different Alg/PAM/CMC-CHs with different AgF contents (AgF-1, AgF-2, AgF-4, AgF-8) showed that AgF-4 had optimal mechanical properties (≈650%). Furthermore, the resistance per tensile strain measurement of Alg/PAM/CMC-CHs confirmed that a visible electrical resistance change occurred in AgF-8 at approximately ≈390% stretching (8.4 Ω to 8.1 kΩ for 2.5 × 2.5 × 20 mm$^3$ sample stretching, GF = ~241). In contrast, AgF-1 exhibited less change in electrical resistance from 0% to ≈400% stretching (4.8 to 25.3 kΩ as the same volume). Finally, electrical performance demonstrations (LED and EMG) using different Alg/PAM/CMC-CHs as conductors were performed. The LED demonstration showed an evident luminous change while stretching the AgF-8 hydrogel, verifying its potential for use as a high-performance strain sensor electrode. The EMG demonstration showed the high resolution of recording signals via conformal contact between the epidermis and AgF-8 through low mechanical properties, verifying its potential for use as a wearable electronics electrode. The suggested conductive composite hydrogel has satisfactory low toughness and high conductive performance. We expect that the Alg/PAM/CMC-CH will be utilized to provide significant benefits in a high-performance wearable strain sensor.

**Author Contributions:** Conceptualization, D.S.; methodology, M.S. and K.P.; software, K.P., H.C. and K.K.; validation, K.P.; formal analysis, M.S. and K.P.; investigation, K.P., H.C. and K.K.; resources, D.S. and K.P.; data curation, D.S. and K.P.; writing—original draft preparation, K.P., H.C.; writing—review and editing, M.S. and D.S.; visualization, K.P., H.C. and K.K. supervision, M.S. and D.S.; project administration, M.S. and D.S.; funding acquisition, M.S. and D.S. All authors have read and agreed to the published version of the manuscript.

**Funding:** This research was supported by the National Research Foundation of Korea (NRF) grant funded by the Korea government (MSIT) (No. 2020R1C1C1005567), Institute of Information & communications Technology Planning & Evaluation (IITP) grant funded by the Korea government (MSIT) (No. 2020-0-00261, Development of low power/low delay/self-power suppliable RF simultaneous information and power transfer system and stretchable electronic epineurium for wireless nerve bypass implementation), and MSIT (Ministry of Science and ICT), Korea, under the ICT Creative Consilience program(IITP-2020-0-01821) supervised by the IITP(Institute for Information & Communications Technology Planning & Evaluation).

**Institutional Review Board Statement:** Not applicable.

**Informed Consent Statement:** Not applicable.

**Data Availability Statement:** The data presented in this study are available in the article.

**Conflicts of Interest:** The funders had no role in the design of the study; in the collection, analyses, or interpretation of data; in the writing of the manuscript, or in the decision to publish the results.

# References

1. Son, D.; Kang, J.; Vardoulis, O.; Kim, Y.; Matsuhisa, N.; Oh, J.Y.; To, J.W.; Mun, J.; Katsumata, T.; Liu, Y.; et al. An integrated self-healable electronic skin system fabricated via dynamic reconstruction of a nanostructured conducting network. *Nat. Nanotech.* **2018**, *13*, 1057–1065. [CrossRef] [PubMed]
2. Kim, D.W.; Song, K.; Seong, D.; Lee, Y.S.; Baik, S.; Song, J.H.; Lee, H.J.; Son, D.; Pang, C. Electrostatic–Mechanical Synergistic In Situ Multiscale Tissue Adhesion for Sustainable Residue-Free Bioelectronics Interfaces. *Adv. Mater.* **2021**, 2105338. [CrossRef] [PubMed]
3. Seo, H.; Han, S.I.; Song, K.; Seong, D.; Lee, K.; Kim, S.H.; Park, T.; Koo, J.H.; Shin, M.; Baac, H.W.; et al. Durable and Fatigue-Resistant Soft Peripheral Neuroprosthetics for In Vivo Bidirectional Signaling. *Adv. Mater.* **2021**, *33*, 2007346. [CrossRef] [PubMed]
4. Song, K.-I.; Seo, H.; Seong, D.; Kim, S.; Yu, K.J.; Kim, Y.-C.; Kim, J.; Kwon, S.J.; Han, H.-S.; Youn, I.; et al. Adaptive self-healing electronic epineurium for chronic bidirectional neural interfaces. *Nat. Commun.* **2020**, *11*, 4195. [CrossRef]
5. Yang, Y.; Wu, M.; Vázquez-Guardado, A.; Wegener, A.J.; Grajales-Reyes, J.G.; Deng, Y.; Wang, T.; Avila, R.; Moreno, J.A.; Minkowicz, S.; et al. Wireless multilateral devices for optogenetic studies of individual and social behaviors. *Nat. Neurosci.* **2021**, *24*, 1035–1045. [CrossRef]
6. Jinno, H.; Yokota, T.; Koizumi, M.; Yukita, W.; Saito, M.; Osaka, I.; Fukuda, K.; Someya, T. Self-powered ultraflexible photonic skin for continuous bio-signal detection via air-operation-stable polymer light-emitting diodes. *Nat. Commun.* **2021**, *12*, 2234. [CrossRef]
7. Huang, Z.; Hao, Y.; Li, Y.; Hu, H.; Wang, C.; Nomoto, A.; Pan, T.; Gu, Y.; Chen, Y.; Zhang, T.; et al. Three-dimensional integrated stretchable electronics. *Nat. Electron.* **2018**, *1*, 473–480. [CrossRef]
8. Kim, D.-H.; Lu, N.; Ma, R.; Kim, Y.-S.; Kim, R.-H.; Wang, S.; Wu, J.; Won, S.M.; Tao, H.; Islam, A.; et al. Epidermal Electronics. *Science* **2011**, *333*, 838–843. [CrossRef]
9. Son, D.; Lee, J.; Qiao, S.; Ghaffari, R.; Kim, J.; Lee, J.E.; Song, C.; Kim, S.J.; Lee, D.J.; Jun, S.W.; et al. Multifunctional wearable devices for diagnosis and therapy of movement disorders. *Nat. Nanotech.* **2014**, *9*, 397–404. [CrossRef]
10. Oh, J.Y.; Rondeau-Gagné, S.; Chiu, Y.-C.; Chortos, A.; Lissel, F.; Wang, G.-J.N.; Schroeder, B.C.; Kurosawa, T.; Lopez, J.; Katsumata, T.; et al. Intrinsically stretchable and healable semiconducting polymer for organic transistors. *Nature* **2016**, *539*, 411–415. [CrossRef]
11. Wang, S.; Xu, J.; Wang, W.; Wang, G.-J.N.; Rastak, R.; Molina-Lopez, F.; Chung, J.W.; Niu, S.; Feig, V.R.; Lopez, J.; et al. Skin electronics from scalable fabrication of an intrinsically stretchable transistor array. *Nature* **2018**, *555*, 83–88. [CrossRef]

12. Kim, S.H.; Baek, G.W.; Yoon, J.; Seo, S.; Park, J.; Hahm, D.; Chang, J.H.; Seong, D.; Seo, H.; Oh, S.; et al. A Bioinspired Stretchable Sensory-Neuromorphic System. *Adv. Mater.* **2021**, *33*, 2104690. [CrossRef]
13. Kim, N.; Lienemann, S.; Petsagkourakis, I.; Alemu Mengistie, D.; Kee, S.; Ederth, T.; Gueskine, V.; Leclère, P.; Lazzaroni, R.; Crispin, X.; et al. Elastic conducting polymer composites in thermoelectric modules. *Nat. Commun.* **2020**, *11*, 1424. [CrossRef]
14. Lee, W.; Yun, H.; Song, J.-K.; Sunwoo, S.-H.; Kim, D.-H. Nanoscale Materials and Deformable Device Designs for Bioinspired and Biointegrated Electronics. *Acc. Mater. Res.* **2021**, *2*, 266–281. [CrossRef]
15. Ohm, Y.; Pan, C.; Ford, M.J.; Huang, X.; Liao, J.; Majidi, C. An electrically conductive silver–polyacrylamide–alginate hydrogel composite for soft electronics. *Nat. Electron.* **2021**, *4*, 185–192.
16. Lin, M.; Zheng, Z.; Yang, L.; Luo, M.; Fu, L.; Lin, B.; Xu, C. A High-Performance, Sensitive, Wearable Multifunctional Sensor Based on Rubber/CNT for Human Motion and Skin Temperature Detection. *Adv. Mater.* **2021**, *34*, 2107309. [CrossRef]
17. Huang, H.; Han, L.; Li, J.; Fu, X.; Wang, Y.; Yang, Z.; Xu, X.; Pan, L.; Xu, M. Super-stretchable, elastic and recoverable ionic conductive hydrogel for wireless wearable, stretchable sensor. *J. Mater. Chem. A* **2020**, *8*, 10291–10300. [CrossRef]
18. Liu, K.; Tran, H.; Feig, V.R.; Bao, Z. Biodegradable and stretchable polymeric materials for transient electronic devices. *MRS Bulletin* **2020**, *45*, 96–102. [CrossRef]
19. Choi, Y.; Park, K.; Choi, H.; Son, D.; Shin, M. Self-Healing, Stretchable, Biocompatible, and Conductive Alginate Hydrogels through Dynamic Covalent Bonds for Implantable Electronics. *Polymers* **2021**, *13*, 1133. [CrossRef]
20. Kim, D.-H.; Xiao, J.; Song, J.; Huang, Y.; Rogers, J.A. Stretchable, Curvilinear Electronics Based on Inorganic Materials. *Adv. Mater.* **2010**, *22*, 2108–2124. [CrossRef]
21. Fan, J.A.; Yeo, W.-H.; Su, Y.; Hattori, Y.; Lee, W.; Jung, S.-Y.; Zhang, Y.; Liu, Z.; Cheng, H.; Falgout, L.; et al. Fractal design concepts for stretchable electronics. *Nat. Commun.* **2014**, *5*, 3266. [CrossRef] [PubMed]
22. Khan, R.; Ilyas, N.; Shamim, M.Z.M.; Khan, M.I.; Sohail, M.; Rahman, N.; Khan, A.A.; Khan, S.N.; Khan, A. Oxide-Based Resistive Switching-Based Devices: Fabrication, Influence Parameters and Applications. *J. Mater. Chem. C* **2021**, *9*, 15755–15788. [CrossRef]
23. Huang, X.; Zhou, X.; Zhou, H.; Zhong, Y.; Luo, H.; Zhang, F. A high-strength self-healing nano-silica hydrogel with anisotropic differential conductivity. *Nano Res.* **2021**, *14*, 2589–2595. [CrossRef]
24. Zhao, L.; Ren, Z.; Liu, X.; Ling, Q.; Li, Z.; Gu, H. A Multifunctional, Self-Healing, Self-Adhesive, and Conductive Sodium Alginate/Poly(vinyl alcohol) Composite Hydrogel as a Flexible Strain Sensor. *ACS Appl. Mater. Interfaces* **2021**, *13*, 11344–11355. [CrossRef] [PubMed]
25. Kadumudi, F.B.; Hasany, M.; Pierchala, M.K.; Jahanshahi, M.; Taebnia, N.; Mehrali, M.; Mitu, C.F.; Shahbazi, M.; Zsurzsan, T.; Knott, A.; et al. The Manufacture of Unbreakable Bionics via Multifunctional and Self-Healing Silk–Graphene Hydrogels. *Adv. Mater.* **2021**, *33*, 2100047. [CrossRef] [PubMed]
26. Wang, J.-J.; Zhang, Q.; Ji, X.-X.; Liu, L.-B. Highly Stretchable, Compressible, Adhesive, Conductive Self-healing Composite Hydrogels with Sensor Capacity. *Chin. J. Polym. Sci.* **2020**, *38*, 1221–1229. [CrossRef]
27. Zheng, H.; Lin, N.; He, Y.; Zuo, B. Self-Healing, Self-Adhesive Silk Fibroin Conductive Hydrogel as a Flexible Strain Sensor. *ACS Appl. Mater. Interfaces* **2021**, *13*, 40013–40031. [CrossRef] [PubMed]
28. Zhang, F.; Xiong, L.; Ai, Y.; Liang, Z.; Liang, Q. Stretchable Multiresponsive Hydrogel with Actuatable, Shape Memory, and Self-Healing Properties. *Adv. Sci.* **2018**, *5*, 1800450. [CrossRef]
29. Wu, M.; Chen, J.; Ma, Y.; Yan, B.; Pan, M.; Peng, Q.; Wang, W.; Han, L.; Liu, J.; Zeng, H. Ultra elastic, stretchable, self-healing conductive hydrogels with tunable optical properties for highly sensitive soft electronic sensors. *J. Mater. Chem. A* **2020**, *8*, 24718–24733. [CrossRef]
30. Sun, J.-Y.; Zhao, X.; Illeperuma, W.R.K.; Chaudhuri, O.; Oh, K.H.; Mooney, D.J.; Vlassak, J.J.; Suo, Z. Highly stretchable and tough hydrogels. *Nature* **2012**, *489*, 133–136. [CrossRef]
31. Kim, S.H.; Kim, Y.; Choi, H.; Park, J.; Song, J.H.; Baac, H.W.; Shin, M.; Kwak, J.; Son, D. Mechanically and electrically durable, stretchable electronic textiles for robust wearable electronics. *RSC Adv.* **2021**, *11*, 22327–22333. [CrossRef]

# Article

# Zinc- and Copper-Loaded Nanosponges from Cellulose Nanofibers Hydrogels: New Heterogeneous Catalysts for the Synthesis of Aromatic Acetals

Laura Riva, Angelo Davide Lotito, Carlo Punta and Alessandro Sacchetti *

Department of Chemistry, Materials, and Chemical Engineering "G. Natta", Politecnico di Milano, 20131 Milan, Italy; laura2.riva@polimi.it (L.R.); angelodavide.lotito@mail.polimi.it (A.D.L.); carlo.punta@polimi.it (C.P.)
* Correspondence: alessandro.sacchetti@polimi.it; Tel.: +39-0223993017

**Abstract:** Herein we report the synthesis of cellulose-based metal-loaded nano-sponges and their application as heterogeneous catalysts in organic synthesis. First, the combination in water solution of TEMPO-oxidized cellulose nanofibers (TOCNF) with branched polyethyleneimine (bPEI) and citric acid (CA), and the thermal treatment of the resulting hydrogel, leads to the synthesis of an eco-safe micro- and nano-porous cellulose nano-sponge (CNS). Subsequently, by exploiting the metal chelation characteristics of CNS, already extensively investigated in the field of environmental decontamination, this material is successfully loaded with Cu (II) or Zn (II) metal ions. Efficiency and homogeneity of metal-loading is confirmed by scanning electron microscopy (SEM) analysis with an energy dispersive X-ray spectroscopy (EDS) detector and by inductively coupled plasma-optical emission spectrometry (ICP-OES) analysis. The resulting materials perform superbly as heterogeneous catalysts for promoting the reaction between aromatic aldehydes and alcohols in the synthesis of aromatic acetals, which play a fundamental role as intermediates in organic synthesis. Optimized conditions allow one to obtain conversions higher than 90% and almost complete selectivity toward acetal products, minimizing, and in some cases eliminating, the formation of carboxylic acid by-products. ICP-OES analysis of the reaction medium allows one to exclude any possible metal-ion release, confirming that catalysis undergoes under heterogeneous conditions. The new metal-loaded CNS can be re-used and recycled five times without losing their catalytic activity.

**Keywords:** nanocellulose hydrogels; cellulose-based nanosponges; heterogeneous catalysis; acetalization; metal-catalyzed reactions; sustainability

## 1. Introduction

The demand for sustainable and environmentally friendly materials is constantly increasing nowadays. The design and development of products and processes by promoting the reduction or possibly even the abatement of the use of dangerous substances, following the guidelines of green chemistry principles [1], encounters the request of the Sustainable Development Goals [2].

Considering these recommendations, in the last years we have developed a new class of nanocellulose-based nanostructured materials which could be easily produced from aqueous dispersions of TEMPO (2,2,6,6-tetramethylpiperidine-1-oxyl)-oxidised cellulose nanofibers (TOCNF) [3], branched polyethyleneimine (bPEI), and citric acid (CA), by thermally treating the resulting hydrogels according to a two-step protocol, namely freeze-drying, followed by heating at about 100 °C [4]. These cellulose-based nanosponges have shown a high micro-porosity (due to the freeze-drying step), but also a nano-porosity derived from the cross-linking between primary amines of bPEI and the carboxylic groups pf TOCNF [5].

Recently, the synthetic protocol has been revised and optimized following an eco-design approach that is an eco-toxicological evaluation of the material in order to guarantee more and more the eco-safety of our final cellulose nanosponges (CNS) [6]. This is due to the fact that these engineered nanomaterials have been primary developed for being in direct contact with the environment, which are particularly efficient in the decontamination of water solutions from organic dyes and heavy metals [4,6,7]. In particular, the latter can be efficiently trapped and coordinated in the CNS network thanks to the chelating action of the several amino-groups present in the nanostructure and deriving from bPEI cross-linker.

More recently the same material has been successfully proposed as an ideal heterogeneous catalyst for promoting Henry and Knoevenagel reactions due to its alkaline properties [8]. In this context, by considering the high affinity of CNS for heavy metals, as demonstrated when exploiting their decontamination action, we envisioned that the same structure could be considered as a biomass-derived support for metal ions, to be used for promoting metal-catalyzed organic reactions under heterogeneous conditions.

The acetalization reaction is one of the most useful protecting methods for carbonyl compounds in multi-step reactions [9]. Moreover, acetals are important compounds with various practical applications, widely used as flavors and fragrances in the cosmetic and food industry [10]. Despite their simple molecular structure, the practical synthesis of acetals is not trivial, since the equilibrium of the reaction is shifted toward the reagents and the yields are often low.

Most methods for acetal synthesis are based on the reaction of a carbonyl compound with alcohols or ortoesters [11] and these reactions are generally catalyzed using p-toluenesulfonic acid, pyridinium salts, triflic acid [12–14] and Lewis acids in their metal chlorides and trifluoromethanesulfonates form, as well as transition metal complexes [15,16]. These homogeneous catalysts show several limitations, which include the problematic work-up procedures, the use of toxic and corrosive reagents, the necessity of neutralization of the strong acid media, and the lack of possibility of recovering and recycling the catalyst [17]. This is of particular importance when contamination of the final products is an issue, as in the case of pharmaceutical substances.

A practical way to prevent these issues is the use of heterogeneous catalysis, in which the metal is supported and immobilized on a solid material, thus allowing for the recovery of the catalyst by simple filtration at the end of the process. For these reactions, different heterogeneous catalysts have been already reported in literature, namely clays [18] and resins [19–21]. However, most of these heterogeneous systems must be used in a high catalyst to substrate ratio and for long reaction times, which favours the formation of side products. At the same time, the use of carbon-based nanostructures as heterogeneous supports for the catalysis of organic reactions has been attracting more and more attention [22,23].

Therefore, we considered the possibility of loading CNS with two different transition metal cations (Zn (II) and Cu (II)), and to use the new metal-organic composites as heterogeneous catalysts for the acetalization reaction.

## 2. Results and Discussion

*2.1. Catalyst Synthesis and Characterization*

The catalyst was synthesized following a previously reported procedure, according to [6,7]. The schematic synthetic procedure is described in Scheme 1.

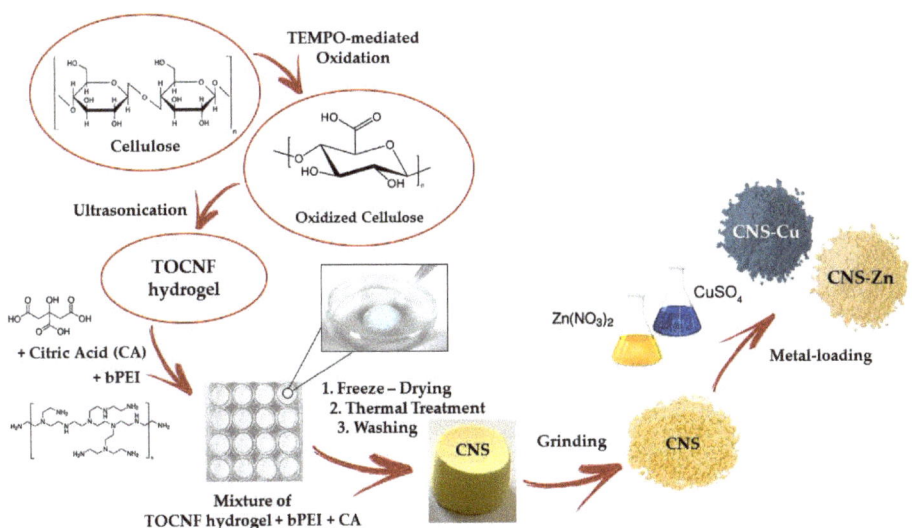

**Scheme 1.** Different production steps for the production of CNS.

Briefly, TOCNF and bPEI were first mixed in deionized water in a 1:2 weight ratio, and CA (18% mol with respect to primary amino groups of bPEI) was added to better fix bPEI in the final network by increasing the number of carboxylic groups. The resulting nanocellulose-based hydrogel was transferred in molds, lyophilized, and then heated in oven at about 100 °C, in order to favour the cross-linking by removing water and promoting the formation of amide bonds between the carboxylic groups of TOCNF and CA and the primary amines of bPEI. Finally, CNS were grinded in a mortar before use, in order to increase the exposed superficial area.

CNS so obtained had been previously characterized by several techniques. First, the formation of amide bonds after thermal treatment had been evidenced by FT-IR analysis [24,25] and also confirmed by $^{13}$C CP-MAS solid-state NMR [4]. Moreover, a high micro-porosity due to the freeze-drying step, with pore sizes in the range of 10–100 µm, had been observed by scanning electron microscopy (SEM) [4], and estimated to be around 70–75% by microcomputed tomography (µ-CT) analysis [24], with trabecular inner structure with an average trabecular thickness of about 30–40 µm and a trabecular separation of about 70–75 µm. Furthermore, nano-porosity in the network had been also revealed by means of small angle neutron scattering (SANS) analysis of water nano-confinement geometries in the sorbent material [5].

Once synthesized, CNS were loaded with Zn (II) and Cu (II) ions, in order to produce the metal-organic composite to be used as heterogeneous catalyst. The loading efficiency was determined by means of ICP-OES and Table 1 shows the percentage in weight of the metals present on the catalyst.

In order to verify the stability of loaded CNS under reaction conditions, we simulated a standard synthetic procedure in a 15 mL vial in absence of reagents and we conducted an ICP-OES analysis on the reaction mixture. The results reported in Table 2 revealed that the amounts of metal released into solution were negligible.

Table 1. ICP-OES analysis result for the sample before the reactions. I, II and III indicate three different samples on which measurements were performed.

| | CNS-Zn | | | | |
|---|---|---|---|---|---|
| | I | II | III | Average | Std. Deviation |
| Zn (w%) | 6.85 | 6.69 | 6.78 | 6.77 | 0.08 |
| | CNS-Cu | | | | |
| | I | II | III | Average | Std. Deviation |
| Cu (w%) | 7.44 | 7.67 | 7.56 | 7.56 | 0.12 |

Table 2. ICP-OES analysis results obtained analyzing the methanol solution after a simulated standard synthetic protocol (2 h, methanol, microwave irradiation, 40 °C). For CNS-Cu 6 mg of catalyst in 15 mL of methanol were used, while for CNS-Zn 45 mg of catalyst in 15 mL of methanol were used (quantities selected according to the optimization tests discussed in Section 3).

| | Cu [mg/L] | Zn [mg/L] |
|---|---|---|
| Test with CNS-Zn | $0.0545 \pm 0.002$ | $0.4613 \pm 0.010$ |
| Test with CNS-Cu | $0.0515 \pm 0.002$ | $0.0883 \pm 0.002$ |

The loading efficacy and homogeneity was also confirmed by SEM-EDS analysis (Figures 1 and 2).

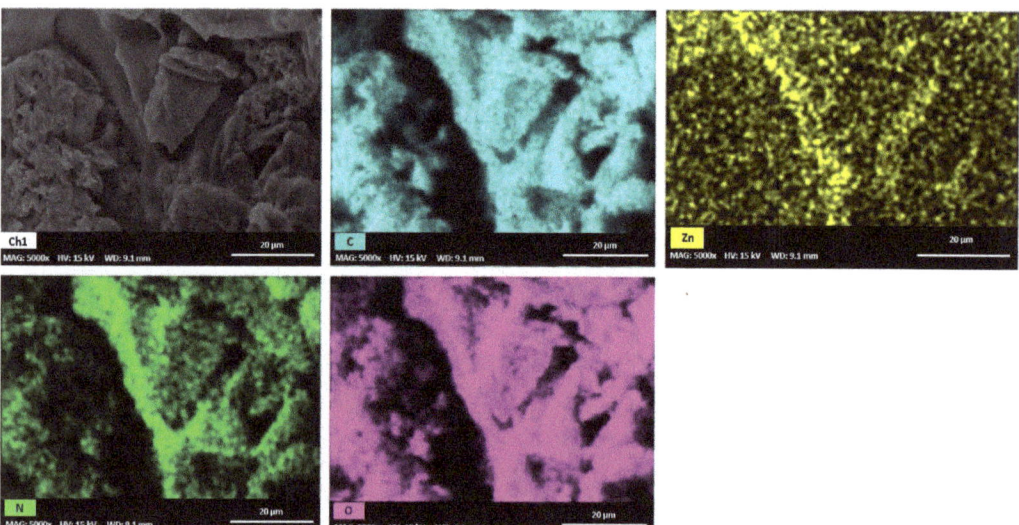

Figure 1. Distribution of elements on CNS-Zn observed with EDS analysis.

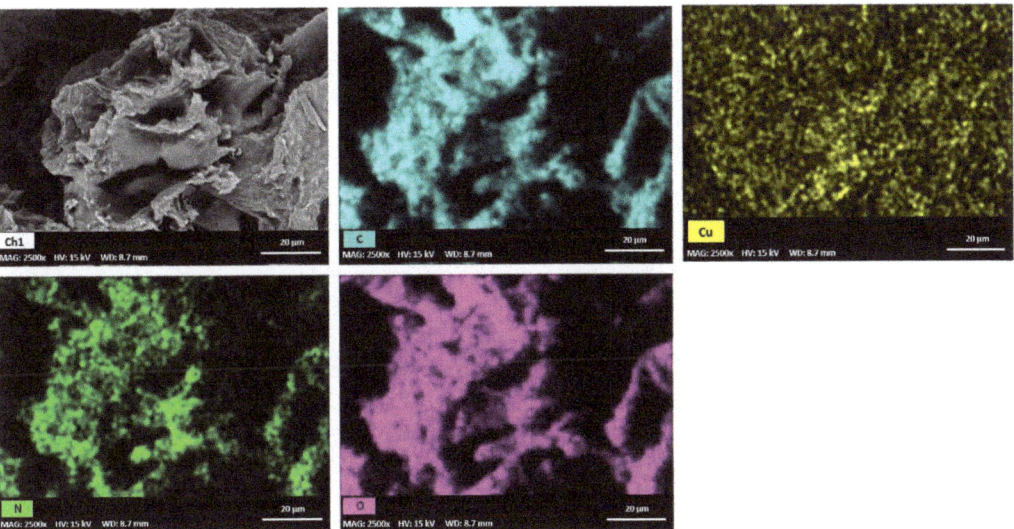

**Figure 2.** Distribution of elements on CNS-Cu observed with EDS analysis.

## 2.2. Optimization of the Reaction Conditions in Methanol

Initial experiments, devoted to the verification of activity of CNS-Cu and CNS-Zn as heterogeneous catalysts and to the optimization of the protocol, were conducted under different operating conditions by reacting p-F-benzaldehyde with methanol under microwave irradiation instead of conventional heating. As expected, the formation of the carboxylic acid product due to the auto-oxidation of the aldehyde in the presence of oxygen could be observed, since the reaction is not conducted in an inert environment. The percentage of carboxylic acid formed is reported in the tables to demonstrate how effective the catalyst is in converting the aldehyde to acetal before its oxidation to carboxylic acid.

In a first set of reactions, the effect of temperature was evaluated (Table 3). Two blank tests in the presence of not-loaded CNS were carried out at 40 °C (*entry 1* of Table 3) and 80 °C (*entry 4* of Table 3), showing that in the absence of metal no formation of any desired product could be observed.

**Table 3.** Optimization of the reaction temperature in methanol [a].

| Entry | T [°C] | Catalyst Type | % Conversion | Selectivity | |
|---|---|---|---|---|---|
| | | | | % 1 | % 2 |
| 1 | 40 | CNS | - | - | - |
| 2 | 40 | CNS-Cu | 42 | 96 | 4 |
| 3 | 40 | CNS-Zn | 98 | 97 | 3 |
| 4 | 80 | CNS | - | - | - |
| 5 | 80 | CNS-Cu | 92 | 90 | 10 |
| 6 | 80 | CNS-Zn | 96 | 76 | 24 |

[a] 50 mg of p-F-benzaldehyde in 2.5 mL of methanol were used, processing the reaction under microwave (MW) irradiation.

Despite the lower conversion values in the case of the Cu-catalyzed reactions, we decided to work at 40 °C when using methanol as a solvent. This choice was taken since working at 40 °C provides advantages in terms of costs and safety. Moreover, by optimizing the amount of catalyst to be used in order to obtain the desired results, we envisioned it could be still possible to reach high conversions also for CNS-Cu. For these reasons, once the temperature was fixed, we progressively reduced the amount of catalyst (Table 4).

**Table 4.** Optimization of the % $w/w$ of CNS-Cu and CNS-Zn [a].

| Entry | Catalyst | % w/w Catalyst | % Conversion | Selectivity % 1 | % 2 |
|---|---|---|---|---|---|
| 1 | CNS-Cu | 35 | 42 | 96 | 4 |
| 2 | CNS-Cu | 15 | 47 | 94 | 6 |
| 3 | CNS-Cu | 2 | 89 | 96 | 4 |
| 4 | CNS-Zn | 35 | 98 | 97 | 3 |
| 5 | CNS-Zn | 15 | 99 | 96 | 4 |
| 6 | CNS-Zn | 2 | 72 | 95 | 5 |

[a] 50 mg of $p$-F-benzaldehyde in 2.5 mL of methanol were used, processing the reaction in MW.

The reactions were repeated starting with 35% $w/w$ catalyst and gradually reducing the quantity of catalyst up to a 2% $w/w$ amount.

The results with **CNS-Cu** showed that, by operating at 40 °C, it was possible to achieve very good conversions (*entry* 3 in Table 4) using only 2% $w/w$ catalyst. Unexpectedly, increasing the amount of CNS-Cu catalyst at T = 40 °C resulted in a lower conversion, whereas at T = 80 °C, high conversion was obtained. We can suggest that at lower temperature, an interaction between the aldehyde and the copper loaded nanosponge is established, thus making the substrate less reactive or available and slowing the reaction. This hypothesis could explain why a lower amount of catalyst is beneficial to the reaction. A possible implication of the backbone CNS material was excluded, since with the use Zinc this effect was not observed. Nonetheless, we decided to not further investigate this very unusual outcome, having reached optimal results with the 2% of catalyst loading.

On the other hand, in the case of CNS-Zn it was necessary to work with 15% $w/w$ catalyst to obtain interesting results. Despite the higher percentage of catalyst, in this case the results were satisfactory also in view of the possibility of recycling it, as it will be discussed later. From the tests conducted, we can therefore conclude that the ideal operating conditions for reactions in methanol were 40 °C and 2% $w/w$ of catalyst for CNS-Cu, and 40 °C and 15% $w/w$ of catalyst for Zn-CNS. In fact, under these conditions, both the catalysts led to a selectivity >95% towards product **1** (acetal) with very high conversions.

### 2.3. Kinetic Tests

The model reaction was also followed along the time, with both CNS-Cu (Table 5) and CNS-Zn (Table 6), in order to analyze the course of the reaction in terms of conversion.

**Table 5.** Results for kinetic tests with CNS-Cu [a].

| Time [min] | % Conversion |
|---|---|
| 30 | 81 |
| 60 | 81 |
| 90 | 88 |
| 120 | 84 |

[a] The reaction was conducted using 50 mg of p-F-benzaldehyde in 2.5 mL of methanol processing the reaction in MW, with a reaction temperature of 40 °C and a 2% w/w of CNS-Cu.

**Table 6.** Results for kinetic tests with CNS-Zn [a].

| Time [min] | % Conversion |
|---|---|
| 30 | 61 |
| 60 | 71 |
| 90 | 81 |
| 120 | 84 |

[a] The reaction was conducted using 50 mg of p-F-benzaldehyde in 2.5 mL of methanol processing the reaction in MW, with a reaction temperature of 40 °C and a 2% w/w of CNS-Zn.

For reactions with CNS-Cu a good conversion could be observed already after 30 min, while for reactions with CNS-Zn it was necessary to extend the time to 120 min to achieve the maximum conversion (Figure 3). Therefore, in order to standardize the reaction time for both metals, 120 min of reaction time was selected as the optimal condition.

**Figure 3.** (A) Conversion trend for reaction with CNS-Cu; (B) conversion trend for the reaction with CNS-Zn.

### 2.4. Scope of the Reaction

With the optimized conditions in hand, we set out to examine the scope of the reaction. First, we extended the transformation to a wider range of carbonyl compounds (Table 7). General reaction is reported in Scheme 2.

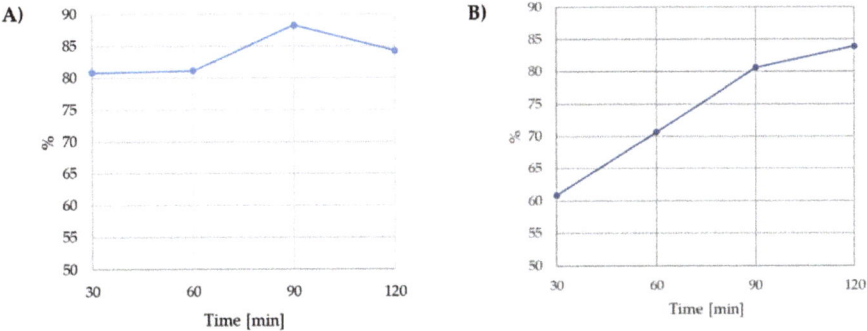

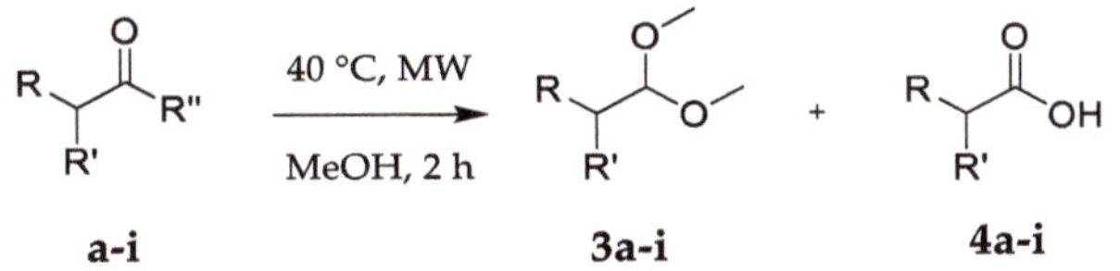

**Scheme 2.** General reaction scheme for the formation of the acetal and the acid product in MeOH.

Table 7. Reactions with different carbonyl compounds using methanol as solvent *.

| Entry | Reagent | Catalyst | % Conversion | Selectivity % 3 | % 4 |
|---|---|---|---|---|---|
| 1 | p-F-benzaldehyde | CNS-Cu | 74 | 1: 93 | 2: 7 |
| 2 | p-Me-benzaldehyde (a) | CNS-Cu | 94 | 3a: 98 | 4a: 2 |
| 3 | m-OMe-benzaldehyde (b) | CNS-Cu | 69 | 3b: 100 | 4b: - |
| 4 | o-OMe-benzaldehyde (c) | CNS-Cu | 93 | 3c: 98 | 4c: 2 |
| 5 | Cyclohexanone (d) | CNS-Cu | 96 | 3d: 100 | 4d: - |
| 6 | Cyclopentanone (e) | CNS-Cu | 55 | 3e: 100 | 4e: - |
| 7 | 2-Naphtaldehyde (f) | CNS-Cu | 86 | 3f: 84 | 4f: 16 |
| 8 | 2-Furaldehyde (g) | CNS-Cu | 52 | 3g: 100 | 4g: - |
| 9 | Benzaldehyde (h) | CNS-Cu | 90 | 3h: 87 | 4h: 13 |
| 10 | p-Cl-benzaldehyde (i) | CNS-Cu | 35 | 3i: 100 | 4i: - |
| 11 | p-F-benzaldehyde | CNS-Zn | 94 | 1: 97 | 2: 3 |
| 12 | p-Me-benzaldehyde (a) | CNS-Zn | 98 | 3a: 100 | 4a: - |
| 13 | m-OMe-benzaldehyde (b) | CNS-Zn | 97 | 3b: 100 | 4b: - |
| 14 | o-OMe-benzaldehyde (c) | CNS-Zn | 99 | 3c: 98 | 4c: 2 |
| 15 | Cyclohexanone (d) | CNS-Zn | 100 | 3d: 100 | 4d: - |
| 16 | Cyclopentanone (e) | CNS-Zn | 55 | 3e: 100 | 4e: - |
| 17 | 2-Naphtaldehyde (f) | CNS-Zn | 87 | 3f: 98 | 4f: 2 |
| 18 | 2-Furaldehyde (g) | CNS-Zn | 49 | 3g: 100 | 4g: - |
| 19 | Benzaldehyde (h) | CNS-Zn | 98 | 3h: 100 | 4h: - |
| 20 | p-Cl-benzaldehyde (i) | CNS-Zn | 37 | 3i: 00 | 4i: - |

* All reactions were conducted in MW at 40 °C for 2 h, using 50 mg of carbonyl compound and 2.5 mL of methanol, with 2% $w/w$ percentage for CNS-Cu and 15% $w/w$ percentage for CNS-Zn.

For the reactions catalyzed by CNS-Cu, selectivity towards product 3 was always almost complete, while conversions were generally quite high, except for some substrates (*entries 6, 8*, and *10* in (Table 7), but in any case, lower than those obtained in the presence of CNS-Zn. Nevertheless, results obtained with CNS-Cu result interesting if we consider that we worked with a very low percentage of catalyst (2% $w/w$). Reactions catalyzed by CNS-Zn showed a high selectivity towards product c too. The conversion in this case turned out to be on average higher than the reactions with CNS-Cu.

Hence, for the choice of catalyst in the case of reactions in methanol it is necessary to decide between higher conversions, thus using CNS-Zn, with the disadvantage of the quantity of catalyst to be used, or to accept slightly lower conversions, but operating with lower quantities of catalyst CNS-Cu.

*2.5. Optimization of the Reaction Conditions in Ethanol*

Finally, we verified if the best conditions developed for methanol could be transferred to ethanol operating with CNS-Cu as catalyst.

In this case we saw a significantly lower conversion in ethanol (Table 8) and consequently we proceeded with the optimization of the reaction by maintaining fixed the percentage amount of the catalyst used (CNS-Cu 2% $w/w$) and varying reaction time and temperature (Table 9).

Table 8. Comparison between the same reaction conditions in methanol and in ethanol [a].

| Solvent | % Conversion | Selectivity (%) % Acetal | % Acid |
|---|---|---|---|
| Methanol | 90 | 1: 87 | 2: 13 |
| Ethanol | 33 | 5: 63 [b] | 2: 37 |

[a] The reaction was conducted using 50 mg of p-F-benzaldehyde in 2.5 mL of solvent processing the reaction in MW, with a reaction temperature of 40 °C, a reaction time of 2 h and a 2% $w/w$ of CNS-Cu. [b] Product 5 is reported in the scheme of Table 9.

**Table 9.** Optimization of the reaction conditions using ethanol as solvent [a].

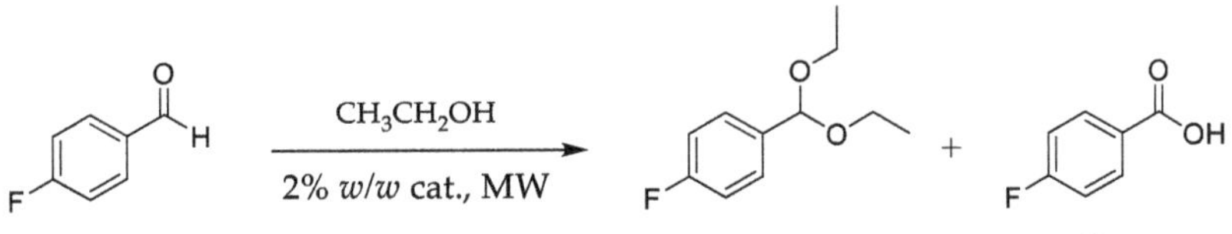

| Entry | T [°C] | t [h] | % Conversion | Selectivity | |
|---|---|---|---|---|---|
| | | | | % 5 | % 2 |
| 1 | 80 | 2 | 61 | 78 | 22 |
| 2 | 80 | 4 | 75 | 82 | 18 |
| 3 | 80 | 6 | 86 | 82 | 18 |
| 4 | 100 | 2 | 68 | 79 | 21 |
| 5 | 100 | 4 | 87 | 80 | 20 |
| 6 | 100 | 6 | 85 | 81 | 19 |

[a] 50 mg of p-F-benzaldehyde in 2.5 mL of ethanol were used, processing the reaction in MW. Catalyst type: CNS-Cu.

The results suggested that by increasing the temperature up to 80 or 100 °C, the selectivity in all cases was about 80% in favour of product **5**. Therefore, in order to choose the best conditions, we focused on only on the results related to conversion.

The operating conditions were selected to be 4 h, 100 °C and 2% w/w catalyst for CNS-Cu. For CNS-Zn, regarding the percentage of catalyst used, we selected the best conditions found in MeOH (15% w/w) and choosing the reaction time and temperature conditions that gave the best performance for CNS-Cu, namely 100 °C for 4 h. Scheme 3 reports the general reaction in ethanol.

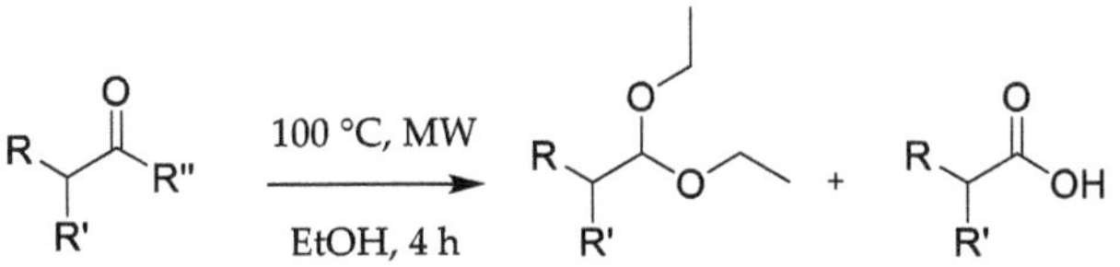

**Scheme 3.** General reaction scheme for the formation of the acetal and the acid product in EtOH.

Once optimized the operating conditions, the reaction was tested on a wide family of carbonyl compounds (Table 10).

The reactions catalyzed by CNS-Cu showed in all cases excellent selectivity, but in most cases the conversion was too low to be considered interesting. The reactions catalyzed by CNS-Zn in addition to excellent selectivity provided high conversions, over 80%, being therefore of greater interest.

Hence, we can conclude that using ethanol as solvent under these operating conditions, unlike in methanol, there is a strong difference in the results obtained with CNS-Cu and CNS-Zn, as only the latter guarantees both high selectivity and high conversion.

Table 10. Reactions with different carbonyl compounds using ethanol as solvent *.

| Entry | Reagent | Catalyst | % Conversion | Selectivity % 6 | Selectivity % 4 |
|---|---|---|---|---|---|
| 1 | p-F-benzaldehyde | CNS-Cu | 68 | 5: 97 | 2: 3 |
| 2 | p-Me-benzaldehyde (a) | CNS-Cu | 50 | 6a: 92 | 4a: 8 |
| 3 | m-OMe-benzaldehyde (b) | CNS-Cu | 12 | 6b: 100 | 4b: - |
| 4 | o-OMe-benzaldehyde (c) | CNS-Cu | 70 | 6c: 98 | 4c: 2 |
| 5 | Cyclohexanone (d) | CNS-Cu | 48 | 6d: 100 | 4d: - |
| 6 | Cyclopentanone (e) | CNS-Cu | 24 | 6e: 100 | 4e: - |
| 7 | 2-Furaldehyde (g) | CNS-Cu | 31 | 6g: 100 | 4g: - |
| 8 | Benzaldehyde (h) | CNS-Cu | 87 | 6h: 80 | 4h: 20 |
| 9 | p-Cl-benzaldehyde (i) | CNS-Cu | 2 | 6i: 100 | 4i: - |
| 10 | p-F-benzaldehyde | CNS-Zn | 84 | 5: 97 | 2: 3 |
| 11 | p-Me-benzaldehyde (a) | CNS-Zn | 76 | 6a: 97 | 4a: 3 |
| 12 | m-OMe-benzaldehyde (b) | CNS-Zn | 80 | 6b: 98 | 4b: 2 |
| 13 | o-OMe-benzaldehyde (c) | CNS-Zn | 89 | 6c: 99 | 4c: 1 |
| 14 | Cyclohexanone (d) | CNS-Zn | 58 | 6d: 100 | 4d: - |
| 15 | Cyclopentanone (e) | CNS-Zn | 31 | 6e: 100 | 4e: - |
| 16 | 2-Furaldehyde (g) | CNS-Zn | 96 | 6g: 97 | 4g: 3 |
| 17 | Benzaldehyde (h) | CNS-Zn | 93 | 6h: 84 | 4h: 16 |
| 18 | p-Cl-benzaldehyde (i) | CNS-Zn | 84 | 6i: 100 | 4i: 0 |

* All reactions were conducted in MW at 100 °C for 4 h, using 50 mg of carbonyl compound and 2.5 mL of EtOH, with the catalyst at 2% w/w percentage for CNS-Cu and at 15% w/w percentage for CNS-Zn.

## 2.6. Catalyst Recycling and Reuse

Referring to the optimized conditions in methanol, we also conducted some tests to verify the possibility to recover and re-use the CNS-metal catalyst at the end of the reaction, by filtering the mixture, drying the solid at about 50 °C for 2 h and then re-using it. Four reusability cycles were performed on both CNS-Cu and CNS-Zn metal-loaded catalysts and results are reported in Tables 11 and 12, respectively, while Figures 4 and 5 show schematically the reusability test results for the same materials.

Table 11. Results of the reusability tests conducted with CNS-Cu [a].

| Entry | T [°C] | t [h] | % Conversion | Selectivity % 1 | Selectivity % 2 |
|---|---|---|---|---|---|
| A | 80 | 2 | 72 | 97 | 3 |
| B | 80 | 4 | 88 | 98 | 2 |
| C | 80 | 6 | 83 | 97 | 3 |
| D | 100 | 2 | 49 | 68 | 32 |

[a] 50 mg of p-F-benzaldehyde in 2.5 mL of methanol were used, processing the reaction in MW. Catalyst type: CNS-Cu % w/w of catalyst: 2%.

**Table 12.** Results of the reusability tests conducted with CNS-Zn [a].

| Entry | T [°C] | t [h] | % Conversion | Selectivity | |
|---|---|---|---|---|---|
| | | | | % 1 | % 2 |
| A | 80 | 2 | 85 | 83 | 17 |
| B | 80 | 4 | 85 | 88 | 12 |
| C | 80 | 6 | 92 | 89 | 11 |
| D | 100 | 2 | 86 | 78 | 22 |

[a] 50 mg of *p*-F-benzaldehyde in 2.5 mL of methanol were used, processing the reaction in MW. Catalyst type: CNS-Zn % *w/w* of catalyst: 15%.

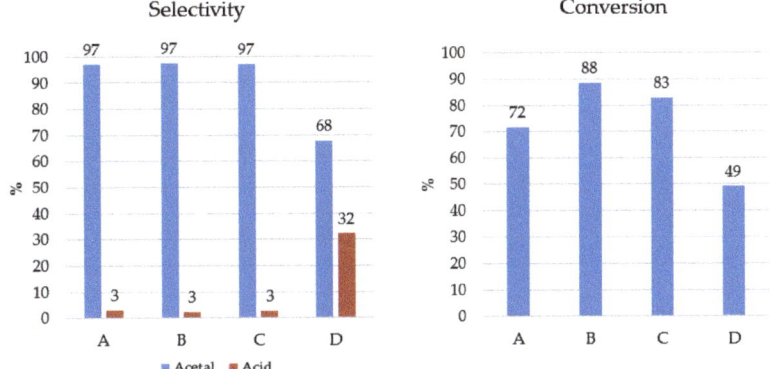

**Figure 4.** Schematic results for the reusability test with CNS-Cu.

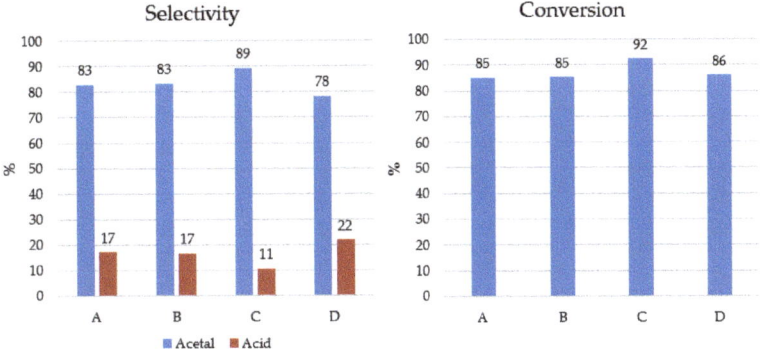

**Figure 5.** Schematic results for the reusability test with CNS-Zn.

Results confirmed that both the catalysts could be reused without any loss of performance in terms of selectivity and conversion, at least for three cycles. In the case of CNS-Cu at the fourth recycling test, there was a decrease in both selectivity and conversion. However, this result could be ascribed to the difficulty in recovering the entire amount of catalyst from the third test, by considering the significantly lower quantities used in this case (2% *w/w* versus 15% *w/w* of CNS-Zn).

## 3. Conclusions

In this work, we investigated the possibility of converting the use of CNS aerogels, obtained by combining TOCNF with bPEI and CA in a unique hydrogel, from a material for water decontamination from heavy metal ions to a totally green heterogeneous support for metal-catalyzed organic reactions. Exploiting the metal chelation characteristics of this aerogel, previously highlighted in the field of environmental remediation, it was possible to load the catalyst with Cu (II) and Zn (II) metal ions and to test the performances of the resulting composites in heterogeneous catalysis for the synthesis of aromatic acetals. The loading efficiency and homogeneity was verified by both ICP-OES and SEM-EDS analysis.

Using methanol as a solvent, we were able to obtain high conversions and good selectivity towards acetal products under optimized conditions, which comprised of operation in a microwave reactor, at low reaction times and temperatures never higher than 100 °C, and with low percentages of the catalysts which, in the case of CNS-Cu, reached a minimum value of 2%.

The scope of the reaction, first tested on $p$-F-benzaldehyde as a starting substrate, was extended to a wider range of aromatic carbonylic compounds, obtaining (in all cases) high yields in the desired acetal products, with in just a few cases tiny amounts of carboxylic acids as by-products. In order to further investigate the general interest of this reaction, ethanol was also used as solvent and reagent. Also in this case, after a required optimization of the reaction conditions, good results were observed both in terms of conversion and selectivity.

The catalysts could be recovered and re-used up to five times without a significant loss in the catalytic activity. In fact, no leaching of metal ions in the reaction medium was observed after ICP-OES analysis of the organic solution.

Due to the high conversions and selectivity observed, the recyclability of the material, and its production starting from sustainable and bio-based renewable sources, the proposed heterogeneous system candidates are much more convenient with respect to both homogeneous and heterogeneous catalysts proposed in the literature for the synthesis of aromatic acetals.

## 4. Materials and Methods

All of the reagents were purchased from Merck (Darmstadt, Germany). Cotton linter was obtained from Bartoli paper factory (Capannori, Lucca, Italy). Deionized water was produced within the laboratories with a Millipore Elix® Deionizer with Progard® S2 ion exchange resins (Merck KGaA, Darmstadt, Germany). All $^1$H-NMR spectra were recorded on a 400 MHz Brüker (Billerica, MA, USA) NMR spectrometer. Microwave reactions were conducted in a Biotage® Initiator+ (Uppsala, Sweden). Other equipment used in the procedures include a Branson SFX250 Sonicator (Emerson Electric Co., Ferguson, MI, USA), a SP Scientific BenchTop Pro Lyophilizer (SP INDUSTRIES, 935 Mearns Road, Warminster, UK), a Büchi Rotavapor® R-124 8 (Flawil, Switzerland) and a Thermotest—Mazzali laboratory oven (Monza, Italy). Scanning electron microscopy (SEM) was performed using a variable-pressure instrument (SEM Cambridge Stereoscan 360) at 100/120 pA with a detector BSD. The operating voltage was 15 kV with an electron beam current intensity of 100 pA. The focal distance was 9 mm. The EDS analysis was performed using a Bruker Quantax 200 6/30 instrument (Billerica, MA, USA). The metal concentrations were measured by ICP-OES atomic emission spectroscopy using a Perkin Elmer Optima 3000 SD spectrometer (Wellesley, MA, USA).

*4.1. TEMPO-Oxidized Cellulose Nanofibers (TOCNFs) Production and Titration*

100 g of cotton linter paper were minced with gradual addition of deionized water. Simultaneously, 2.15 g of tetramethyl-piperidine-$N$-oxide (TEMPO) and 15.42 g of KBr were dissolved in 2 L of deionized water. Once the paper was homogeneously blended with water, the solution was transferred in the reaction keg and water was added in order to obtain a total volume of 5.7 L. Then, 437 mL of 12% $w/v$ NaClO aqueous solution were

dripped in the reaction mixture, keeping the pH above 10.5–11.0 with dropwise addition of NaOH 4 M. The suspension was left stirring overnight and after 12–16 h the oxidized cellulose was acidified with HCl 12 N until a pH of 1–2 to induce the aggregation of the cellulose fibers and their easy separation from water. The oxidized cellulose was then filtered on a Buchner funnel and washed with deionized water until neutrality and then titrated for the estimation of the obtained concentration of carboxylic acids on the cellulose structure (oxidation degree). This latter was performed by titration of –COOH groups with NaOH 0.1 N using phenolphthalein as colorimetric indicator. The oxidation degree was estimated to be around 1.5 $mmol_{COOH}/g_{TOCNF}$.

### 4.2. Synthesis of CNS

3.5 g of TOCNF were suspended in 140 mL of deionized water, in order to obtain a 2.5% $w/v$ solution. Granular NaOH (5.25 mmol, 0.210 g) was added, and the suspension was ultrasonicated to further promote the separation of the nanofibers, obtaining a transparent mixture. This latter was then acidified with 1 N HCl (20 mL), filtered and washed until neutral pH. The residual water content was calculated and the remaining water necessary to obtain a 2.8–3% $w/v$ TOCNF solution was split up in three quotas, in order to re-suspend the cellulose and dissolve the cross-linking polymer (bPEI 25 kDa) and the co-reticulant agent CA, in three separate batches. The amount of the cross-linking polymer (7 g) was calculated as double of the initial weight of TOCNF and 1.8 g of co-reticulant were used. Once dissolved in water, the cross-linker and the co-reticulant agents were slowly added to the TOCNF solution, while continuously stirring up to obtain a white and homogeneous hydrogel, which was placed in 24-wells well-plates and quickly frozen at −35 °C. After 24 h, the well-plates were moved to the freeze-dryer. At the end of the process, white cylindrical-shaped spongy aerogels were obtained. They were removed from the wells and placed in an oven, at the initial temperature of 55 °C. The temperature was then slowly raised up to 102 °C and then kept constant for 16 h. At the end of the thermal treatment, CNS were washed 6 times with deionized water (6 × 100 mL) and the last time with 97% ethanol solution (1 × 50 mL).

### 4.3. Metal Loading on CNS

The uploading of the metals on CNS's bulk was carried out with aqueous solutions of $Zn(NO_3)_2$ (prepared dissolving 250 g of $Zn(NO_3)_2$ in 250 mL deionized water) and $CuSO_4$ (prepared dissolving 80 g of $CuSO_4$ in 250 mL deionized water). 1 g of CNS was stirred for 4 h in 250 mL of Zn (II) and Cu (II) solutions. At the end of the sorption process, CNS were filtered off and then washed once with deionized water (500 mL), obtaining Zn-loaded (CNS-Zn) and Cu-loaded (CNS-Cu) catalysts.

### 4.4. General Procedure for Catalytic Synthesis of Acetals

The catalyst (CNS-Zn or CNS-Cu), the reagents and finally the alcohol as solvent were added in a 5 mL microwave vial, equipped with a magnetic stirrer, in the quantities reported in Tables 3–12. The reactions were conducted in a microwave reactor in a time ranging from 2 to 6 h, with a temperature comprised between 40 °C and 100 °C, at atmospheric pressure.

### 4.5. Products Characterization

All conversion and selectivity data were calculated by $^1$H-NMR analysis on the crude (see Supplementary Materials). For a better accuracy, the percentage of the acid by-product in the mixture, often present in very low amounts, was determined by GC-MS analysis.

The $^1$H-NMR spectra of known products **1**, **3a–i** and **7a–h**, are in agreement with the literature references as reported in the Supplementary Materials.

Product **2** was recovered and purified according to the following procedure in order to demine the isolated yield and to compare that with the value of calculated by $^1$H-NMR with internal standard. The catalyst was removed by filtration and the alcoholic solvent was removed under vacuum. The obtained crude was dissolved in 15 mL of ethyl acetate and washed with deionized water (3 × 15 mL). The extracted organic layer was dried with sodium sulfate and the solvent was removed under vacuum by means of a Rotavapor®. Flash column chromatography was performed to purify the product by using 10–12 mm silica gel packing. The eluent was chosen in order to move the desired components to $R_f$ 0.35 on analytical TLC. The selected eluent was a mixture of hexane and ethyl acetate in a 9:1 ratio.

The unknown product **5** was isolated according to procedure previously reported for product **2**. Flash column chromatography was performed to purify product **5** by using 10–12 mm silica gel packing and the selected eluent was a mixture of hexane and ethyl acetate in a 95:5 ratio. The characterization of new compound **5** is here reported.

1-(diethoxymethyl)-4-fluorobenzene **5**. $^1$H NMR (400 MHz, Chloroform-$d$) δ 7.45 (dd, $J$ = 8.6, 5.6 Hz, 2H), 7.03 (t, $J$ = 8.8 Hz, 2H), 5.48 (s, 1H), 3.60 (dq, $J$ = 9.8, 7.2 Hz, 2H), 3.52 (dq, $J$ = 9.5, 7.1 Hz, 2H), 1.23 (t, $J$ = 7.1 Hz, 6H). $^{13}$C NMR (101 MHz, Chloroform-$d$) δ 162.7 (d, $J$ = 246.2 Hz), 135.1 (d, $J$ = 3.2 Hz), 128.4 (d, $J$ = 8.1 Hz, 2C), 114.9 (d, $J$ = 21.4 Hz, 2C), 100.9, 60.9, 15.1. Elemental analysis for $C_{11}H_{15}FO_2$: C, 66.65%; H, 7.63%; F, 9.58%; O, 16.14%. Found C, 66.12%; H, 7.65%.

**Supplementary Materials:** The following supporting information can be downloaded at: https://www.mdpi.com/article/10.3390/gels8010054/s1. $^1$H NMR Spectra of the crude products for all the known compounds used to determine the product yield. Figure S1: $^1$H NMR spectrum of product **1** in CDCl$_3$; Figure S2: $^1$H NMR spectrum of product **3a** in CDCl$_3$; Figure S3: $^1$H NMR spectrum of product **3b** in CDCl$_3$; Figure S4: $^1$H NMR spectrum of product **3c** in CDCl$_3$; Figure S5: $^1$H NMR spectrum of product **3d** in CDCl$_3$; Figure S6: $^1$H NMR spectrum of product **3e** in CDCl$_3$; Figure S7: $^1$H NMR spectrum of product **3f** in CDCl$_3$; Figure S8: $^1$H NMR spectrum of product **3g** in CDCl$_3$; Figure S9: $^1$H NMR spectrum of product **3h** in CDCl$_3$; Figure S10: $^1$H NMR spectrum of product **3i** in CDCl$_3$; Figure S11: $^1$H NMR spectrum of product **6a** in CDCl$_3$; Figure S12: $^1$H NMR spectrum of product **6b** in CDCl$_3$; Figure S13: $^1$H NMR spectrum of product **6c** in CDCl$_3$; Figure S14: $^1$H NMR spectrum of product **6d** in CDCl$_3$; Figure S15: $^1$H NMR spectrum of product **6e** in CDCl$_3$; Figure S16: $^1$H NMR spectrum of product **6g** in CDCl$_3$; Figure S17: $^1$H NMR spectrum of product **6h** in CDCl$_3$; Figure S18: $^1$H NMR spectrum of product **6i** in CDCl$_3$; Figure S19: $^1$H NMR spectrum of product **5** in CDCl$_3$; Figure S20: $^{13}$C-APT NMR spectrum of product **5** in CDCl$_3$; Figure S21: $^{13}$C NMR spectrum of product **5** in CDCl$_3$; Figure S22: COSY NMR spectrum of product **5** in CDCl$_3$; Figure S23: HSQC NMR spectrum of product **5** in CDCl$_3$.

**Author Contributions:** Conceptualization, C.P. and A.S.; methodology, L.R. and A.D.L.; validation, L.R. and A.D.L.; formal analysis, investigation, L.R. and A.D.L.; resources, A.S. and C.P.; data curation, A.S.; writing—original draft preparation, L.R.; writing—review and editing, C.P.; supervision, A.S. and C.P. All authors have read and agreed to the published version of the manuscript.

**Funding:** This research received no external funding.

**Institutional Review Board Statement:** Not applicable.

**Informed Consent Statement:** Not applicable.

**Conflicts of Interest:** The authors declare no conflict of interest.

# References

1. Anastas, P.T.; Warner, J.C. *Green Chemistry: Theory and Practice. Green Chemistry: Theory and Practice*; Oxford University Press: New York, NY, USA, 1998.
2. Hák, T.; Janoušková, S.; Moldan, B. Sustainable Development Goals: A need for relevant indicators. *Ecol. Indic.* **2016**, *60*, 565–573. [CrossRef]
3. Isogai, A.; Saito, T.; Fukuzumi, H. TEMPO-oxidized cellulose nanofibers. *Nanoscale* **2011**, *3*, 71–85. [CrossRef] [PubMed]
4. Melone, L.; Rossi, B.; Pastori, N.; Panzeri, W.; Mele, A.; Punta, C. TEMPO-Oxidized Cellulose Cross-Linked with Branched Polyethyleneimine: Nanostructured Adsorbent Sponges for Water Remediation. *Chempluschem* **2015**, *80*, 1408–1415. [CrossRef]
5. Paladini, G.; Venuti, V.; Almásy, L.; Melone, L.; Crupi, V.; Majolino, D.; Pastori, N.; Fiorati, A.; Punta, C. Cross-linked cellulose nano-sponges: A small angle neutron scattering (SANS) study. *Cellulose* **2019**, *26*, 9005–9019. [CrossRef]

6. Fiorati, A.; Grassi, G.; Graziano, A.; Liberatori, G.; Pastori, N.; Melone, L.; Bonciani, L.; Pontorno, L.; Punta, C.; Corsi, I. Eco-design of nanostructured cellulose sponges for sea-water decontamination from heavy metal ions. *J. Clean. Prod.* **2020**, *246*, 119009. [CrossRef]
7. Riva, L.; Pastori, N.; Panozzo, A.; Antonelli, M.; Punta, C. Nanostructured cellulose-based sorbent materials for water decontamination from organic dyes. *Nanomaterials* **2020**, *10*, 1570. [CrossRef]
8. Riva, L.; Punta, C.; Sacchetti, A. Co-Polymeric Nanosponges from Cellulose Biomass as Heterogeneous Catalysts for amine-catalyzed Organic Reactions. *ChemCatChem* **2020**, *12*, 6214–6222. [CrossRef]
9. Greene, T.W. *Greene's Protective Groups in Organic Synthesis*; Peter, G.M.E., Ed.; John Wiley & Sons, Inc.: Hoboken, NJ, USA, 2014; ISBN 9781118905074.
10. Bickers, D.R.; Calow, P.; Greim, H.A.; Hanifin, J.M.; Rogers, A.E.; Saurat, J.H.; Sipes, I.G.; Smith, R.L.; Tagami, H. The safety assessment of fragrance materials. *Regul. Toxicol. Pharmacol.* **2003**, *37*, 218–273. [CrossRef]
11. Smith, B.M.; Graham, A.E. Indium triflate mediated acetalization of aldehydes and ketones. *Tetrahedron Lett.* **2006**, *47*, 9317–9319. [CrossRef]
12. Dauben, W.G.; Look, G.C.; Gerdes, J.M. Organic Reactions at High Pressure. Conversion of Cyclic Alkanones and Enones to 1,3-Dioxolanes. *J. Org. Chem.* **1986**, *51*, 4964–4970. [CrossRef]
13. Howard, E.G.; Lindsey, R.V. The Chemistry of Some 5-m-Dithianones and Dithiacycloalkanones. *J. Am. Chem. Soc.* **1960**, *82*, 158–164. [CrossRef]
14. Fieser, L.F.; Stevenson, R. Cholesterol and Companions. IX. Oxidation of Δ5-Cholestene-3-one with Lead Tetraacetate. *J. Am. Chem. Soc.* **1954**, *76*, 1728–1733. [CrossRef]
15. Janus, E. Lewis acids immobilized in ionic liquid-Application for the acetal synthesis. *Polish. J. Chem. Technol.* **2013**, *15*, 78–80. [CrossRef]
16. Krompiec, S.; Penkala, M.; Szczubiałka, K.; Kowalska, E. Transition metal compounds and complexes as catalysts in synthesis of acetals and orthoesters: Theoretical, mechanistic and practical aspects. *Coord. Chem. Rev.* **2012**, *256*, 2057–2095. [CrossRef]
17. Shimizu, K.I.; Hayashi, E.; Hatamachi, T.; Kodama, T.; Kitayama, Y. SO3H-functionalized silica for acetalization of carbonyl compounds with methanol and tetrahydropyranylation of alcohols. *Tetrahedron Lett.* **2004**, *45*, 5135–5138. [CrossRef]
18. Kumar, H.M.S.; Reddy, B.V.S.; Mohanty, P.K.; Yadav, J.S. Clay catalyzed highly selective O-alkylation of primary alcohols with orthoesters. *Tetrahedron Lett.* **1997**, *38*, 3619–3622. [CrossRef]
19. Patwardhan, S.A.; Dev, S. Amberlyst-15, a superior catalyst for the preparation of enol ethers and acetals. *Synthesis* **1974**, *5*, 348–349. [CrossRef]
20. Kawabata, T.; Mizugaki, T.; Ebitani, K.; Kaneda, K. Highly efficient heterogeneous acetalization of carbonyl compounds catalyzed by a titanium cation-exchanged montmorillonite. *Tetrahedron Lett.* **2001**, *20*, 131–135. [CrossRef]
21. Sinhamahapatra, A.; Sutradhar, N.; Ghosh, M.; Bajaj, H.C.; Panda, A.B. Mesoporous sulfated zirconia mediated acetalization reactions. *Appl. Catal. A Gen.* **2011**, *402*, 87–93. [CrossRef]
22. Soni, J.; Sethiya, A.; Agarwal, S. The Role of Carbon-based Solid Acid Catalysts in Organic Synthesis. In *Advances in Organic Synthesis*; Bentham Science Publisher: Sharjah, United Arab Emirates, 2021; pp. 235–291.
23. Zhai, Y.; Zhu, Z.; Dong, S. Carbon-Based Nanostructures for Advanced Catalysis. *ChemCatChem* **2015**, *7*, 2806–2815. [CrossRef]
24. Fiorati, A.; Turco, G.; Travan, A.; Caneva, E.; Pastori, N.; Cametti, M.; Punta, C.; Melone, L. Mechanical and drug release properties of sponges from cross-linked cellulose nanofibers. *Chempluschem* **2017**, *82*, 848–858. [CrossRef] [PubMed]
25. Paladini, G.; Venuti, V.; Crupi, V.; Majolino, D.; Fiorati, A.; Punta, C. FTIR-ATR analysis of the H-bond network of water in branched polyethyleneimine/TEMPO-oxidized cellulose nano-fiber xerogels. *Cellulose* **2020**, *27*, 8605–8618. [CrossRef]

MDPI
St. Alban-Anlage 66
4052 Basel
Switzerland
Tel. +41 61 683 77 34
Fax +41 61 302 89 18
www.mdpi.com

*Gels* Editorial Office
E-mail: gels@mdpi.com
www.mdpi.com/journal/gels

www.ingramcontent.com/pod-product-compliance
Lightning Source LLC
LaVergne TN
LVHW070508100526
838202LV00014B/1811